Emergent Micro- and Nanomaterials for Optical, Infrared, and Terahertz Applications

Driven by continuing pursuits in device miniaturization and performance improvement, emergent micro- and nanomaterials hold the keys to enabling next-generation technologies in optical, infrared, and terahertz applications, owing to their unique properties and strong responses in these frequency bands. Development of these fascinating materials has triggered a number of opportunities in the applied sciences, and some have even made their impact in the market. *Emergent Micro- and Nanomaterials for Optical, Infrared, and Terahertz Applications* reviews state-of-the-art developments in various emergent materials and their implementation in applications such as sensors, waveplates, communications, and light sources, among others. The book discusses the similarities, advantages, and limitations and offers a comparative of each material. This volume:

- Covers all emergent materials (natural and artificial) that are promising for optical, infrared, and terahertz applications
- Comparatively analyzes these materials, elucidating their unique advantages, limitations, and application scopes
- Provides an up-to-date record on achievements and progress in cutting-edge optical, infrared, and terahertz applications
- Offers a comprehensive overview to connect multidisciplinary fields, such as materials, physics, and optics, to serve as a basis for future progress

This book is a valuable reference for engineers, researchers, and students in the areas of materials and optics, as well as physics, and will benefit both junior- and senior-level researchers.

Emerging Materials and Technologies

Series Editor: Boris I. Kharissov

Nanomaterials in Manufacturing Processes
Dhiraj Sud, Anil Kumar Singla, and Munish Kumar Gupta

Advanced Materials for Wastewater Treatment and Desalination
A.F. Ismail, P.S. Goh, H. Hasbullah, and F. Aziz

Green Synthesized Iron-Based Nanomaterials: Application and Potential Risk
Piyal Mondal and Mihir Kumar Purkait

Polymer Nanocomposites in Supercapacitors
Soney C George, Sam John and Sreelakshmi Rajeevan

Polymers Electrolytes and Their Composites for Energy Storage/ Conversion Devices
Edited by Achchhe Lal Sharma, Anil Arya and Anurag Gaur

Hybrid Polymeric Nanocomposites from Agricultural Waste
Sefiu Adekunle Bello

Photoelectrochemical Generation of Fuels
Edited by Anirban Das, Gyandeshwar Kumar Rao and Kasinath Ojha

Emergent Micro- and Nanomaterials for Optical, Infrared, and Terahertz Applications
Edited by Song Sun, Wei Tan, and Su-Huai Wei

Gas Sensors: Manufacturing, Materials, and Technologies
Edited by Ankur Gupta, Mahesh Kumar, Rajeev Kumar Singh and Shantanu Bhattacharya

Environmental Biotechnology
Fundamentals to Modern Techniques
Sibi G

Emerging Two Dimensional Materials and Applications
Edited by Arun Kumar Singh, Ram Sevak Singh, and Anar Singh

For more information about this series, please visit:
www.routledge.com/Emerging-Materials-and-Technologies/book-series/CRCEMT

Emergent Micro- and Nanomaterials for Optical, Infrared, and Terahertz Applications

Edited by
Song Sun
Wei Tan
Su-Huai Wei

CRC Press
Taylor & Francis Group
Boca Raton London New York

CRC Press is an imprint of the
Taylor & Francis Group, an **informa** business

First edition published 2023
by CRC Press
6000 Broken Sound Parkway NW, Suite 300, Boca Raton, FL 33487-2742

and by CRC Press
4 Park Square, Milton Park, Abingdon, Oxon, OX14 4RN

CRC Press is an imprint of Taylor & Francis Group, LLC

ISBN: 978-1-032-06505-2 (hbk)
ISBN: 978-1-032-06507-6 (pbk)
ISBN: 978-1-003-20260-8 (ebk)

DOI: 10.1201/9781003202608

Typeset in Times
by Apex CoVantage, LLC

Contents

Preface... vii

Editors.. ix

Contributors xi

Chapter 1   Introduction to Optical, Infrared, and Terahertz
Frequency Bands 1

Song Sun, Wei Tan, and Su-Huai Wei

Chapter 2   Theory of Electromagnetic Fields....................................... 23

Yuriy Akimov

Chapter 3   Theoretical Design of Nanomaterials for Optical and Terahertz
Applications..61

*Xiao Jiang, Jiafeng Xie, Zhou Li, Bing Huang,
and Su-Huai Wei*

Chapter 4   Plasmonic Materials and Their Applications 83

Jinfeng Zhu, Yinong Xie, and Yuan Gao

Chapter 5   Artificial Metamaterials, Metasurfaces, and Their Applications 121

Dacheng Wang

Chapter 6   Low Loss Dielectric Materials and Their Applications....................153

Song Sun

Chapter 7   Chiral Metamaterials and Their Applications 179

Yidong Hou, Xuannan Wu, and Xiu Yang

Chapter 8   Emerging Two-Dimensional Materials and
Their Applications in Detection of Polarized Light........................ 207

Xiao Luo, Qing Liu, Huidong Yin, and Fucai Liu

Chapter 9   Phase Change Materials 239

Qingyang Du

Chapter 10 Magnetic and Spintronic Materials and
Their Applications..261

Zheng Feng and Wei Tan

Chapter 11 Soft and Flexible Materials and Their Applications 285

Yongbiao Wan, Zhiguang Qiu, and Chuan Fei Guo

Chapter 12 Piezoelectric Materials and Their Applications...............................329

Taiping Zhang and Qilin Hua

Chapter 13 Hybrid Perovskite Materials and Their Applications.......................363

Ru Li and Xue Liu

Chapter 14 Near-Infrared Organic Materials for Biological Applications...........393

Qi-Wei Zhang and Yang Tian

Index...425

Preface

Driven by the continuing pursuits in device miniaturization and performance improvement, emergent micro-nanomaterials hold the keys for empowering the next generation technologies in the optical, infrared, and terahertz applications, because they possess unique properties and have strong responses in their corresponding frequency bands. Some of these materials have been artificially engineered to manipulate the electromagnetic waves at the subwavelength scale. These fascinating materials have triggered a number of research opportunities in the applied sciences and even made their impact in the market.

Due to their great potentials and enormous opportunities, many materials have been proposed and studied for applications in the optical to terahertz bands, demonstrating a diversity of novel behaviors and underlying physical mechanisms. The difference between these materials grants their own operating bands, advantages, limitations, and applications, therefore, research about these materials is generally conducted in different communities in different ways. However, these materials share a more unified perspective based on *electromagnetics* because optical, infrared, and terahertz waves are just different frequency bands of electromagnetic waves, and thus, they are governed by the same *materials science,* which controls the fundamental material properties as well as their associated applications. A comparative analysis of various materials will strengthen the understanding of each material and clearly distinguish one from the others. In addition, the recent development in hybrid material systems also requires knowledge from multiple disciplines. A general overview that simultaneously covers these emergent materials in parallel is required, and this can promote advances not only in any individual field, but also in potential multidisciplinary, multi-material systems.

As there are already plenty of books about a particular material or application, this book is aimed to provide a different perspective by holistically examining various emergent materials that are extensively researched for optical, infrared, and terahertz applications, including plasmonic, dielectric nanophotonic, spintronic, soft, and low dimensional materials. Serving as a general overview and guideline, this book offers a pivot that connects many other specific fields. This will be an important and convenient resource for students and junior researchers who are still seeking their research directions. But it is an equally important resource to help experienced researchers and engineers learn about other competing materials and technologies.

This book contains 14 chapters, organized as follows. Chapter 1 introduces the background of optical, infrared, and terahertz radiation, including their properties and major applications. Chapter 2 presents the fundamental theory of electromagnetic waves in homogenous/inhomogeneous media from an electromagnetic perspective. Meanwhile, Chapter 3 provides the material engineering approaches for optical, infrared, and terahertz applications from a materials science perspective. Jointly, Chapters 2 and 3 describe the principle of wave-matter interaction from both *wave* and *matter* points. After that, Chapters 4 through 6 review the state-of-the-art developments in plasmonic materials, artificial metamaterials and metasurfaces, and dielectric nanophotonics, respectively, illustrating fascinating results in tailoring electromagnetic waves via delicately designed micro-nanoscale objects. Chapter 7 overviews the chiral materials and structures for manipulating circularly polarized waves. Chapter 8 discusses various two-dimensional materials for polarized light detection. Chapter 9 reviews the development of phase-changing materials for dynamic control of photonic and terahertz devices. Chapter 10 covers the emergent magnetic and spintronic materials and their associated applications. Chapter 11 overviews various soft and flexible materials for optoelectronic applications. Chapter 12 reviews the piezoelectric materials for sensing and modulating photonics and optoelectronic devices. Chapter 13 introduces perovskite material systems and their associated applications. Finally, Chapter 14 discusses novel organic materials for infrared imaging and therapy. The similarities, advantages, and limitations of each material will be compared and discussed.

We gratefully acknowledge the support from the Microsystem and Terahertz Research Center (MTRC) at China Academy of Engineering Physics (CAEP) and the Beijing Computational Science Research Center (CSRC). We also acknowledge all the contributors: Yuriy Akimov, Xiao Jiang, Jiafeng Xie, Zhou Li, Bing Huang, Jinfeng Zhu, Yinong Xie, Yuan Gao, Dacheng Wang, Yidong Hou, Xuannan Wu, Xiu Yang, Xiao Luo, Qing Liu, Huidong Ying, Fucai Liu, Qingyang Du, Zheng Feng, Yongbiao Wan, Zhiguang Qiu, Chuan Fei Guo, Taiping Zhang, Qilin Hua, Ru Li, Xue Liu, Qi-Wei Zhang and Yang Tian, without whose significant effort and hard work, it is not possible that this book could be published. Song Sun would like to give thanks to his family, especially his wife Yexi Chen. He also dedicates this book to commemorate his father.

We hope this book will serve as a basis for future progress in relevant fields and possibly nurture some new ideas and interdisciplinary research directions. We also hope this book will be a valuable reference for engineers, scientists, and students in the areas of materials, optical, infrared, and terahertz in general, and attract more fresh blood to join these communities and make these areas continually booming.

Song Sun
Wei Tan
Su-Huai Wei

Editors

Song Sun, PhD, is an associate professor at the Microsystem and Terahertz Research Center, China Academy of Engineering Physics. He earned a BS (first class honor, 2009) and a PhD (2013) in microelectronics at Nanyang Technological University, Singapore. Prof. Sun started his career at the Institute of High Performance Computing, Agency for Science Technology and Research, Singapore in 2013 and joined the Microsystem and Terahertz Research Center, China, in 2017. His expertise is in theoretical physics and computational electromagnetics, with a strong background in a vacuum electronics, plasmonics, nanophotonics, metamaterials, and metasurfaces. Prof. Sun has authored or co-authored over 70 peer-reviewed journal papers, conference proceedings, and book chapters. He serves as an active member in IEEE, ACS, and the Chinese Optical Society (senior member).

Wei Tan, PhD, is an associate professor and director of the Terahertz Physics Department in the Microsystem and Terahertz Research Center, China Academy of Engineering Physics. He earned a BS (2007) and a PhD (2012) in condensed matter physics at Tongji University, China. Prof. Tan started his career at the Beijing Computational Science Research Center, China as a postdoctoral researcher in 2012 and joined the Microsystem and Terahertz Research Center, China, in 2014. His expertise is in theoretical physics and the modeling of electromagnetic wave-matter interactions, with a strong background in terahertz physics, semiconductor physics, photonic crystals, and metamaterials. Prof. Tan has authored or co-authored over 40 peer-reviewed journal papers and conference proceedings, including an invited review article and a perspective article on the topic of spintronic terahertz emitters.

Su-Huai Wei, PhD, is a chair professor and head of the Materials and Energy Division of the Beijing Computational Science Research Center (CSRC). He earned a BS in physics at Fudan University in 1981 and a PhD at the College of William and Mary in 1985. He joined the National Renewable Energy Laboratory (NREL) in 1985 and was a laboratory fellow and manager of the Theoretical Materials Science Group before he joined the CSRC in 2015. His research focuses on developing electronic structure theory of semiconductor compounds, alloys, and nanomaterials for optoelectronic and energy-related applications. He has published more than 500 papers in leading scientific journals, including more than 70 in *Physical Review Letters*. He is a fellow of both of the American Physical Society and the Materials Research Society.

Contributors

Yuriy Akimov
A*STAR Institute of High Performance
    Computing
Singapore

Qingyang Du
Department of Electrical Engineering
Xiamen University
Xiamen, Fujian, China

Zheng Feng
CAEP Microsystem and Terahertz
    Research Center
and
CAEP Institute of Electronic
    Engineering
Chengdu, Sichuan, China

Yuan Gao
Institute of Electromagnetics and
    Acoustics
and
Key Laboratory of Electromagnetic
    Wave Science and Detection
    Technology
Xiamen University
Xiamen, Fujian, China

Chuan Fei Guo
Department of Materials Science and
    Engineering
Southern University of Science and
    Technology
Shenzhen, Guangdong, China

Yidong Hou
College of Physics
Sichuan University
Chengdu, Sichuan, China

Qilin Hua
Beijing Institute of Nanoenergy and
    Nanosystems
Chinese Academy of Sciences
and
School of Nanoscience and Technology
University of Chinese Academy of
    Sciences
Beijing, China

Bing Huang
Beijing Computational Science
    Research Center
Beijing, China

Xiao Jiang
Beijing Computational Science
    Research Center
Beijing, China

Ru Li
College of Optoelectronic Engineering
Chongqing University
Chongqing, China

Zhou Li
GBA Branch of Aerospace Information
    Research Institute
Guangzhou, Guangdong, China

Fucai Liu
Yangtze Delta Region Institute
    (Huzhou)
and
School of Optoelectronic Science and
    Engineering
University of Electronic Science and
    Technology of China
Chengdu, Sichuan, China

Qing Liu
Yangtze Delta Region Institute
(Huzhou)
and
School of Optoelectronic Science and
Engineering
University of Electronic Science and
Technology of China
Chengdu, Sichuan, China

Xue Liu
College of Optoelectronic
Engineering
Chongqing University
Chongqing, China

Xiao Luo
Yangtze Delta Region Institute
(Huzhou)
and
School of Optoelectronic Science and
Engineering
University of Electronic Science and
Technology of China
Chengdu, Sichuan, China

Zhiguang Qiu
School of Electronics and Information
Technology
and
State Key Lab of Opto-Electronic
Materials and Technologies
Sun Yat-sen University
Guangzhou, Guangdong, China

Song Sun
CAEP Microsystem and Terahertz
Research Center
and
CAEP Institute of Electronic
Engineering
Chengdu, Sichuan, China

Wei Tan
CAEP Microsystem and Terahertz
Research Center
and
CAEP Institute of Electronic
Engineering
Chengdu, Sichuan, China

Yang Tian
Shanghai Key Laboratory of
Green Chemistry and
Chemical Processes
School of Chemistry and Molecular
Engineering
East China Normal University
Shanghai, China

Yongbiao Wan
CAEP Microsystem and Terahertz
Research Center
and
CAEP Institute of Electronic
Engineering
Chengdu, Sichuan, China

Dacheng Wang
CAEP Microsystem and Terahertz
Research Center
and
CAEP Institute of Electronic
Engineering
Chengdu, Sichuan, China

Su-Huai Wei
Beijing Computational Science
Research Center
Beijing, China

Xuannan Wu
College of Physics
Sichuan University
Chengdu, Sichuan, China

Jiafeng Xie
GBA Branch of Aerospace Information
    Research Institute
Guangzhou, Guangdong, China

Yinong Xie
Institute of Electromagnetics and
    Acoustics
and
Key Laboratory of Electromagnetic Wave
    Science and Detection Technology
Xiamen University
Xiamen, Fujian, China

Xiu Yang
College of Physics
Sichuan University
Chengdu, Sichuan, China

Huidong Yin
Yangtze Delta Region Institute
    (Huzhou)
and
School of Optoelectronic Science and
    Engineering
University of Electronic Science and
    Technology of China
Chengdu, Sichuan, China

Qi-Wei Zhang
Shanghai Key Laboratory of Green
    Chemistry and Chemical Processes
School of Chemistry and Molecular
    Engineering
East China Normal University
Shanghai, China

Taiping Zhang
CAEP Microsystem and Terahertz
    Research Center
and
CAEP Institute of Electronic
    Engineering
Chengdu, Sichuan, China

Jinfeng Zhu
Institute of Electromagnetics and
    Acoustics
and
Key Laboratory of Electromagnetic
    Wave Science and Detection
    Technology
Xiamen University
Xiamen, Fujian, China

1 Introduction to Optical, Infrared, and Terahertz Frequency Bands

Song Sun, Wei Tan, and Su-Huai Wei

CONTENTS

1.1 Introduction to Electromagnetic Waves 1
1.2 Brief Overview of Optical Frequency Bands 2
    1.2.1 Ultraviolet Regime 2
        1.2.1.1 Categories of UV Radiation 2
        1.2.1.2 Sources of UV Radiation 4
        1.2.1.3 Applications of UV Light 6
    1.2.2 Visible Light Regime 8
        1.2.2.1 Various Color Categories 8
        1.2.2.2 Sources of Visible Light 8
        1.2.2.3 Applications of Visible Light 11
1.3 Brief Overview on Infrared Frequency Band 13
    1.3.1 Categories of Infrared Radiations 13
    1.3.2 Sources of Infrared Radiations 13
    1.3.3 Applications of Infrared Light 14
1.4 Brief Overview of the Terahertz Frequency Band 16
    1.4.1 Sources of Terahertz Radiation 17
    1.4.2 Applications of Terahertz Radiations 18
1.5 Summary 20
Acknowledgements 20
Bibliography 20

1.1 INTRODUCTION TO ELECTROMAGNETIC WAVES

In physics, electromagnetic waves (EM) are synchronized oscillations of electric and magnetic fields propagating through space. Depending on the oscillation frequencies, the electromagnetic waves can be divided into many bands including gamma rays, X-rays, optical waves, infrared waves, terahertz waves, microwaves, and radio frequency waves, as shown in Figure 1.1. Although each band differs from other bands in terms of effects on matter and applications, their propagation processes and wave-matter interaction mechanisms are generally governed by Maxwell's equations. The fundamental EM wave-matter interaction theory will be introduced in

DOI: 10.1201/9781003202608-1

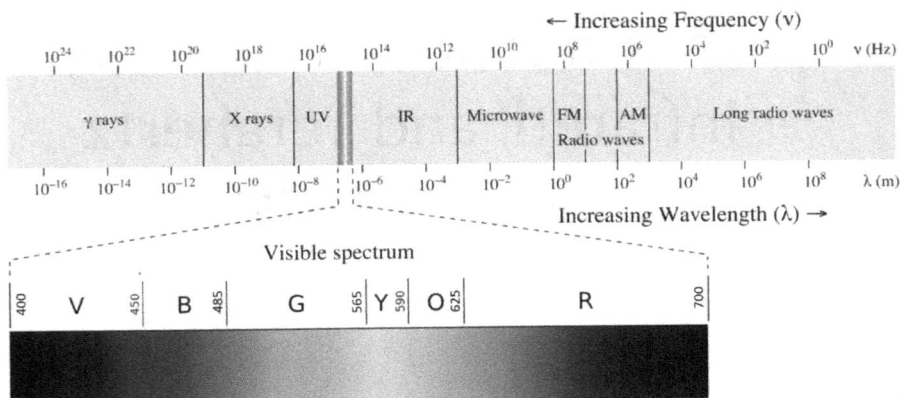

FIGURE 1.1   Different electromagnetic frequency bands.

Chapter 2. This book mainly focuses on three intermediate frequency bands: optical, infrared, and terahertz regimes. Unlike short-wavelength gamma rays and X-rays, these three bands are non-ionizing radiation since their photon energies are generally not sufficient to ionize atoms or break chemical bonds of molecules. The effects of these radiations on matter are caused primarily by heating effects. Unlike long wavelength microwaves and radio frequency waves with well-understood properties and matured applications, the wave-matter interactions in these three bands are less explored and many applications are yet to be developed. Hereafter, a brief introduction to the backgrounds of these three frequency bands is presented.

1.2   BRIEF OVERVIEW OF OPTICAL FREQUENCY BANDS

In a general sense, the optical frequency band accounts for both ultraviolet (UV) and visible regimes. The UV regime is the electromagnetic radiation with wavelengths from 10 nm (~ 30 PHz) to 400 nm (~ 750 THz), longer than X-rays while shorter than visible light. On the other hand, the visible regime (often termed light or visible light) covers wavelengths from 400 nm to 700 nm (~ 430 THz), longer than UV radiation while shorter than infrared waves. Besides this simple difference in wavelength, the UV and visible bands demonstrate unique phenomena when interacting with materials, thereby enriching a vast variety of applications. Hereafter, we shall briefly introduce some key features of these two frequency domains and their associated applications.

1.2.1   Ultraviolet Regime

1.2.1.1   Categories of UV Radiation

UV radiation was first discovered in 1801 by German physicist Johann Wilhelm Ritter when he found that invisible rays just beyond the violet end could quickly darken the silver chloride-soaked paper [1]. After years of research, a family of UV radiation was identified and can be divided into eight sub-types as summarized in Table 1.1, depending on their wavelength ranges as well as electromagnetic properties [2]. Note that there is some overlap among various sub-types. In particular, the first three categories can be distinguished by their absorptivity in atmosphere, where UV-A (also named soft UV or long-wave UV) is not absorbed by the ozone layer, UV-B (also called intermediate UV) is mostly absorbed by the ozone layer and UV-C (also named hard UV or short-wave UV) is completely absorbed by the ozone layer. This is because UV-C as well as even shorter wavelength UV light consists of high energy ionization radiation that could be absorbed by the oxygen to produce ozone.

The next four categories are defined by wavelength ranges together with some special functionalities. N-UV is the wavelength band that is visible to some birds, insects, and fishes. M-UV and F-UV are similar to the UV-B and UV-C, but with slightly different boundaries. E-UV (also called X-UV) is extremely high energy ionization radiation that results in a completely different wave-matter interaction mechanism. It is known that ultrahigh energy photons with wavelength smaller than 30 nm could empower effective interaction with inner-shell electrons and nuclei of atoms, whereas those with larger wavelengths mainly interact with outer valence electrons. This property makes E-UV strongly absorbed by most natural substances, which could be utilized for solar imaging in MSSTA (multi-spectral-solar-telescope array) and NIXT (normal-incidence-X-ray-telescope) sounding rockets to observe E-UV sources in space (e.g., coronal star) and to study the composition of the interstellar medium using E-UV spectroscopy.

The last category is termed V-UV, which is the spectral combination of E-UV and F-UV. This particular UV band is strongly absorbed by oxygen molecules in air, whereas 150 nm–200 nm wavelength UV radiation could pass through nitrogen.

TABLE 1.1
Spectral Categories of Ultraviolet Radiation

Category	Abbreviation	Wavelength (nm)	Frequency (PHz)
Ultraviolet-A	UV-A	315–400	0.75–0.95
Ultraviolet-B	UV-B	280–315	0.95–1.07
Ultraviolet-C	UV-C	100–280	1.07–3
Near ultraviolet	N-UV	300–400	0.75–1
Middle ultraviolet	M-UV	200–300	1–1.5
Far ultraviolet	F-UV	122–200	1.5–2.46
Extreme ultraviolet	E-UV	10–121	2.46–30
Vacuum ultraviolet	V-UV	10–200	1.5–30

Source: Materials are adapted from ISO-standard ISO-21348 [2]

Therefore, it is often exploited in those instruments operating in oxygen-free environment (e.g., gas chamber filled with pure nitrogen) without the need of expensive vacuum pump, for instance, UV immersion photolithograph equipment at 193 nm in the semiconductor industry [3].

1.2.1.2 Sources of UV Radiation

Though UV radiation is unperceivable to human eyes (blocked by lens and cornea of eyes), it is widely presented in this universe. Theoretically, any object with a sufficiently high temperature could emit a broadband of electromagnetic waves including UV light based on the black body radiation. In reality, the sun emits UV radiation at all wavelength bands mentioned above, which contributes approximately 10% of the total solar irradiance as shown in Figure 1.2 [4]. Note that the solar irradiance at the top of atmosphere is different to that at the earth ground surface, since the atmosphere blocks more than 70% of UV radiation (i.e., only most of UV-A and part of UV-B could reach the earth ground depending on the cloud thickness).

Besides hot planets, various artificial UV sources are developed including lamps, light-emitting diodes (LED) and lasers [5]. There are several kinds of lamps that could emit UV radiations. The first type is called black light or Wood's lamp, which produces long wave UV radiation based on either fluorescent or incandescent mechanism. Black light is commonly used in the situation where no visible light is needed, for instance, to trace fluorescence-tagged biochemical substance or to detect counterfeit money. The second type is the short-wave UV lamp, which typically emits short-wave UV-C light with peaks at 253.7 nm and 185 nm due to the mercury vapor filled within the lamp. These short-wave UV lamps are extensively utilized for disinfection purpose (also named germicidal lamps) in biomedical, food and water industries. The third type is the gas-discharge lamp, which could empower flexible UV radiation at various spectral

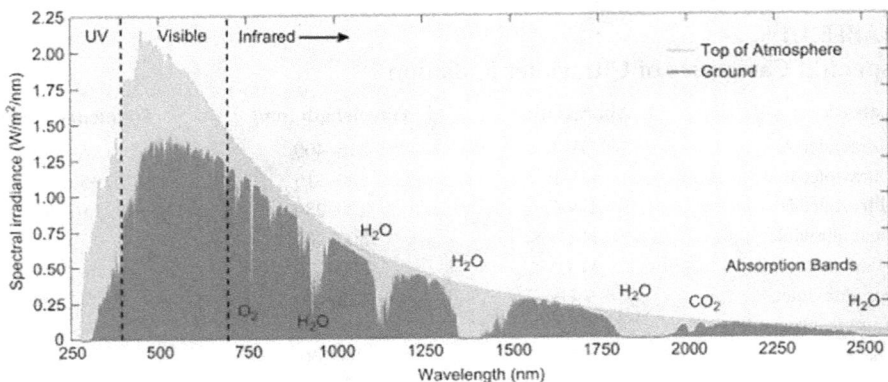

FIGURE 1.2 Solar irradiance spectrum as a function of wavelength.

lines depending on the gas types containing in the tube. The most commonly used gas-discharge lamps are neon lamp, deuterium arc lamp, xenon arc lamp, and mercury-xenon arc lamp, which cover the whole UV-A/UV-B/UV-C bands. The fourth type is the metal-halide lamp offering a high intensity white radiation via a mixture of gaseous mercury and metal halide (e.g., sodium iodide), since the metal ion could disassociate from the halide compound during the operation and produce additional emission power. Lastly, excimer lamp is a quasi-mono-chromatic UV source originated from the spontaneous emission of excimer mol-ecules, which spans over a wide range of UV spectra depending on the molecule types [6].

Besides the lamp bulbs, solid-state LED devices are also developed to facilitate smaller, cheaper, longer lifetime and more energy efficient UV sources [7]. LEDs emit UV radiation based on the electron-hole recombination mechanism, whose emission wavelength depends on the bandgaps of semiconductor materials. With the development of high quality AlGaN materials, UV LEDs with various emis-sion wavelengths can be manufactured by varying Al/Ga fraction which covers a wide range of UV-A (365 nm for GaN) to UV-C regime (210 nm for AlN). Other wide bandgap semiconductor materials such as diamond (235 nm) and boron nitride (215 nm) can also be used for deep UV LEDs. Compared to conventional UV bulbs, LED devices cause limited environmental contamination since no hazardous gas (e.g., mercury) is used, and have a faster switching time since there is no time delay due to filament warm up.

In general, the UV radiations from lamps or LEDs are incoherent (i.e., random polarization and no determinant phase), and laser devices are made for coherent UV emission. Nitrogen gas laser typically generates UV beams at 337 nm and 357 nm by electronically excite the nitrogen molecule. Excimer lasers could yield a wide spectral range of UV lasers depending on the types of excimer molecules. In particular, ArF laser emits UV at 193 nm and is routinely used in photolithography for integrated circuit manufacture. Up to present, the lowest UV wavelength pro-duced by the excimer laser is 126 nm via Ar_2^* and 96 nm via Ne_2^* [8]. Furthermore, solid state semiconductor lasers directly emitting UV lights are created based on the cerium-doped crystals, neodymium-doped fluoride fibers or some wide band-gap semiconductors (e.g. GaN etc), with typical emission wavelengths around UV-A regime. Note that the cutting-edge semiconductor technology has demon-strated that AlGaN quantum wells could achieve lasing in UV-C regime (~ 250 nm) [9]. To access shorter UV wavelength, a convenient method is to adopt nonlin-ear frequency conversion technique. For instance, the third and fourth harmonic generations from Nd:YAG laser producing 355 nm and 266 nm laser emission wavelengths. Applying a more general four-wave-mixing principle, tunable laser emission extending a wide range of V-UV band (75 nm ~ 200 nm) can be achieved via mixing photons from ArF laser and another tunable visible or near-infrared laser.

The aforementioned UV sources are difficult to access the E-UV domain. An indirect approach is to use excimer laser to excite a very hot plasma. The electron transition of hot plasma naturally radiates E-UV at 13.5 nm, which is crucial for photolithograph in semiconductor industry. On the other hand, free electron laser based on expensive synchrotron facilities could generate UV radiation at all wavelength, including E-UV down to 10 nm.

1.2.1.3   Applications of UV Light

There are many applications harvesting the benefits of UV light. Figure 1.3 briefly introduces some of them.

Human health issue. UV radiation is a double-edged blade to human health [10]. On one hand, UV-B could promote human body to produce vitamin D, which is critical for life. In addition, UV rays can also treat certain skin conditions such as psoriasis and jaundice for therapy purpose. On the other hand, overexposure to UV light could cause damage to skins and eyes. It is found that all UV bands could damage collagen fibers and accelerate aging of skin. Specifically, the high energy UV-C and UV-B radiations are capable of directly damaging DNA and eventually lead to deadly skin cancer. Exposure to UV-C and UV-B also damages the cornea, lens and retina of eye and causes photokeratitis. Even the low energy UV-A radiation is found to cause indirect damage to DNA due to some harmful reactive chemical intermediates. These potential risks trigger a serious concern on ozone depletion as well as a safety debate on sunbath and sun tanning bed.

Sterilization and disinfection. Since high energy UV radiation could damage DNA, it could be naturally applied for sterilization and disinfection of microorganisms in biology laboratories, medical facilities, food processing and waste water treatments. Mercury vapor lamp emitting at the peak of germicidal effectiveness curve (~ 265 nm) is the most commonly used UV source, while UV-C LED gradually gains its popularity due to its tunable emission wavelength and monochromatic nature.

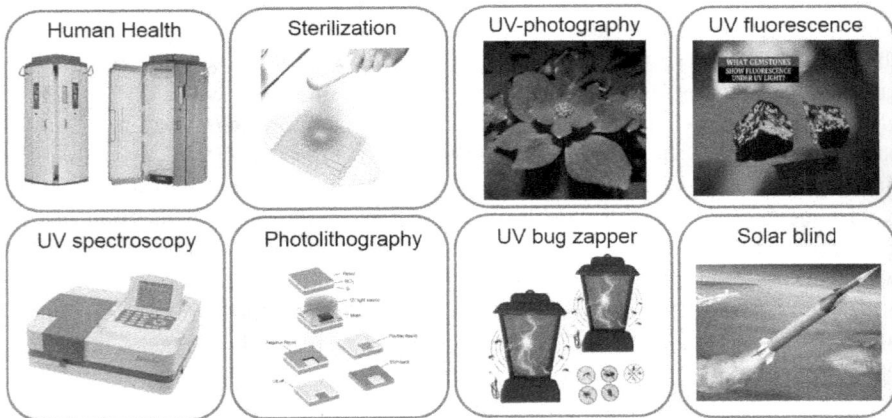

FIGURE 1.3   Various applications of UV radiation.

Photography. To take a UV photography, the target is illuminated directly with a UV source and the reflected signal is collected via a specially developed lens together with a UV-passing filter. UV photograph serves a number of scientific, medical, forensic and artistic purposes, for example, revealing deterioration of art works or structures, diagnosing certain skin conditions, and even analyzing the composition and temperature of the interstellar medium.

UV induced fluorescence. This technique applies a UV source to excite fluorescence dyes which emits visible or infrared radiation. UV-induced fluorescence is widely used in forensic, biochemistry and non-destructive testing, for example, tracing artifacts in archaeological sites, detecting counterfeited goods, documents or currency, and revealing surface defects or magnetic leakage fields. It also have been used in paints, papers, and textiles to enhance color or to produce special effects.

Analyzing bio-chemical species. Many bio-chemical compounds possess distinctive fluorescence under UV illumination. Therefore, UV source is an investigating tool at the crime scene to identify body fluids such as semen, blood and saliva. It is also frequently used in various industries requiring sanitary compliance (e.g. hotel industry), since many organic molecules show traced signals under UV light. UV-Visible multispectral analyzer is adopted to read illegible documents by distinguishing ink from paper. It is also used in chemical industry to analyze the structures of chemical compounds or to monitor environmental contaminations of hazardous gases (e.g., sulfur compounds etc) in fossil-fired power plants.

Photolithography. Sophisticated UV source is an essential component for fine resolution photolithograph, which is of great significance in the semiconductor industry. As a rule of thumb, the minimum linewidth of photolithography is proportional to the wavelength of optical source. A smaller linewidth generally results in a smaller transistor size, which is critical to fulfill Moore's law in building more compact and more powerful integrated circuit. Currently, 193 nm UV sources are commonly used for fabrication of semiconductor devices and circuits, while 13.5 nm UV sources are available for extremely fine resolution lithograph.

Biology. Some of the birds, reptiles and insects can see UV light and use it as a communication system for sex recognition and mating behavior. Many insects rely on UV emission from celestial objects as references for flight navigation. An UV trap (so called bug zapper) could disrupt the navigation processes and attract small flies, eventually eliminating them with electric shock.

National defense. An important task of national defense involves scanning the sky for missiles and other airborne threats below the atmosphere, but the bright flare of an incoming rocket might be overwhelmed by the sun during the day. The high temperature flare emits a broadband of electromagnetic waves including UV with wavelength shorter than 280 nm, which could be used to trace missile flares without being affected by the sun radiation (atmosphere blocks this high energy UV from the sun, which is called solar blind). The solar blind regime has also been proposed for high speed free space communication.

TABLE 1.2
Spectral Color Regimes

Color	Wavelength (nm)	Frequency (THz)
Violet	400–450	670–790
Blue	450–485	620–670
Cyan	485–500	600–620
Green	500–565	530–600
Yellow	565–590	510–530
Orange	590–625	480–510
Red	625–700	400–430

1.2.2 VISIBLE LIGHT REGIME

1.2.2.1 Various Color Categories

Visible light is the electromagnetic radiation that can be conceived as different colors by human eyes and brains, which spans from 400 nm to 700 nm as shown in Table 1.2. There are two things worthy to be mentioned: 1) Table 1.2 does not contain all the colors that the human visual system can distinguish. Unsaturated colors such as pink or magenta, are absent in color spectrum since they consist of multiple wavelengths; 2) the boundaries for visible light is not very strict. Some individual is capable of perceiving UV light down to 310 nm.

Although it can be directly seen by human eyes, the discovery of visible light takes quite a long time [11]. Early in the 13th century, Roger Bacon had already proposed that the rainbow in the sky is formed under a similar principle to light passing through crystal. In 1665, Isaac Newton had shown, for the first time that, a prism could disassemble and reassemble white light. At an appropriate incident angle, the prism could bend visible light and each color is redirected to a slightly different angle depending on its wavelength. He was also the first one to use the word "spectrum" and initially divided the spectrum into six colors (red, orange, yellow, green, blue, and violet). He then proposed to add "indigo" as the seventh color, but later suggested that it merely a shadow of blue and violet ("indigo" proposed by Newton actually refers to "blue" in modern color spectrum, and his "blue" refers to "cyan"). Until the 19th century, the boundaries of visible light became more definite since the invisible UV and infrared light were measured by Johann Wilhelm Ritter and William Herschel respectively. In 1802, Thomas Young first obtained the wavelengths of different colors, and connected the visible spectrum to the color vision of human in Young-Helmholtz theory [12].

1.2.2.2 Sources of Visible Light

There are mainly five types of visible light sources that we frequently encounter in our ordinary lives: natural sources like the sun or fire, incandescent lamps, LED, laser, as well as visible fluorescence. Similar as UV, objects with adequate temperature could emit visible light according to the black body radiation. The concept of "color temperature" is introduced to characterize the equivalent temperature of a visible light source by modelling it as an ideal black body radiator [13]. In general, color temperatures beyond 5000 K are called "cool" colors and appear to be white or blue (e.g., sun), while those below 3000 K are named "warm" color and appear to be yellow or red (e.g., match or candle flame). The radiation of sun closely resembles a black body radiation with an effective color temperature at about 5780 K. The visible light contributes approximately 40% of the total solar irradiance (see Figure 1.2). Note that the sun may appear various colors such as red, orange, yellow or white depending on its position in the sky, which essentially results from the scattering of sunlight in the atmosphere instead of changing on the color temperature. Two things are worthy to mention: 1) the meaning of color temperature is completely different from the context of conventional temperature, and so are the word "cool" and "warm" used here; 2) for light sources not based on thermal radiations (e.g., LED and laser), their emission does not follow the black body spectra. Correlated color temperature (CCT) is commonly adopted for these sources, which most closely matches human color perception.

An incandescent lamp converts thermal energies into visible photons with a filament enclosed in a glass bulb [14]. Electric current flows through the filament to heat it up until it glows, which mimics a black body radiator at the same temperature. It was first discovered in 1761 by Ebenezer Kinnersley that heating a metal wire could produce incandescence. In 1802, Humphry Davy created an incandescent light by passing electric current through a platinum strip. In 1838, Marcellin Jobard invented the first incandescent light bulb with a vacuum atmosphere using a carbon filament, and two years later, Warren de la Rue first implemented platinum as the filament in a vacuum bulb. In 1859, Moses G. Farmer built an electric incandescent light bulb using a platinum filament and filed a patent, but it was sold to Thomas Edison later. In 1878, Thomas Edison started his own research on incandescent lamps and demonstrated the first successful test in 1879. After testing many filament materials and improving the designs, he discovered that a carbonized bamboo filament could last more than 1200 hours and eventually paved the way for commercialization. Nowadays, very cheap incandescent lamps are available with three major color options of incandescent bulbs for consumers depending on the input voltages and filament materials: soft white (2700 K–3000 K), cool white (3500 K–4100 K) and day light (5000 K–6500 K).

The main drawback for incandescent lamp is that its thermal radiation always leads to a continuous spectrum, and auxiliary device such as filter is needed to obtain monochromatic light. In addition, the emission of incandescent lamp is omnidirectional, which could not satisfy the demands for unidirectional radiations in applications such as wireless communication and wearable electronics. To overcome these obstacles, semiconductor LED devices are invented based on electroluminescence mechanism [15]. Although the first LED was invented to radiate infrared light at 900 nm in 1961, the first visible LED emitting red light was demonstrated shortly after that via a GaAsP tunneling junction. In 1971, the first green LED was created with zinc-doped GaN material by Pankove and Miller. In 1972, the first yellow LED is invented with a combination of two GaP chips (one red and one green) by M. George Craford. He also significantly improved the brightness of red and red-orange LEDs, with AlGaAs and AlInGaP material systems respectively. Until 1993, high-brightness blue LED was demonstrated by Shuji Nakamura based on a GaN system after the major breakthrough of GaN epitaxial growth and p-type doping. Since then, all the individual colors in the visible spectrum are completely covered. White light could thus be resembled by mixing the three primary colors (red, green, blue) of LEDs or using phosphor materials (e.g., cerium doped YAG crystal) excited by blue LED. Nowadays, more advanced LED technologies, such as organic LED or quantum dot LED, have been developed to obtain better illumination performances. More details can be found in a recent review [16].

Fluorescence spontaneous emission is another type of visible light source, which is commonly considered as the basis for laser stimulated emission. Unlike LED, which is mainly driven by electric power, the fluorescence emission could be excited by a variety of power sources including biochemical, optical, mechanical, and electrical sources. Nature itself contains many fluorescence species such as firefly, jelly fish and coral, whose fluorescence emission result from the chemical reactions of the organisms (also called bioluminescence). Other than that, organic fluorescence dyes, artificial quantum dots (QD), nitrogen-vacancy center (NV), atomic emissions are found to emit visible light under electric or optical stimulus at different wavelengths and bandwidths [17]. The rich variety of fluorophores thus enable a plethora of emergent applications such as biochemistry, optoelectronic device and system, high resolution display, and optical communication, among the others. In addition, certain long afterglow materials demonstrate extremely long lifetime fluorescence emissions (also called phosphorescence) that they can continue to glow-in-dark even after the excitation source is switched off.

Because of its incoherent radiation nature, LED, or fluorescence spontaneous emission in general cannot have very high intensity or brightness. Laser devices are made for applications requiring highly monochromatic light, stable and high output power as well as highly directional radiation. The theoretical framework of stimulated emission process of electromagnetic radiation was first proposed by Albert Einstein in 1917, and experimentally demonstrated 30 years later by Willis E. Lamb and Robert C. Retherford. In 1953, the stimulated emission was first utilized to amplify the microwave radiation which is now termed as maser. In 1955, Nikolay Basov and Aleksandr Prokhorov proposed optical pumping of a multi-level system for carrier population inversion, which is now the mainstream approach for continuous laser operation. In 1957, Charles H. Townes and Arthur L. Schawlow started their pioneer research on optical maser in infrared and visible regime. In 1957 and 1958, Gordon Gould, Aleksandr Prokhorov and Charles H. Townes independently proposed the idea of using open resonator for light amplifying and named it as laser for non-microwave device. The first functioning laser was invented in 1960 by Theodore H. Maiman using the flashlamp-pumped ruby crystal, which yielded pulsed red laser emission at 694 nm. Until 1962, the first semiconductor laser diode based on GaAs was invented in the near-infrared band by Robert N. Hall, and later in the visible regime by Nick Holonyak

Nowadays, various types of visible light laser devices are developed including gas lasers (e.g., He-Ne laser emits 543 nm, 594 nm, 605 nm, 612 nm and 633 nm), semiconductor laser diode (e.g., GaN laser emits 400 nm ~ 480 nm; AlGaInP laser emits 630 nm ~ 645 nm), nonlinear laser system (e.g., second harmonic generation from 1064 nm Nd laser), tunable visible light dye laser (e.g., rhodamine 6G emits 540 nm ~ 680 nm) and supercontinuum free electron laser. By mixing various colors, the first white light laser was made in 2015 based on a mixture of zinc, cadmium, sulfur, and selenium [18].

1.2.2.3 Applications of Visible Light

A few typical applications using visible light are introduced in the following (see Figure 1.4):

Natural phenomena. When we see that the grass is green, the hair is black or the flower is red, it is because these objects could reflect some specific colors into our eyes while being absorptive or transparent for other colors. Some exotic natural phenomena are also results of interference, diffraction, scattering of visible light. For instance, the rainbow on the sky is due to the reflection and refraction of sunlight by water droplets in the atmosphere. In addition, some biological processes are also affected by visible light, e.g., the photomorphogenesis process of plant growth.

Spectroscopy. UV-Vis multispectral spectroscopy is developed to study the electronic transition properties of atoms and molecules, which is routinely used in analytical chemistry to quantitively determine the types and concentrations of different analytes, such as metal ions, highly conjugated organic compounds, and biological macromolecules. It is also commonly used in the fields of optics and materials to measure the thickness and optical properties of thin films, which could give the complex refractive index dispersion of a given film based on the measured reflectance spectra via Forouhi-Bloomer equations.

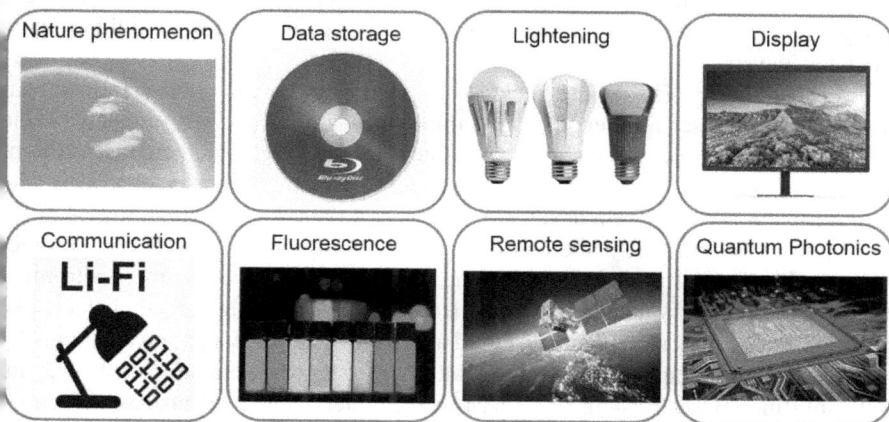

FIGURE 1.4 Various applications of visible light.

Data storage. The Blue-ray technology in the computer, film, and gaming industries exploits blue laser of 405 nm to read and write the disc, which is able to store several hours of high-definition video. Recently, ultra HD Blu-ray discs (4K Blu-ray) and players have become available, allowing a storage capacity of 100 GB. More advanced technologies including using UV laser or Holographic Versatile Disc (HVD) are under development, which might further enhance the storage capacity to a few TB in a single disc.

Lightening. Long time ago, the sunlight is the only visible light source that mankind can rely on. The invention of incandescent lamp frees human from their dependencies on sunlight and allows them working, studying, and playing even after the night fall, greatly enriching people's lives. Nowadays, more advanced LED devices are gradually adopted to replace the conventional lamps in many applications including aviation lighting, automotive headlamps, advertising, general lighting, traffic signals, and medical devices, among the others.

Display and photography. With the development of monochromatically visible light sources as well as detectors, high-definition color display and photography are achievable, which have been widely used in computer and projector monitor, digital camera, and smart phones. Continuous efforts are spent to create miniaturized visible sources (e.g., micro-LED and nano-LED), in order to further reduce the pixel size and increase the resultant image resolution.

Communication and positioning. Visible light communication (VLC) utilizes fluorescence lamp or LED to transmit signal at a high speed over a short distance, which could be beneficial for applications such as the internet-of-things (IoTs) and indoor Li-Fi technology. In addition, visible light indoor positioning is feasible by using LED as the beacon which transmit the location signal via the VLC approach.

Fluorescence imaging. Visible fluorophores are frequently used as markers to label organic compounds such as protein and DNA, whose emission intensity is proportional to the concentration of fluorophore. As a complementary to the UV-Vis absorption spectroscopy, fluorescence spectroscopy is a popular instrument especially in the medical and biochemical industry. Combining with modern nanotechnology, detection and dynamic tracing of single molecule labeled with fluorophore can be achieved.

Remote sensing. Visible laser device has been used for remote sensing applications (so-called laser altimetry). For example, it could be utilized to calculate the elevation of earth's polar ice sheets, or to measure the heights and characteristic of clouds, or to estimate the heights and structures of canopy of forests, or to sense the distribution of aerosols from dust storms and forest fires.

Quantum optics. Lastly, there is an important property of light called particle-wave duality, indicating that the light can not only be described by the electromagnetic wave, but also by quanta of light called photon. When lights mainly behave like particles, exotic optical phenomena could arise which can no longer be explained by the classic electromagnetism but the quantum physics, e.g., entanglement, vacuum Rabi-splitting, Bose-Einstein condensates, etc. Interesting readers may refer to a classic text book for more details [19].

TABLE 1.3
Sub-division of Infrared Regime

Category	Wavelength (μm)	Frequency (THz)
Near-infrared (NIR)	0.7– 1.4	214–430
Short-wavelength infrared (SWIR)	1.4–3	100–214
Middle-wavelength infrared (MWIR)	3–8	37–100
Long-wavelength infrared (LWIR)	8–15	20–37
Far-infrared (FIR)	15–1000	0.3–20

1.3   BRIEF OVERVIEW ON INFRARED FREQUENCY BAND

First discovered by William Herschel in 1800, the infrared (IR) regime is now defined as the electromagnetic wave within frequency range of 700 nm (430 THz) to 1 mm (300 GHz), which is sandwiched between the visible and microwave bands. IR radiation is ubiquitously presented almost everywhere since the black body radiation from any object near room temperature is at infrared region. The universal existence of IR radiation is thus heavily utilized in industrial, scientific, military, and medical applications.

1.3.1   CATEGORIES OF INFRARED RADIATIONS

Table 1.3 gives a commonly used sub-division scheme for IR radiations. NIR (also called IR-A region) is defined by water absorption, and plays an important role in fiber optical communication, night vision equipment and NIR spectroscopy. SWIR (also called IR-B region) is also related to water absorption and often used for long-distance telecommunications. MWIR (also called thermal infrared band) is the band where the infrared signature of aircraft's engine exhaust plume lies within, and has important military applications. LWIR (also called thermal imaging band) is frequently used to obtain a completely passive image of objects with temperatures slightly higher than room temperature. It is only based on the thermal emission from objects and does not require any external light source. FIR is typically used in astronomy to observe interstellar gases where new stars are formed. Sometimes it is also used for human body detection as well as thermal therapy. The MWIR, LWIR and FIR together are called IR-C region. Note that the long-wavelength end of FIR overlaps with the terahertz band.

The boundaries between different IR categories are not strict, which may vary depending on the applications. For example, the astronomers in NASA use a different scheme where NIR is from 0.7 μm to 2.5 μm, middle-infrared (MIR) is from 2.5 μm to 25 μm, and FIR is from 25 μm to 1000 μm. The ISO defines NIR from 0.7 μm to 3 μm, MIR from 3 μm to 50 μm, and FIR from 50 μm to 1000 μm [20].

1.3.2   SOURCES OF INFRARED RADIATIONS

In general, there are three types of infrared sources that are commonly used: thermal radiation sources, IR LEDs, and IR lasers. Any object with appropriate temperature could emit thermal radiation in IR frequency, which has a maximum emission wavelength that is inversely proportional to the absolute temperature of object, in accordance with Wien's displacement law. The concept of emissivity is used to characterize the capability of infrared emission of an object, where that with higher emissivity appear hotter. Sun emits at all IR wavelength, which contributes 50% of total solar emission power. Even the UV or visible light from solar irradiance could be reabsorbed by objects on earth and reemitted as infrared waves. Incandescent lamps or fires at high temperatures could generate continuous emission spectra in UV, visible, and infrared regimes, where the infrared radiation accounts 75% of the total emitted power. Note that the glass bulb of incandescent lamp would block IR radiation with wavelength larger than 5 μm, and therefore it is commonly used as NIR and MIR source. In addition, gas-discharging lamps such as xenon and neon could generate strong NIR radiation during the gas-discharging process, and is commonly served for infrared communication purpose. Objects at room temperature emit radiation mostly in the LWIR band (8 μm ~ 25 μm).

Early in the 1950s, infrared emission from various semiconductor alloys (e.g., GaAs, GaSb, InP, SiGe) had been discovered, and the first LED was invented at 900 nm in 1961 [15]. Today, the most commonly used IR LEDs are based on GaAs with the central emission peak within NIR region (typically around 830 nm ~ 980 nm). In addition, with the development in crystal growth technology, MIR LEDs are now available based on heterostructures of III-V materials (e.g., InGaAs, InAsSb) with emission wavelengths of 1.4 μm ~ 7 μm, which are particularly useful for hazardous gas monitoring (e.g., CH_4, CO_2). Infrared LEDs are widely used for remote control (e.g., television) where an infrared beam carrying instructions in binary code is sent from the remote control to the targeted device. They are also utilized in the security systems. For instance, security camera uses infrared LED to emit appropriated amount of infrared light according to the light level in the recorded area, enabling infrared imaging at night.

To obtain high power coherent IR radiation, various kinds of infrared laser sources are developed [21]. In NIR region, solid state lasers based on doped crystals (e.g., Nd:YAG, Nd:YVO$_4$, Yb:YAG emit at 1064 nm) or semiconductors (e.g., GaAs emits around 900 nm) are most commonly used. In MIR region, gas discharging lasers (e.g., CO emits around 6 μm ~ 8 μm and CO_2 emits at 10.6 μm), doped crystal lasers (e.g., Co:MgF$_2$ emits around 1.7 μm ~ 2.5 μm, Cr^{2+}:ZnSe emits around 3.5 μm), and fiber lasers (e.g., Er-doped fiber emits at 2.8 μm, Ho-doped fiber emit around 3 μm ~ 3.9 μm) are frequently encountered. In the FIR region, quantum cascade laser (QCL) based on a repeat stack of semiconductor multi-quantum well heterostructures is invented to emit IR laser radiation in MIR and FIR bands, from 2.6 μm to 250 μm (can extends to 355 μm under an external magnetic field). Other than that, p-Ge semiconductor (emit at 40 μm ~ 400 μm) and methanol molecule (emit at all FIR wavelength) are newly discovered for FIR lasing [22]. Lastly, nonlinear frequency conversion and free-electron laser can also generate IR radiations.

1.3.3 Applications of Infrared Light

Several common applications with IR radiations are described in the following (see Figure 1.5) [23,24]:

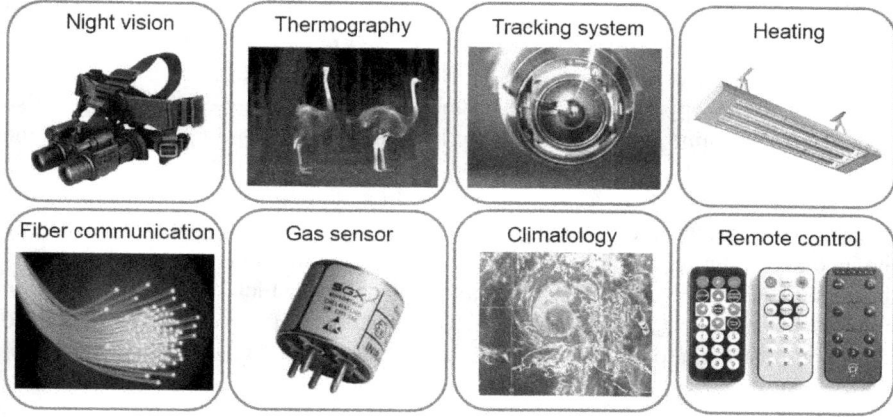

FIGURE 1.5 Various applications of infrared waves.

Night vision. IR radiation is heavily used in night vision equipment where the visible light is insufficient, increasing in-the-dark visibility. Night vision devices operate through a process involving the conversion of infrared light into electrons that are then amplified by a chemical and electrical process and then converted back into visible light.

Thermography. Also named thermal imaging, IR radiation is used to measure the temperature (or emissivity) of an object. Since IR radiation is emitted by all objects based on their temperatures, thermography allows people to visualize the object and its environment without visible light. It was mainly used for military and industrial applications, but now is also available for civil applications, e.g., infrared camera on car.

Tracking. Also known as infrared homing, infrared tracking refers to a passive missile guidance system which tracks the infrared emission from a hot target (e.g., engines of vehicle or aircraft) with temperature higher than the background.

Heating and cooling. IR radiation could be a very efficient method to control the temperature of an object. On one hand, IR radiation could be a deliberate heating source, which is used in several civil (e.g., saunsa, de-icing, cooking, or grilling, etc.) and industry applications (e.g., curing of coating, annealing, plastic welding, etc.). On the other hand, it can also be used as a radiative cooling method to cool buildings or other systems, in particular for LWIR band since it can escape into space through the atmosphere.

Communications. Today, the optical fiber communication systems are operated at NIR region where the least dispersion wavelengths (1330 nm and 1550 nm) of silica fiber locate within. In addition, free space infrared communication uses IR laser is sometimes used in the urban area to reduce the cost. Other than that, data transmission among computers and handphones based on IR LEDs are popular for indoor communication in high population density area. Lastly, IR radiation is the most common way for remote control to command appliances.

Spectroscopy. Since the chemical bonds of many molecules vibrates at infrared frequencies, IR vibration spectroscopy can be used to detect these species or study the composition of organic compounds. Both NIR and MIR bands are commonly used for spectroscopy. Similar as the visible light source, IR spectroscopy can also be used to determine the complex refractive index dispersion of a thin film in the infrared region.

Meteorology, climatology, and astronomy. In meteorology, weather satellites use thermal or infrared images to estimate the cloud heights and types, or to calculate the land and water temperatures. In climatology, IR radiation is monitored to reflect the energy exchange between the earth and atmosphere, which provides information on climate change (e.g., global warming). In astronomy, infrared telescope is applied to measure the radiation heats from various kinds of stars (e.g., imbedded stars, protostars) in our galaxy.

Environmental science. Many hazardous gas molecules possess chemical bond signatures in the infrared frequencies, for example, CO_2, CO, CH_4, SO_2, etc. Therefore, infrared sensors are usually deployed to monitor these harmful gases for personal and environmental safety.

Biological effects. Many animals use infrared as vision or sensing systems. For example, pit viper has a pair of infrared sensory on its head to detect the temperature of prey. Other organisms that have thermoreceptive organs include pythons, boas, bat, just to name a few. Other than that, fish use NIR to capture prey and for phototactic swimming orientation. Some fungi require near-infrared light for ejection.

1.4   BRIEF OVERVIEW OF THE TERAHERTZ FREQUENCY BAND

At the long wavelength end of FIR regime, there exists a special electromagnetic wave band called terahertz (also called T-ray, T-wave or T-light) with frequencies from 0.3 THz to 10 THz, or equivalently from 1 mm to ~30 μm in wavelength (sometimes it also refers to the frequency range from 0.1 THz to 10 THz). Falling at the transition region between infrared and microwave domains, terahertz wave shares some properties from each of them. Similar to microwaves, terahertz radiation can penetrate many non-conducting materials including clothing, paper, wood, plastic, and ceramics. Although the penetration depth is smaller than that of microwave, terahertz radiation could generate images with higher resolution due to its shorter wavelength, which makes it potentially suitable for non-destructive testing. Similar to infrared wave, terahertz radiation is strongly absorbed by certain gas and liquid molecules, which makes it capable for spectroscopic analyzing. In particular, terahertz wave could penetrate some distance into human tissue in a non-ionizing manner, which makes it a potential replacement for medical X-rays. These unique properties of terahertz band have driven the development of novel technologies in many applications. Note that some communities such as astronomy also name terahertz wave as submillimeter wave. These two concepts are essentially identical.

1.4.1 SOURCES OF TERAHERTZ RADIATION

Both natural and artificial sources are found to emit terahertz radiation [25]. In nature, any object with temperature higher than two kelvins is able to generate terahertz wave, according to the black body radiation. Though the natural thermal emissions are typically weak, they are very important for characterizing the cold cosmic dust in the interstellar clouds. At present, there are several telescope systems operating in terahertz band such as James Clerk Maxwell Telescope and the one located at Herschel Space Observatory.

Besides, various artificial sources are available for terahertz generation, including thermal lamps, electronic sources, lasers, optically pumped sources and mechanically excited sources. The first type of terahertz source is the thermal lamp (e.g., mercury lamp) which also operates based on black body radiation and is able to generate a broadband continuous emission spectrum at high temperature. For this type of source, the terahertz radiation is usually at the long wavelength end of the thermal emission spectrum with a very limited power efficiency. Therefore, it is more often used in UV, visible and NIR regimes instead of terahertz band.

The second type is the electronic terahertz source, which can be further divided into two sub-types: vacuum electronic and solid-state electronic sources. For the former type, travelling wave tube (e.g., backward wave oscillator, extended-interaction klystron oscillator) is commonly used to amplify electromagnetic signal from microwave to terahertz regimes. Other than that, vacuum particle accelerators (e.g., gyrotron, synchrotron, free electron laser) could also generate a broadband electromagnetic wave including terahertz band. The main advantage of vacuum electronic source is the high terahertz output power, but at a cost of bulky and expensive equipment. For the latter type, semiconductor diodes (e.g., Schottky-barrier diode) and transistors (e.g., BiCMOS, HEMT) are widely used for frequency multiplication up to the THz regime, and negative differential resistance devices (e.g., Gunn diode, resonant-tunneling diode) are employed as THz oscillators for direct THz wave generation. Up to date, the multiplier based on Schottky-barrier diode can operated at the frequencies up to ~ 3 THz, while the oscillator based on resonant-tunneling diode can generate THz wave up to ~ 2 THz. Recently, a novel solid-state terahertz source based on layered superconductor $Bi_2Sr_2CaCu_2O_8$ was developed. The advantage of solid-state electronic source relies on its high compactness and high integration level.

Next, various types of lasers could produce coherent terahertz radiations. QCL could directly emit FIR and terahertz wave up to 355 μm, which is a very compact terahertz source that are popularly used. The main drawback of QCL is the requirement of cryogenic cooling system. In addition, certain molecular gas lasers can emit terahertz radiation at discrete frequency based on some molecular rotation states. For example, CO_2 pumped methanol laser can yield terahertz at 2.5 THz, but with a very low conversion efficiency. Furthermore, some semiconductor lasers (e.g., p-Ge) could generate terahertz wave up to 400 μm.

Optically pumped source produces terahertz radiation by exciting certain target material or structure with pulsed or continuous laser. Perhaps the most commonly used optically pumped source is the photoconductive switch (also called Auston switch) illuminated by femtosecond laser pulse, whose accelerated photocurrent causes the charge carriers to radiate picosecond terahertz pulse with frequency from 0.3 THz to 3 THz. Photoconductive switch can also be operated in a continuous-wave mode via difference frequency generation of two laser beams. In this case, photoconductive switches are also called photomixers. Some nonlinear crystals (e.g., ZnTe, GaP, $LiNbO_3$) are also frequently adopted to produce pulsed terahertz radiation with frequency from 0.5 THz—3 THz, which results from the optical rectification principle—a second order nonlinear process. Other than that, ferromagnetic and antiferromagnetic materials are recently discovered to generate terahertz radiation (also named spintronic terahertz source) under femtosecond laser illumination, which is essentially attributed to the spin-charge conversion mechanism. This novel type of magnetic material will be discussed in Chapter 10.

Finally, terahertz radiation can also be achieved via mechanical excitation. It was recently report that the process of peeling adhesive tape from the roll could emit unpolarized terahertz wave from 1 THz–20 THz, which is essentially attributed to the tribocharging of adhesive tape and subsequently discharge, eventually resulting in bremsstrahlung. In addition, surface formation could also induce terahertz radiation (also named terahertz surfoluminescence), which is originated from the ultrafast separation of charge within a newly formed surface. Nevertheless, the present mechanically excited terahertz source suffers from limited output power, incoherent and unpolarized radiation, which is still in its infancy and not commonly used in practice.

1.4.2 APPLICATIONS OF TERAHERTZ RADIATIONS

The unique properties of terahertz radiation could enable various applications as shown in Figure 1.6, including different imaging systems, wireless communication, and particle accelerator [26].

Medical imaging. Compared to the ionizing X-ray imaging, the low photon energy of terahertz radiation in general does not damage living tissues and DNA. At certain frequencies, terahertz wave can penetrate several millimeters of tissue with low water content (e.g., fat, teeth) and reflect back, which makes it suitable for non-invasive and painless medical imaging. Since it is strongly absorbed by water molecule, terahertz radiation can clearly differentiate the tissue content from water content, providing solid evidence for medical diagnosis and surgery. During the outbreaking of COVID-19, terahertz spectroscopy and imaging have also been proposed as a rapid screening tool.

Security. Because the terahertz wave can penetrate normal fabrics and plastics, it can be used for remote screening to uncover concealed weapons on a person, avoiding the intrusion of human body privacy. This is of particular interest for security industry, because many materials of interest (e.g., chemicals, drugs, explosives) have unique "fingerprints" in the terahertz range. The first compact THz camera prototype was developed in 2004 by a company called ThruVision which successfully demonstrated the capabilities to image guns and explosive

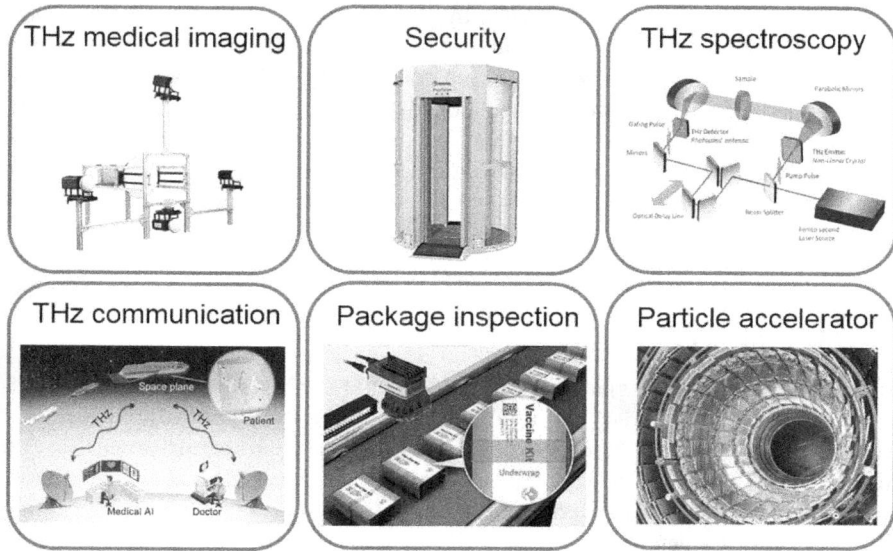

FIGURE 1.6   Various applications of terahertz wave.

under clothing. Today, mature THz security screening system is already available in the market.

Spectroscopy. Terahertz spectroscopy could provide valuable information in the fields of astronomy, materials science, and biochemistry. In particular, the terahertz time-domain spectroscopy (THz-TDS) is able to analyze samples that are opaque in the visible and near-infrared regimes. Though it requires the sample to be thin, THz-TDS yields coherent pulse radiation which contains abundant information in a broadband frequency range. In addition, terahertz waves are commonly used to study material properties in high magnetic fields, since the Larmor frequencies of electron spins are in the terahertz band.

Ultrahigh-speed wireless communication. The next-generation (6G) communication is proposed to offer peak data rate in the range of 100 Gb/s–1 Tb/s. The meet the requirement, the THz band was announced to be open for experimental use for 6G wireless communication networks. Back in 2012, a team of Japanese researchers reported a new record of 3 Gb/s for wire data transmission using terahertz wave, and most recently, THz communication experiments with a data rate of ~ 100 Gb/s have already been demonstrated using diverse technologies (both electronics- and photonics-based configurations), which shows great potential for ultrahigh-speed wireless communication. Furthermore, by using vacuum electronic amplifier for high output power of ~ 2 W, THz communication distance has been extended to over 20 km with a data rate of 5 Gb/s [27].

Manufacture process monitoring. Terahertz radiation can be potentially utilized to inspect packaged goods since the traits of plastics and cardboards are transparent at THz frequencies. This is highly demanded in the manufacturing industry for quality control and process monitoring. In general, terahertz imaging systems could result in a higher resolution than scanning acoustic microscopes but lower resolution than X-ray systems. Nevertheless, the non-ionizing nature of terahertz wave is preferred for certain goods such as semiconductor electronic chips or living tissues.

Particle accelerator. A new types of particle accelerator called "beam driven dielectric wakefield accelerators" (DWAs) operates in the terahertz band, which pushes the plasma breakdown threshold for surface electric fields into the multi-GV/m range. Interesting readers may refer to Ref. [28] for more detailed explanation

1.5   SUMMARY

So far, we have taken a glance at the fundamental concepts and associated applications of optical, infrared, and terahertz bands. In these intermediate frequency regimes, many materials demonstrate exotic properties which are different to those in microwave and radio frequency regimes. For example, materials that are simply perfect electric conductors (PEC) or transparent in microwave regime, could be highly dispersive and frequency-dependent in the optical, infrared, and terahertz regimes, exhibiting very interesting features such as semimetal, semiconductor, plasmonic, etc. Furthermore, the discovery of newly emergent materials such as 2D materials, artificial metamaterials, organic materials, piezoelectric materials, among the others, demonstrate unique wave-matter interaction phenomena in optical, infrared, and terahertz bands which might enable some paradigm-shift technologies or even trigger a completely new range of applications. In the following chapters, a systematic review on these emergent materials and their opportunities in optical, infrared, and terahertz bands will be discussed.

ACKNOWLEDGEMENTS

This work is supported by the National Natural Science Foundation of China (NSFC) (Nos. 62005256, 12088101, 11991060, U1930402) and China Academy of Engineering Physics Innovation and Development Fund (No. CX20200011).

BIBLIOGRAPHY

[1] Hockberger, P. E. 2002. A history of ultraviolet photobiology for humans, animals and microorganisms. *Photochem. Photobiol.* 76:561–579.
[2] *ISO-standard 21348:2007 space environment (natural and artificial)—Process for determining solar irradiances.* https://www.iso.org/obp/ui/#iso:std:iso:21348:ed-1:v1:en (accessed July 09, 2021).
[3] Paetzel, R., Albrecht, H. S., Lokai, P., Zshocke, W., Schmidt, T., Bragin, I., Schroeder, T., Reusch, C., and Spratte, S. 2003. Excimer lasers for superhigh NA 193-nm lithography. *SPIE Proceeding of Microlithography—Optical Microlithography XVI*, vol. 5040. https://doi.org/10.1117/12.485344
[4] United States Department of Energy, National Renewable Energy Laboratory. *Solar spectral irradiance: ASTM G-173s.* https://www.nrel.gov/grid/solar-resource/spectra.html (accessed July 13, 2021).

[5] Wikipedia. *Ultraviolet.* https://en.wikipedia.ahau.cf/wiki/Ultraviolet#Photography (accessed July 14, 2021).

[6] Rhodes, C. K. 1984. *Excimer lasers.* Springer.

[7] Muramoto, Y., Kimura, M., and Nouda, S. 2014. Development and future of ultraviolet light-emitting diodes: UV-LED will replace the UV lamp. *Semicond. Sci. Technol.* 29:084004.

[8] Neeser, S., Schumann, M., and Langhoff, H. 1996. Improved gain for the Ar_2^* excimer laser at 126 nm. *Appl. Phys. B* 63:103–105.

[9] Shan, M. C., Zhang, Y., Tran, T. B., Jiang, J. A., Long, H. L., Zheng, Z. H., Wang, A. G., Guo, W., Ye, J. C., Chen, C. Q., Dai, J. N., and Li, X. H. 2019. Deep UV laser at 249 nm based GaN quantum wells. *ACS Photon.* 6:2387–2391.

[10] Young, A. R. 2006. Acute effects of UVR on human eyes and skin. *Prog. Biophys. Mol. Bio.* 92:80–85.

[11] Wikipedia. *Visible spectrum.* https://en.wikipedia.ahau.cf/wiki/Visible_spectrum (accessed July 14, 2021).

[12] Young, T. 1802. Bakerian lecture: On the theory of light and colours. *Phil. Trans. R. Soc. Lond.* 92:12–48.

[13] Choudhury, A. K. R. 2014. *Principles of colour and appearance measurement: Object appearance, colour perception and instrumental measurement.* Woodhead Publishing.

[14] Sell, K. 2001. *Revolution in lamps: A chronicle of 50 years of progress.* 2nd edition. River Publishers.

[15] Schubert, E. F. 2012. *Light-emitting diode.* 2nd edition. Cambridge University Press.

[16] Ma, L., Yu, P., Wang, W. H., Kuo, H. C., Govorov, A. O., Sun, S., and Wang, Z. M. 2021. Nanoantenna-enhanced light-emitting diodes: Fundamental and recent progress. *Laser Photon. Rev.* 15:2000367.

[17] Resch-Genger, U., Grabolle, M., Cavaliere-Jaricot, S., Nitschke, R., and Nann, T. 2008. Quantum dots versus organic dyes as fluorescent labels. *Nat. Methods* 5:763–775.

[18] Silfvast, W. T. 1996. *Laser fundamentals.* Cambridge University Press.

[19] Scully, M. O., and Zubairy, M. S. 1997. *Quantum optics.* Cambridge University Press.

[20] *ISO-standard 20473:2007 Optics and photonics—spectral bands.* https://www.iso.org/standard/39482.html (accessed August 04, 2021).

[21] Vodopayanov, K. L. 2020. *Laser-based infrared sources and applications.* Wiley Publishing Group.

[22] Zerbetto, S. C., and Vasconcellos, E. C. C. 1994. Far infrared laser lines produced by methanol and its isotopic species: A review. *Int. J. Infrared Milli.* 15:889–933.

[23] Bramson, M. A. 1968. *Infrared radiation—a handbook for applications.* Springer.

[24] Wikipedia. *Infrared.* https://en.wikipedia.iwiki.uk/wiki/Infrared (accessed August 07, 2021).

[25] Lewis, R. A. 2014. A review of terahertz sources. *J. Phys. D: Appl. Phys.* 47:374001.

[26] Wikipedia. *Terahertz radiation.* https://en.wikipedia.ahau.cf/wiki/Terahertz_radiation#cite_note-BBC-44 (accessed August 13, 2021).

[27] Wu, Q., Lin, C., Lu, B., Miao, L., Hao, X., Wang, Z., Jiang, Y., Lei, W., Den, X., Chen, H., Yao, J., and Zhang, J. 2017. A 21 km 5 Gbps real time wireless communication system at 0.14 THz. *42nd International Conference on Infrared, Millimeter, and Terahertz Waves (IRMMW-THz).* No. 17259056. https://ieeexplore.ieee.org/document/8066870 (accessed July 21, 2022).

[28] Thompson, M. C., Badakov, H., Cook, A. M., et al. 2008. Breakdown limits on giga-volt-per-meter electron-beam-driven wakefields in dielectric structures. *Phys. Rev. Lett.* 100:214801.

2 Theory of Electromagnetic Fields

Yuriy Akimov

CONTENTS

2.1 Electromagnetic Fields in a Vacuum..24
    2.1.1 Microscopic Maxwell's Equations 24
    2.1.2 Sources of Electromagnetic Fields 25
    2.1.3 Transfer of Electromagnetic Energy.............................27
    2.1.4 Transfer of Electromagnetic Momentum....................30
    2.1.5 Excitation of Electromagnetic Fields...........................32
    2.1.6 Eigenmode Resonances 34
2.2 Electromagnetic Fields in a Homogeneous Isotropic Medium.......38
    2.2.1 Microscopic and Macroscopic Descriptions of a Medium 38
    2.2.2 Macroscopic Maxwell's Equations..............................40
    2.2.3 Macroscopic Transfer of Energy and Momentum 40
    2.2.4 Material Equations...41
    2.2.5 Temporal and Spatial Dispersion.................................45
    2.2.6 Local Response Approximation....................................46
2.3 Electromagnetic Fields in an Inhomogeneous Medium 52
    2.3.1 Inhomogeneous Medium Response 52
    2.3.2 Excitation of Electromagnetic Fields in an Inhomogeneous Medium 53
    2.3.3 Piecewise Homogeneous Media 55
    2.3.4 Boundary Conditions 56
    2.3.5 Induced Surface Charge and Surface Current..............57
    2.3.6 Dissipation in Piecewise Homogeneous Media...........59
2.4 Summary...60
Bibliography 60

In this chapter, we will consider fundamentals of classical theory of electromagnetic fields. In Section 2.1, we will overview microscopic properties of electromagnetic fields excited in a vacuum. Macroscopic description of electromagnetic fields propagating in a homogeneous isotropic medium will be given in Section 2.2. In Section 2.3, we will discuss properties of electromagnetic fields excited in inhomogeneous structures. The chapter will conclude with a summary in Section 2.4.

DOI: 10.1201/9781003202608-2

2.1 ELECTROMAGNETIC FIELDS IN A VACUUM

2.1.1 MICROSCOPIC MAXWELL'S EQUATIONS

In classical electrodynamics, excitation of electromagnetic fields in a vacuum is described by *microscopic Maxwell's equations* (free of any statistical averaging):

$$\nabla \times \mathbf{B}(t,\mathbf{r}) = \mu_0 \left[\varepsilon_0 \frac{\partial \mathbf{E}(t,\mathbf{r})}{\partial t} + \mathbf{J}(t,\mathbf{r}) \right], \tag{2.1}$$

$$\nabla \times \mathbf{E}(t,\mathbf{r}) = -\frac{\partial \mathbf{B}(t,\mathbf{r})}{\partial t}, \tag{2.2}$$

where $\mathbf{E}(t,\mathbf{r})$, $\mathbf{B}(t,\mathbf{r})$, and $\mathbf{J}(t,\mathbf{r})$ are the *electric field, magnetic induction,* and *current density*; $\varepsilon_0 = 8.85418782 \cdot 10^{-12}$ F/m and $\mu_0 = 4\pi \cdot 10^{-7}$ H/m are the electric and magnetic constants. These equations describe how currents with a given density $\mathbf{J}(t,\mathbf{r})$ excite the fields $\mathbf{E}(t,\mathbf{r})$ and $\mathbf{B}(t,\mathbf{r})$. By applying divergence to the right- and left-hand sides of these equations and by taking into account the identity

$$\nabla \cdot [\nabla \times \mathbf{F}(t,\mathbf{r})] \equiv 0,$$

we can get additional properties of the vector fields $\mathbf{E}(t,\mathbf{r})$, $\mathbf{B}(t,\mathbf{r})$, and $\mathbf{J}(t,\mathbf{r})$:

$$0 = \mu_0 \left[\varepsilon_0 \frac{\partial \nabla \cdot \mathbf{E}(t,\mathbf{r})}{\partial t} + \nabla \cdot \mathbf{J}(t,\mathbf{r}) \right], \tag{2.3}$$

$$0 = -\frac{\partial \nabla \cdot \mathbf{B}(t,\mathbf{r})}{\partial t}. \tag{2.4}$$

Equation (2.3) can be rewritten as the continuity equation that describes charge conservation,

$$\frac{\partial \rho(t,\mathbf{r})}{\partial t} + \nabla \cdot \mathbf{J}(t,\mathbf{r}) = 0, \tag{2.5}$$

where $\rho(t,\mathbf{r})$ represents the *charge density* that defines according to Eq. (2.3) the divergence of electric field:

$$\nabla \cdot \mathbf{E}(t,\mathbf{r}) = \frac{\rho(t,\mathbf{r})}{\varepsilon_0}. \tag{2.6}$$

Equation (2.4) gives us the divergence relation for magnetic induction,

$$\nabla \cdot \mathbf{B}(t,\mathbf{r}) = 0. \tag{2.7}$$

Mathematically speaking, divergence of a vector field reveals sources and sinks of the field. Therefore, the divergence relations are important in understanding of physical origins of fields. From them, we can conclude that charges create electric fields

Time-varying charges give rise to currents. But what is the origin of magnetic induction? To answer this question, we do a bit of vector analysis.

2.1.2 Sources of Electromagnetic Fields

Helmholtz decomposition is the powerful tool of vector analysis that will help us clarify structure and origin of electromagnetic field. This decomposition forms the fundamental theorem of vector calculus [1, 2]. According to it, any twice continuously differentiable vector field $\mathbf{F}(t,\mathbf{r})$ localized in space can be uniquely decomposed into *transverse* (solenoidal), $\mathbf{F}_t(t,\mathbf{r})$, and *longitudinal* (conservative), $\mathbf{F}_l(t,\mathbf{r})$, components:

$$\mathbf{F}(\mathbf{r}) = \mathbf{F}_t(\mathbf{r}) + \mathbf{F}_l(\mathbf{r}), \tag{2.8}$$

that exhibit

$$\nabla \cdot \mathbf{F}_t(\mathbf{r}) = 0, \qquad \nabla \times \mathbf{F}_l(\mathbf{r}) = 0. \tag{2.9}$$

The transverse and longitudinal components are orthogonal in Hilbert space

$$\int_0^\infty \mathbf{F}_t(\mathbf{r}) \cdot \mathbf{F}_l(\mathbf{r}) d^3\mathbf{r} = 0 \tag{2.10}$$

and can be obtained from $\mathbf{F}(\mathbf{r})$ as follows:

$$\mathbf{F}_t(\mathbf{r}) = \frac{1}{4\pi} \nabla \times \int_0^\infty \frac{\nabla' \times \mathbf{F}(\mathbf{r}')}{|\mathbf{r} - \mathbf{r}'|} d^3\mathbf{r}', \tag{2.11}$$

$$\mathbf{F}_l(\mathbf{r}) = -\frac{1}{4\pi} \nabla \int_0^\infty \frac{\nabla' \cdot \mathbf{F}(\mathbf{r}')}{|\mathbf{r} - \mathbf{r}'|} d^3\mathbf{r}', \tag{2.12}$$

where $\nabla' = \partial / \partial \mathbf{r}'$.

If we do Helmholtz decomposition of the vector fields $\mathbf{E}(t,\mathbf{r})$, $\mathbf{B}(t,\mathbf{r})$, and $\mathbf{J}(t,\mathbf{r})$, we can split Maxwell's equations into the four equations:

$$\nabla \times \mathbf{B}_t(t,\mathbf{r}) = \mu_0 \left[\varepsilon_0 \frac{\partial \mathbf{E}_t(t,\mathbf{r})}{\partial t} + \mathbf{J}_t(t,\mathbf{r}) \right], \tag{2.13}$$

$$\nabla \times \mathbf{E}_t(t,\mathbf{r}) = -\frac{\partial \mathbf{B}_t(t,\mathbf{r})}{\partial t}, \tag{2.14}$$

$$0 = \mu_0 \left[\varepsilon_0 \frac{\partial \mathbf{E}_l(t,\mathbf{r})}{\partial t} + \mathbf{J}_l(t,\mathbf{r}) \right], \tag{2.15}$$

$$0 = -\frac{\partial \mathbf{B}_l(t,\mathbf{r})}{\partial t}. \tag{2.16}$$

These equations reveal that excitation of electromagnetic fields goes in two ways: (i) *nonlocal* excitation of transverse fields by transverse currents and (ii) *local* excitation of longitudinal fields by longitudinal currents. These processes go independently, though the current densities $\mathbf{J}_t(t,\mathbf{r})$ and $\mathbf{J}_l(t,\mathbf{r})$ are linked by $\mathbf{J}(t,\mathbf{r})$.

Applying the divergence operator to Eqs. (2.15) and (2.16), we get the following relations:

$$\nabla \cdot \mathbf{J}_l(t,\mathbf{r}) = -\frac{\partial \rho(t,\mathbf{r})}{\partial t}, \tag{2.17}$$

$$\nabla \cdot \mathbf{E}_l(t,\mathbf{r}) = \frac{\rho(t,\mathbf{r})}{\varepsilon_0}, \tag{2.18}$$

$$\nabla \cdot \mathbf{B}_l(t,\mathbf{r}) = 0. \tag{2.19}$$

These relations demonstrate that charges give rise to only *longitudinal* electric field, while time-varying charges contribute to only *longitudinal* current density. All transverse fields, including $\mathbf{J}_t(t,\mathbf{r})$, $\mathbf{E}_t(t,\mathbf{r})$ and $\mathbf{B}_t(t,\mathbf{r})$, have zero divergence by default and, therefore, do not possess sources and sinks. Does this mean that all transverse fields do not have their physical origins? No, zero divergence *does not* mean that physical origin of the field does not exist. As soon as the field exists, its physical origin exists too, following the causality principle. In particular, Eqs. (2.13) and (2.14) clearly say that the cause of fields $\mathbf{E}_t(t,\mathbf{r})$ and $\mathbf{B}_t(t,\mathbf{r})$ is $\mathbf{J}_t(t,\mathbf{r})$.

To clarify the physical origin of transverse fields, let us consider an elementary particle bearing a charge q and an intrinsic magnetic dipole moment $\mathbf{m}(t)$. Within the relativity theory, elementary particles cannot have finite dimensions [3] and, hence, are treated as point-like with the use of the Dirac delta function $\delta[\mathbf{r}-\mathbf{r}_0(t)]$:

$$\rho(t,\mathbf{r}) = q\delta\left[\mathbf{r} - \mathbf{r}_0(t)\right], \tag{2.20}$$

where $\mathbf{r}_0(t)$ is the trajectory of the particle. The current density of the particle,

$$\mathbf{J}(t,\mathbf{r}) = q\frac{d\mathbf{r}_0(t)}{dt}\delta\left[\mathbf{r} - \mathbf{r}_0(t)\right] + \nabla \times \mathbf{m}(t)\delta\left[\mathbf{r} - \mathbf{r}_0(t)\right], \tag{2.21}$$

represents the coupled state of the transverse $\mathbf{J}_t(t,\mathbf{r})$ and longitudinal $\mathbf{J}_l(t,\mathbf{r})$ current densities:

$$\mathbf{J}_l(t,\mathbf{r}) = \frac{q}{4\pi}\left(\frac{d\mathbf{r}_0(t)}{dt} \cdot \nabla\right)\frac{\mathbf{r} - \mathbf{r}_0(t)}{\left|\mathbf{r} - \mathbf{r}_0(t)\right|^3}, \tag{2.22}$$

$$\mathbf{J}_t(t,\mathbf{r}) = \nabla \times \left(\frac{q}{4\pi}\frac{d\mathbf{r}_0(t)}{dt} \times \frac{\mathbf{r} - \mathbf{r}_0(t)}{\left|\mathbf{r} - \mathbf{r}_0(t)\right|^3} + \mathbf{m}(t)\delta\left[\mathbf{r} - \mathbf{r}_0(t)\right]\right). \tag{2.23}$$

Following this decomposition, a point charge q results in both transverse and longitudinal currents, while a point magnetic dipole moment $\mathbf{m}(t)$ gives rise to transverse current only. Finally, these currents excite all transverse and longitudinal fields. In other words, the physical origins of longitudinal fields $\mathbf{J}_l(t,\mathbf{r})$ and $\mathbf{E}_l(t,\mathbf{r})$ are charges [to be specific, moving charges for $\mathbf{J}_l(t,\mathbf{r})$ and any charges for $\mathbf{E}_l(t,\mathbf{r})$], while the physical origins of transverse fields $\mathbf{J}_t(t,\mathbf{r})$, $\mathbf{E}_t(t,\mathbf{r})$ and $\mathbf{B}_t(t,\mathbf{r})$ are intrinsic magnetic moments and moving charges.

Note, intrinsic magnetic moments are responsible for so-called magnetic properties of matter. They are fully given by transverse current with $\mathbf{J}_t(t,\mathbf{r})$. Moving charges also can contribute to $\mathbf{J}_t(t,\mathbf{r})$ causing magnetic effects similar to those of intrinsic magnetic moments. Their contribution is commonly treated as an orbital (charge-induced) magnetic moment. Thus, moving charges have a broader spectrum of effects—they are responsible for both so-called electric properties of matter given by $\mathbf{J}_l(t,\mathbf{r})$ and magnetic ones described by $\mathbf{J}_t(t,\mathbf{r})$.

2.1.3 TRANSFER OF ELECTROMAGNETIC ENERGY

Now let us explore energy transfer by electromagnetic fields. For this, we use Maxwell's equations to calculate the quantity

$$\varepsilon_0 \mathbf{E}(t,\mathbf{r}) \cdot \frac{\partial \mathbf{E}(t,\mathbf{r})}{\partial t} + \frac{1}{\mu_0} \mathbf{B}(t,\mathbf{r}) \cdot \frac{\partial \mathbf{B}(t,\mathbf{r})}{\partial t} =$$

$$\frac{1}{\mu_0} \mathbf{E}(t,\mathbf{r}) \cdot \left[\nabla \times \mathbf{B}(t,\mathbf{r}) \right] - \frac{1}{\mu_0} \mathbf{B}(t,\mathbf{r}) \cdot \left[\nabla \times \mathbf{E}(t,\mathbf{r}) \right] - \mathbf{E}(t,\mathbf{r}) \cdot \mathbf{J}(t,\mathbf{r}).$$

If we take into account the vector identity

$$\mathbf{E}(t,\mathbf{r}) \cdot [\nabla \times \mathbf{B}(t,\mathbf{r})] - \mathbf{B}(t,\mathbf{r}) \cdot [\nabla \times \mathbf{E}(t,\mathbf{r})] = \nabla \cdot [\mathbf{E}(t,\mathbf{r}) \times \mathbf{B}(t,\mathbf{r})],$$

we can rewrite this equation in the following form:

$$\frac{\varepsilon_0}{2} \frac{\partial}{\partial t} \left[\mathbf{E}^2(t,\mathbf{r}) + c^2 \mathbf{B}^2(t,\mathbf{r}) \right] + \nabla \cdot \frac{1}{\mu_0} [\mathbf{E}(t,\mathbf{r}) \times \mathbf{B}(t,\mathbf{r})] = -\mathbf{J}(t,\mathbf{r}) \cdot \mathbf{E}(t,\mathbf{r}) \quad (2.24)$$

that describes the power balance between an electromagnetic field and a current. In the right-hand side of this equation, we have the *power density* spent by the current on excitation of the electromagnetic field,

$$a(t,\mathbf{r}) = -\mathbf{J}(t,\mathbf{r}) \cdot \mathbf{E}(t,\mathbf{r}). \quad (2.25)$$

The left-hand side represents different mechanisms for that power expenditure. The first one is increase of the *energy density*

$$w(t,\mathbf{r}) = \frac{\varepsilon_0}{2} \left[\mathbf{E}^2(t,\mathbf{r}) + c^2 \mathbf{B}^2(t,\mathbf{r}) \right] \quad (2.26)$$

of electric field $\mathbf{E}(t,\mathbf{r})$ and magnetic induction $\mathbf{B}(t,\mathbf{r})$ over time. The second mechanism is the radiation of electromagnetic energy out. It is given by the divergence of the *Poynting vector*:

$$\mathbf{S}(t,\mathbf{r}) = \frac{1}{\mu_0}\mathbf{E}(t,\mathbf{r}) \times \mathbf{B}(t,\mathbf{r}), \qquad (2.27)$$

being the energy flux density of the electromagnetic field. The power balance of these mechanisms is known as the *Poynting theorem*,

$$\frac{\partial w(t,\mathbf{r})}{\partial t} + \nabla \cdot \mathbf{S}(t,\mathbf{r}) = a(t,\mathbf{r}), \qquad (2.28)$$

that describes how energy of electromagnetic field transfers in time and space.

As we saw above, electromagnetic fields are independently excited by transverse and longitudinal currents. The energy transfers accompanying these processes occur separately too. They can be written in the form of Poynting theorem as well:

$$\frac{\partial w_{t,l}(t,\mathbf{r})}{\partial t} + \nabla \cdot \mathbf{S}_{t,l}(t,\mathbf{r}) = a_{t,l}(t,\mathbf{r}). \qquad (2.29)$$

The energy density $w_t(t,\mathbf{r})$, power flux $\mathbf{S}_t(t,\mathbf{r})$, and current power density $a_t(t,\mathbf{r})$ of transverse fields can be derived in the way similar to total fields, but from Eqs. (2.13) and (2.14):

$$w_t(t,\mathbf{r}) = \left(\varepsilon_0/2\right)\left[\mathbf{E}_t^2(t,\mathbf{r}) + c^2\mathbf{B}_t^2(t,\mathbf{r})\right],$$

$$\mathbf{S}_t(t,\mathbf{r}) = \mu_0^{-1}\mathbf{E}_t(t,\mathbf{r}) \times \mathbf{B}_t(t,\mathbf{r}),$$

$$a_t(t,\mathbf{r}) = -\mathbf{E}_t(t,\mathbf{r}) \cdot \mathbf{J}_t(t,\mathbf{r}),$$

while the longitudinal counterparts are given by Eqs. (2.15) and (2.16):

$$w_l(t,\mathbf{r}) = (\varepsilon_0/2)\mathbf{E}_l^2(t,\mathbf{r}),$$

$$\mathbf{S}_l(t,\mathbf{r}) \equiv 0,$$

$$a_l(t,\mathbf{r}) = -\mathbf{E}_l(t,\mathbf{r}) \cdot \mathbf{J}_l(t,\mathbf{r}).$$

Note that the subscripts l,t in Eq. (2.29) stay not for the Helmholtz components of the respective quantities, but for the quantities given by transverse and longitudinal fields $\{\mathbf{E}(t,\mathbf{r}),\mathbf{B}(t,\mathbf{r})\}_{l,t}$. As any longitudinal fields possess $\mathbf{B}_l(t,\mathbf{r}) \equiv 0$, they always result in $\mathbf{S}_l(t,\mathbf{r}) \equiv 0$ and, thus, *do not* transfer their energy. This feature is confirmed by *local* excitation of longitudinal fields. In contrast to them, transverse fields feature $\mathbf{S}_t(t,\mathbf{r}) \not\equiv 0$ and, hence, are able to transfer their energy.

As for the transfer of total energy of electromagnetic field, it goes in a quite complicated manner following the interference of longitudinal and transverse fields, with

$$w(t,\mathbf{r}) = w_t(t,\mathbf{r}) + w_l(t,\mathbf{r}) + w_i(t,\mathbf{r}),$$

$$\mathbf{S}(t,\mathbf{r}) = \mathbf{S}_t(t,\mathbf{r}) + \mathbf{S}_l(t,\mathbf{r}) + \mathbf{S}_i(t,\mathbf{r}),$$

$$a(t,\mathbf{r}) = a_t(t,\mathbf{r}) + a_l(t,\mathbf{r}) + a_i(t,\mathbf{r}),$$

where the interference terms are given by

$$w_i(t,\mathbf{r}) = \varepsilon_0 \mathbf{E}_t(t,\mathbf{r}) \cdot \mathbf{E}_l(t,\mathbf{r}),$$

$$\mathbf{S}_i(t,\mathbf{r}) = \mu_0^{-1} \mathbf{E}_l(t,\mathbf{r}) \times \mathbf{B}_t(t,\mathbf{r}),$$

$$a_i(t,\mathbf{r}) = -\mathbf{E}_t(t,\mathbf{r}) \cdot \mathbf{J}_l(t,\mathbf{r}) - \mathbf{E}_l(t,\mathbf{r}) \cdot \mathbf{J}_t(t,\mathbf{r}).$$

According to this interference, both transverse and longitudinal fields participate in transfer of total energy. It is given by two power fluxes $\mathbf{S}_t(t,\mathbf{r})$ and $\mathbf{S}_i(t,\mathbf{r})$ resulted from the transfer by pure transverse fields and from the transfer caused by interference of transverse and longitudinal fields. The necessary condition for transfer of total energy is nonzero *transverse* field.

Although the energy transfer of total field obeys quite complicated interference of Helmholtz fields, its integral characteristics are free of the interference effects. As transverse and longitudinal fields are orthogonal in Hilbert space, the respective integral contributions of the terms $w_i(t,\mathbf{r})$ and $a_i(t,\mathbf{r})$ always vanish, so that the total energy of electromagnetic field and the total power can be reduced to

$$\int_0^\infty w(t,\mathbf{r}) \mathrm{d}^3\mathbf{r} = \int_0^\infty \left[w_t(t,\mathbf{r}) + w_l(t,\mathbf{r}) \right] \mathrm{d}^3\mathbf{r},$$

$$\int_0^\infty a(t,\mathbf{r}) \mathrm{d}^3\mathbf{r} = \int_0^\infty \left[a_t(t,\mathbf{r}) + a_l(t,\mathbf{r}) \right] \mathrm{d}^3\mathbf{r}.$$

Eventually, the integral form of Poynting theorem (2.28) for the total energy stored in the entire space can be written in the following form:

$$\frac{\partial}{\partial t} \int_0^\infty \left[w_t(t,\mathbf{r}) + w_l(t,\mathbf{r}) \right] \mathrm{d}^3\mathbf{r} + \int_0^\infty \nabla \cdot \left[\mathbf{S}_t(t,\mathbf{r}) + \mathbf{S}_i(t,\mathbf{r}) \right] \mathrm{d}^3\mathbf{r}$$

$$= \int_0^\infty \left[a_t(t,\mathbf{r}) + a_l(t,\mathbf{r}) \right] \mathrm{d}^3\mathbf{r}, \tag{2.30}$$

where the second integral corresponds to the power radiated out by electromagnetic fields at infinity. If we separately consider energy transfer by transverse and longitudinal fields (2.29), we get

$$\frac{\partial}{\partial t} \int_0^\infty w_{t,l}(t,\mathbf{r}) \mathrm{d}^3\mathbf{r} + \int_0^\infty \nabla \cdot \mathbf{S}_{t,l}(t,\mathbf{r}) \mathrm{d}^3\mathbf{r} = \int_0^\infty a_{t,l}(t,\mathbf{r}) \mathrm{d}^3\mathbf{r}.$$

Substituting it into Eq. (2.30), we arrive at

$$\int\limits_{0}^{\infty} \nabla \cdot \mathbf{S}_i \left(t, \mathbf{r}\right) d^3 \mathbf{r} = 0,$$

demonstrating the interference-free radiation of energy out of the system.

2.1.4 TRANSFER OF ELECTROMAGNETIC MOMENTUM

Electromagnetic fields transfer not only energy, but momentum too. Equation of momentum transfer can be derived, if we use Maxwell's equations to calculate the following quantity:

$$\varepsilon_0 \frac{\partial \mathbf{E}(t, \mathbf{r})}{\partial t} \times \mathbf{B}(t, \mathbf{r}) + \varepsilon_0 \mathbf{E}(t, \mathbf{r}) \times \frac{\partial \mathbf{B}(t, \mathbf{r})}{\partial t} = -\varepsilon_0 \mathbf{E}(t, \mathbf{r}) \times [\nabla \times \mathbf{E}(t, \mathbf{r})]$$

$$-\frac{1}{\mu_0} \mathbf{B}(t, \mathbf{r}) \times [\nabla \times \mathbf{B}(t, \mathbf{r})] - \mathbf{J}(t, \mathbf{r}) \times \mathbf{B}(t, \mathbf{r}).$$

By using the vector identity

$$\mathbf{F}(t, \mathbf{r}) \times [\nabla \times \mathbf{F}(t, \mathbf{r})] = \frac{1}{2} \nabla \mathbf{F}^2 (t, \mathbf{r}) - [\mathbf{F}(t, \mathbf{r}) \cdot \nabla] \mathbf{F}(t, \mathbf{r})$$

we can rewrite the equation above in the following form:

$$\varepsilon_0 \frac{\partial}{\partial t} \mathbf{E}(t, \mathbf{r}) \times \mathbf{B}(t, \mathbf{r}) = -\varepsilon_0 \mathbf{E}(t, \mathbf{r}) \nabla \cdot \mathbf{E}(t, \mathbf{r}) - \frac{1}{\mu_0} \mathbf{B}(t, \mathbf{r}) \nabla \cdot \mathbf{B}(t, \mathbf{r})$$

$$-\nabla \cdot \left\{ \varepsilon_0 \left[\mathbf{E}(t, \mathbf{r}) \mathbf{E}(t, \mathbf{r}) - \overset{\leftrightarrow}{\mathbf{I}} \frac{\mathbf{E}^2 (t, \mathbf{r})}{2} \right] + \mu_0^{-1} \left[\mathbf{B}(t, \mathbf{r}) \mathbf{B}(t, \mathbf{r}) - \overset{\leftrightarrow}{\mathbf{I}} \frac{\mathbf{B}^2 (t, \mathbf{r})}{2} \right] \right\}$$

$$-\mathbf{J}(t, \mathbf{r}) \times \mathbf{B}(t, \mathbf{r}),$$

where $\overset{\leftrightarrow}{\mathbf{I}}$ is the second order unit tensor. Taking into account the divergence relations for $\mathbf{E}(t, \mathbf{r})$ and $\mathbf{B}(t, \mathbf{r})$, we arrive at the equation

$$\varepsilon_0 \frac{\partial}{\partial t} \mathbf{E}(t, \mathbf{r}) \times \mathbf{B}(t, \mathbf{r}) + \nabla \cdot \left\{ \varepsilon_0 \left[\mathbf{E}(t, \mathbf{r}) \mathbf{E}(t, \mathbf{r}) - \overset{\leftrightarrow}{\mathbf{I}} \frac{\mathbf{E}^2 (t, \mathbf{r})}{2} \right] \right.$$

$$\left. +\mu_0^{-1} \left[\mathbf{B}(t, \mathbf{r}) \mathbf{B}(t, \mathbf{r}) - \overset{\leftrightarrow}{\mathbf{I}} \frac{\mathbf{B}^2 (t, \mathbf{r})}{2} \right] \right\} \qquad (2.31)$$

$$= -\rho(t, \mathbf{r}) \mathbf{E}(t, \mathbf{r}) - \mathbf{J}(t, \mathbf{r}) \times \mathbf{B}(t, \mathbf{r}),$$

describing force balance between particles and an electromagnetic field. The right hand side of Eq. (2.31) is the *force density* of the particles acting on the electromagnetic field,

$$\mathbf{f}(t,\mathbf{r}) = -\rho(t,\mathbf{r})\mathbf{E}(t,\mathbf{r}) - \mathbf{J}(t,\mathbf{r}) \times \mathbf{B}(t,\mathbf{r}). \tag{2.32}$$

This force is opposite to the Lorentz force, with which the electromagnetic field acts on the particles. In the left-hand side of this equation, we have two types of response of the electromagnetic field to the particles. The first response is increase of the *momentum density* of the electromagnetic field,

$$\mathbf{p}(t,\mathbf{r}) = \varepsilon_0 \mathbf{E}(t,\mathbf{r}) \times \mathbf{B}(t,\mathbf{r}) = \frac{\mathbf{S}(t,\mathbf{r})}{c^2}, \tag{2.33}$$

over time. The second one is spatial transfer of the momentum given by the divergence of the *Maxwell stress tensor*,

$$\overleftrightarrow{\mathbf{T}}(t,\mathbf{r}) = \varepsilon_0 \left[\mathbf{E}(t,\mathbf{r})\mathbf{E}(t,\mathbf{r}) - \overleftrightarrow{\mathbf{I}} \frac{\mathbf{E}^2(t,\mathbf{r})}{2} \right] + \mu_0^{-1} \left[\mathbf{B}(t,\mathbf{r})\mathbf{B}(t,\mathbf{r}) - \overleftrightarrow{\mathbf{I}} \frac{\mathbf{B}^2(t,\mathbf{r})}{2} \right]. \tag{2.34}$$

Finally, the *momentum transfer equation* can be written in the following form:

$$\frac{\partial \mathbf{p}(t,\mathbf{r})}{\partial t} + \nabla \cdot \overleftrightarrow{\mathbf{T}}(t,\mathbf{r}) = \mathbf{f}(t,\mathbf{r}). \tag{2.35}$$

As transverse and longitudinal fields are excited independently, transfers of their momenta go separately too. Their equations can be written as follows:

$$\frac{\partial \mathbf{p}_{l,t}(t,\mathbf{r})}{\partial t} + \nabla \cdot \overleftrightarrow{\mathbf{T}}_{l,t}(t,\mathbf{r}) = \mathbf{f}_{l,t}(t,\mathbf{r}). \tag{2.36}$$

The momentum density $\mathbf{p}_t(t,\mathbf{r})$, Maxwell stress tensor $\overleftrightarrow{\mathbf{T}}_t(t,\mathbf{r})$, and particle force density $\mathbf{f}_t(t,\mathbf{r})$ for transverse components can be derived in the same way as for the total field, but from Eqs. (2.13) and (2.14):

$$\mathbf{p}_t(t,\mathbf{r}) = \varepsilon_0 \mathbf{E}_t(t,\mathbf{r}) \times \mathbf{B}_t(t,\mathbf{r}),$$

$$\overleftrightarrow{\mathbf{T}}_t(t,\mathbf{r}) = \varepsilon_0 \left[\mathbf{E}_t(t,\mathbf{r})\mathbf{E}_t(t,\mathbf{r}) - \overleftrightarrow{\mathbf{I}} \frac{\mathbf{E}_t^2(t,\mathbf{r})}{2} \right] + \mu_0^{-1} \left[\mathbf{B}_t(t,\mathbf{r})\mathbf{B}_t(t,\mathbf{r}) - \overleftrightarrow{\mathbf{I}} \frac{\mathbf{B}_t^2(t,\mathbf{r})}{2} \right],$$

$$\mathbf{f}_t(t,\mathbf{r}) = -\mathbf{J}_t(t,\mathbf{r}) \times \mathbf{B}_t(t,\mathbf{r}),$$

while the longitudinal counterparts are given by Eqs. (2.15) and (2.16):

$$\mathbf{p}_l(t,\mathbf{r}) \equiv 0,$$

$$\overleftrightarrow{\mathbf{T}}_l(t,\mathbf{r}) = \varepsilon_0 \left[\mathbf{E}_l(t,\mathbf{r})\mathbf{E}_l(t,\mathbf{r}) - \overleftrightarrow{\mathbf{I}} \frac{\mathbf{E}_l^2(t,\mathbf{r})}{2} \right],$$

$$\mathbf{f}_l(t,\mathbf{r}) = -\rho(t,\mathbf{r})\mathbf{E}_l(t,\mathbf{r}).$$

Note, longitudinal fields possess $\mathbf{p}_l(t,\mathbf{r}) \equiv 0$ and, thus, *do not* bear any momentum. Only transverse fields can have nonzero momentum, $\mathbf{p}_t(t,\mathbf{r}) \not\equiv 0$.

As for the transfer of total momentum of electromagnetic field, it follows the interference of longitudinal and transverse fields with

$$\mathbf{p}(t,\mathbf{r}) = \mathbf{p}_t(t,\mathbf{r}) + \mathbf{p}_l(t,\mathbf{r}) + \mathbf{p}_i(t,\mathbf{r}),$$

$$\overleftrightarrow{\mathbf{T}}(t,\mathbf{r}) = \overleftrightarrow{\mathbf{T}}_t(t,\mathbf{r}) + \overleftrightarrow{\mathbf{T}}_l(t,\mathbf{r}) + \overleftrightarrow{\mathbf{T}}_i(t,\mathbf{r}),$$

$$\mathbf{f}(t,\mathbf{r}) = \mathbf{f}_t(t,\mathbf{r}) + \mathbf{f}_l(t,\mathbf{r}) + \mathbf{f}_i(t,\mathbf{r}),$$

where the interference terms are given by

$$\mathbf{p}_i(t,\mathbf{r}) = \varepsilon_0 \mathbf{E}_l(t,\mathbf{r}) \times \mathbf{B}_t(t,\mathbf{r}),$$

$$\overleftrightarrow{\mathbf{T}}_i(t,\mathbf{r}) = \varepsilon_0 \left[\mathbf{E}_t(t,\mathbf{r})\mathbf{E}_l(t,\mathbf{r}) + \mathbf{E}_l(t,\mathbf{r})\mathbf{E}_t(t,\mathbf{r}) - \overleftrightarrow{\mathbf{I}}\mathbf{E}_t(t,\mathbf{r}) \cdot \mathbf{E}_l(t,\mathbf{r}) \right],$$

$$\mathbf{f}_i(t,\mathbf{r}) = -\rho(t,\mathbf{r})\mathbf{E}_t(t,\mathbf{r}) - \mathbf{J}_l(t,\mathbf{r}) \times \mathbf{B}_t(t,\mathbf{r}).$$

According to this interference, both transverse and longitudinal fields contribute to the momentum of total field. They are given by two momentum densities $\mathbf{p}_t(t,\mathbf{r}) = \mathbf{S}_t(t,\mathbf{r})/c^2$, $\mathbf{p}_i(t,\mathbf{r}) = \mathbf{S}_i(t,\mathbf{r})/c^2$ that come from pure transverse fields and from interference of the Helmholtz fields.

In contrast to the energy, the momentum of total field stored in the entire space remains governed by the interference of transverse and longitudinal fields:

$$\frac{\partial}{\partial t} \int_0^\infty \left[\mathbf{p}_t(t,\mathbf{r}) + \mathbf{p}_i(t,\mathbf{r}) \right] \mathrm{d}^3\mathbf{r} +$$

$$\int_0^\infty \nabla \cdot \left[\overleftrightarrow{\mathbf{T}}_t(t,\mathbf{r}) + \overleftrightarrow{\mathbf{T}}_l(t,\mathbf{r}) + \overleftrightarrow{\mathbf{T}}_i(t,\mathbf{r}) \right] \mathrm{d}^3\mathbf{r} = \int_0^\infty \left[\mathbf{f}_t(t,\mathbf{r}) + \mathbf{f}_l(t,\mathbf{r}) + \mathbf{f}_i(t,\mathbf{r}) \right] \mathrm{d}^3\mathbf{r}$$

with the particular integral relations

$$\frac{\partial}{\partial t} \int_0^\infty \mathbf{p}_{t,l}(t,\mathbf{r})\mathrm{d}^3\mathbf{r} + \int_0^\infty \nabla \cdot \overleftrightarrow{\mathbf{T}}_{t,l}(t,\mathbf{r})\mathrm{d}^3\mathbf{r} = \int_0^\infty \mathbf{f}_{t,l}(t,\mathbf{r})\mathrm{d}^3\mathbf{r}$$

for the pure fields and

$$\frac{\partial}{\partial t} \int_0^\infty \mathbf{p}_i(t,\mathbf{r})\mathrm{d}^3\mathbf{r} + \int_0^\infty \nabla \cdot \overleftrightarrow{\mathbf{T}}_i(t,\mathbf{r})\mathrm{d}^3\mathbf{r} = \int_0^\infty \mathbf{f}_i(t,\mathbf{r})\mathrm{d}^3\mathbf{r}$$

for their interference.

2.1.5 EXCITATION OF ELECTROMAGNETIC FIELDS

Owing to the linearity of Maxwell's equations with respect to $\mathbf{E}(t,\mathbf{r})$, $\mathbf{B}(t,\mathbf{r})$, and $\mathbf{J}(t,\mathbf{r})$, we can do the Fourier transform of all these fields to the frequency space,

$$\mathbf{F}(t,\mathbf{r}) = \frac{1}{(2\pi)^{1/2}} \int_{-\infty}^{\infty} d\omega\, \mathbf{F}(\omega,\mathbf{r})e^{-i\omega t},$$

and reduce (2.1) and (2.2) to

$$\nabla \times \mathbf{B}(\omega,\mathbf{r}) = -i\omega\mu_0\varepsilon_0 \mathbf{E}(\omega,\mathbf{r}) + \mu_0 \mathbf{J}(\omega,\mathbf{r}), \qquad (2.37)$$

$$\nabla \times \mathbf{E}(\omega,\mathbf{r}) = i\omega\mathbf{B}(\omega,\mathbf{r}), \qquad (2.38)$$

where ω is the angular frequency. For Helmholtz fields, these equations can be split into the following four equations:

$$\nabla \times \mathbf{B}_t(\omega,\mathbf{r}) = -i\omega\mu_0\varepsilon_0 \mathbf{E}_t(\omega,\mathbf{r}) + \mu_0 \mathbf{J}_t(\omega,\mathbf{r}),$$

$$\nabla \times \mathbf{E}_t(\omega,\mathbf{r}) = i\omega\mathbf{B}_t(\omega,\mathbf{r}),$$

$$-i\omega\varepsilon_0 \mathbf{E}_l(\omega,\mathbf{r}) + \mathbf{J}_l(\omega,\mathbf{r}) = 0,$$

$$i\omega\mathbf{B}_l(\omega,\mathbf{r}) = 0.$$

Magnetic induction field demonstrates different behaviors for transverse and longitudinal parts:

$$\mathbf{B}_t(\omega,\mathbf{r}) = -\frac{i}{\omega}\nabla \times \mathbf{E}_t(\omega,\mathbf{r}) \neq 0, \qquad (2.39)$$

$$\mathbf{B}_l(\omega,\mathbf{r}) \equiv 0, \qquad (2.40)$$

following the fundamental distinction of the Helmholtz fields. As for electric fields, they also show dissimilar dynamics:

$$\mathbf{E}_t(\omega,\mathbf{r}) = \hat{G}_t^{-1}(\omega)\mathbf{J}_t(\omega,\mathbf{r}), \qquad (2.41)$$

$$\mathbf{E}_l(\omega,\mathbf{r}) = \hat{G}_l^{-1}(\omega)\mathbf{J}_l(\omega,\mathbf{r}), \qquad (2.42)$$

where $\hat{G}_{t,l}^{-1}(\omega)$ are the inverse operators of

$$\hat{G}_t(\omega) = i\varepsilon_0\omega\left[1 + k_0^{-2}(\omega)\nabla^2\right], \qquad (2.43)$$

$$\hat{G}_l(\omega) = i\varepsilon_0\omega \qquad (2.44)$$

with $k_0(\omega) = \omega\sqrt{\varepsilon_0\mu_0}$. Note that the Helmholtz electric fields are excited independently by $\mathbf{J}_{t,l}(\omega,\mathbf{r})$ and experience different resonances.

2.1.6   EIGENMODE RESONANCES

To explore resonances of electromagnetic fields, we consider a linear vector equation

$$\hat{G}(\omega)\mathbf{M}(\omega,\mathbf{r}) = \mathbf{F}(\omega,\mathbf{r}),$$

where the linear operator $\hat{G}(\omega)$ describes a process of excitation of a vector field $\mathbf{M}(\omega,\mathbf{r})$ by a vector field $\mathbf{F}(\omega,\mathbf{r})$. As a linear operator, $\hat{G}(\omega)$ is characterized with its own eigenvalues $G_n(\omega)$ and eigenfields $\tilde{\mathbf{M}}_n(\omega,\mathbf{r})$:

$$\hat{G}(\omega)\tilde{\mathbf{M}}_n(\omega,\mathbf{r}) = G_n(\omega)\tilde{\mathbf{M}}_n(\omega,\mathbf{r}).$$

The eigenvalues $G_n(\omega)$ can have a number of complex roots ω_{nm},

$$G_n(\omega_{nm}) = 0.$$

The roots ω_{nm} are called *eigenfrequencies* and define the conditions when the operator $\hat{G}(\omega)$ gives the null output field for a nonzero input field $\tilde{\mathbf{M}}_n(\omega,\mathbf{r})$,

$$\hat{G}(\omega_{nm})\tilde{\mathbf{M}}_n(\omega_{nm},\mathbf{r}) = 0.$$

The respective solutions given by the eigenfrequencies ω_{nm} and eigenfields $\tilde{\mathbf{M}}_n(\omega_{nm},\mathbf{r})$ are called *eigenmodes* of the operator $\hat{G}(\omega)$.

In the vicinity of eigenfrequencies, $\omega \approx \mathrm{Re}\,\omega_{nm}$, the excited field $\mathbf{M}(\omega,\mathbf{r})$ can experience resonant growth, if (i) ω_{nm} does not stay far away from the real axis ω, i.e. when $|\mathrm{Im}\,\omega_{nm}| \ll |\mathrm{Re}\,\omega_{nm}|$, (ii) the exciting field is nonzero in this frequency range, $\mathbf{F}(\omega,\mathbf{r}) \neq 0$, and (iii) the spatial profile of $\mathbf{F}(\omega,\mathbf{r})$ is properly chosen to follow the distribution of $\tilde{\mathbf{M}}_n(\omega_{nm},\mathbf{r})$ for $\mathbf{M}(\omega,\mathbf{r})$. Under these conditions, the excited field $\mathbf{M}(\omega,\mathbf{r})$ can have resonance on eigenmode of the operator $\hat{G}(\omega)$ caused by smallness of its eigenvalue $G_n(\omega)$ in the vicinity of $\mathrm{Re}\,\omega_{nm}$.

According to the eigenmode theory, the efficiency of resonant excitation follows the accuracy of matching of the field characteristics (distribution and frequency) for excited and eigen fields. This accuracy is controlled by $\mathbf{F}(\omega,\mathbf{r})$. If for two different fields $\mathbf{F}(\omega,\mathbf{r})$, the matching conditions are satisfied, we can see resonances of $\mathbf{M}(\omega,\mathbf{r})$ around the same frequencies given by the operator $\hat{G}(\omega)$.

Applying this theory to transverse and longitudinal electric fields with their $\hat{G}_t(\omega)$ and $\hat{G}_l(\omega)$, we can conclude that resonant frequencies of $\mathbf{E}_t(\omega,\mathbf{r})$ and $\mathbf{E}_l(\omega,\mathbf{r})$ are different. If we seek the eigenfields of $\hat{G}_t(\omega)$ and $\hat{G}_l(\omega)$ in the form of plane waves,

$$\tilde{\mathbf{E}}(\omega_m(\mathbf{k}),\mathbf{k},\mathbf{r}) = \mathbf{E}_0(\omega_m(\mathbf{k}),\mathbf{k})e^{i\mathbf{k}\cdot\mathbf{r}},$$

where $\mathbf{k}$ is the real wavevector (used as a continuous parameter instead of the integer index n), we can get the following dispersion relations:

$$G_t(\omega_m(\mathbf{k}),\mathbf{k}) = i\varepsilon_0 \frac{\omega_m^2(\mathbf{k}) - k^2 c^2}{\omega_m(\mathbf{k})} = 0,$$

$$G_l\left(\omega_m(\mathbf{k}),\mathbf{k}\right) = i\varepsilon_0\omega_m(\mathbf{k}) = 0,$$

that define the eigenfrequencies of transverse and longitudinal eigenmodes. For the transverse eigenmodes, we get

$$\omega_m(\mathbf{k}) = \pm ck, \tag{2.45}$$

while for longitudinal ones, we obtain

$$\omega_m(\mathbf{k}) = 0. \tag{2.46}$$

The transverse eigenmodes are known as *photons*. They are the elementary radiation eigenmodes, whose eigenfields propagate in a vacuum at the fixed phase velocity $\omega_m(\mathbf{k})/k$ equal to the speed of light. The longitudinal eigenmodes are static and known as *virtual photons*. As $\omega_m(\mathbf{k})$ of all these eigenmodes are real, the resonances on them can be seen as divergences in Fourier images,

$$\mathbf{F}(\omega,\mathbf{k}) = \frac{1}{(2\pi)^2} \int\limits_{-\infty}^{\infty} dt \int\limits_{0}^{\infty} d^3\mathbf{r}\, \mathbf{F}(t,\mathbf{r}) e^{-i(\mathbf{k}\cdot\mathbf{r}-\omega t)},$$

for the excited fields,

$$\mathbf{E}_t(\omega,\mathbf{k}) = G_t^{-1}(\omega,\mathbf{k})\mathbf{J}_t(\omega,\mathbf{k}), \tag{2.47}$$

$$\mathbf{E}_l(\omega,\mathbf{k}) = G_l^{-1}(\omega,\mathbf{k})\mathbf{J}_l(\omega,\mathbf{k}). \tag{2.48}$$

Eventually, these divergences appear in the total work A done by currents on excitation of electromagnetic fields. To calculate it, we do a bit of math,

$$A = \int\limits_{-\infty}^{\infty} dt \int\limits_{0}^{\infty} d^3\mathbf{r}\, a(t,\mathbf{r}) = -\int\limits_{-\infty}^{\infty} dt \int\limits_{0}^{\infty} d^3\mathbf{r}\, \mathbf{E}(t,\mathbf{r})\cdot\mathbf{J}(t,\mathbf{r})$$

$$= -\frac{1}{(2\pi)^4} \int\limits_{-\infty}^{\infty} dt \int\limits_{0}^{\infty} d^3\mathbf{r} \iint\limits_{-\infty}^{\infty} d\omega' d\omega'' \iint\limits_{0}^{\infty} d^3\mathbf{k}' d^3\mathbf{k}'' e^{i\left[(\mathbf{k}'+\mathbf{k}'')\cdot\mathbf{r} - (\omega'+\omega'')t\right]}.$$

$$\left[\mathbf{E}_t\left(\omega',\mathbf{k}'\right)\cdot\mathbf{J}_t\left(\omega'',\mathbf{k}''\right) + \mathbf{E}_l\left(\omega',\mathbf{k}'\right)\cdot\mathbf{J}_l\left(\omega'',\mathbf{k}''\right)\right].$$

With the use of the Dirac delta function,

$$\delta(\omega) = \frac{1}{2\pi} \int\limits_{-\infty}^{\infty} e^{-i\omega t}\, dt,$$

we can rewrite this equation as follows:

$$A = -\mathrm{Re}\int_{-\infty}^{\infty}d\omega\int_{0}^{\infty}d^3\mathbf{k}\left[\mathbf{E}_t\left(\omega,\mathbf{k}\right)\cdot\mathbf{J}_t^*\left(\omega,\mathbf{k}\right)+\mathbf{E}_l\left(\omega,\mathbf{k}\right)\cdot\mathbf{J}_l^*\left(\omega,\mathbf{k}\right)\right],\qquad(2.49)$$

where we took into account that any real field $\mathbf{F}(t,\mathbf{r})$ features $\mathbf{F}(-\omega,-\mathbf{k})=\mathbf{F}^*(\omega,\mathbf{k})$. This gives us the spectral density of work done by the current at angular frequency ω and wavevector $\mathbf{k}$:

$$a\left(\omega,\mathbf{k}\right) = -\mathrm{Re}\left[G_t^{-1}\left(\omega,\mathbf{k}\right)\left|\mathbf{J}_t\left(\omega,\mathbf{k}\right)\right|^2 + G_l^{-1}\left(\omega,\mathbf{k}\right)\left|\mathbf{J}_l\left(\omega,\mathbf{k}\right)\right|^2\right].\qquad(2.50)$$

As we can see, transverse and longitudinal currents contribute

$$a_t\left(\omega,\mathbf{k}\right) = \mathrm{Re}\left[\frac{\mathrm{i}}{\varepsilon_0}\frac{\omega}{\omega^2 - k^2 c^2}\left|\mathbf{J}_t\left(\omega,\mathbf{k}\right)\right|^2\right],\qquad(2.51)$$

$$a_l\left(\omega,\mathbf{k}\right) = \mathrm{Re}\left[\frac{\mathrm{i}}{\omega\varepsilon_0}\left|\mathbf{J}_l\left(\omega,\mathbf{k}\right)\right|^2\right]\qquad(2.52)$$

to $a(\omega,\mathbf{k})$. Note, these contributions are zero for any ω and $\mathbf{k}$ except the singular lines $\omega = \omega_m(\mathbf{k})$. In other words, transverse and longitudinal currents do not spend anything on excitation of electromagnetic fields except the fields on the singular lines given by $\omega = \omega_m(\mathbf{k})$. Amongst the entire $(\omega,\mathbf{k})$ space, only these singular lines contribute to the spectral work $a(\omega,\mathbf{k})$. To get those contributions, we use the Landau rule [7] of adding +i0 (following the causality principle) to all resonant frequencies,

$$a_t\left(\omega,\mathbf{k}\right) = -\frac{\left|\mathbf{J}_t\left(\omega,\mathbf{k}\right)\right|^2}{\varepsilon_0}\mathrm{Im}\left[\lim_{\omega\to-kc+\mathrm{i}0}\frac{\omega}{\omega^2-k^2c^2}+\lim_{\omega\to kc+\mathrm{i}0}\frac{\omega}{\omega^2-k^2c^2}\right],$$

$$a_l\left(\omega,\mathbf{k}\right) = -\frac{\left|\mathbf{J}_l\left(\omega,\mathbf{k}\right)\right|^2}{\varepsilon_0}\mathrm{Im}\lim_{\omega\to+\mathrm{i}0}\frac{1}{\omega}.$$

With the use of Sokhotsky's formula,

$$\lim_{\varepsilon\to+0}\frac{1}{x\pm\mathrm{i}\varepsilon} = \mathcal{P}\left(\frac{1}{x}\right)\mp\mathrm{i}\pi\delta\left(x\right),$$

where $\mathcal{P}$ denotes the Cauchy principal value, we obtain

$$a_t\left(\omega,\mathbf{k}\right) = \frac{\pi}{2\varepsilon_0}\left|\mathbf{J}_t\left(\omega,\mathbf{k}\right)\right|^2\left[\delta\left(\omega+kc\right)+\delta\left(\omega-kc\right)\right],\qquad(2.53)$$

$$a_l\left(\omega,\mathbf{k}\right) = \frac{\pi}{\varepsilon_0}\left|\mathbf{J}_l\left(\omega,\mathbf{k}\right)\right|^2\delta\left(\omega\right).\qquad(2.54$$

Finally, the works done by transverse and longitudinal currents on excitation of electromagnetic fields in a vacuum are given by

$$A_t = \frac{\pi}{\varepsilon_0} \int_0^\infty d^3k \left| J_t \left(ck, \mathbf{k} \right) \right|^2, \tag{2.55}$$

$$A_l = \frac{\pi}{\varepsilon_0} \int_0^\infty d^3k \left| J_l \left(0, \mathbf{k} \right) \right|^2. \tag{2.56}$$

These are the total losses incurred by currents on excitation of eigenfields with $\omega = \omega_m(\mathbf{k})$. As follows from Poynting theorem (2.29), these losses go into pumping of the eigenfields' energy and into radiation of the energy by eigenfields out of the system. The total energy of electromagnetic fields pumped into all transverse and longitudinal fields are

$$W_{t,l} = \int_{-\infty}^{\infty} dt \int_0^\infty d^3\mathbf{r} \frac{\partial w_{t,l}(t,\mathbf{r})}{\partial t} = \int_{-\infty}^{\infty} d\omega \int_0^\infty d^3k \, w_{t,l}(\omega,\mathbf{k}), \tag{2.57}$$

where the spectral densities of electromagnetic fields energy are given by

$$w_t(\omega,\mathbf{k}) = \mathrm{Re}\left[i\omega\varepsilon_0 \left[\left| \mathbf{E}_t(\omega,\mathbf{k}) \right|^2 - c^2 \left| \mathbf{B}_t(\omega,\mathbf{k}) \right| \right] \right]$$

$$= \mathrm{Re}\left[\frac{i}{\varepsilon_0} \frac{\omega}{\omega^2 - k^2 c^2} \left| \mathbf{J}_t(\omega,\mathbf{k}) \right|^2 \right],$$

$$w_l(\omega,\mathbf{k}) = \mathrm{Re}\left[i\omega\varepsilon_0 \left| \mathbf{E}_l(\omega,\mathbf{k}) \right|^2 \right] = \mathrm{Re}\left[\frac{i}{\omega\varepsilon_0} \left| \mathbf{J}_l(\omega,\mathbf{k}) \right|^2 \right].$$

Note $w_t(\omega,\mathbf{k})$ and $w_t(\omega,\mathbf{k})$ coincide with expressions (2.51) and (2.52) for $a_t(\omega,\mathbf{k})$ and $a_l(\omega,\mathbf{k})$. Eventually, we get that the currents' works completely go into energy of eigenmodes:

$$W_t = A_t, \quad W_l = A_l, \tag{2.58}$$

meaning that *the net energy radiated out of the system is zero*,

$$W_{\mathrm{rad}} = \int_{-\infty}^{\infty} dt \int_0^\infty \nabla \cdot \mathbf{S}_t(t,\mathbf{r}) d^3\mathbf{r} = 0. \tag{2.59}$$

Highlight, the total energy stored in electromagnetic fields is zero for all frequencies except the eigenfrequencies $\omega_m(\mathbf{k})$. This means that neither transverse nor longitudinal currents do a work on excitation of those fields. They spend their energy on excitation of eigenmodes only. In practice, however, $\mathbf{J}_l(0,\mathbf{k}) = 0$, as follows from the continuity equation,

$$\mathbf{J}_l(\omega,\mathbf{k}) = \frac{\omega\mathbf{k}}{k^2} \rho(\omega,\mathbf{k}), \tag{2.60}$$

making the contributions A_l and W_l zero.

2.2 ELECTROMAGNETIC FIELDS IN A HOMOGENEOUS ISOTROPIC MEDIUM

2.2.1 MICROSCOPIC AND MACROSCOPIC DESCRIPTIONS OF A MEDIUM

In Section 2.1.2, we discussed the microscopic current of an elementary particle. For the ith elementary particle, it can be written as follows:

$$\mathbf{J}_i(t,\mathbf{r}) = \frac{\partial \mathbf{P}_{i,l}(t,\mathbf{r})}{\partial t} + \nabla \times \left(\mathbf{P}_{i,l}(t,\mathbf{r}) \times \frac{d\mathbf{r}_i(t)}{dt} + \mathbf{m}_i(t)\delta\left[\mathbf{r} - \mathbf{r}_i(t)\right] \right), \quad (2.61)$$

where $\mathbf{P}_{i,l}(t,\mathbf{r})$ is the longitudinal field defined by the charge q_i and trajectory $\mathbf{r}_i(t)$ of the particle,

$$\mathbf{P}_{i,l}(t,\mathbf{r}) = -\frac{q_i}{4\pi} \frac{\mathbf{r} - \mathbf{r}_i(t)}{\left|\mathbf{r} - \mathbf{r}_i(t)\right|^3}. \quad (2.62)$$

To specify $\mathbf{J}_i(t,\mathbf{r})$, we should know the trajectory of the particle, $\mathbf{r}_i(t)$, and dynamics of its intrinsic magnetic moment, $\mathbf{m}_i(t)$. For N elementary particles composing the medium, the current density can be rewritten with two fields $\mathbf{P}(t,\mathbf{r})$ and $\mathbf{M}(t,\mathbf{r})$,

$$\mathbf{J}(t,\mathbf{r}) = \frac{\partial \mathbf{P}(t,\mathbf{r})}{\partial t} + \nabla \times \mathbf{M}(t,\mathbf{r}), \quad (2.63)$$

where

$$\mathbf{P}(t,\mathbf{r}) = \sum_{i=1}^{N} \left[\mathbf{P}_{i,l}(t,\mathbf{r}) - \nabla \times \left(\mathbf{r}_i(t) \times \mathbf{P}_{i,l}(t,\mathbf{r}) \right) \right], \quad (2.64)$$

$$\mathbf{M}(t,\mathbf{r}) = \sum_{i=1}^{N} \left[\left(\mathbf{r}_i(t) \times \frac{\partial \mathbf{P}_{i,l}(t,\mathbf{r})}{\partial t} \right) + \mathbf{m}_i(t)\delta\left[\mathbf{r} - \mathbf{r}_i(t)\right] \right]. \quad (2.65)$$

These expressions demonstrate independence of the fields $\mathbf{P}(t,\mathbf{r})$ and $\mathbf{M}(t,\mathbf{r})$. Following them, we need to know the positions $\mathbf{r}_i(t)$ of all particles to specify $\mathbf{P}(t,\mathbf{r})$ at a given point of time, while for $\mathbf{M}(t,\mathbf{r})$ we need to know not only the particles' positions, but their velocities $d\mathbf{r}_i(t)/dt$ and intrinsic magnetic moments $\mathbf{m}_i(t)$ too.

Microscopic description of current requires us to know how all the particles composing the medium move in electromagnetic fields. With increasing number of elementary particles N, the task of current density calculation becomes very complicated and troublesome, as the total number of particles in solids reaches $\sim 10^{30}$ per 1 m^3. To simplify microscopic description of a medium, the *macroscopic* approach is introduced [4, 5], where all physical fields are statistically averaged in space over small volumes. The size of those volumes, a_{macro}, is much smaller compared to the wavelength, but much larger than the lattice constant. Such averaging allows us to neglect all fluctuations that appear at atomic scales, below a_{macro}, and describe the medium from the macroscopic point of view, where particle currents are not discrete, but continuously distributed in space.

As microscopic Maxwell's equations are linear to the microscopic fields $\mathbf{J}(t,\mathbf{r})$, $\mathbf{E}(t,\mathbf{r})$ and $\mathbf{B}(t,\mathbf{r})$, the macroscopic field averaging keeps the equations unchanged. However, this averaging affects the description of currents' fields $\mathbf{P}(t,\mathbf{r})$ and $\mathbf{M}(t,\mathbf{r})$, which can no longer be functions of the particles' trajectories, but are some functions of the excited macroscopic fields $\mathbf{E}(t,\mathbf{r})$ and $\mathbf{B}(t,\mathbf{r})$. The macroscopic fields $\mathbf{P}(t,\mathbf{r})$ and $\mathbf{M}(t,\mathbf{r})$ are independent of each other, as follows from their microscopic definition given by Eqs. (2.64) and (2.65), and describe different electromagnetic effects.

Another important step in macroscopic description is introduction of a charge-neutral medium by dividing the total macroscopic charge density $\rho(t,\mathbf{r})$ into two parts:

$$\rho(t,\mathbf{r}) = \rho_{\text{med}}(t,\mathbf{r}) + \rho_{\text{ext}}(t,\mathbf{r}), \tag{2.66}$$

where sign-varying $\rho_{\text{med}}(t,\mathbf{r})$ describes fully compensated total charge of the medium,

$$\int \rho_{\text{med}}(t,\mathbf{r}) d^3\mathbf{r} = 0,$$

and the second part, $\rho_{\text{ext}}(t,\mathbf{r})$, represents uncompensated charges external to the medium,

$$\int \rho_{\text{ext}}(t,\mathbf{r}) d^3\mathbf{r} \neq 0.$$

Similarly, we can decompose the total macroscopic current density $\mathbf{J}(t,\mathbf{r})$,

$$\mathbf{J}(t,\mathbf{r}) = \mathbf{J}_{\text{med}}(t,\mathbf{r}) + \mathbf{J}_{\text{ext}}(t,\mathbf{r}). \tag{2.67}$$

Then, by writing the medium's current density as

$$\mathbf{J}_{\text{med}}(t,\mathbf{r}) = \frac{\partial \mathbf{P}_{\text{med}}(t,\mathbf{r})}{\partial t} + \nabla \times \mathbf{M}_{\text{med}}(t,\mathbf{r}), \tag{2.68}$$

we get the physical meaning of the field $\mathbf{P}_{\text{med}}(t,\mathbf{r})$—this is the *polarization field* describing the density of medium's charges,

$$\rho_{\text{med}}(t,\mathbf{r}) = -\nabla \cdot \mathbf{P}_{\text{med}}(t,\mathbf{r}). \tag{2.69}$$

It is non-zero inside the medium and vanishes outside of it,

$$\int_{V_{\text{med}}} \nabla \cdot \mathbf{P}_{\text{med}}(t,\mathbf{r}) d^3\mathbf{r} = \oint_{S_{\text{med}}} \mathbf{P}_{\text{med}}(t,\mathbf{r}) d\mathbf{S} = 0.$$

As for the field $\mathbf{M}_{\text{med}}(t,\mathbf{r})$, it is the *magnetization field*. The current given by $\nabla \times \mathbf{M}_{\text{med}}(t,\mathbf{r})$ is independent of charge variation and contributed by intrinsic and orbital magnetic moments of the medium's particles.

2.2.2 Macroscopic Maxwell's Equations

Introduction of the macroscopically defined fields allows us to rewrite the Maxwell's equations in the following *macroscopic* form:

$$\nabla \times \mathbf{H}(t,\mathbf{r}) = \frac{\partial \mathbf{D}(t,\mathbf{r})}{\partial t} + \mathbf{J}_{\text{ext}}(t,\mathbf{r}), \tag{2.70}$$

$$\nabla \times \mathbf{E}(t,\mathbf{r}) = -\frac{\partial \mathbf{B}(t,\mathbf{r})}{\partial t}, \tag{2.71}$$

with the divergence relations given by

$$\nabla \cdot \mathbf{D}(t,\mathbf{r}) = \rho_{\text{ext}}(t,\mathbf{r}), \tag{2.72}$$

$$\nabla \cdot \mathbf{B}(t,\mathbf{r}) = 0, \tag{2.73}$$

where $\mathbf{D}(t,\mathbf{r})$ and $\mathbf{H}(t,\mathbf{r})$ are the auxiliary fields called the *electric displacement* and *magnetic field*, introduced to account for the polarization and magnetization of the charge-compensated medium,

$$\mathbf{D}(t,\mathbf{r}) = \varepsilon_0 \mathbf{E}(t,\mathbf{r}) + \mathbf{P}_{\text{med}}(t,\mathbf{r}), \tag{2.74}$$

$$\mathbf{H}(t,\mathbf{r}) = \frac{1}{\mu_0} \mathbf{B}(t,\mathbf{r}) - \mathbf{M}_{\text{med}}(t,\mathbf{r}). \tag{2.75}$$

To clarify the name of field $\mathbf{H}(t,\mathbf{r})$, we regroup the terms in Maxwell's equations as follows:

$$\nabla \times \mathbf{H}(t,\mathbf{r}) = \varepsilon_0 \frac{\partial \mathbf{E}(t,\mathbf{r})}{\partial t} + \frac{\partial \mathbf{P}_{\text{med}}(t,\mathbf{r})}{\partial t} + \mathbf{J}_{\text{ext}}(t,\mathbf{r}),$$

$$\nabla \times \mathbf{E}(t,\mathbf{r}) = -\mu_0 \frac{\partial \mathbf{H}(t,\mathbf{r})}{\partial t} - \mu_0 \frac{\partial \mathbf{M}_{\text{med}}(t,\mathbf{r})}{\partial t}.$$

The fields $\mathbf{E}(t,\mathbf{r})$ and $\mathbf{H}(t,\mathbf{r})$ appear in these equations symmetrically. Following this symmetry, the field $\mathbf{H}$ is commonly called as the magnetic field, by analogy with the electric field $\mathbf{E}$, although it is actually an auxiliary quantity.

2.2.3 Macroscopic Transfer of Energy and Momentum

Statistical averaging of the fields does not change the form of Maxwell's equations. As such, macroscopic transfer of energy and momentum remain obeying Eqs. (2.28) and (2.35) with formal change of the microscopic fields to the macroscopic ones. Separation of the medium's and external currents in macroscopic description divides the source terms in the energy and momentum transfer equations:

$$\frac{\partial w(t,\mathbf{r})}{\partial t} + \nabla \cdot \mathbf{S}(t,\mathbf{r}) - a_{\text{med}}(t,\mathbf{r}) = a_{\text{ext}}(t,\mathbf{r}), \tag{2.76}$$

$$\frac{\partial \mathbf{p}(t,\mathbf{r})}{\partial t} + \nabla \cdot \overset{\leftrightarrow}{\mathbf{T}}(t,\mathbf{r}) - \mathbf{f}_{\text{med}}(t,\mathbf{r}) = \mathbf{f}_{\text{ext}}(t,\mathbf{r}), \tag{2.77}$$

where

$$a_{\text{med}}(t,\mathbf{r}) = -\mathbf{E}(t,\mathbf{r}) \cdot \mathbf{J}_{\text{med}}(t,\mathbf{r}), \tag{2.78}$$

$$a_{\text{ext}}(t,\mathbf{r}) = -\mathbf{E}(t,\mathbf{r}) \cdot \mathbf{J}_{\text{ext}}(t,\mathbf{r}) \tag{2.79}$$

and

$$\mathbf{f}_{\text{med}}(t,\mathbf{r}) = -\rho_{\text{med}}(t,\mathbf{r})\mathbf{E}(t,\mathbf{r}) - \mathbf{J}_{\text{med}}(t,\mathbf{r}) \times \mathbf{B}(t,\mathbf{r}), \tag{2.80}$$

$$\mathbf{f}_{\text{ext}}(t,\mathbf{r}) = -\rho_{\text{ext}}(t,\mathbf{r})\mathbf{E}(t,\mathbf{r}) - \mathbf{J}_{\text{ext}}(t,\mathbf{r}) \times \mathbf{B}(t,\mathbf{r}). \tag{2.81}$$

The medium-related terms describe additional mechanisms for spending of the power and force of the external currents, which pump the entire system. In particular, the power density

$$-a_{\text{med}}(t,\mathbf{r}) = \mathbf{E}(t,\mathbf{r}) \cdot \frac{\partial \mathbf{P}_{\text{med}}(t,\mathbf{r})}{\partial t} + \frac{\partial \left[\mathbf{M}_{\text{med}}(t,\mathbf{r}) \cdot \mathbf{B}(t,\mathbf{r}) \right]}{\partial t}$$
$$-\mathbf{B}(t,\mathbf{r}) \cdot \frac{\partial \mathbf{M}_{\text{med}}(t,\mathbf{r})}{\partial t} + \nabla \cdot \left[\mathbf{M}_{\text{med}}(t,\mathbf{r}) \times \mathbf{E}(t,\mathbf{r}) \right]$$

describes additional expenditures for (i) increase of polarization field $\partial \mathbf{P}_{\text{med}}(t,\mathbf{r})/\partial t$ by electric field $\mathbf{E}(t,\mathbf{r})$, (ii) increase of magnetic interaction energy $\mathbf{M}_{\text{med}}(t,\mathbf{r}) \cdot \mathbf{B}(t,\mathbf{r})$, (iii) decrease of magnetization field $\partial \mathbf{M}_{\text{med}}(t,\mathbf{r})/\partial t$ by magnetic induction $\mathbf{B}(t,\mathbf{r})$, and (iv) radiation of the power flux $\mathbf{M}_{\text{med}}(t,\mathbf{r}) \times \mathbf{E}(t,\mathbf{r})$. As for the force density

$$-\mathbf{f}_{\text{med}}(t,\mathbf{r}) = \frac{\partial}{\partial t} \left[\mathbf{P}_{\text{med}}(t,\mathbf{r}) \times \mathbf{B}(t,\mathbf{r}) \right] - \left[\nabla \cdot \mathbf{P}_{\text{med}}(t,\mathbf{r}) \right]$$
$$\mathbf{E}(t,\mathbf{r}) + \mathbf{P}_{\text{med}}(t,\mathbf{r}) \times \left[\nabla \times \mathbf{E}(t,\mathbf{r}) \right] - \mathbf{B}(t,\mathbf{r}) \times \left[\nabla \times \mathbf{M}_{\text{med}}(t,\mathbf{r}) \right],$$

it goes for increase of the polarization-induced momentum density $\mathbf{P}_{\text{med}}(t,\mathbf{r}) \times \mathbf{B}(t,\mathbf{r})$ and the force of electromagnetic fields acting on the medium's fields.

2.2.4 MATERIAL EQUATIONS

Macroscopic Maxwell's equations describe generation of electromagnetic field by external currents in terms of two pairs of electric $\{\mathbf{E}(t,\mathbf{r}),\mathbf{D}(t,\mathbf{r})\}$ and magnetic $\{\mathbf{B}(t,\mathbf{r}),\mathbf{H}(t,\mathbf{r})\}$ fields. However, they do not form a closed set of equations, until we provide *material relations* for the response of the charge-neutral medium to electric and magnetic fields. These relations are given by field-dependent functions for the polarization $\mathbf{P}_{\text{med}}(t,\mathbf{r}) = \mathbf{P}_{\text{med}}(\mathbf{E}(t,\mathbf{r}))$ and magnetization $\mathbf{M}_{\text{med}}(t,\mathbf{r}) = \mathbf{M}_{\text{med}}(\mathbf{B}(t,\mathbf{r}))$ vectors that finally result in material relations for the auxiliary fields $\mathbf{D}(t,\mathbf{r}) = \mathbf{D}(\mathbf{E}(t,\mathbf{r}))$

and $\mathbf{H}(t,\mathbf{r}) = \mathbf{H}(\mathbf{B}(t,\mathbf{r}))$. In general, these relations are nonlinear. However, for not high $\mathbf{E}(t,\mathbf{r})$ and $\mathbf{B}(t,\mathbf{r})$ fields, they can be approximated with linear functions. It is the so-called *linear electrodynamics* approach. In this approximation, the response of the medium at a given point $\mathbf{r}$ and moment t is assumed to be a linear function of electromagnetic fields at any point r' taken at all preceding moments $t' < t$ in accordance with the causality principle,

$$P_{\text{med},i}\left(t,\mathbf{r}\right) = \varepsilon_0 \int_{-\infty}^{t} dt' \int_{0}^{\infty} d^3\mathbf{r}' \chi_{ij}^{e}\left(t,t',\mathbf{r},\mathbf{r}'\right) E_j\left(t',\mathbf{r}'\right), \qquad (2.82)$$

$$M_{\text{med},i}\left(t,\mathbf{r}\right) = \int_{-\infty}^{t} dt' \int_{0}^{\infty} d^3\mathbf{r}' \chi_{ij}^{m}\left(t,t',\mathbf{r},\mathbf{r}'\right) H_j\left(t',\mathbf{r}'\right). \qquad (2.83)$$

Here, we write the dependence $\mathbf{M}_{\text{med}}(\mathbf{H})$ instead of $\mathbf{M}_{\text{med}}(\mathbf{B})$ following the linearity of medium response.

The tensors $\chi_{ij}^{e}(t,t',\mathbf{r},\mathbf{r}')$ and $\chi_{ij}^{m}(t,t',\mathbf{r},\mathbf{r}')$ in Eqs. (2.82) and (2.83) characterize the efficiency of the material response transfer from one point of space and time to another. Macroscopic averaging of the fields over a_{macro} larger than the lattice constant allows us to treat the medium response uniform in space. As for the response in time, it is uniform following the homogeneity of time. Finally, the functions χ_{ij}^{e}, χ_{ij}^{m} can be treated as dependent on the differences $t - t'$ and $\mathbf{r} - \mathbf{r}'$,

$$P_{\text{med},i}(t,\mathbf{r}) = \varepsilon_0 \int_{-\infty}^{t} dt' \int_{0}^{\infty} d^3\mathbf{r}' \chi_{ij}^{e}\left(t - t',\mathbf{r} - \mathbf{r}'\right) E_j\left(t',\mathbf{r}'\right), \qquad (2.84)$$

$$M_{\text{med},i}(t,\mathbf{r}) = \int_{-\infty}^{t} dt' \int_{0}^{\infty} d^3\mathbf{r}' \chi_{ij}^{m}\left(t - t',\mathbf{r} - \mathbf{r}'\right) H_j\left(t',\mathbf{r}'\right). \qquad (2.85)$$

By performing the Fourier transform of $\mathbf{P}_{\text{med}}$, $\mathbf{M}_{\text{med}}$, $\mathbf{E}$, $\mathbf{H}$ from the $(t,\mathbf{r})$ space to the $(\omega,\mathbf{k})$ space, we get the material relations for polarization and magnetization fields,

$$P_{\text{med},i}(\omega,\mathbf{k}) = \varepsilon_0 \chi_{ij}^{e}(\omega,\mathbf{k}) E_j(\omega,\mathbf{k}), \qquad (2.86)$$

$$M_{\text{med},i}(\omega,\mathbf{k}) = \chi_{ij}^{m}(\omega,\mathbf{k}) H_j(\omega,\mathbf{k}). \qquad (2.87)$$

Here, $\chi_{ij}^{e}(\omega,\mathbf{k})$ and $\chi_{ij}^{m}(\omega,\mathbf{k})$ are the tensors of *complex electric and magnetic susceptibility* given by

$$\chi_{ij}^{e}(\omega,\mathbf{k}) = \int_{0}^{\infty} dt_1 \int_{-\infty}^{\infty} d^3\mathbf{r}_1 \chi_{ij}^{e}\left(t_1,\mathbf{r}_1\right) e^{-i\left(\mathbf{k}\cdot\mathbf{r}_1 - \omega t_1\right)}, \qquad (2.88)$$

$$\chi_{ij}^{m}(\omega,\mathbf{k}) = \int_{0}^{\infty} dt_1 \int_{-\infty}^{\infty} d^3\mathbf{r}_1 \chi_{ij}^{m}\left(t_1,\mathbf{r}_1\right) e^{-i\left(\mathbf{k}\cdot\mathbf{r}_1 - \omega t_1\right)}, \qquad (2.89)$$

where $t_1 = t - t'$ and $\mathbf{r}_1 = \mathbf{r} - \mathbf{r}'$.

As for the fields $\mathbf{D}(\omega,\mathbf{k})$ and $\mathbf{B}(\omega,\mathbf{k})$, we get the following material relations:

$$D_i(\omega,\mathbf{k}) = \varepsilon_0 \varepsilon_{ij}(\omega,\mathbf{k}) E_j(\omega,\mathbf{k}), \tag{2.90}$$

$$B_i(\omega,\mathbf{k}) = \mu_0 \mu_{ij}(\omega,\mathbf{k}) H_j(\omega,\mathbf{k}), \tag{2.91}$$

where the tensors of *complex dielectric permittivity* $\varepsilon_{ij}(\omega,\mathbf{k})$ and *magnetic permeability* $\mu_{ij}(\omega,\mathbf{k})$ are defined as

$$\varepsilon_{ij}(\omega,\mathbf{k}) = 1 + \chi_{ij}^e(\omega,\mathbf{k}), \tag{2.92}$$

$$\mu_{ij}(\omega,\mathbf{k}) = 1 + \chi_{ij}^m(\omega,\mathbf{k}). \tag{2.93}$$

The tensors $\chi_{ij}^{e,m}(\omega,\mathbf{k})$ and $\chi_{ij}^m(\omega,\mathbf{k})$ are the Fourier images of the real-valued tensors $\chi_{ij}^e(t,\mathbf{r})$ and $\chi_{ij}^m(t,\mathbf{r})$. This brings the following symmetry property for $\chi_{ij}^{e,m}(\omega,\mathbf{k})$:

$$\chi_{ij}^{e,m}(-\omega,-\mathbf{k}) = \left[\chi_{ij}^{e,m}(\omega,\mathbf{k}) \right]^*.$$

Also, the tensors $\chi_{ij}^{e,m}(\omega,\mathbf{k})$ are analytic in the complex upper half-plane of ω and vanish faster than $1/|\omega|$ as $|\omega| \to \infty$ [6]. As such, their real and imaginary parts obey the Kramers-Kronig relations:

$$\mathrm{Re}\left[\chi_{ij}^{e,m}(\omega,\mathbf{k}) \right] = \frac{1}{\pi} P \int\limits_{-\infty}^{\infty} \frac{\mathrm{Im}\left[\chi_{ij}^{e,m}(\omega,\mathbf{k}) \right]}{\omega' - \omega} d\omega',$$

$$\mathrm{Im}\left[\chi_{ij}^{e,m}(\omega,\mathbf{k}) \right] = -\frac{1}{\pi} P \int\limits_{-\infty}^{\infty} \frac{\mathrm{Re}\left[\chi_{ij}^{e,m}(\omega,\mathbf{k}) \right]}{\omega' - \omega} d\omega'.$$

As for the tensors of $\varepsilon_{ij}(\omega,\mathbf{k})$ and $\mu_{ij}(\omega,\mathbf{k})$, they exhibit similar symmetry property:

$$\varepsilon_{ij}(-\omega,-\mathbf{k}) = \varepsilon_{ij}^*(\omega,\mathbf{k}),$$

$$\mu_{ij}(-\omega,-\mathbf{k}) = \mu_{ij}^*(\omega,\mathbf{k}),$$

and the Kramers-Kronig relations:

$$\mathrm{Re}\,\varepsilon_{ij}(\omega,\mathbf{k}) - \delta_{ij} = \frac{1}{\pi} P \int\limits_{-\infty}^{\infty} \frac{\mathrm{Im}\,\varepsilon_{ij}(\omega,\mathbf{k})}{\omega' - \omega} d\omega',$$

$$\mathrm{Im}\,\varepsilon_{ij}(\omega,\mathbf{k}) = -\frac{1}{\pi} P \int\limits_{-\infty}^{\infty} \frac{\mathrm{Re}\,\varepsilon_{ij}(\omega,\mathbf{k}) - \delta_{ij}}{\omega' - \omega} d\omega',$$

$$\mathrm{Re}\,\mu_{ij}(\omega,\mathbf{k}) - \delta_{ij} = \frac{1}{\pi} P \int\limits_{-\infty}^{\infty} \frac{\mathrm{Im}\,\mu_{ij}(\omega,\mathbf{k})}{\omega' - \omega} d\omega',$$

$$\mathrm{Im}\,\mu_{ij}(\omega,\mathbf{k}) = -\frac{1}{\pi} P \int\limits_{-\infty}^{\infty} \frac{\mathrm{Re}\,\mu_{ij}(\omega,\mathbf{k}) - \delta_{ij}}{\omega' - \omega} d\omega'.$$

Next, we assume that the medium is isotropic. For such a medium, the response should be identical in any direction. This require the medium to have zero electric and magnetic dipole moments

$$\int \mathbf{J}_{med}\left(t,\mathbf{r}\right)d^3\mathbf{r} = 0, \quad \int \mathbf{r} \times \mathbf{J}_{med}\left(t,\mathbf{r}\right)d^3\mathbf{r} = 0.$$

Under these conditions, $\varepsilon_{ij}(\omega,\mathbf{k})$ and $\mu_{ij}(\omega,\mathbf{k})$ can be composed of the unit tensor δ_{ij} and the tensor k_ik_j, as they are the only two tensors of the second rank formed of the wavevector $\mathbf{k}$. In this case, we have

$$\varepsilon_{ij}(\omega,\mathbf{k}) = \left(\delta_{ij} - \frac{k_ik_j}{k^2}\right)\varepsilon_t(\omega,\mathbf{k}) + \frac{k_ik_j}{k^2}\varepsilon_l(\omega,\mathbf{k}), \tag{2.94}$$

$$\mu_{ij}(\omega,\mathbf{k}) = \left(\delta_{ij} - \frac{k_ik_j}{k^2}\right)\mu_t(\omega,\mathbf{k}) + \frac{k_ik_j}{k^2}\mu_l(\omega,\mathbf{k}). \tag{2.95}$$

Thus, among nine components of each tensor $\varepsilon_{ij}(\omega,\mathbf{k})$ and $\mu_{ij}(\omega,\mathbf{k})$, only two components are independent. These components are $\varepsilon_t(\omega,\mathbf{k})$ and $\varepsilon_l(\omega,\mathbf{k})$ for $\varepsilon_{ij}(\omega,\mathbf{k})$, or $\mu_t(\omega,\mathbf{k})$ and $\mu_l(\omega,\mathbf{k})$ for $\mu_{ij}(\omega,\mathbf{k})$. The meaning of those tensor components gets clear, if we write $\mathbf{D}(\omega,\mathbf{k})$ and $\mathbf{B}(\omega,\mathbf{k})$ in the vector form,

$$\mathbf{D}(\omega,\mathbf{k}) = \varepsilon_0\left[\varepsilon_t(\omega,\mathbf{k})\mathbf{E}_t(\omega,\mathbf{k}) + \varepsilon_l(\omega,\mathbf{k})\mathbf{E}_l(\omega,\mathbf{k})\right],$$

$$\mathbf{B}(\omega,\mathbf{k}) = \mu_0\left[\mu_t(\omega,\mathbf{k})\mathbf{H}_t(\omega,\mathbf{k}) + \mu_l(\omega,\mathbf{k})\mathbf{H}_l(\omega,\mathbf{k})\right],$$

where the subscripts t and l denote the Helmholtz components of the fields,

$$\mathbf{F}_t(\omega,\mathbf{k}) = \frac{\mathbf{k} \times [\mathbf{F}(\omega,\mathbf{k}) \times \mathbf{k}]}{k^2}, \tag{2.96}$$

$$\mathbf{F}_l(\omega,\mathbf{k}) = \frac{\mathbf{k} \cdot [\mathbf{F}(\omega,\mathbf{k}) \cdot \mathbf{k}]}{k^2}. \tag{2.97}$$

Following these expressions, $\varepsilon_l(\omega,\mathbf{k})$ and $\mu_l(\omega,\mathbf{k})$ give the medium response to longitudinal electric and magnetic fields, while $\varepsilon_t(\omega,\mathbf{k})$ and $\mu_t(\omega,\mathbf{k})$ describe the medium response to transverse electric and magnetic fields,

$$\mathbf{D}_t(\omega,\mathbf{k}) = \varepsilon_0\varepsilon_t(\omega,\mathbf{k})\mathbf{E}_t(\omega,\mathbf{k}), \quad \mathbf{D}_l(\omega,\mathbf{k}) = \varepsilon_0\varepsilon_l(\omega,\mathbf{k})\mathbf{E}_l(\omega,\mathbf{k}),$$

$$\mathbf{B}_t(\omega,\mathbf{k}) = \mu_0\mu_t(\omega,\mathbf{k})\mathbf{H}_t(\omega,\mathbf{k}), \quad \mathbf{B}_l(\omega,\mathbf{k}) = \mu_0\mu_l(\omega,\mathbf{k})\mathbf{H}_l(\omega,\mathbf{k}).$$

Note, the longitudinal magnetic induction is always zero, $\mathbf{B}_l(\omega,\mathbf{k}) \equiv 0$. The longitudinal part of magnetization does not affect excitation of electromagnetic fields and, thus, can be set to zero. Then, the longitudinal magnetic field is zero too, $\mathbf{H}_l(\omega,\mathbf{k}) = -\mathbf{M}_{med, l}(\omega,\mathbf{k}) = 0$. Finally, $\mu_l(\omega,\mathbf{k})$ can be chosen arbitrary as not making any sense.

2.2.5   TEMPORAL AND SPATIAL DISPERSION

Medium response given in terms of two tensors $\varepsilon_{ij}(\omega,\mathbf{k})$ and $\mu_{ij}(\omega,\mathbf{k})$ is frequency and wavevector-dependent. This means that medium responses differently to external excitations with different ω and $\mathbf{k}$. As a result, any electromagnetic pulse injected into the medium will disperse in time and space. Such media with frequency and wavevector dependences are called *dispersive*. The frequency dependencies of the tensors $\varepsilon_{ij}(\omega,\mathbf{k})$ and $\mu_{ij}(\omega,\mathbf{k})$ describe the *temporal dispersion* of electromagnetic fields, while their wavevector dependencies give the *spatial dispersion*.

Temporal dispersion arises due to the inertia and friction of the medium particles. They make polarization and magnetization inertial to electric and magnetic fields. As a result, the medium response at a moment t depends on the field values at all preceding moments $t' \le t$. The time interval $\tau = t - t'$, over which the previous history still has a significant effect, is defined by the characteristic frequencies ω_P for polarization effects and ω_M for magnetization ones. For electromagnetic fields oscillating at a very high frequency $|\omega| \gg \omega_P, \omega_M$, the medium particles do not have enough time to form any significant polarization and magnetization, leading to

$$\varepsilon_{ij}\left(\omega \rightarrow \pm\infty, \mathbf{k}\right) = \delta_{ij}, \quad \mu_{ij}\left(\omega \rightarrow \pm\infty, \mathbf{k}\right) = \delta_{ij}.$$

However, at frequencies below or close to the characteristic frequencies, the temporal dispersion gets stronger and should be taken into account.

In contrast to temporal dispersion, spatial one comes from nonlocality of the medium response. Physically, it is given by dependence of polarization and magnetization fields $\mathbf{P}(t,\mathbf{r})$, $\mathbf{M}(t,\mathbf{r})$ on electric and magnetic fields $\mathbf{E}(t,\mathbf{r}')$, $\mathbf{H}(t,\mathbf{r}')$ in the vicinity of the point $\mathbf{r}$. The region over which the nonlocality takes place is defined by the characteristic length $a_s = a_{\text{macro}} + a_{\text{dyn}}$ given by the macroscopic field averaging and particle dynamics. The length a_{macro} is the inherent error of the macroscopic approach that does not allow us to look at smaller scales, but is necessary to fulfill the condition of medium's uniformity. The length a_{dyn} is given by the average traveling distance of the electrons at the highest occupied orbital

$$a_{\text{dyn}} = \sqrt{\frac{2E_{\text{HOMO}}}{m\omega^2}},$$

where E_{HOMO} and m are the energy and effective mass of those electrons.

As a_{dyn} decreases with the frequency very fast, the response of typical solids starting from the infrared range is highly localized ($ka_s \approx ka_{\text{macro}} \ll 1$) with negligibly small spatial dispersion. It allows us to treat $\varepsilon_{ij}(\omega,\mathbf{k})$ and $\mu_{ij}(\omega,\mathbf{k})$ independent of the wavevector $\mathbf{k}$, when

$$\varepsilon_t(\omega,\mathbf{k}) = \varepsilon_t(\omega), \quad \mu_t(\omega,\mathbf{k}) = \mu_t(\omega), \tag{2.98}$$

$$\varepsilon_l(\omega,\mathbf{k}) = \varepsilon_l(\omega), \quad \mu_l(\omega,\mathbf{k}) = \mu_l(\omega). \tag{2.99}$$

Thus, uniform isotropic solids are well described at infrared frequencies and above within the *local response* approximation, where the material relations

$$\mathbf{D}_{t,l}(\omega,\mathbf{k}) = \varepsilon_0 \varepsilon_{t,l}(\omega)\mathbf{E}_{t,l}(\omega,\mathbf{k}), \tag{2.100}$$

$$\mathbf{B}_{t,l}(\omega,\mathbf{k}) = \mu_0 \mu_{t,l}(\omega)\mathbf{H}_{t,l}(\omega,\mathbf{k}) \tag{2.101}$$

are given by two functions of *complex scalar dielectric permittivity*

$$\varepsilon_{t,l}(\omega) = \varepsilon'_{t,l}(\omega) + i\varepsilon''_{t,l}(\omega)$$

and two functions of *complex scalar magnetic permeability*

$$\mu_{t,l}(\omega) = \mu'_{t,l}(\omega) + i\mu''_{t,l}(\omega).$$

These functions obey the Kramers-Kronig relations:

$$\varepsilon'_{t,l}(\omega) - 1 = \frac{1}{\pi} \mathcal{P} \int_{-\infty}^{\infty} \frac{\varepsilon''_{t,l}(\omega)}{\tilde{\omega} - \omega} d\tilde{\omega}, \quad \varepsilon''_{t,l}(\omega) = -\frac{1}{\pi} \mathcal{P} \int_{-\infty}^{\infty} \frac{\varepsilon'_{t,l}(\omega) - 1}{\tilde{\omega} - \omega} d\tilde{\omega},$$

$$\mu'_{t,l}(\omega) - 1 = \frac{1}{\pi} \mathcal{P} \int_{-\infty}^{\infty} \frac{\mu''_{t,l}(\omega)}{\tilde{\omega} - \omega} d\tilde{\omega}, \quad \mu''_{t,l}(\omega) = -\frac{1}{\pi} \mathcal{P} \int_{-\infty}^{\infty} \frac{\mu'_{t,l}(\omega) - 1}{\tilde{\omega} - \omega} d\tilde{\omega},$$

and feature the following symmetry:

$$\varepsilon'_{t,l}(-\omega) = \varepsilon'_{t,l}(\omega), \quad \varepsilon''_{t,l}(-\omega) = -\varepsilon''_{t,l}(\omega),$$

$$\mu'_{t,l}(-\omega) = \mu'_{t,l}(\omega), \quad \mu''_{t,l}(-\omega) = -\mu''_{t,l}(\omega).$$

The symmetry relations reveal the even behavior of $\varepsilon'_{t,l}(\omega)$ and $\mu'_{t,l}(\omega)$, as well as the odd behavior of $\varepsilon''_{t,l}(\omega)$ and $\mu''_{t,l}(\omega)$. In Section 2.2.6, we will see that excitation of electromagnetic fields within the local response approximation is affected by $\varepsilon_l(0)$. Following the symmetry relations, $\varepsilon''_{t,l}(\omega)$ change their signs at $\omega = 0$ passing through zero or infinity. The former is observed for non-conductive media, while the latter takes place for conductive materials.

2.2.6 LOCAL RESPONSE APPROXIMATION

Local response approximation is the most widely used description of medium response. Within this approximation, macroscopic Maxwell's equations are written in the $(\omega,\mathbf{r})$ space as follows:

$$\nabla \times \mathbf{H}(\omega,\mathbf{r}) = -i\omega\mathbf{D}(\omega,\mathbf{r}) + \mathbf{J}_{\text{ext}}(\omega,\mathbf{r}), \tag{2.102}$$

$$\nabla \times \mathbf{E}(\omega,\mathbf{r}) = i\omega\mathbf{B}(\omega,\mathbf{r}), \tag{2.103}$$

where the material relations are given in terms of the scalar functions $\varepsilon_{t,l}(\omega)$ and $\mu_{t,l}(\omega)$,

$$\mathbf{D}_{t,l}(\omega,\mathbf{r}) = \varepsilon_0 \varepsilon_{t,l}(\omega)\mathbf{E}_{t,l}(\omega,\mathbf{r}), \tag{2.104}$$

$$\mathbf{B}_{t,l}(\omega,\mathbf{r}) = \mu_0 \mu_{t,l}(\omega)\mathbf{H}_{t,l}(\omega,\mathbf{r}). \tag{2.105}$$

For Helmholtz fields, the Maxwell's equations split into the following four equations:

$$\nabla \times \mathbf{H}_t(\omega,\mathbf{r}) = -i\omega \mathbf{D}_t(\omega,\mathbf{r}) + \mathbf{J}_{\text{ext},t}(\omega,\mathbf{r}),$$

$$\nabla \times \mathbf{E}_t(\omega,\mathbf{r}) = i\omega \mathbf{B}_t(\omega,\mathbf{r}),$$

$$-i\omega \mathbf{D}_l(\omega,\mathbf{r}) + \mathbf{J}_{\text{ext},l}(\omega,\mathbf{r}) = 0,$$

$$i\omega \mathbf{B}_l(\omega,\mathbf{r}) = 0.$$

Following these, the magnetic induction fields look similar to those in the microscopic description:

$$\mathbf{B}_t(\omega,\mathbf{r}) = -\frac{i}{\omega}\nabla \times \mathbf{E}_t(\omega,\mathbf{r}) \neq 0, \tag{2.106}$$

$$\mathbf{B}_l(\omega,\mathbf{r}) \equiv 0, \tag{2.107}$$

while the electric fields are substantially different,

$$\mathbf{E}_t(\omega,\mathbf{r}) = \hat{G}_t^{-1}(\omega)\mathbf{J}_{\text{ext},t}(\omega,\mathbf{r}), \tag{2.108}$$

$$\mathbf{E}_l(\omega,\mathbf{r}) = \hat{G}_l^{-1}(\omega)\mathbf{J}_{\text{ext},l}(\omega,\mathbf{r}), \tag{2.109}$$

where $\hat{G}_{t,l}^{-1}(\omega)$ account for the medium response:

$$\hat{G}_t(\omega) = i\omega\varepsilon_0 \frac{\varepsilon_t(\omega)\mu_t(\omega)k_0^2(\omega) + \nabla^2}{\mu_t(\omega)k_0^2(\omega)}, \tag{2.110}$$

$$\hat{G}_l(\omega) = i\omega\varepsilon_0\varepsilon_l(\omega) \tag{2.111}$$

with $k_0 = \omega\sqrt{\varepsilon_0\mu_0}$ as in microscopic description. We see that transverse and longitudinal fields are still excited independently by their respective currents. Excitation of $\mathbf{E}_t(\omega,\mathbf{r})$ and $\mathbf{B}_t(\omega,\mathbf{r})$ by $\mathbf{J}_t(\omega,\mathbf{r})$ is still nonlocal, while excitation of $\mathbf{E}_l(\omega,\mathbf{r})$ by $\mathbf{J}_l(\omega,\mathbf{r})$ remains local.

Resonances of the excited fields in local response approximation are governed by the medium's $\varepsilon_{t,l}(\omega),\mu_{t,l}(\omega)$. By writing the eigenfields in the form of plane waves,

$$\tilde{\mathbf{E}}\left(\omega_m(\mathbf{k}),\mathbf{k},\mathbf{r}\right) = \mathbf{E}_0\left(\omega_m(\mathbf{k}),\mathbf{k}\right)e^{i\mathbf{k}\cdot\mathbf{r}},$$

where the real wavevector **k** plays the role of eigenmode parameter, we obtain the following dispersion relations:

$$G_t\big(\omega_m(\mathbf{k}),\mathbf{k}\big) = i\varepsilon_0\,\frac{\varepsilon_t\big(\omega_m(\mathbf{k})\big)\mu_t\big(\omega_m(\mathbf{k})\big)\omega_m^2(\mathbf{k}) - c^2 k^2}{\mu_t\big(\omega_m(\mathbf{k})\big)\omega_m(\mathbf{k})} = 0,$$

$$G_l\big(\omega_m(\mathbf{k}),\mathbf{k}\big) = i\omega_m(\mathbf{k})\varepsilon_0\varepsilon_l\big(\omega_m(\mathbf{k})\big) = 0,$$

which can be reduced to

$$\varepsilon_t\big(\omega_m(\mathbf{k})\big)\mu_t\big(\omega_m(\mathbf{k})\big)\omega_m^2(\mathbf{k}) - c^2 k^2 = 0 \qquad (2.112)$$

for transverse eigenmodes and to

$$\omega_m(\mathbf{k})\varepsilon_l\big(\omega_m(\mathbf{k})\big) = 0 \qquad (2.113)$$

for longitudinal ones.

As we can see, the dispersion relation of transverse eigenmodes is the modified relation of photons. Now, it contains the additional factor $\varepsilon_t(\omega_m(\mathbf{k}))\mu_t(\omega_m(\mathbf{k}))$ that changes the eigenfrequency of photons and brings additional solutions for $\omega_m(\mathbf{k})$. These eigenmodes are called *polaritons*[1]. For complex functions $\varepsilon_t(\omega)$ and $\mu_t(\omega)$, eigenfrequencies of polaritons are complex. They do not cause divergences of excited field, as was in the case of photons.

As for longitudinal eigenmodes, their dispersion relation is the modified relation of virtual photons with the complex factor $\varepsilon_l(\omega)$. This factor brings new complex eigenfrequencies $\omega_m(\mathbf{k})$. These new eigenmodes are called *phonons*, *plasmons*, and *excitons*, depending on the medium type. As for the virtual photons with $\omega_m(\mathbf{k}) = 0$, their presence or absence in the system is defined by the status of medium's conduction. In particular, if we introduce medium's conductivities $\sigma_{t,l}(\omega)$ as the complex coefficients between the excited electric fields and the medium's current density

$$\mathbf{J}_{\text{med}}(\omega,\mathbf{r}) = \sigma_l(\omega)\mathbf{E}_l(\omega,\mathbf{r}) + \sigma_t(\omega)\mathbf{E}_t(\omega,\mathbf{r}),$$

we arrive at the direct relation between the conductivities $\sigma_{t,l}(\omega)$ and dielectric permittivities $\varepsilon_{t,l}(\omega)$ of the medium:

$$\sigma_{t,l}(\omega) = -i\omega\big[\varepsilon_{t,l}(\omega) - 1\big],$$

that gives us

$$\lim_{\omega \to 0}\omega\varepsilon_l(\omega) = i\sigma_l(0).$$

---

1 Depending on the medium type, polaritons can be specified as phonon-polaritons, plasmon-polaritons, or exciton-polaritons.

This limit tells us that $\omega_m(\mathbf{k}) = 0$ can be a solution of Eq. (2.113), but only for a non-conductive medium with $\sigma_l(0) = 0$. For a conductive medium with $\sigma_l(0) \neq 0$, virtual photons are no longer solutions of the dispersive relation. Thus, we can expect to see the divergence of the excited fields at $\omega = 0$ for non-conductive media and no singularity for conductive media.

Resonances on the medium eigenmodes can be seen in Fourier images of the excited fields $\mathbf{E}_{t,l}(\omega,\mathbf{k})$:

$$\mathbf{E}_t(\omega,\mathbf{k}) = G_t^{-1}(\omega,\mathbf{k})\mathbf{J}_{ext,t}(\omega,\mathbf{k}), \tag{2.114}$$

$$\mathbf{E}_l(\omega,\mathbf{k}) = G_l^{-1}(\omega,\mathbf{k})\mathbf{J}_{ext,l}(\omega,\mathbf{k}). \tag{2.115}$$

As the medium eigenfrequencies $\omega_m(\mathbf{k})$ are complex [except $\omega_m(\mathbf{k}) = 0$ for nonconductive media], the strength of the resonances appears dependent on the imaginary part of $\omega_m(\mathbf{k})$. The smaller the $\left|\omega_m''(\mathbf{k})\right|$ compared to the $\left|\omega_m'(\mathbf{k})\right|$, the stronger the resonance. For eigenmodes with $\left|\omega_m''(\mathbf{k})\right| \geq \left|\omega_m'(\mathbf{k})\right|$, the resonance strength may be too small to be distinguished as a separate peak.

Note, imaginary parts of $\varepsilon_{t,l}(\omega)$ and $\mu_t(\omega)$ define the total work done by external currents on excitation of electromagnetic fields:

$$A_{ext} = \int\limits_{-\infty}^{\infty} dt \int\limits_{0}^{\infty} d^3\mathbf{r}\, a_{ext}(t,\mathbf{r}) =$$

$$-\mathrm{Re} \int\limits_{0}^{\infty} d^3\mathbf{k} \int\limits_{-\infty}^{\infty} d\omega \left[\mathbf{E}_t(\omega,\mathbf{k}) \cdot \mathbf{J}_{ext,t}^*(\omega,\mathbf{k}) + \mathbf{E}_l(\omega,\mathbf{k}) \cdot \mathbf{J}_{ext,l}^*(\omega,\mathbf{k}) \right].$$

This work is separately contributed by external transverse and longitudinal currents,

$$A_{ext,t} = -\mathrm{Re} \int\limits_{0}^{\infty} d^3\mathbf{k} \int\limits_{-\infty}^{\infty} d\omega\, G_t^{-1}(\omega,\mathbf{k}) \left|\mathbf{J}_{ext,t}(\omega,\mathbf{k})\right|^2,$$

$$A_{ext,l} = -\mathrm{Re} \int\limits_{0}^{\infty} d^3\mathbf{k} \int\limits_{-\infty}^{\infty} d\omega\, G_l^{-1}(\omega,\mathbf{k}) \left|\mathbf{J}_{ext,l}(\omega,\mathbf{k})\right|^2.$$

The contribution by external transverse current has no singularities in contrast to microscopic description,

$$A_{ext,t} = \int\limits_{0}^{\infty} d^3\mathbf{k} \int\limits_{-\infty}^{\infty} d\omega \frac{\mu_0\omega\left[\mu_t''(\omega)k^2 + \varepsilon_t''(\omega)\left|\mu_t(\omega)\right|^2 k_0^2(\omega)\right]}{\left|\varepsilon_t(\omega)\mu_t(\omega)k_0^2(\omega) - k^2\right|^2} \left|\mathbf{J}_{ext,t}(\omega,\mathbf{k})\right|^2. \tag{2.117}$$

As for $A_{\text{ext},l}$, we should separately consider conductive and non-conductive media. For a conductive medium, we do not have any singularity on the frequency axis, resulting in

$$A_{\text{ext},l} = \int_0^\infty d^3k \int_{-\infty}^\infty d\omega \frac{\varepsilon_l''(\omega)}{\varepsilon_0 \omega |\varepsilon_l(\omega)|^2} |J_{\text{ext},l}(\omega, k)|^2 . \tag{2.118}$$

A non-conductive medium brings a singularity at $\omega = 0$ and leads to

$$A_{\text{ext},l} = \mathcal{P}\int_0^\infty d^3k \int_{-\infty}^\infty d\omega \frac{\varepsilon_l''(\omega)}{\varepsilon_0 \omega |\varepsilon_l(\omega)|^2} |J_{\text{ext},l}(\omega, k)|^2$$

$$+ \frac{\pi \varepsilon_l'(0)}{\varepsilon_0 |\varepsilon_l(0)|^2} \int_0^\infty d^3k |J_{\text{ext},l}(0, k)|^2 . \tag{2.119}$$

As we can see, $\varepsilon_{t,l}''(\omega)$ and $\mu_t''(\omega)$ define the spectral work done by external currents on excitation of electromagnetic fields at frequency ω.

Similar to microscopic description, some part of work A_{ext} done by external currents goes into pumping of electromagnetic energy W:

$$W = \int_{-\infty}^\infty dt \int_0^\infty d^3r \frac{\partial w(t, \mathbf{r})}{\partial t} = \int_{-\infty}^\infty d\omega \int_0^\infty d^3k \left[w_t(\omega, \mathbf{k}) + w_l(\omega, \mathbf{k}) \right], \tag{2.120}$$

where the spectral densities $w_{t,l}(\omega, \mathbf{k})$ are given by

$$w_t(\omega, \mathbf{k}) = \operatorname{Re}\left[i\omega \varepsilon_0 \left[|E_t(\omega, \mathbf{k})|^2 - c^2 |B_t(\omega, \mathbf{k})|^2 \right] \right],$$

$$w_l(\omega, \mathbf{k}) = \operatorname{Re}\left[i\omega \varepsilon_0 |E_l(\omega, \mathbf{k})|^2 \right]. \tag{2.121}$$

Obviously,

$$W_t = 0, \quad W_l = 0$$

for a conductive medium and

$$W_t = 0, \quad W_l = \frac{\pi}{\varepsilon_0 |\varepsilon_l(0)|^2} \int_0^\infty d^3k |J_{\text{ext},l}(0, k)|^2 \tag{2.122}$$

for a non-conductive one.

The rest of work done goes into dissipation by the medium and radiation of energy out of the system. The energy dissipated can be calculated as follows:

$$Q = -\int_{-\infty}^{\infty} dt \int_{0}^{\infty} d^3\mathbf{r}\, a_{med}(t,\mathbf{r}) = \mathrm{Re}\int_{0}^{\infty} d^3\mathbf{k} \int_{-\infty}^{\infty} d\omega \left[\begin{array}{l} \mathbf{E}_t(\omega,\mathbf{k}) \cdot \mathbf{J}_{med,t}^*(\omega,\mathbf{k}) \\ +\mathbf{E}_l(\omega,\mathbf{k}) \cdot \mathbf{J}_{med,l}^*(\omega,\mathbf{k}) \end{array} \right]. \quad (2.123)$$

This energy is separately contributed by transverse and longitudinal fields. The transverse fields give

$$Q_t = \int_{0}^{\infty} d^3\mathbf{k} \int_{-\infty}^{\infty} d\omega \frac{\mu_0 \omega \left[\mu_t''(\omega)k^2 + \varepsilon_t''(\omega)|\mu_t(\omega)|^2 k_0^2(\omega) \right]}{\left| \varepsilon_t(\omega)\mu_t(\omega)k_0^2(\omega) - k^2 \right|^2} |\mathbf{J}_{ext,t}(\omega,\mathbf{k})|^2, \quad (2.124)$$

while the longitudinal fields contribute

$$Q_l = \int_{0}^{\infty} d^3\mathbf{k} \int_{-\infty}^{\infty} d\omega \frac{\varepsilon_l''(\omega)}{\varepsilon_0 \omega |\varepsilon_l(\omega)|^2} |\mathbf{J}_{ext,l}(\omega,\mathbf{k})|^2 \quad (2.125)$$

in a conductive medium and

$$Q_l = \mathcal{P}\int_{0}^{\infty} d^3\mathbf{k} \int_{-\infty}^{\infty} d\omega \frac{\varepsilon_l''(\omega)}{\varepsilon_0 \omega |\varepsilon_l(\omega)|^2} |\mathbf{J}_{ext,l}(\omega,\mathbf{k})|^2$$
$$+ \frac{\pi\left[\varepsilon_l'(0)-1\right]}{\varepsilon_0 |\varepsilon_l(0)|^2} \int_{0}^{\infty} d^3\mathbf{k} |\mathbf{J}_{ext,l}(0,\mathbf{k})|^2 \quad (2.126)$$

in a non-conductive one. As the dissipated energy Q should always be positive regardless of $\mathbf{J}_{ext,t}(\omega,\mathbf{k})$, we can obtain the following requirements for $\varepsilon_{t,l}''(\omega)$ and $\mu_t''(\omega)$

$$\omega\varepsilon_{t,l}''(\omega) \geq 0, \quad \omega\mu_t''(\omega) \geq 0$$

valid for both conductive and non-conductive media. Defined in this way the imaginary parts result in non-negative A_{ext}, W and Q. Moreover, they are linked by the relation

$$A_{ext} = W + Q,$$

which means that *the excited electromagnetic fields do not radiate the energy out of the system,*

$$W_{rad} = \int_{-\infty}^{\infty} dt \int_{0}^{\infty} \nabla \cdot \mathbf{S}_t(t,r) d^3\mathbf{r} = 0. \quad (2.127)$$

Highlight, the work done by external currents at all frequencies completely goes into compensation of the medium dissipation except the singular frequency $\omega = 0$ for non-conductive media. This means that neither transverse nor longitudinal external currents do a work on excitation of those fields. As for the singular frequency $\omega = 0$ of non-conductive media, its contributions to A_{ext}, W, and Q in practice vanish due to zero static component $\mathbf{J}_{ext,l}(0, \mathbf{k})$ that follows from

$$\mathbf{J}_{ext,l}(\omega, \mathbf{k}) = \frac{\omega \mathbf{k}}{k^2} \rho_{ext}(\omega, \mathbf{k}), \qquad (2.128)$$

given by the continuity equation.

2.3 ELECTROMAGNETIC FIELDS IN AN INHOMOGENEOUS MEDIUM

2.3.1 INHOMOGENEOUS MEDIUM RESPONSE

When we described excitation of electromagnetic fields in a medium, we assumed that the medium is homogeneous. If it is inhomogeneous, or structurally composed by different materials, we no longer can describe the medium polarization and magnetization fields with Eqs. (2.84) and (2.85). In this case,

$$P_{med,i}(t, \mathbf{r}) = \varepsilon_0 \int_{-\infty}^{t} dt' \int_{0}^{\infty} d^3 \mathbf{r}' \chi_{ij}^{e}(t - t', \mathbf{r}, \mathbf{r}') E_j(t', \mathbf{r}'),$$

$$M_{med,i}(t, \mathbf{r}) = \int_{-\infty}^{t} dt' \int_{0}^{\infty} d^3 \mathbf{r}' \chi_{ij}^{m}(t - t', \mathbf{r}, \mathbf{r}') H_j(t', \mathbf{r}').$$

These relations can be written in a shorter form:

$$\mathbf{P}_{med}(t, \mathbf{r}) = \varepsilon_0 \hat{\chi}^{e} \mathbf{E}(t, \mathbf{r}), \qquad (2.129)$$

$$\mathbf{M}_{med}(t, \mathbf{r}) = \hat{\chi}^{m} \mathbf{H}(t, \mathbf{r}), \qquad (2.130)$$

where $\hat{\chi}^{e}$ and $\hat{\chi}^{m}$ are the linear integral operators transforming the fields $\mathbf{E}(t, \mathbf{r})$ and $\mathbf{H}(t, \mathbf{r})$ in time and space following the tensors $\chi_{ij}^{e}(t - t', \mathbf{r}, \mathbf{r}')$ and $\chi_{ij}^{m}(t - t', \mathbf{r}, \mathbf{r}')$.

By performing the Fourier transform from the $(t, \mathbf{r})$ space to the $(\omega, \mathbf{r})$ space, we rewrite the material relations,

$$\mathbf{P}_{med}(\omega, \mathbf{r}) = \varepsilon_0 \hat{\chi}^{e}(\omega) \mathbf{E}(\omega, \mathbf{r}), \qquad (2.131)$$

$$\mathbf{M}_{med}(\omega, \mathbf{r}) = \hat{\chi}^{m}(\omega) \mathbf{H}(\omega, \mathbf{r}), \qquad (2.132)$$

with the use of frequency-dependent operators $\hat{\chi}^e(\omega)$ and $\hat{\chi}^m(\omega)$, which describe the following linear integral transformations:

$$P_{\text{med},i}(\omega,\mathbf{r}) = \varepsilon_0 \int_0^\infty d^3\mathbf{r}'\, \chi_{ij}^e(\omega,\mathbf{r},\mathbf{r}')E_j(\omega,\mathbf{r}'),$$

$$M_{\text{med},i}(\omega,\mathbf{r}) = \int_0^\infty d^3\mathbf{r}'\, \chi_{ij}^m(\omega,\mathbf{r},\mathbf{r}')H_j(\omega,\mathbf{r}'),$$

where $\chi_{ij}^e(\omega,\mathbf{r},\mathbf{r}')$ and $\chi_{ij}^m(\omega,\mathbf{r},\mathbf{r}')$ are given by

$$\chi_{ij}^e(\omega,\mathbf{r},\mathbf{r}') = \int_0^\infty dt_1 \chi_{ij}^e(t_1,\mathbf{r},\mathbf{r}')e^{i\omega t_1}, \tag{2.133}$$

$$\chi_{ij}^m(\omega,\mathbf{r},\mathbf{r}') = \int_0^\infty dt_1 \chi_{ij}^m(t_1,\mathbf{r},\mathbf{r}')e^{i\omega t_1}, \tag{2.134}$$

with $t_1 = t - t'$.

Frequency-dependent operators of electric and magnetic susceptibility allow us to write the material relations for $\mathbf{D}(\omega,\mathbf{r})$ and $\mathbf{B}(\omega,\mathbf{r})$:

$$\mathbf{D}(\omega,\mathbf{r}) = \varepsilon_0\hat{\varepsilon}(\omega)\mathbf{E}(\omega,\mathbf{r}), \tag{2.135}$$

$$\mathbf{B}(\omega,\mathbf{r}) = \mu_0\hat{\mu}(\omega)\mathbf{H}(\omega,\mathbf{r}), \tag{2.136}$$

where the operators of dielectric permittivity and magnetic permeability are defined as follows:

$$\hat{\varepsilon}(\omega) = 1 + \hat{\chi}^e(\omega), \tag{2.137}$$

$$\hat{\mu}(\omega) = 1 + \hat{\chi}^m(\omega). \tag{2.138}$$

These operators generally account for inhomogeneity and anisotropy of the medium.

2.3.2 Excitation of Electromagnetic Fields in an Inhomogeneous Medium

In contrast to homogeneous media, whose response was described with tensors $\varepsilon_{ij}(\omega,\mathbf{k})$ and $\mu_{ij}(\omega,\mathbf{k})$, inhomogeneous media are given with linear operators $\hat{\varepsilon}(\omega)$ and $\hat{\mu}(\omega)$. This description results in the following form of macroscopic Maxwell's equations:

$$\nabla \times \mathbf{H}(\omega,\mathbf{r}) = -i\omega\varepsilon_0\hat{\varepsilon}(\omega)\mathbf{E}(\omega,\mathbf{r}) + \mathbf{J}_{\text{ext}}(\omega,\mathbf{r}), \tag{2.139}$$

$$\nabla \times \mathbf{E}(\omega,\mathbf{r}) = i\omega\mu_0\hat{\mu}(\omega)\mathbf{H}(\omega,\mathbf{r}). \tag{2.140}$$

To perform Helmholtz decomposition of Maxwell's equations, we need to decompose the material relations first. For that, we introduce Helmholtz parts of the operators $\hat{\varepsilon}(\omega)$ and $\hat{\mu}(\omega)$ as follows:

$$\mathbf{D}_t(\omega,\mathbf{r}) = \varepsilon_0 \left[\hat{\varepsilon}_{tt}(\omega)\mathbf{E}_t(\omega,\mathbf{r}) + \hat{\varepsilon}_{tl}(\omega)\mathbf{E}_l(\omega,\mathbf{r}) \right],$$

$$\mathbf{D}_l(\omega,\mathbf{r}) = \varepsilon_0 \left[\hat{\varepsilon}_{lt}(\omega)\mathbf{E}_t(\omega,\mathbf{r}) + \hat{\varepsilon}_{ll}(\omega)\mathbf{E}_l(\omega,\mathbf{r}) \right],$$

$$\mathbf{B}_t(\omega,\mathbf{r}) = \mu_0 \left[\hat{\mu}_{tt}(\omega)\mathbf{H}_t(\omega,\mathbf{r}) + \hat{\mu}_{tl}(\omega)\mathbf{H}_l(\omega,\mathbf{r}) \right],$$

$$\mathbf{B}_l(\omega,\mathbf{r}) = \mu_0 \left[\hat{\mu}_{lt}(\omega)\mathbf{H}_t(\omega,\mathbf{r}) + \hat{\mu}_{ll}(\omega)\mathbf{H}_l(\omega,\mathbf{r}) \right].$$

Now, we can do decomposition of Maxwell's equations:

$$\nabla \times \mathbf{H}_t(\omega,\mathbf{r}) = -i\omega\varepsilon_0 \left[\hat{\varepsilon}_{tt}(\omega)\mathbf{E}_t(\omega,\mathbf{r}) + \hat{\varepsilon}_{tl}(\omega)\mathbf{E}_l(\omega,\mathbf{r}) \right] + \mathbf{J}_{ext,t}(\omega,\mathbf{r}),$$

$$\nabla \times \mathbf{E}_t(\omega,\mathbf{r}) = i\omega\mu_0 \left[\hat{\mu}_{tt}(\omega)\mathbf{H}_t(\omega,\mathbf{r}) + \hat{\mu}_{tl}(\omega)\mathbf{H}_l(\omega,\mathbf{r}) \right],$$

$$0 = -i\omega\varepsilon_0 \left[\hat{\varepsilon}_{lt}(\omega)\mathbf{E}_t(\omega,\mathbf{r}) + \hat{\varepsilon}_{ll}(\omega)\mathbf{E}_l(\omega,\mathbf{r}) \right] + \mathbf{J}_{ext,l}(\omega,\mathbf{r}),$$

$$0 = i\omega\mu_0 \left[\hat{\mu}_{lt}(\omega)\mathbf{H}_t(\omega,\mathbf{r}) + \hat{\mu}_{ll}(\omega)\mathbf{H}_l(\omega,\mathbf{r}) \right].$$

We can see that medium's inhomogeneity makes the four equations generally coupled with the operators of dielectric permittivity and magnetic permeability.

Solution for excited electric fields can be written in the following operator form:

$$\mathbf{E}_t(\omega,\mathbf{r}) = \hat{G}_t^{-1}(\omega) \left[\mathbf{J}_{ext,t}(\omega,\mathbf{r}) - \hat{\varepsilon}_{tl}(\omega)\hat{F}_l^{-1}(\omega)\mathbf{J}_{ext,l}(\omega,\mathbf{r}) \right], \tag{2.141}$$

$$\mathbf{E}_l(\omega,\mathbf{r}) = \hat{G}_l^{-1}(\omega) \left[\mathbf{J}_{ext,l}(\omega,\mathbf{r}) - \hat{\varepsilon}_{lt}(\omega)\hat{F}_t^{-1}(\omega)\mathbf{J}_{ext,t}(\omega,\mathbf{r}) \right], \tag{2.142}$$

where the operators $\hat{G}_t(\omega)$ and $\hat{G}_l(\omega)$ are given by

$$\hat{G}_t(\omega) = i\omega\varepsilon_0 \left[\hat{F}_t(\omega) - \hat{\varepsilon}_{tl}(\omega)\hat{F}_l^{-1}(\omega)\hat{\varepsilon}_{lt}(\omega) \right],$$

$$\hat{G}_l(\omega) = i\omega\varepsilon_0 \left[\hat{F}_l(\omega) - \hat{\varepsilon}_{lt}(\omega)\hat{F}_t^{-1}(\omega)\hat{\varepsilon}_{tl}(\omega) \right],$$

with

$$\hat{F}_t(\omega) = \hat{\varepsilon}_{tt}(\omega) - k_0^{-2}(\omega)\nabla \times \left[\hat{\mu}_{tt}(\omega) - \hat{\mu}_{tl}(\omega)\hat{\mu}_{ll}(\omega)\hat{\mu}_{lt}(\omega) \right]^{-1} \nabla \times$$

and

$$\hat{F}_l(\omega) = \hat{\varepsilon}_{ll}(\omega).$$

Now, the excitation is generally *nonlocal* for both Helmholtz groups of fields caused by their coupling through the medium response. As follows from Eqs. (2.141) and (2.142), this coupling is realized through the operators $\hat{\varepsilon}_{tl}(\omega)$ and $\hat{\varepsilon}_{lt}(\omega)$. If either of them results in zero, the coupling between transverse and longitudinal fields disappears making excitation of the longitudinal field fully local.

The operators $\hat{G}_t(\omega)$ and $\hat{G}_l(\omega)$ possess their own eigenmodes. Their eigenfields can formally be written as $\ker[\hat{G}_t(\omega_m)]$ and $\ker[\hat{G}_l(\omega_m)]$. They exist at different eigenfrequencies ω_m, which are generally complex, but can be real in some cases. At frequencies $\omega = \omega_m'$, we can observe resonant growth of the excited fields over the entire space, if the conditions of resonant excitation given in Section 2.1.6 are fulfilled. Finally, these resonances can be seen in the work done by external currents $A = A_t + A_l$:

$$A_t = -\mathrm{Re}\int_0^\infty d^3\mathbf{r}\int_{-\infty}^\infty d\omega\, \mathbf{J}_{\mathrm{ext},t}^*(\omega,\mathbf{r})\cdot \mathbf{E}_t(\omega,\mathbf{r}),$$

$$A_l = -\mathrm{Re}\int_0^\infty d^3\mathbf{r}\int_{-\infty}^\infty d\omega\, \mathbf{J}_{\mathrm{ext},l}^*(\omega,\mathbf{r})\cdot \mathbf{E}_l(\omega,\mathbf{r}),$$

or in the energy dissipated in the medium $Q = Q_t + Q_l$:

$$Q_t = \mathrm{Re}\int_0^\infty d^3\mathbf{r}\int_{-\infty}^\infty d\omega\, \mathbf{J}_{\mathrm{med},t}^*(\omega,\mathbf{r})\cdot \mathbf{E}_t(\omega,\mathbf{r}),$$

$$Q_l = \mathrm{Re}\int_0^\infty d^3\mathbf{r}\int_{-\infty}^\infty d\omega\, \mathbf{J}_{\mathrm{med},l}^*(\omega,\mathbf{r})\cdot \mathbf{E}_l(\omega,\mathbf{r}).$$

2.3.3 PIECEWISE HOMOGENEOUS MEDIA

Now, let us consider the most frequent macroscopic inhomogeneity—an interface between two media. We assume that every medium is homogeneous and isotropic far away from the interface, where they can be described with their bulk scalar permittivities $\varepsilon_{1,2}(\omega)$ and permeabilities $\mu_{1,2}(\omega)$:

$$\varepsilon_{1t}(\omega) = \varepsilon_{1l}(\omega) = \varepsilon_1(\omega), \quad \mu_{1t}(\omega) = \mu_{1l}(\omega) = \mu_1(\omega), \tag{2.143}$$

$$\varepsilon_{2t}(\omega) = \varepsilon_{2l}(\omega) = \varepsilon_2(\omega), \quad \mu_{2t}(\omega) = \mu_{2l}(\omega) = \mu_2(\omega). \tag{2.144}$$

Close to the interface, scalar permittivity and permeability are inhomogeneous, given by $\varepsilon(\omega,\mathbf{r})$ and $\mu(\omega,\mathbf{r})$. They smoothly vary from $\varepsilon_1(\omega)$ and $\mu_1(\omega)$ to $\varepsilon_2(\omega)$ and $\mu_2(\omega)$ within a finite transition layer, properties of which are defined by microscopic composition of the interface. The width of the transition layer is given by $a_{\mathrm{trans}} = a_{\mathrm{micro}} + a_{\mathrm{macro}}$, where a_{micro} is the microscopic size of the inhomogeneity, which typically varies in solids from one to tens of nanometers depending on the type of interface, and a_{macro} is the scale of statistical averaging in macroscopic description.

The minimal width of the transition layer is given by a_{macro}, which is the smallest recognizable length in the macroscopic description. Thus, any microscopic interface is macroscopically continuous. Even an ideal sharp interface with a step-like microscopic profile and $a_{\text{micro}} = 0$ exhibits macroscopically continuous transition with $a_{\text{trans}} = a_{\text{micro}}$. Additional factors such as interface composition, roughness, and diffuseness only increase a_{trans} and make the transition layer thicker.

At the same time, the typical values of a_{trans} are often much below the wavelength, $a_{\text{trans}}k \ll 1$. It allows us to model such transitions as macroscopic discontinuities between piecewise uniform media. Following this criterion, more complex structures composed of several uniform domains can be considered as piecewise homogeneous, if the size of every domain is much larger than a_{trans}. In this case, the Maxwell's equations can be solved separately for every homogeneous domain with the uniform material equations,

$$\mathbf{D}_j(\omega, \mathbf{r}) = \varepsilon_0 \varepsilon_j(\omega) \mathbf{E}_j(\omega, \mathbf{r}), \tag{2.145}$$

$$\mathbf{B}_j(\omega, \mathbf{r}) = \mu_0 \mu_j(\omega) \mathbf{H}_j(\omega, \mathbf{r}), \tag{2.146}$$

where $\varepsilon_j(\omega)$ and $\mu_j(\omega)$ are the scalar complex permittivity and permeability of the jth domain. Finally, the fields in adjoining domains can be seamed with the respective boundary conditions.

2.3.4   BOUNDARY CONDITIONS

The boundary conditions for piecewise homogeneous structures can be derived from the macroscopic Maxwell's equations. To get them, we consider an arbitrary field $\mathbf{F}(\mathbf{r})$ discontinuous at an interface given by $\mathbf{r} = \mathbf{r}_\Omega$. If this field describes a real physical quantity, then it can experience only finite jumps that can be calculated with the following integrals

$$\mathbf{n}_\Omega \times \left[\mathbf{F}_2(\mathbf{r}_\Omega) - \mathbf{F}_1(\mathbf{r}_\Omega) \right] = \int_{\mathbf{n}_\Omega \cdot \mathbf{r}_\Omega - 0}^{\mathbf{n}_\Omega \cdot \mathbf{r}_\Omega + 0} \nabla \times \mathbf{F}(\mathbf{r}) \, d(\mathbf{n}_\Omega \cdot \mathbf{r}),$$

$$\mathbf{n}_\Omega \cdot \left[\mathbf{F}_2(\mathbf{r}_\Omega) - \mathbf{F}_1(\mathbf{r}_\Omega) \right] = \int_{\mathbf{n}_\Omega \cdot \mathbf{r}_\Omega - 0}^{\mathbf{n}_\Omega \cdot \mathbf{r}_\Omega + 0} \nabla \cdot \mathbf{F}(\mathbf{r}) \, d(\mathbf{n}_\Omega \cdot \mathbf{r}),$$

where the integration is performed across the interface in the direction of the unit vector $\mathbf{n}_\Omega$ perpendicular to the interface Ω and pointing from medium 1 to medium 2.

If we perform integration of Maxwell's equations (2.70)–(2.71) across the interface between two media, we get

$$\mathbf{n}_\Omega \times \left[\mathbf{H}_2(t, \mathbf{r}_\Omega) - \mathbf{H}_1(t, \mathbf{r}_\Omega) \right] = \mathbf{I}_{\text{ext}}(t, \mathbf{r}_\Omega), \tag{2.147}$$

$$\mathbf{n}_\Omega \times \left[\mathbf{E}_2 \left(t, \mathbf{r}_\Omega \right) - \mathbf{E}_1 \left(t, \mathbf{r}_\Omega \right) \right] = 0, \tag{2.148}$$

where $\mathbf{I}_{\text{ext}}(t,\mathbf{r}_\Omega)$ is the *surface density of external current* at the interface,

$$\mathbf{I}_{\text{ext}} \left(t, \mathbf{r}_\Omega \right) = \int_{\mathbf{n}_\Omega \cdot \mathbf{r}_\Omega - 0}^{\mathbf{n}_\Omega \cdot \mathbf{r}_\Omega + 0} \mathbf{J}_{\text{ext}} \left(t, \mathbf{r} \right) \mathbf{d} \left(\mathbf{n}_\Omega \cdot \mathbf{r} \right). \tag{2.149}$$

Thus, tangential components of $\mathbf{H}(t,\mathbf{r})$ can experience discontinuity at the interface brought by an *external surface current*. At the same time, the tangential components of $\mathbf{E}(t,\mathbf{r})$ are always continuous.

Similarly, we can integrate divergence relations (2.72)–(2.73) and obtain

$$\mathbf{n}_\Omega \cdot \left[\mathbf{D}_2 \left(t, \mathbf{r}_\Omega \right) - \mathbf{D}_1 \left(t, \mathbf{r}_\Omega \right) \right] = \sigma_{\text{ext}} \left(t, \mathbf{r}_\Omega \right), \tag{2.150}$$

$$\mathbf{n}_\Omega \cdot \left[\mathbf{B}_2 \left(t, \mathbf{r}_\Omega \right) - \mathbf{B}_1 \left(t, \mathbf{r}_\Omega \right) \right] = 0, \tag{2.151}$$

where $\sigma_{\text{ext}}(t,\mathbf{r}_\Omega)$ is the *surface density of external charge* at the interface,

$$\sigma_{\text{ext}} \left(t, \mathbf{r}_\Omega \right) = \int_{\mathbf{n}_\Omega \cdot \mathbf{r}_\Omega - 0}^{\mathbf{n}_\Omega \cdot \mathbf{r}_\Omega + 0} \rho_{\text{ext}} \left(t, \mathbf{r} \right) \mathbf{d} \left(\mathbf{n}_\Omega \cdot \mathbf{r} \right). \tag{2.152}$$

Thus, the normal component of $\mathbf{D}(t,\mathbf{r})$ is discontinuous at the interface in the presence of an *external surface charge*, while the normal component of $\mathbf{B}(t,\mathbf{r})$ is always continuous.

2.3.5 Induced Surface Charge and Surface Current

Boundary conditions (2.147) and (2.150) can be rewritten for the fields $\mathbf{B}(t,\mathbf{r})$ and $\mathbf{E}(t,\mathbf{r})$ as follows:

$$\mathbf{n}_\Omega \times \left[\mathbf{B}_2 \left(t, \mathbf{r}_\Omega \right) - \mathbf{B}_1 \left(t, \mathbf{r}_\Omega \right) \right] = \mu_0 \left[\mathbf{I}_{\text{med}} \left(t, \mathbf{r}_\Omega \right) + \mathbf{I}_{\text{ext}} \left(t, \mathbf{r}_\Omega \right) \right], \tag{2.153}$$

$$\mathbf{n}_\Omega \cdot \left[\mathbf{E}_2 \left(t, \mathbf{r}_\Omega \right) - \mathbf{E}_1 \left(t, \mathbf{r}_\Omega \right) \right] = \frac{1}{\varepsilon_0} \left[\sigma_{\text{med}} \left(t, \mathbf{r}_\Omega \right) + \sigma_{\text{ext}} \left(t, \mathbf{r}_\Omega \right) \right], \tag{2.154}$$

where $\sigma_{\text{med}}(t,\mathbf{r}_\Omega)$ and $\mathbf{I}_{\text{med}}(t,\mathbf{r}_\Omega)$ are the *surface densities of induced charge and current* at the interface between the two media,

$$\sigma_{\text{med}} \left(t, \mathbf{r}_\Omega \right) = \int_{\mathbf{n}_\Omega \cdot \mathbf{r}_\Omega - 0}^{\mathbf{n}_\Omega \cdot \mathbf{r}_\Omega + 0} \rho_{\text{med}} \left(t, \mathbf{r} \right) \mathbf{d} \left(\mathbf{n}_\Omega \cdot \mathbf{r} \right), \tag{2.155}$$

$$\mathbf{I}_{\text{med}}\left(t,\mathbf{r}_{\Omega}\right) = \int_{\mathbf{n}_{\Omega}\cdot\mathbf{r}_{\Omega}-0}^{\mathbf{n}_{\Omega}\cdot\mathbf{r}_{\Omega}+0} \mathbf{J}_{\text{med}}\left(t,\mathbf{r}\right)\text{d}\left(\mathbf{n}_{\Omega}\cdot\mathbf{r}\right). \tag{2.156}$$

Thus, the tangential components of $\mathbf{B}(t,\mathbf{r})$ and normal component of $\mathbf{E}(t,\mathbf{r})$ are discontinuous at the interface even in the absence of external surface charge and surface current. This discontinuity is brought by different polarization and magnetization properties of the two media, which result in appearance of *induced surface charge* and *induced surface current*.

Note, the boundary conditions for electromagnetic fields can be extended to the frequency domain with a simple change of t to ω:

$$\mathbf{n}_{\Omega}\times\left[\mathbf{E}_2\left(\omega,\mathbf{r}_{\Omega}\right)-\mathbf{E}_1\left(\omega,\mathbf{r}_{\Omega}\right)\right] = 0, \tag{2.157}$$

$$\mathbf{n}_{\Omega}\cdot\left[\mathbf{E}_2\left(\omega,\mathbf{r}_{\Omega}\right)-\mathbf{E}_1\left(\omega,\mathbf{r}_{\Omega}\right)\right] = \frac{1}{\varepsilon_0}\left[\sigma_{\text{med}}\left(\omega,\mathbf{r}_{\Omega}\right)+\sigma_{\text{ext}}\left(\omega,\mathbf{r}_{\Omega}\right)\right], \tag{2.158}$$

$$\mathbf{n}_{\Omega}\times\left[\mathbf{B}_1\left(\omega,\mathbf{r}_{\Omega}\right)-\mathbf{B}_1\left(\omega,\mathbf{r}_{\Omega}\right)\right] = \mu_0\left[\mathbf{I}_{\text{med}}\left(\omega,\mathbf{r}_{\Omega}\right)+\mathbf{I}_{\text{ext}}\left(\omega,\mathbf{r}_{\Omega}\right)\right], \tag{2.159}$$

$$\mathbf{n}_{\Omega}\cdot\left[\mathbf{B}_2\left(\omega,\mathbf{r}_{\Omega}\right)-\mathbf{B}_1\left(\omega,\mathbf{r}_{\Omega}\right)\right] = 0, \tag{2.160}$$

where

$$\sigma_{\text{med, ext}}\left(\omega,\mathbf{r}_{\Omega}\right) = \int_{\mathbf{n}_{\Omega}\cdot\mathbf{r}_{\Omega}-0}^{\mathbf{n}_{\Omega}\cdot\mathbf{r}_{\Omega}+0} \rho_{\text{med,ext}}\left(\omega,\mathbf{r}\right)\text{d}\left(\mathbf{n}_{\Omega}\cdot\mathbf{r}\right), \tag{2.161}$$

$$\mathbf{I}_{\text{med,ext}}\left(\omega,\mathbf{r}_{\Omega}\right) = \int_{\mathbf{n}_{\Omega}\cdot\mathbf{r}_{\Omega}-0}^{\mathbf{n}_{\Omega}\cdot\mathbf{r}_{\Omega}+0} \mathbf{J}_{\text{med,ext}}\left(\omega,\mathbf{r}\right)\text{d}\left(\mathbf{n}_{\Omega}\cdot\mathbf{r}\right). \tag{2.162}$$

The density of induced charge and current in a piecewise homogeneous medium can be written with step-like susceptibilities $\chi^e(\omega,\mathbf{r})$ and $\chi^m(\omega,\mathbf{r})$:

$$\rho_{\text{med}}\left(\omega,\mathbf{r}\right) = -\nabla\cdot\mathbf{P}_{\text{med}}\left(\omega,\mathbf{r}\right) = -\varepsilon_0\nabla\cdot\left[\chi^e\left(\omega,\mathbf{r}\right)\mathbf{E}\left(\omega,\mathbf{r}\right)\right],$$

$$\mathbf{J}_{\text{med}}\left(\omega,\mathbf{r}\right) = -i\omega\mathbf{P}_{\text{med}}\left(\omega,\mathbf{r}\right)+\nabla\times\mathbf{M}_{\text{med}}\left(\omega,\mathbf{r}\right)$$
$$= -i\omega\varepsilon_0\chi^e\left(\omega,\mathbf{r}\right)\mathbf{E}\left(\omega,\mathbf{r}\right)+\nabla\times\left[\chi^m\left(\omega,\mathbf{r}\right)\mathbf{H}\left(\omega,\mathbf{r}\right)\right].$$

Finally, the surface charge and surface current induced at the interface between two media are given by

$$\sigma_{\text{med}}\left(\omega,\mathbf{r}_{\Omega}\right) = -\mathbf{n}_{\Omega}\cdot\left[\mathbf{P}_{\text{med,2}}\left(\omega,\mathbf{r}_{\Omega}\right)-\mathbf{P}_{\text{med,1}}\left(\omega,\mathbf{r}_{\Omega}\right)\right]$$
$$= -\varepsilon_0\mathbf{n}_{\Omega}\cdot\left[\chi_2^e\left(\omega\right)\mathbf{E}_2\left(\omega,\mathbf{r}_{\Omega}\right)-\chi_1^e\left(\omega\right)\mathbf{E}_1\left(\omega,\mathbf{r}_{\Omega}\right)\right], \tag{2.163}$$

$$\mathbf{I}_{med}\left(\omega,\mathbf{r}_{\Omega}\right) = \mathbf{n}_{\Omega} \times \left[\mathbf{M}_{med,2}\left(\omega,\mathbf{r}_{\Omega}\right) - \mathbf{M}_{med,1}\left(\omega,\mathbf{r}_{\Omega}\right)\right]$$

$$= \mathbf{n}_{\Omega} \times \left[\chi_2^m\left(\omega\right)\mathbf{H}_2\left(\omega,\mathbf{r}_{\Omega}\right) - \chi_1^m\left(\omega\right)\mathbf{H}_1\left(\omega,\mathbf{r}_{\Omega}\right)\right]. \tag{2.164}$$

These expressions demonstrate that surface charge appears as a result of discontinuous polarization field, while surface current is caused by discontinuous magnetization field.

2.3.6 DISSIPATION IN PIECEWISE HOMOGENEOUS MEDIA

As we saw above, all currents induced in a piecewise homogeneous medium can effectively be divided into two groups: (i) *volume currents* induced inside every domain and (ii) *surface currents* induced at the domains' interfaces. Following this division, we can write the volume density of induced currents as follows,

$$\mathbf{J}_{med}\left(\omega,\mathbf{r}\right) = \mathbf{J}_{med}^V\left(\omega,\mathbf{r}\right) + \mathbf{J}_{med}^S\left(\omega,\mathbf{r}\right). \tag{2.165}$$

Then, the volume current induced inside the domains is given by the volume density

$$\mathbf{J}_{med}^V\left(\omega,\mathbf{r}\right) = -i\omega\varepsilon_0[\varepsilon(\omega,\mathbf{r})\mu(\omega,\mathbf{r}) - 1]\mathbf{E}(\omega,\mathbf{r}) + \chi^m(\omega,\mathbf{r})\mathbf{J}_{ext}^V(\omega,\mathbf{r}), \tag{2.166}$$

while the surface current induced at the domains' interfaces is given by the volume density

$$\mathbf{J}_{med}^S\left(\omega,\mathbf{r}\right) = \nabla\chi^m(\omega,\mathbf{r}) \times \mathbf{H}(\omega,\mathbf{r}) + \chi^m(\omega,\mathbf{r})\mathbf{J}_{ext}^S\left(\omega,\mathbf{r}\right). \tag{2.167}$$

In a medium, induced currents lead to dissipation losses. These losses can be calculated as

$$Q\left(\omega\right) = \text{Re} \int_0^\infty \mathbf{E}^*\left(\omega,\mathbf{r}\right) \cdot \mathbf{J}_{med}\left(\omega,\mathbf{r}\right)d^3\mathbf{r},$$

where the integration is performed over the entire space. In a case of piecewise homogeneous medium, these losses appear as *volume* and *surface absorption* of electromagnetic energy:

$$Q\left(\omega\right) = \int_0^\infty \left[q^V\left(\omega,\mathbf{r}\right) + q^S\left(\omega,\mathbf{r}\right)\right]d^3\mathbf{r}. \tag{2.168}$$

The volume absorption is characterized by the loss density

$$q^V\left(\omega,\mathbf{r}\right) = \omega\varepsilon_0\text{Im}\left[\varepsilon\left(\omega,\mathbf{r}\right)\mu\left(\omega,\mathbf{r}\right)\left|\mathbf{E}\left(\omega,\mathbf{r}\right)\right|^2\right]$$

$$+ \text{Re}\left[\chi^m\left(\omega,\mathbf{r}\right)\mathbf{E}^*\left(\omega,\mathbf{r}\right) \cdot \mathbf{J}_{ext}^V\left(\omega,\mathbf{r}\right)\right], \tag{2.169}$$

while the surface absorption is given by

$$q^{S}(\omega,\mathbf{r}) = -\mathrm{Re}\left[\nabla\chi^{m}(\omega,\mathbf{r})\cdot\mathbf{E}^{*}(\omega,\mathbf{r})\times\mathbf{H}(\omega,\mathbf{r})\right]$$
$$+\mathrm{Re}\left[\chi^{m}(\omega,\mathbf{r})\mathbf{E}^{*}(\omega,\mathbf{r})\cdot\mathbf{J}_{\mathrm{ext}}^{S}(\omega,\mathbf{r})\right]. \qquad (2.170)$$

Note, the second terms in $q^{V}(\omega,\mathbf{r})$ and $q^{S}(\omega,\mathbf{r})$ contribute to volume and surface absorption only for those domains and interfaces, which contain the external current. These terms describe *direct* volume/surface heating by the external current. For the domains and interfaces not containing the external current, the second terms do not contribute to volume/surface absorption. In this case, $q^{V}(\omega,\mathbf{r})$ and $q^{S}(\omega,\mathbf{r})$ are completely given by the first terms, describing *indirect* volume/surface heating by the external current at a distance.

Following the theory of eigenmodes, the volume and surface absorption rates appear simultaneously resonant on the same eigenmodes, if the conditions of resonant excitation listed in Section 2.1.6 are fulfilled. This is a general feature of eigenmode resonances enabling global enhancement of excited fields over the entire space, including all domains and interfaces.

2.4 SUMMARY

In this chapter, we have reviewed the classical theory of electromagnetic fields. We have discussed the main properties of fields excited in a vacuum. We have demonstrated that excitation of electromagnetic fields proceeds in two ways following the fundamental distinctions of the Helmholtz fields. These distinctions become more pronounced in the presence of a homogeneous isotropic medium, where excitation of the Helmholtz fields is accompanied by different resonant effects. In inhomogeneous media, behavior of the Helmholtz fields becomes even more complicated and coupled through the medium inhomogeneity. In a piecewise homogeneous media, this results in distinct volume and surface effects such as absorption of electromagnetic fields and direct/indirect heating of the medium.

BIBLIOGRAPHY

[1] I. S. Gradshteyn and I. M. Ryzhik, *Tables of Integrals, Series, and Products*, 6th ed. (Academic Press, 2000).
[2] G. B. Arfken and H. J. Weber, *Mathematical Methods for Physicists*, 6th ed. (Elsevier Academic Press, 2005).
[3] L. D. Landau and E. M. Lifshitz, *The Classical Theory of Fields*, 4th ed. (Butterworth-Heinemann, 1980).
[4] L. D. Landau, E. M. Lifshitz and L. P. Pitaevskii, *Electrodynamics of Continuous Media* 2nd ed. (Butterworth-Heinemann, 1984).
[5] J. D. Jackson, *Classical Electrodynamics*, 2nd ed. (Wiley, 1975).
[6] L. D. Landau and E. M. Lifshitz, *Statistical Physics*, 2nd ed. (ButterworthHeinemann 1969).
[7] E. M. Lifshitz and L. P. Pitaevskii, *Physical Kinetics* (Butterworth-Heinemann, 1981).

3 Theoretical Design of Nanomaterials for Optical and Terahertz Applications

Xiao Jiang, Jiafeng Xie, Zhou Li, Bing Huang, and Su-Huai Wei

CONTENTS

3.1 Introduction 61
3.2 Brief Overview of Applications of 2D Optical Materials 62
    3.2.1 Graphene 62
    3.2.2 Transition Metal Di-Chalcogenides 64
    3.2.3 Hexagonal Boron Nitride 66
    3.2.4 Layered Black Phosphorus 68
    3.2.5 2D Perovskite Oxides and Organic-Inorganic Structures 68
3.3 Applications of Nonlinear Optical Materials 70
    3.3.1 Second Harmonic Generation 70
    3.3.2 Third Harmonic Generation 72
    3.3.3 Giant Nonlinear Enhancement Based on Double Plasmonic
            Resonance 73
3.4 Applications of THz Materials 74
    3.4.1 Photonic Green's Function, Optical Conductivity, and
            Electron-Phonon Interaction 74
    3.4.2 THz Driven Topological Phase Transition 76
    3.4.3 Other THz Related Interaction 77
3.5 Summary and Outlook 77
Bibliography 78

3.1 INTRODUCTION

Optical properties have been one of the most fascinating and functional aspects of any nanomaterial. They are generally customized by altering parameters such as particle size, shape, surface characteristics, and various other variables [1]. Major application fields based on optical properties include light emission and detection, solar cells, photocatalysis, photoelectronic and imaging, and biosensing. The basic understanding of the fundamental optical properties and related spectroscopic techniques can help distinguish the

DOI: 10.1201/9781003202608-3

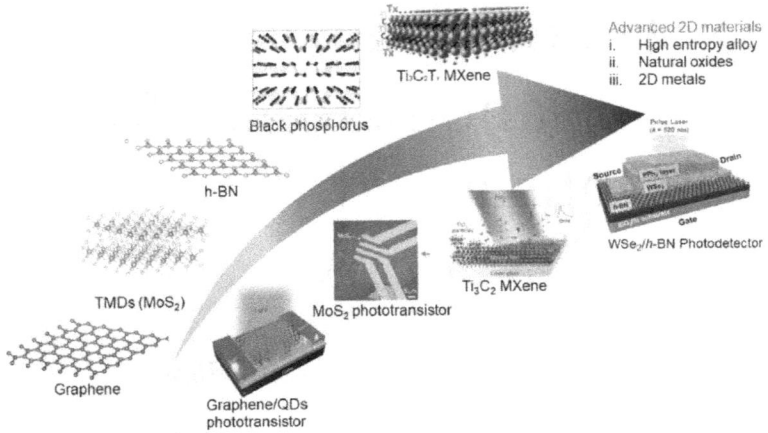

FIGURE 3.1   Overview of different types of 2D materials and their optoelectronic applications.
Source: Adapted from Ref. [87]. Copyright 2012 American Chemical Society.

nanomaterials. Reviews exist on the discussion of linear optics, nonlinear optics, light-emitting, photodetection, and anisotropy in 2D materials, as shown in Figure 3.1 [1–3]. A recent review was published discussing the tunable properties of 2D materials [4]. Graphene's optical properties are quite unique, with band structure having van Hove-like singularities. Chiral symmetry also exists for the quasiparticles, which helps fix the direction of pseudospin to be parallel for electrons or antiparallel for holes [5], whereas in monolayer TMDCs, optical absorption is dominated by direct transitions between valence band (VB) and conduction band (CB) states around the K and K′ points. Direct band-to-band transitions in 2D are characterized by a step function-like spectrum originated from the energy-independent joint-density-of-states and transition matrix elements near parabolic band edges [2]. Later, hexagonal boron nitride (hBN) was studied for light-matter interactions at the atomic scale [6] whose atomic structure is folded, resulting in high anisotropy of phonons, photons, and electrons [7]. Here, in this chapter, we have focused on the recent technological innovations in 2D materials such as h-BN, BP, and perovskite oxides. A wide variety of optical device applications, recent theoretical first-principle calculations, and future advancements in device applications have been discussed.

3.2   BRIEF OVERVIEW OF APPLICATIONS OF 2D OPTICAL MATERIALS

3.2.1   GRAPHENE

Graphene is one of the pioneers in the field of 2D materials. Light-matter interaction in graphene gives rise to exciting optical properties and has impacted the fields of optoelectronics, nanoelectronics, and nonlinear optics. The linear absorption of graphene shows an absorption band centered at ~260 nm due to π-π* transition of electrons in π-conjugated sp2 carbon core [8]. Graphene does not show any visible luminescence. Previously, several researchers have studied the use of graphene in ultrafast and

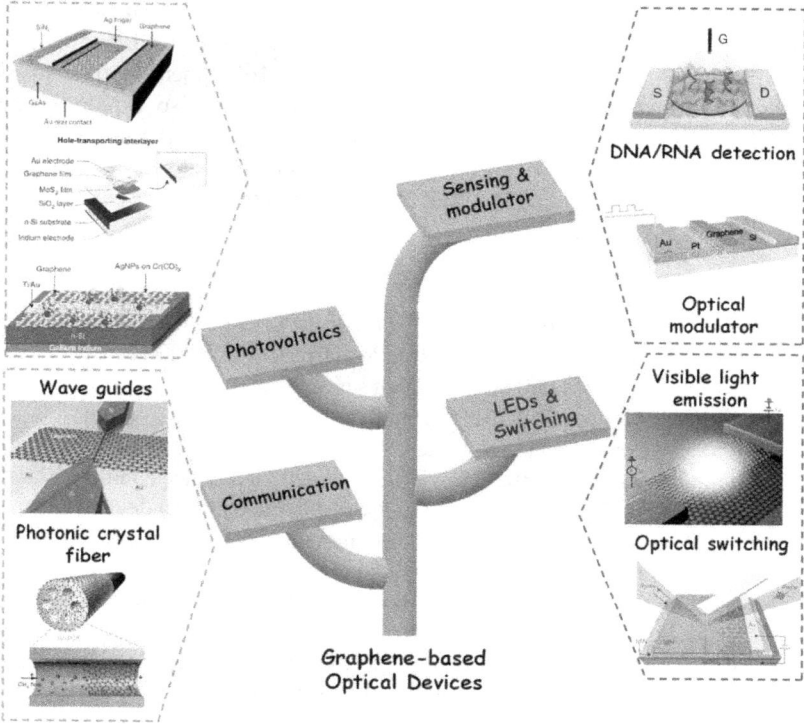

FIGURE 3.2   Schematic representation of graphene-based optical device applications.
Source: Adapted and modified from Ref. [13]. No permission required.

efficient optical switching, solar cell, optical modulators, plasmonic devices, transparent light emitters, exhibiting low operation voltage, ultrafast optical communications, and state-of-the-art photodetectors [5,9,10], as shown in Figure 3.2. It is believed that absorption property in the UV region may have huge applications in solid-state lighting [11]. Recently, optical switching gained massive attention as it overcame the limitations of electrical switching. Notami *et al.* greatly enhanced the nonlinear absorption of graphene loaded with plasmonic waveguides and achieved ultrafast optical switching with switching energy and time of 35 fJ and 260 fs, respectively, connected to conventional waveguides for use in integrated circuits [12]. Studies exist in the field of biological molecule sensing of nucleic acids using deformed graphene by FET-based biosensors [13]. In another study, computational simulations revealed "electrical hot spots" in the sensing channel, which reduces the charge screening. Recently, meter-scale level optical fiber was demonstrated by a combination of graphene and photonic crystal fiber, and it shows broadband response and significant modulation depth under a low gate voltage [14]. Recent studies of composite graphene and W- and Mo-based TMD heterostructures are reported to form electroluminescent systems with linewidths approaching homogeneous limits near the THz rate. The spatially localized hot electrons (~2,800 K) resulted in a 1000-fold enhancement in thermal radiation efficiency [15]. Distinct nonlinear nano-optical properties of graphene are also well-reported, with enhanced broadband four-wave mixing response in fs nanoimaging being revealed. The strong electron-electron interaction was also recently studied [16]. Other graphitic structures like reduced graphene oxide (rGO) infused nanofluid were also explored for their optical filtrations of solar energy with thermal efficiencies reaching up to 30%. They influence the functionality of hybrid solar cells [17]. In UV-Vis absorption spectra measurements, the hydrophobic nature of rGO is found to improve the dispersion stability.

3.2.2 TRANSITION METAL DI-CHALCOGENIDES

2D transition metal di-chalcogenides (TMDs) consist of over 40 compounds with the general formula of MX_2 (X = S, Se, and Te). Primarily group VI TMDs are extensively studied, and their associated optical applications have been demonstrated. Some of them include MoS_2, WS_2, $MoSe_2$, ReS_2, $MoTe_2$, and WTe_2, synthesized by various approaches [11]. Their reduced dimensionalities make 2D TMDs fascinating with strong light-matter interaction and enhanced optical properties. Varied lateral sizes and the number of layers are observed in 2D TMDs. From the literature, the bandgap decreased by ~ 0.3 eV–0.35 eV as the layer number increased from monolayer to bulk in Mo- and W-based TMDs, showing weak dependence of optical bandgap on layer number [18].

Sulfides. The optical absorption of exfoliated molybdenum sulfide (MoS_2) layer has been investigated by several researchers, and the major excitonic absorption peaks are observed at higher wavelength regions [19]. The layered dependent optical reflectance spectra ReS_2 are presented in Figure 3.3(a). It shows that the peak position changes with layer number. Temperature effects varied the optical bandgaps of the ReS_2 films (10 layers); the bandgap varied from 1.36 eV (303 K) to 1.38 eV (383 K). Theoretical predictions showed similar results where the bandgap increased from 1.32 eV to 1.40 eV. Energy level degeneracy was attributed to the weaker coupling between the Re 5d orbital and S 3p orbital, leading to the smaller energy level splitting with increased temperature [20].

MoS_2 monolayer has a stable, gate tunable optical response at room temperature (RT) near excitonic transition [21]. It also shows strong excitonic photoluminescence (PL) in the higher wavelength region (Figure 3.3(b)). Hybrid nanostructures of MoS_2 show enhancement in polarization near exciton binding energies [22]. Other TMD alloys of Mo-, like $Mo_xW_{1-x}Se_2$ and $WS_{2y}Se_{2(1-y)}$, exhibit enhanced PL emission in the monolayers which is due to deep trap states produced by vacancies that promote the emission of excitons and trions [23]. Theoretical calculations also confirmed the experimental results, where MoS_2 with four different morphology-controlled plasmonic nanoparticles was studied. Furthermore, the plasmonic strain reduces the bandgap by 32 times and enhanced photoresponse due to massive hot electron injection [24]. An asymmetric Fabry–Perot cavity was formed based on a hybrid structure of MoS_2/hBN/ Au/SiO_2 by vertical stacking, whose PL intensity is two orders of magnitude larger than that of monolayer MoS_2. The strong absorption was justified from photonic localization on the top of the microcavity [25]. Monolayer WS_2 onto exfoliated graphite by high-temperature chemical vapor deposition (CVD) showed a single excitonic PL peak with a Lorentzian profile at RT and an 8 meV bandgap at 79 K. In a similar study, temperature-dependent PL spectra of WS_2 on different substrates was analyzed. Similarly, PL emission spectra of MoS_2 and WS_2 were also studied on two different substrates, and it

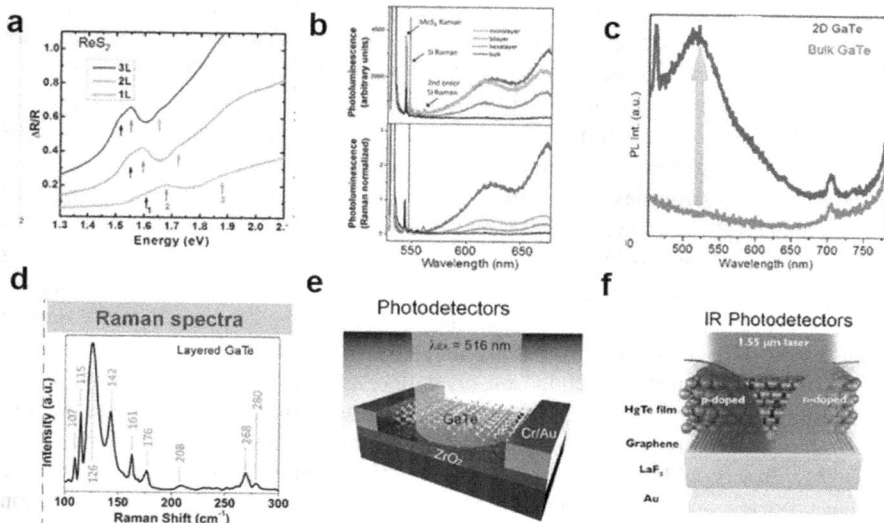

FIGURE 3.3 (a) Layer-dependent reflectance spectra of ReS_2.

Source: Adapted from Ref. [20]. Copyright 2018 American Chemical Society.

(b) PL emission spectra of monolayer MoS_2 with Raman signal.

Source: Adapted from Ref. [22]. Copyright 2011 American Chemical Society.

(c) Comparison PL emission spectrum of bulk and 2D GaTe.

Source: Adapted and modified from Ref. [88]. Copyright 2021 American Chemical Society.

(d) Raman spectra of GaTe.

Source: Adapted from Ref. [31]. Copyright 2016 American Chemical Society.

(e) GaTe photodetector.

Source: Adapted from Ref. [34]. Copyright 2018 American Chemical Society.

(f) IR photodetector using HgTe.

Source: Adapted from Ref. [26]. Copyright 2020 American Chemical Society.

is observed that with hBN, the PL emission becomes narrower [26]. Studies with multi-atom doped ZnS exhibit high fluorescence efficiency of ~ 62% in the visible region.

Selenides. Some of the 2D selenides have applications in a wide range of optical devices. The bandgap transition from direct to indirect was noted when the thickness of the layers (L) is reduced below 6 nm. When L decreases below ~10 nm, the PL intensity decreases by a factor > 10. This value is significantly larger than that for any other luminescent material. At this thickness, the optical bandgap was found to be 1.44 eV–1.47 eV [27]. An encapsulated InSe device offers high quality and ambient-stable mobility of 30 $cm^2V^{-1}s^{-1}$–120 $cm^2V^{-1}s^{-1}$ compared to ~ 1 $cm^2V^{-1}s^{-1}$ of un-encapsulated devices. For complete hBN encapsulation to GaSe, PL with a photoresponsivity of 84.2 AW^{-1} (at 405 nm) was observed [28]. Based on the anisotropic nonlinear behavior of the material, a SnSe-based all-optical switch was proposed. The nonlinear optical response was polarization-dependent, and an unexpectedly high on/off ratio was achieved. In another study, the SnSe-decorated nonlinear device in fiber lasers with ultrashort mode-locked pulses at 1.5 µm and 2.0 µm was fabricated. Additionally, studies indicated that the selenide materials could serve as good saturable absorbers for lasers in the broadband area [29].

Tellurides. 2D tellurides are seldom explored for their optical properties compared to other chalcogenides. Layered Si_2Te_3 and $Mn-Si_2Te_3$ have been studied for their high-pressure optical phonon behaviors. Raman modes in $Mn-Si_2Te_3$ show phonon stiffening and softening, suggesting negative linear compressibility [30]. In-plane optical anisotropy and RT PL spectrum in the visible range were observed in the 2D GaTe sample (Figure 3.3(c)). A GaTe multilayer study showed weak anisotropy in the visible range, and the Raman intensity depended on crystalline orientation (Figure 3.3(d)). These results suggest high photoresponsivity and the possibility of generating a large number of dangling bonds, providing recombination sites for carriers in low dimensional structures [31]. Based on the density function theory (DFT), monolayer GaTe gave unique optical properties because the anisotropic layer could affect bandgap and absorption coefficient [32]. The optical absorption in the visible region with high electron and hole mobilities was observed for the monolayer GeTe. It also shows a larger bandgap, which is strain-tunable, compared to its bulk form [33]. GaTe nanoflakes also showed enhanced performance as photodetectors as depicted in Figure 3.3(e), exhibiting better responsivity and illuminating properties [34]. An enhanced photodetection range and faster response time were noted with a combination of HgTe and graphene structure as shown in the Figure 3.3(f). The structure constructed on LaF_3 substrate offers high gate tunablity and possible charge carrier polarities in graphene and HgTe [26].

3.2.3 Hexagonal Boron Nitride

With highly dispersive surface phonon-polariton modes, hexagonal boron nitride (hBN) is a natural hyperbolic material. Raman spectra of mono-, bi-, and trilayer BN show a characteristic peak of phonon mode, analogous to the G peak of graphene. A progressive weaker peak is observed as layer number decreases, which are ~ 50 times smaller than that of graphene's G peak in monolayer BN under the same measurement conditions as seen in Figure 3.4(a). Strain effects on hBN are pretty interesting. The phonon frequency shifts are due to compressive strain in hBN (Figure 3.4(b)). Figure 3.4(c) depicts monolayer graphene on hBN film with the details mentioned in the figure. Also, Figure 3.4(d) shows the permittivity of hBN, and the result shows a possibility of hyperbolicity [35]. A Van der Waals (vdW) WS_2/MoS_2 heterostructure on hBN flake showed excitonic optical responses. This was mapped to the presence of several valleys in the electronic structure. At 1.3 eV–1.7 eV (at RT), PL emission was observed for WS_2/MoS_2 heterostructure, which is generally absent in bare WS_2 or MoS_2 monolayer. The theoretical analysis concluded that the PL peaks originated in monolayer hBN were due to the following reasons: (I) direct K–K interlayer excitons, (II) indirect Q–Γ interlayer excitons, and (III) indirect K–Γ interlayer excitons. The level alignment was also calculated and had good agreement with experimental PL spectra [6]. Figure 3.4(e) illustrates local strain measurement in annealed hBN flakes of ~ 6 nm thickness on a SiO_2/Si substrate. The first-principle calculation was carried

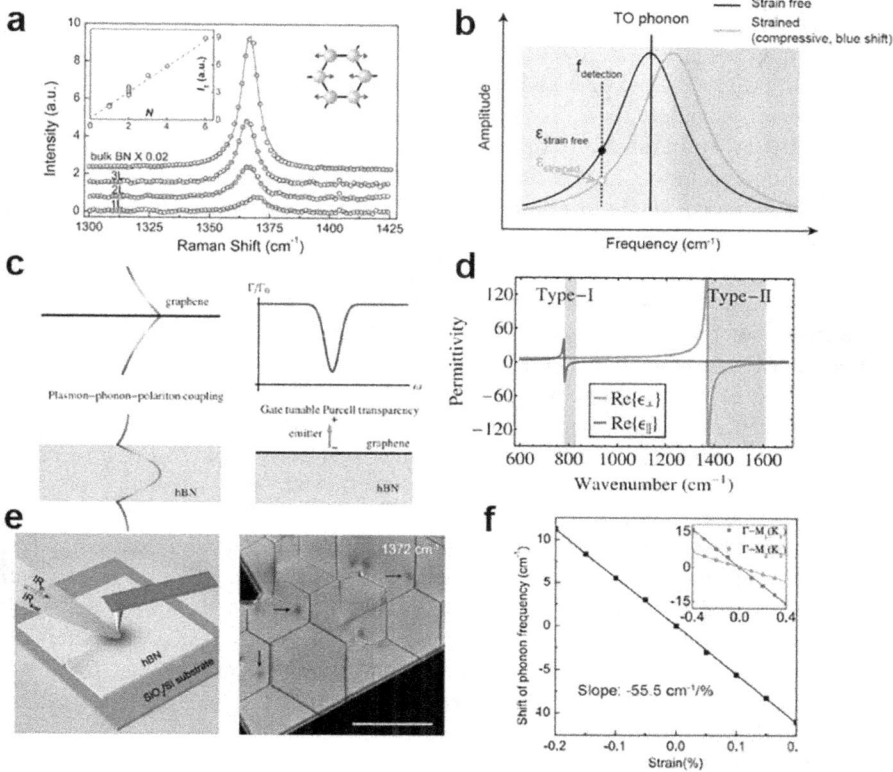

FIGURE 3.4 (a) Raman spectra of hBN. Inset shows the changes in integrated intensity with the layer number. The picture in the top right side shows phonon mode for Raman peak. (b) Illustration of compressive strain induced phonon frequency shift and near-field IR contrast (c) The interaction of graphene plasmon with hBN phonon (left) and emission of a nearby dipole into hybrid modes (right). (d) Permittivity of hBN. (e) The near-field IR measurement technique using hBN. (f) Theoretical results of phonon frequency as a function of isotropic strain.

Source: (a), (b), (e), and (f) are adopted from Ref. [36]. Copyright 2019 American Chemical Society. (c) and (d) are adopted from Ref. [35]. Copyright 2015 American Chemical Society.

out by assuming isotropic biaxial strain, where hBN transverse optical (TO) phonon frequency shift was also observed (Figure 3.4(f)) [36]. Theoretical and experimental optical properties of heterostructures of hBN with other 2D materials are also studied. Theoretically, studies of graphene-hBN heterostructure were conducted by Kumar *et al.* regarding the interaction of graphene plasmon and hBN phonon.

In graphene-hBN heterostructures, we observed hybridization of plasmon and phonon, as well as the reduction of the group velocity of light in the IR region, and this was because of strong plasmon dipole–dipole coupling. Tuning the coupling strength can open a new area for light manipulation and detection in the mid-IR window. Plasmon-phonon polaritons in these heterostructures were also studied [37].

3.2.4 LAYERED BLACK PHOSPHORUS

Black phosphorus (BP) was investigated as 2D material recently, offering high carrier mobility and thickness-dependent direct bandgap. Multilayer BP sheets under periodic stress modulate their optoelectronic properties, yielding quantum confinement and better strain tunability than the TMD counterparts. BP finds a wide range of applications, including photovoltaics and optoelectronics [38]. Optical conductivity was studied for a single layer and bulk BP at varied applied strain along zigzag (σ_{ZZ}) and armchair (σ_{AC}) directions. Similarly, wavelength and pulse-dependent nonlinear optical properties of BP nanosheets have been studied by several researchers. Figure 3.5(a) shows absorption spectra of the BP dispersed in ethanol [13]. Reports of encapsulation technique usage showed reduced exciton binding energy by 70% in monolayer BP and elimination of the bound exciton in the four-layer BP structure. This changes the nature of the excited states and absorption spectrum.

Size-dependent nonlinear optical response of BP nanosheets synthesized by liquid penetrant examination (LPE) for nanosecond laser pulses was reported recently. Results showed that the nonlinear absorption coefficient of BP depended on laser intensity and lateral flake dimension. Other major application areas of BP include switchable electronic circuits. A switchable gate voltage led to switchable optical linear dichroism. Figure 3.5(b) shows GW quasiparticle of stacked BP, and Figure 3.5(c) shows 90° twisted BP with a 2-fold degeneracy [39]. Enhanced photoresponsivity of a BP-based photodetector with a high photocurrent ratio (~ 8.7) was obtained. Thus, BP has potential applications in telecommunication, sensing, and IR polarimetry imaging. The current-rectifying behavior can be in the heterojunction of BP/MoS_2 photodetector by tuning the gate voltage and forward-to-reverse bias current ratio exceeding 10^3. Figure 3.5(d) illustrates the anisotropic crystal structure of BP [40]. Strain-induced anisotropy resulted in electron–phonon interaction behavior in strained BP, and a high PL lifetime of BP as suitable candidates for live-cell imaging has been studied. Figure 3.5(e) shows a representation of electronic band structures of BP with different thicknesses [41]. Figure 3.5(f) shows photocurrent (IPh) and photoresponsivity (R) of the junction under 1.55 μm light illumination at the bias voltage $V_{ds} = 3$ V. Figure 3.5(g) shows the photoresponse of the junction at a different voltage [42]. Theoretical and experimental anisotropic studies of the material also exist due to atomic vibrations at increased temperatures as well as in RT.

3.2.5 2D PEROVSKITE OXIDES AND ORGANIC-INORGANIC STRUCTURES

Perovskites are high entropy oxides with multiple cation Wyckoff positions and find a wide range of applications in optical and electronic devices. The freestanding 2D monolayer of perovskite is a tunable wide bandgap semiconducting material. Theoretical studies on some important perovskite oxides (*e.g.*, $SrTiO_3$, $LaAlO_3$, and $KTaO_3$) were compared with graphene and MoS_2 monolayer for their optical properties. Organic light-emitting diodes were also fabricated using calcium niobate ($CaNbO_3$) nanosheets. With a wide bandgap of ~ 3.5 eV, $CaNbO_3$ has been used as electron transport layers (ETLs) and electron injection layers (EILs). The operational lifetime of the devices was exceptional, with high luminance. 2D $CsPb_2Br_5$ exhibits enhanced performance in optoelectronic devices such as white LEDs. The packaged

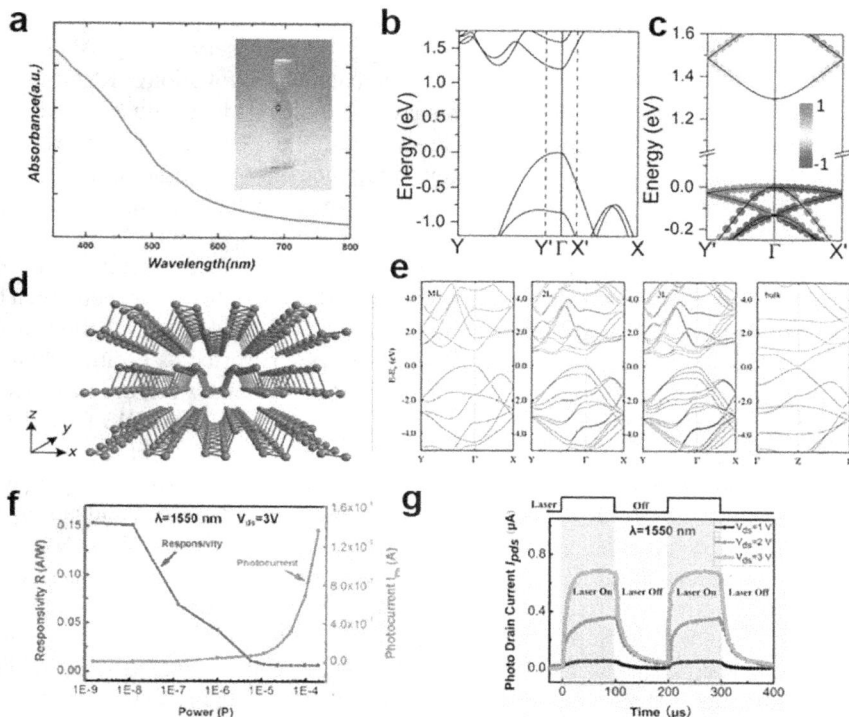

FIGURE 3.5 (a) UV-Vis absorption spectrum of the BP. Inset shows the digital photograph of the BP dispersion.

Source: Adapted and modified from Ref. [89]. Copyright 2017 American Chemical Society.

(b) GW quasiparticle band structure of naturally stacked bilayer BP and (c) 90° twisted bilayer BP.

Source: Adapted and modified from Ref. [39]. Copyright 2016 American Chemical Society.

(d) Crystal structure of BP.

Source: Adapted and modified from Ref. [90], no permission required.)

(e) Calculated electronic band structures of BP with different thicknesses.

Source: Adapted and modified from Ref. [41]. Copyright 2016 American Chemical Society.

(f) The photocurrent (IPh) and photoresponsivity (R) of BP and (g) time-resolved photoresponse of the junction photodiode.

Source: Adapted and modified from Ref. [42]. Copyright 2016 American Chemical Society.

WLED is (0.33, 0.33), indicating that it emits white light with high color rendering index of $\sim$ 94%, far superior compared to other reports. 2D perovskites have also found applications as photodetectors. An optimized photodetector CsPbBr$_3$/Au exhibited photoresponsivity of 41.0 AW^{-1} with a specific detectivity of 1.67 × 10^{12} Jones under an incident of 232 µW/cm^2. It is estimated that the charge recombination rate constants for 2D perovskite solar cells are larger than the 3D compound.

The charge-carrier recombination in the planar device architecture was particularly noted for $(BA)_2(MA)_4Pb_5I_{16}$. The bandgap change was not observed for $[TBA_xH_{1-x}]$ + $[Ca2Nb3O10]^-$ nanosheets obtained through intercalation–exfoliation of $KCa_2Nb_3O_{10}$. The structure was preserved after delamination, and the bandgap values helped analyzing the electronic structure. Due to the varied optical bandgap and accelerated interfacial charge transfer process, many studies on perovskite structures aiding other applications such as photocatalysis and visible photoelectrochemical oxidation have been explored in recent years. The UV-Vis absorption spectrum of $CH_3NH_3PbI_3$ with interband optical transitions between VB and CB has been detected recently [43]. The bandgap of similar structures of the layered perovskite material obtained by various other synthesis methods were found to be ~ 3.53 eV. Complex halide structures like $(CH_3(CH_2)_3NH_3)_2(CH_3NH_3)_4Pb_5I_{16}$ have been reported as exciting materials for the fabrication of solar cells. Other layered perovskites were used as LEDs. The large bandgap and reduced reflectivity of ~ 0.014, 0.013, and 0.013 for $SrTiO_3$, $LaAlO_3$, and $KTaO_3$ at zero frequency were estimated. This implies that zero absorption is responsible for the transparency of the materials in the low-frequency region. The peak reflectivity value is larger than that of graphene (< 0.001, bandgap of 0.5 eV–1.2 eV) and comparable to those of MoS_2 (~ 0.09, bandgap of 2.8 eV).

The most recent advances in the 2D organic-inorganic halide perovskites have become competitive materials for efficient solar energy harvesting. The optimization of printed spiral coils (PSCs) involves tuning of structures, composition, and defect passivation in perovskite absorbers, the device structure, and also the interface modifications. After the first report of 2D perovskites as absorbers, the recent advancements have reached ~ 18% efficiency [44]. One of the most recent studies of organic-inorganic hybrid perovskite was done with 2D hybrid lead bromides, $(C_7H_{18}N_2)PbBr_4$ and $(C_9H_{22}N_2)PbBr_4$; their possible use in the optoelectronic field was discussed briefly. Their optical bandgap lies in between 2.76 eV and 2.78 eV, respectively, for $(C_7H_{18}N_2)$ $PbBr_4$ and $(C_9H_{22}N_2)PbBr_4$. Along with it, we observe broad photoluminescent spectra that originate from free and self-trapped excitons. The exciton energy levels were controlled through chain length variations which influenced the material optical properties. In another study, the high tunability in the state-of-the-art hybrid Ruddlesden-Popper perovskites (RPPs) was discussed. As compared to their bulk counterpart, the 2D layered RPPs showed improved environmental stability to external light, stress, and humidity [45]. This was attributed to the absence of ion migration in 2D RPPs and hence improving the retention capacity and better performance in optical devices.

3.3 APPLICATIONS OF NONLINEAR OPTICAL MATERIALS

3.3.1 Second Harmonic Generation

Second harmonic generation (SHG) from 2D layered-materials (2DLM) has received much scientific interest due to its potential applications in active photonic nanodevices. Indeed, SHG is forbidden in free-standing or pristine graphene, BP, even-layer TMDCs due to its centrosymmetric property [46]. However, SHG from these materials has been investigated both theoretically and experimentally by breaking the inversion symmetry via an external excitation, such as electrical excitation (field/current/charge), doping, and structural variations, or by placing them onto a substrate. Thus,

Dean *et al.* have demonstrated SHG from exfoliated single layer (SL) and multi-layer (ML) graphene films on oxidized Si (001) substrate at 800 nm in the femtosecond regime [47]. The SHG emission effect was confirmed by the quadratic dependence of SHG intensity ($I_{2\omega}$) upon the pump intensity (I_ω), namely $I_\omega \propto I_\omega^2$.

The SHG from suspended SL and bilayer (BL) graphene has also been observed and attributed to long-range curvature fluctuations of the suspended graphene. This assumption was validated by the analysis of 2D spatially resolved Raman and SHG mapping. In particular, SHG studies were extensively done on monolayer TMCs due to the inherently noncentrosymmetric nature of these materials and their large optical nonlinearities. Thus, it has been reported that TMCs have large second-order nonlinear susceptibility, more than an order of 1 nm/V [48]. This is several orders of magnitude larger than that of most dielectric materials [17, 34]. The most in-depth studied TMDCs are MoS$_2$ [48], MoSe$_2$, WS$_2$ [49], and WSe$_2$ [50]. Thus, Kumar *et al.* have observed very strong SHG emission from SL MoS$_2$ nanosheets prepared by mechanical and chemical exfoliation methods, with excitation at 810 nm. The estimated effective bulk-like second-order susceptibility $\chi^{(2)}$, defined as the ratio between the surface second-order susceptibility and the layer thickness, is ~ 10^5 pm/V for mechanical exfoliated monolayer MoS$_2$ and 5×10^3 pm/V for the CVD-grown films. Similar values were reported in other studies of SL WSe$_2$ at 816nm ($\chi^{(2)} = 5 \times 10^3$ pm/V) and WS$_2$ at 832 nm ($\chi^{(2)} = 4.5 \times 10^3$ pm/V) [51].

It has also been observed that SHG from TMDCs is highly dependent on the number of stacked layers, excitation wavelength, and structural inhomogeneity. MoS$_2$ with SL or odd number of layers, in which Mo atoms are sandwiched between two S atoms and arranged in a trigonal prismatic lattice, belongs to the D_{3h}^3 space group and is a noncentrosymmetric material; thus, SHG is allowed. When stacked in an even number of layers, the inversion symmetry is restored (in this case, the material belongs to the D_{3d}^3 space group), which results in the vanishing of the SHG [46].

Figure 3.6(a) shows the layer-dependent SHG emission from MoS$_2$ nanosheets at 1560 nm excitation [46]. The results show that odd-layer sheets have strong SHG signal, whereas even layers of MoS$_2$ exhibit vanishing or very weak SHG emission. It has been found that the SHG emission from an odd number of layers is almost the

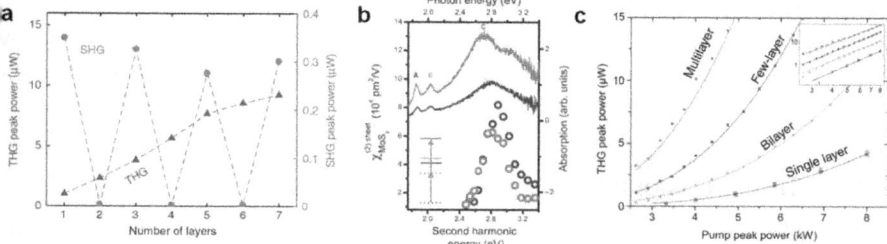

FIGURE 3.6 (a) Layer-dependent SHG and THG emission from MoS$_2$ [46]. (b) Linear and SHG emission spectra of monolayer (blue) and trilayer (green) MoS$_2$ showing the resonance enhancement at exciton band [54]. (c) THG emission from various graphene layers as a function of pump power [60].

Source: (b) and (c) Copyright 2017 American Chemical Society.

same for nonresonant SHG emitted photons, whereas in the resonant case (the energy of SHG photons is larger than the bandgap) a significant reduction in SHG (*e.g.*, five layers of MoS_2 show nearly eight times less SHG than the SL) is observed with increasing the odd number of layers [52]. This dependence is attributed to the reabsorption of SHG photons and interlayer coupling in MoS_2 nanosheets. Similar results are observed for hBN nanosheets under 810 nm excitation. In contrast, the quadratic dependence of SHG on the number of layers has been observed for the TMDCs of spiral WS_2 nanosheets due to the broken symmetry from the twisted structures.

Stronger SHG has been reported for resonance excitation and excitation at edges of 2DLMs. Figure 3.6(b) shows the resonant feature of SHG emission with C peaks of both monolayer and trilayer MoS_2 [53]. It has been suggested that the resonant enhancement is due to the increased electronic states at the C peak. Similar to graphene, TMDCs with even number of layers also exhibit SHG when the inversion symmetries are broken using external sources, including charge, plasmonic hot carriers, and specific structures, such as heterostructures, spirals, and pyramids [54].

What's more, the role of weak interlayer coupling in the second harmonic generation (SHG) effects of two-dimensional van der Waals (vdW) systems has also been studied theoretically [55]. We take homobilayer MoS_2/MoS_2 and heterobilayer $MoS_2/MoSe_2$ as typical examples, and have systemically investigated their SHG susceptibilities $\chi^{(2)}$ as a function of interlayer hopping strength (t_{int}) using first-principles calculations. For the $\chi^{(2)}_{yyy}(0;0,0)$ of both MoS_2/MoS_2 and $MoS_2/MoSe_2$, although the increase of tint can increase the intensities of interlayer optical transitions (IOT), the increased band repulsion around the point can eventually decrease their $\chi^{(2)}_{yyy}(0;0,0)$ values; the larger the tint, the smaller the $\chi^{(2)}_{yyy}(0;0,0)$. For the $|\chi^{(2)}_{yyy}(-2\omega;\omega,\omega)|$ spectra of $MoS_2/MoSe_2$ in the low photon-energy region, their peak values are very sensitive to the variable tint, due to the strong tint-dependent IOT dominating in the band edge; the larger the tint, the larger the $\left|\chi^{(2)}_{yyy}(-2\omega;\omega,\omega)\right|$. For the $\left|\chi^{(2)}_{yyy}(-2\omega;\omega,\omega)\right|$ of MoS_2/MoS_2 in the high photon-energy region, comparing to the $MoS_2/MoSe_2$, their peak values will decrease in a much more noticeable way as the tint increases, due to the larger reduction of band-nesting effect. Our study not only can successfully explain the puzzling experimental observations for the different SHG responses in different bilayer transition metal dichalcogenides under variable tint, but also may provide a general understanding for designing controllable SHG effects in the vdW systems.

3.3.2 THIRD HARMONIC GENERATION

All 2DLMs exhibit third harmonic generation (THG) and four wave mixing (FWM) to various degrees irrespective of their symmetry properties. The use of THG for material characterization is a particularly efficient investigative tool for centrosymmetric materials, including graphene, BP, and even/ML TMDCs [56]. THG and FWM in graphene-based materials have been investigated both theoretically and experimentally [57]. Thus, it has been revealed that the effective bulk-like third-order nonlinear susceptibility, $\chi^{(3)}$, strongly depends on the excitation wavelength, experimental conditions, and sample preparation method.

Strong THG from mechanically exfoliated single layer (SL) and few layer (FL) graphitic films on oxidized Si substrate was demonstrated by Kumar et al. with excitation of 1720 nm wavelength upon normal incidence. The measured values of $\chi^{(3)}$ were of the order of 10^{-16} m^2/V^2. One order of magnitude larger value of $\chi^{(3)}$ (10^{-15} m^2/V^2) was measured by Saynatjoki et al. for SL graphene with reference to the substrate (Si/SiO$_2$) at 1550 nm. Similar values were reported by Hendry et al. over the wavelength range of 760 nm to 840 nm using the FWM technique. Enhanced and broadband electrically tunable THG from graphene samples have also been experimentally demonstrated [58].

There are several theoretical predictions of large THG with $\chi^{(3)}$ of the order of 10^{-12} m^2/V^2; the large nonlinearity is attributed to substrate effects, heterostructures, dopants, and strong plasmonic near-field enhancements [59]. Layer-dependent tunable THG from BP nanosheets was reported near telecom wavelength [60]; the measured value of $\chi^{(3)}$ is about 10^{-19} m^2/V^2. Moreover, strong excitonic resonantly enhanced THG from FL BP was observed by Rodrigues et al.; the measured THG emission is three orders of magnitude larger than that of SL graphene under similar experimental conditions [61].

Studies of THG in MoS$_2$ [58], MoSe$_2$, WSe$_2$, WS$_2$, ReS$_2$, GaSe [62], and GaTe [63] have been performed by several groups. In contrast to the on-off response of SHG, THG exhibits a gradual increase with the number of layers (Figure 3.6(a)). The measured $\chi^{(3)}$ of SL to FL MoS$_2$ nanosheets in the wavelength range of 1560 nm to 1950 nm was of the order of 10^{-19} m^2/V^2 [58]. Saynatjoki et al. and Woodward et al. have measured the THG from SL MoS$_2$ and compared it to that of SL graphene under similar experimental conditions. It was found that the THG efficiency in MoS$_2$ is larger by a factor of three to four than that in graphene [46]. Moreover, THG in SL to FL of ReS$_2$, GaTe, and GaSe was reported to be larger by one to two orders of magnitude than that in SL MoS$_2$; the measured $\chi^{(3)}$ are about 10^{-18} m^2/V^2 to 10^{-16} m^2/V^2. Other class of 2DLMs (perovskites) have also been found to possess a large THG response. Thus, Abdelwahab et al., have measured $\chi^{(3)}$ of 1.12×10^{-17} m^2/V^2 for 2D lead halide Ruddlesden-Popper perovskites under resonant exciton excitation (at 1675 nm). It was found that the maximum THG conversion efficiency (0.006%) is more than five orders of magnitude larger than those of BP and MoS$_2$. Moreover, it has been shown that there is an optimum number of layers for which one achieves maximum THG (Figure 3.6(c)). Also, a quadratic dependence of THG on the number of layers, for less than 15 layers, has been reported in graphene, TMCs (e.g., MoS$_2$, WSe$_2$, and GaSe) [46], and BP. A decrease in THG was observed for a large number of layers and attributed to the phase mismatch or depletion of fundamental and/or THG signal by the sample absorption and reflection [57].

3.3.3 GIANT NONLINEAR ENHANCEMENT BASED ON DOUBLE PLASMONIC RESONANCE

In this section, we present a recently proposed mechanism to enhance THG in graphene nanoribbons (GNRs) based on a so-called double-resonance plasmon effect [64]. The geometrical structure is a 1D graphene grating with period Λ and width of graphene ribbons W. Graphene structures are assumed to be located at $z = z_s$ and placed on a substrate with relative permittivity ε_s (for specificity, assumed to be glass, $\varepsilon_s = 2.25$).

The mechanism of THG enhancement can be understood as follows. By varying the width of graphene ribbons, one can engineer the spectral resonances of the grating to ensure that the resonance wavelength of the fundamental plasmon coincides with the wavelength of the incoming beam, whereas the resonance wavelength of one of the higher-order plasmons is exactly a third of the resonance wavelength of the fundamental plasmon mode. Under these circumstances, the diffraction grating will be efficiently excited at the fundamental frequency, which will lead to a strong field enhancement at this frequency and radiate effectively at the TH, as there is another plasmon resonance at this wavelength. In effect, such a diffraction grating would act as a highly effective receiver at the fundamental frequency and a strong emitter (efficient antenna) at the TH.

Using the numerical method GS-RCWA, the dispersion map of the linear optical response of graphene ribbons is calculated. The results suggest that it is indeed possible to engineer an optical diffraction grating with the desired property, because it can be seen that there are certain values of the graphene ribbon width W, for which a graphene plasmon mode exists at both FF and TH wavelengths. To be more specific, for $W = 85$ nm, graphene diffraction grating supports a fundamental plasmon mode at the fundamental frequency corresponding to $\lambda_{FF} = 9.03$ μm and a third-order plasmon mode at $\lambda_{TH} = \lambda_{FF}/3 = 3.01$ μm.

Importantly, the double-resonance condition is fulfilled when graphene ribbons are in relatively close proximity to their nearest neighbors. Therefore, although the excitation of localized surface plasmons on graphene ribbons plays the major role in the observed enhancement of the TH intensity, the optical coupling between neighboring ribbons and other diffractive effects could affect as well the optical response of the graphene structure. What's more, a further enhancement of the TH intensity occurs when the double-resonance condition holds. Thus, it was determined by choosing the wavelength of the incident beam to be equal to the resonance wavelength of the fundamental plasmon and varying the width of the ribbons. It can be seen that a maximum intensity of the TH is achieved when $W = 85$ nm, that is for the width at which there are plasmons at both fundamental frequency and TH.

3.4 APPLICATIONS OF THZ MATERIALS

3.4.1 Photonic Green's Function, Optical Conductivity, and Electron-Phonon Interaction

Electromagnetic waves propagate in space and time, follow the rules of Maxwell equations. From the early days when Dirac wrote down the Dirac equation in quantum mechanics, many similarities with Maxwell equations were found. However, one difference remains, quantum mechanics requires the energy of electromagnetic waves to be quantified, while in classical Maxwell equations we could not find the Planck constant. To rewrite the Maxwell equations into a Dirac-like form, we could define the Maxwell Hamiltonian H_{Max}, which is determined from the dielectric property, magnetic permeability, and magneto-electric property of a specific material. In this way, the Planck constant could be added. The photonic Green's function is defined as the inverse of $(z-H_{Max})$, where z is the frequency variable. The photonic Green's function appears in the relation between the electric field and the electric current. The optical conductivity tensor σ appears in the opposite way $J = \sigma E$, which could be calculated from the Kubo formula. Kubo formula is one kind of the fluctuation-dissipation theorem, which is the key idea in many physical problems, such as Einstein's Brownian particle, Nyquist's formula, Callen-Welton relation and Casimir forces.

The electron-phonon interaction renormalizes quasiparticle dynamics and leads to important observable changes in electronic properties which illustrate the effects of many body renormalizations not captured in single particle theories. For the Dirac electrons in graphene an example, features observed in the density of states (DOS) and in the dispersion curves measured in angular resolved photo emission spectroscopy have been interpreted as phonon structure. The DOS of electrons in graphene coupled to a phonon in an external magnetic field were calculated and found that coupling to an Einstein mode broadens the Landau levels and radically alters the DOS by introducing a new set of peaks at energies [65,66], and the phenomenon were confirmed by David's and Adam's experiment [67,68].

The large renormalization effects from electron-phonon interaction is clearly seen in the modifications to the DOS. The four phonon structures are at $\omega = \pm \omega_E$, $-\mu - \omega_E - \Delta$ and $-\mu - \omega_E + \Delta$ and $-\mu - \omega_E + \Delta$, where ω_E is the phonon frequency, μ is the chemical potential, and Δ is the gap. By comparison, the corresponding boson structures in the real part of the dynamic longitudinal optical conductivity $Re(\sigma_{xx}(\omega))$ is more closely related to a convolution of two DOS factors. For example, gapless Dirac fermions ($\Delta = 0$) and gapped case with $\Delta = 20$ meV. Bare band with a small residual scattering will broaden out the intraband Drude peak, which is large only at small ω and is centered at $\omega = 0$. The bare chemical potential is 35 meV and the onset of a second absorption band from the interband transitions starts at $\omega = 2\mu$. These transitions continue up to large energies and provide the so-called universal background. In our units for $Re(\sigma_{xx}(\omega))$ which is $e^2/\hbar$, this background [69,70] has a height of 1/16. For $\Delta = 0$ this height is almost unaffected by the electron-phonon interaction. However, the appearance of the Holstein processes above $\omega = \omega_E$ which provides significant phonon assisted absorption in the photon region above ω_E and below the main interband absorption edge at 2μ. The other feature to be noted is that, for the correlated case, the onset of the interband transitions has moved to lower energies and is now at twice the value of the interacting chemical potential. For the gapped fermion case another important element appears. As is well known [71] when $\Delta = 0$ there is a peak in the interband transitions just above the threshold energy which persists up to a few Δ above the threshold before the value of the background is reestablished at its universal value. In the clean limit we see that, just above the interband onset, the conductivity is larger than its universal value of $e^2/16\hbar$ which is the bare band result. We note that, with electron-phonon, the magnitude of the absorption in the region of the edge still remains above the universal background value but now there is also a small phonon structure highlighted. No such structure is seen for $\Delta = 0$. It is the variation with energy of the background (in the presence of a finite gap) which allows for the phonon structure to be revealed.

The probability of occupation of the state k at zero temperature is denoted by $n(k)$ and in the bare band picture is a step function $\theta(k_F - k)$. For $T = 0.1$ K, with a weak electron-phonon coupling $g = 10$ and the gap $\Delta = 20$ meV, $n(k)$ is considerably reduced from value one throughout the occupied states. It still has a finite discontinuous jump at $k = k_F$ (Fermi wave-vector) with correlation tail for $k > k_F$. The out-of-plane spin $S_z(k)$ shows the qualitatively similar effect. There are finite tails beyond this momentum which are entirely due to many body correlation effects that go beyond a bare band description. They can be taken as the representative of other correlation effect such as those due to electron-electron interactions rather than electron-phonon interaction.

3.4.2 THz Driven Topological Phase Transition

According to optical response theory, optical susceptibility plays a key role in light-matter interactions. It is a function of optical frequency ω, thus the response of material can be divided into four regimes [72]: low frequency regime, absorption region, reflection regime, and transparent regime. For low frequency within terahertz, the far infrared optics is strongly and directly coupled with phonons. And it would introduce a damage-free method to control the material behaviors though lattice vibrations. Owing to the coherent phonon, monolayer transition metal dichalcogenides are predicted to show ultrafast topological phase transitions under terahertz light [73]. Indeed, a light-induced topological phase transitions in $MoTe_2$, from trivial hexagonal to the topological insulator distorted phase, are observed experimentally. And this metastable phase persists indefinitely over months after THz exposure. Combining ultralow frequency Raman spectroscopy with first principles calculations, Z. Li and coworkers investigated the phonon-assisted electronic states modulation of few-layer $PdSe_2$ at terahertz frequencies. Two distinctive types of coherent phonon excitations could couple preferentially to different types of electronic excitations: the intralayer (4.3 THz) mode to carriers and the interlayer (0.35 THz) mode to excitons.

In addition, charge density wave (CDW) is another mechanics that induces light- and THz-driven topological phase transition. Both theoretical and experimental evidences show that strong k-dependent EPC and CDW phase [74–76]. TMDs exhibit a rich set of Peierls-like CDW orders [77]. Based on x-ray diffraction and the first-principles calculations, Y. Liu $et\ al.$ investigated the nature of charge density waves and superconductivity in 1T-$TaSe_2$-$x$$Te_x$ ($0 \le x \le 2$) [78]. Doping-induced disordered distribution of Se/Te suppresses CDWs in 1T-$TaSe_2$. Moreover, K. Tanimura studied the femtosecond-laser excitation-induced CDW phase in 1T-TaS_2 [79]. And coherent-phonon spectroscopy results show that, together with the amplitude mode of CDW with a frequency of 2.41 THz, two other modes with frequencies of 2.34 and 2.07 THz are excited in the photoexcited commensurate CDW phase at several tens of picoseconds after excitation. Recently, G. Marini and coworkers studied the CDW orders in $MoTe_2$ and WTe_2 with ultrafast optical pumping [80]. Besides TMDs, many other materials show good performances in optical and THz frequencies with strong EPC. S. Krylow $et\ al.$ studied ultrafast structural relaxation dynamics of femtosecond laser-excited graphene using ab-initio molecular dynamics simulations including EPC [81]. Recent studies of light-modulated EPC explored the photoexcited graphene during the ultrafast photocarrier dynamics [82]. The graphene hot-electron bolometer detectors with superconducting electrodes gives an electrical NEP of 15 fW/Hz$^{0.5}$ and a dynamic range of 47 dB at 0.3 K.

3.4.3 OTHER THz RELATED INTERACTION

Three-dimensional topological insulator Bi_2Se_3 is an excellent thermoelectric material with excellent thermoelectric coefficient. And topological surface state is observed in thin films [83] and Fe/Mn-doped crystal [84,85]. Recently, terahertz emission based on ultrafast photothermoelectric effect is observed Dirac semimetal-lic Cd_3As_2, and when a weak magnetic field (~ 0.4 T) is applied, the response clearly indicates an order of magnitude enhancement on transient photothermoelectric current generation compared to the photo-Seebeck effect. Valley-Hall and spin-Hall effect are observed in monolayer TMDs, and the intrinsic spin Hall conductivity is one order of magnitude larger than that in inversion-symmetric bulk states [86].

3.5 SUMMARY AND OUTLOOK

The optical properties of various nanomaterials have been discussed in this chapter. Though graphene and 2D TMDs shows excellent optoelectronic properties in various areas, several modifications are needed for the enhancement of optoelectronics applications in modern electronics.

Making heterojunctions is another way that can be explored with 2D materials. To fine-tune the optical properties of the 2D materials, the formation of heterojunction provides excellent flexibility. Several methods have been used to make heterojunctions. For enhancing the device performance, making an interface with continuous band alignment, optimizing the carrier numbers, etc., can be employed. The 3D printing technology can be utilized to build 2D materials optoelectronics circuits to tune the properties of the device. The heterojunction of 2D materials will be formed layer by layer using 3D printing. For fabrication of stacked heterostructure of few atomic thicknesses of each layer, 3D printing method, CVD, epitaxial growth technique, and so forth can be done. These kinds of vertically stacked heterostructure can be a potential candidate for functional integrated optical devices such as photovoltaic cells, phototransistors, photodetectors, LEDs, and optical sensors. Also, the high sensitivity to the environmental changes of 2D materials makes them promising candidates for biosensing applications. Modification of electronic structure is a useful technique that can improve the optoelectronic properties of the 2D materials. To improve the performance of the 2D-2D heterojunction in optoelectronic devices, electron-hole transport phenomena will be control by engineering the interfacial band structure. The interfacial charge and energy transfer play an essential role for optoelectronic devices. Therefore, optimization of several parameters, such as bandgap offset at the interface, thickness of spacer, laser excitation power, charge separation, and transport rate, is very important for future optoelectronics devices using 2D materials. Generally, 2D materials also show high electron mobility which increases the efficiency of the photo carriers. Therefore, during the fabrication of heterojunction devices, the selection of appropriate 2D materials is very important. Additionally, to modify the device structures, 3D printing technology is one of the promising approaches. In the case of 3D printed based hybrid optical devices, tuning the layer number, charge transport modification, alloying, and so forth could improve the efficiency of the fabricated heterojunction. We hope the present work will contribute to further works on these exciting materials.

BIBLIOGRAPHY

[1] Xia, F., Wang, H., Xiao, D., Dubey, M., and Ramasubramaniam, A. 2014. Two-dimensional material nanophotonics. *Nat. Photon.* 8:899–907.

[2] Mak, K. F., and Shan, J. 2016. Photonics and optoelectronics of 2D semiconductor transition metal dichalcogenides. *Nat. Photon.* 10:216–226.

[3] Guo, B., Xiao, Q. L., Wang, S. H., and Zhang, H. 2019. 2D layered materials: Synthesis, nonlinear optical properties, and device applications. *Laser Photon. Rev.* 13:1800327.

[4] Ma, Q., Ren, G., Xu, K., and Ou, J. Z. 2020. Tunable optical properties of 2D materials and their applications. *Adv. Opt. Mater.* 9:2001313.

[5] Grigorenko, A. N., Polini, M., and Novoselov, K. S. 2012. Graphene plasmonics. *Nat. Photon.* 6:749–758.

[6] Latini, S., Winther, K. T., Olsen, T., and Thygesen, K. S. 2017. Interlayer excitons and band alignment in MoS_2/hBN/WSe_2 van der Waals heterostructures. *Nano Lett.* 17:938–945.

[7] Qiao, J., Kong, X., Hu, Z. X., Yang, F., and Ji, W. 2014. High-mobility transport anisotropy and linear dichroism in few-layer black phosphorus. *Nat. Commun.* 5:4475.

[8] Pramanik, A., Biswas, S., Tiwary, C. S., Sarkar, R., and Kumbhakar, P. 2018. Colloidal n-doped graphene quantum dots with tailored luminescent downshifting and detection of UVA radiation with enhanced responsivity. *ACS Omega.* 3:16260–16270.

[9] Liu, M., Yin, X., Ulin-Avila, E., Geng, B., Zentgraf, T., Ju, L., Wang, F., and Zhang, X. 2011. A graphene-based broadband optical modulator. *Nature.* 474:64–67.

[10] Gan, X., Shiue, R.-J., Gao, Y., Meric, I., Heinz, T. F., Shepard, K., Hone, J., Assefa, S., and Englund, D. 2013. Chip-integrated ultrafast graphene photodetector with high responsivity. *Nat. Photon.* 7:883–887.

[11] Chen, H., Liu, T., Su, Z., Shang, L., and Wei, G. 2018. 2D transition metal dichalcogenide nanosheets for photo/thermo-based tumor imaging and therapy. *Nanoscale Horiz.* 3:74–89.

[12] Ono, M., Hata, M., Tsunekawa, M., Nozaki, K., Sumikura, H., Chiba, H., and Notomi, M. 2019. Ultrafast and energy-efficient all-optical switching with graphene-loaded deep-subwavelength plasmonic waveguides. *Nat. Photon.* 14:37–43.

[13] Hwang, M. T., Heiranian, M., Kim, Y., et al. 2020. Ultrasensitive detection of nucleic acids using deformed graphene channel field effect biosensors. *Nat. Commun.* 11:1543.

[14] Chen, K., Zhou, X., Qiao, R., et al. 2019. Graphene photonic crystal fibre with strong and tunable light–matter interaction. *Nat. Photon.* 13:754–759.

[15] Kim, Y. D., Kim, H., Cho, Y., et al. 2015. Bright visible light emission from graphene. *Nat. Nanotech.* 10:676–681.

[16] Jiang, T., Kravtsov, V., Tokman, M., Belyanin, A., and Raschke, M. B. 2019. Ultrafast coherent nonlinear nanooptics and nanoimaging of graphene. *Nat. Nanotech.* 14:838–843.

[17] Abdelrazik, A. S., Tan, K. H., Aslfattahi, N., Saidur, R., and Al-Sulaiman, F. A. 2020. Optical properties and stability of water-based nanofluids mixed with reduced graphene oxide decorated with silver and energy performance investigation in hybrid photovoltaic thermal solar systems. *Int. J. Energy Res.* 44:11487–11508.

[18] Synnatschke, K., Cieslik, P. A., Harvey, A., Castellanos-Gomez, A., Tian, T., Shih, C.-J., Chernikov, A., Santos, E. J. G., Coleman, J. N., and Backes, C. 2019. Length- and thickness-dependent optical response of liquid-exfoliated transition metal dichalcogenides. *Chem. Mater.* 31:10049–10062.

[19] Vikraman, D., Akbar, K., Hussain, S., Yoo, G., Jang, J.-Y., Chun, S.-H., Jung, J., and Park, H. J. 2017. Direct synthesis of thickness-tunable MoS_2 quantum dot thin layers: Optical, structural and electrical properties and their application to hydrogen evolution. *Nano Energy.* 35:101–114.

[20] Zhao, K., Huang, F., Dai, C.-M., Li, W., Chen, S.-Y., Jiang, K., Huang, Y.-P., Hu, Z., and Chu, J. 2018. Temperature dependence of phonon modes, optical constants, and optical band gap in two-dimensional ReS_2 films. *J. Phys. Chem. C* 122:29464–29469.

[21] Kravets, V. G., Wu, F., Auton, G. H., Yu, T., Imaizumi, S., and Grigorenko, A. N. 2019. Measurements of electrically tunable refractive index of MoS_2 monolayer and its usage in optical modulators. *NPJ 2D Mater. Appl.* 3:36.

[22] Eda, G., Yamaguchi, H., Voiry, D., Fujita, T., Chen, M., and Chhowalla, M. 2011. Erratum: Photoluminescence from chemically exfoliated MoS_2. *Nano Lett.* 11:5111–5116.

[23] Sun, Y., Fujisawa, K., Lin, Z., Lei, Y., Mondschein, J. S., Terrones, M., and Schaak, R. E. 2017. Low-temperature solution synthesis of transition metal dichalcogenide alloys with tunable optical properties. *J. Am. Chem. Soc.* 139:11096–11105.

[24] Sriram, P., Wen, Y. P., Manikandan, A., et al. 2020. Enhancing quantum yield in strained MoS_2 bilayers by morphology-controlled plasmonic nanostructures toward superior photodetectors. *Chem. Mater.* 32:2242–2252.

[25] Wang, Q. X., Guo, J., Ding, Z. J., Qi, D. Y., Jiang, J. Z., Wang, Z., Chen, W., Xiang, Y., Zhang, W., and Wee, A. T. S. 2017. Fabry-Perot cavity-enhanced optical absorption in ultrasensitive tunable photodiodes based on hybrid 2D materials. *Nano Lett.* 17:7593–7598.

[26] Noumbe, U. N., Greboval, C., Livache, C., et al. 2020. Reconfigurable 2D/0D p-n graphene/HgTe nanocrystal heterostructure for infrared detection. *ACS Nano.* 14: 4567–4576.

[27] Mudd, G. W., Svatek, S. A., Ren, T. H., et al. 2013. Tuning the bandgap of exfoliated InSe nanosheets by quantum confinement. *Adv. Mater.* 25:5714–5718.

[28] Zhang, C. X., Ouyang, H., Miao, R. L., et al. 2019. Anisotropic nonlinear absorption: Anisotropic nonlinear optical properties of a SnSe flake and a novel perspective for the application of all-optical switching. *Adv. Opt. Mater.* 7:1900631.

[29] Wang, Z., Li, F., Guo, J., Ma, C. Y., Song, Y. F., He, Z., Liu, J., Zhang, Y. P., Li, D. L., and Zhang, H. 2020. Facile synthesis of 2D tin selenide for near- and mid-infrared ultrafast photonics applications. *Adv. Opt. Mater.* 8:1902183.

[30] Johnson, V. L., Anilao, A., and Koski, K. J. 2019. Pressure-dependent phase transition of 2D layered silicon telluride (Si_2Te_3) and manganese intercalated silicon telluride. *Nano Res.* 12:2373–2377.

[31] Huang, S. X., Tatsumi, Y. K., Ling, X., et al. 2016. In-plane optical anisotropy of layered gallium telluride. *ACS Nano.* 10:8964–8972.

[32] Abed Al-Abbas, S. S., Muhsin, M. K., and Jappor, H. R. 2018. Tunable optical and electronic properties of gallium telluride monolayer for photovoltaic absorbers and ultraviolet detectors. *Chem. Phys. Lett.* 713:46–51.

[33] Qiao, M., Chen, Y. L., Wang, Y., and Li, Y. F. 2018. The germanium telluride monolayer: A two dimensional semiconductor with high carrier mobility for photocatalytic water splitting. *J. Mater. Chem. A* 6:4119–4125.

[34] Kang, J., Sangwan, V. K., Lee, H.-S., Liu, X., and Hersam, M. C. 2018. Solution-processed layered gallium telluride thin-film photodetectors. *ACS Photon.* 5:3996–4002.

[35] Kumar, A., Low, T., Fung, K. H., Avouris, P., and Fang, N. X. 2015. Tunable light-matter interaction and the role of hyperbolicity in graphene-hBN system. *Nano Lett.* 15:3172–3180.

[36] Lyu, B., Li, H., Jiang, L., et al. 2019. Phonon polariton-assisted infrared nanoimaging of local strain in hexagonal boron nitride. *Nano Lett.* 19:1982–1989.

[37] Jia, Y., Zhao, H., Guo, Q., Wang, X., Wang, H., and Xia, F. 2015. Tunable plasmon-phonon polaritons in layered graphene-hexagonal boron nitride heterostructures. *ACS Photon.* 2:907–912.

[38] Quereda, J., San-Jose, P., Parente, V., Vaquero-Garzon, L., Molina-Mendoza, A. J., Agrait, N., Rubio-Bollinger, G., Guinea, F., Roldan, R., and Castellanos-Gomez, A. 2016. Strong modulation of optical properties in black phosphorus through strain-engineered rippling. *Nano Lett.* 16:2931–2937.

[39] Cao, T., Liu, Z., Qiu, D. Y., and Louie, S. G. 2016. Gate switchable transport and optical anisotropy in 90° twisted bilayer black phosphorus. *Nano Lett.* 16:5542–5546.

[40] Villegas, C. E., Rocha, A. R., and Marini, A. 2016. Anomalous temperature dependence of the band gap in black phosphorus. *Nano Lett.* 16:5095–5101.

[41] Lu, J., Yang, J., Carvalho, A., Liu, H., Lu, Y., and Sow, C. H. 2016. Light-matter interactions in phosphorene. *Acc. Chem. Res.* 49:1806–1815.

[42] Ye, L., Li, H., Chen, Z., and Xu, J. 2016. Near-infrared photodetector based on MoS_2/black phosphorus heterojunction. *ACS Photon.* 3:692–699.

[43] Li, P., et al. 2017. Two-dimensional $CH_3NH_3PbI_3$ perovskite nanosheets for ultrafast pulsed fiber lasers. *ACS Appl. Mater. Interfaces.* 9:12759–12765.

[44] Luo, T., Wang, C., Pan, Z., Jin, C., Fu, Z., and Jin, Y. 2019. Maternal polystyrene microplastic exposure during gestation and lactation altered metabolic homeostasis in the dams and their F_1 and F_2 offspring. *Environ. Sci. Technol.* 53:10978–10992.

[45] Tsai, H., Nie, W., Blancon, J., et al. 2016. High-efficiency two-dimensional Ruddlesden-Popper perovskite solar cells. *Nature.* 536:312–316.

[46] Saynatjoki, A., Karvonen, L., Rostami, H., et al. 2017. Ultra-strong nonlinear optical processes and trigonal warping in MoS_2 layers. *Nat. Commun.* 8:893.

[47] Dean, J. J., and van Driel, H. M. 2009. Second harmonic generation from graphene and graphitic films. *Appl. Phys. Lett.* 95:261910.

[48] Kumar, N., Najmaei, S., Cui, Q., Ceballos, F., Ajayan, P. M., Lou, J., and Zhao, H. 2013. Second harmonic microscopy of monolayer MoS_2. *Phys. Rev. B* 87:161403(R).

[49] Lin, X., Liu, Y., Wang, K., Wei, C., Zhang, W., Yan, Y., Li, Y. J., Yao, J., and Zhao, Y. S. 2018. Two-dimensional pyramid-like WS_2 layered structures for highly efficient edge second-harmonic generation. *ACS Nano.* 12:689–696.

[50] Wang, G., Marie, X., Gerber, I., Amand, T., Lagarde, D., Bouet, L., Vidal, M., Balocchi, A., and Urbaszek, B. 2015. Giant enhancement of the optical second-harmonic emission of WSe_2 monolayers by laser excitation at exciton resonances. *Phys. Rev. Lett.* 114:097403.

[51] Janisch, C., Wang, Y., Ma, D., Mehta, N., Elias, A. L., Perea-Lopez, N., Terrones, M., Crespi, V., and Liu, Z. 2014. Extraordinary second harmonic generation in tungsten disulfide monolayers. *Sci. Rep.* 4:5530.

[52] Li, Y., Rao, Y., Mak, K. F., You, Y., Wang, S., Dean, C. R., and Heinz, T. F. 2013. Probing symmetry properties of few-layer MoS_2 and h-BN by optical second-harmonic generation. *Nano Lett.* 13:3329–3333.

[53] Trolle, M. L., Tsao, Y.-C., Pedersen, K., and Pedersen, T. G. 2015. Observation of excitonic resonances in the second harmonic spectrum of MoS_2. *Phys. Rev. B* 92:161409(R).

[54] Fan, X., Jiang, Y., Zhuang, X., et al. 2017. Broken symmetry induced strong nonlinear optical effects in spiral WS_2 nanosheets. *ACS Nano.* 11:4892–4898.

[55] Jiang, X., Kang, L., and Huang, B. 2022. Role of interlayer coupling in second harmonic generation in bilayer transition metal dichalcogenides. *Phys. Rev. B* 105:045415.

[56] Wang, R., Chien, H.-C., Kumar, J., Kumar, N., Chiu, H.-Y., and Zhao, H. 2014. Third-harmonic generation in ultrathin films of MoS_2. *ACS Appl. Mater. Interfaces.* 6:314–318.

[57] Hong, S.-Y., Dadap, J. I., Petrone, N., Yeh, P.-C., Hone, J., and Osgood, R. M. 2013. Optical third-harmonic generation in graphene. *Phys. Rev. X* 3:021014.

[58] Soavi, G., Wang, G., Rostami, H., et al. 2018. Broadband, electrically tunable third-harmonic generation in graphene. *Nat. Nanotech.* 13:583–588.

[59] Zhang, M., Li, G., and Li, L. 2014. Graphene nanoribbons generate a strong third-order nonlinear optical response upon intercalating hexagonal boron nitride. *J. Mater. Chem. C* 2:1482–1488.

[60] Autere, A., Ryder, C. R., Saynatjoki, A., et al. 2017. Rapid and large-area characterization of exfoliated black phosphorus using third-harmonic generation microscopy. *J. Phys. Chem. Lett.* 8:1343–1350.

[61] Rodrigues, M. J., de Matos, C. J., Ho, Y. W., Peixoto, H., de Oliveira, R. E., Wu, H. Y., Neto, A. H., and Viana-Gomes, J. 2016. Black phosphorus: Resonantly increased optical frequency conversion in atomically thin black phosphorus. *Adv. Mater.* 28:10692.

[62] Zhou, X., Cheng, J., Zhou, Y., Cao, T., Hong, H., Liao, Z. M., Wu, S. W., Peng, H. L., Liu, K. H., and Yu, D. P. 2015. Strong second-harmonic generation in atomic layered GaSe. *J. Am. Chem. Soc.* 137:7994–7997.

[63] Susoma, J., Karvonen, L., Säynätjoki, A., Mehravar, S., Norwood, R. A., Peyghambarian, N., Kieu, K., Lipsanen, H., and Riikonen, J. 2016. Second and third harmonic generation in few-layer gallium telluride characterized by multiphoton microscopy. *Appl. Phys. Lett.* 108:073103.

[64] You, J. W., You, J., Weismann, M., and Panoiu, N. C. 2017. Double-resonant enhancement of third-harmonic generation in graphene nanostructures. *Philos. Trans. A: Math. Phys. Eng. Sci.* 375:20160313.

[65] Pound, A., Carbotte, J. P., and Nicol, E. J. 2011. Effects of electron-phonon coupling on Landau levels in graphene. *Phys. Rev. B* 84:085125.

[66] Li, Z., and Carbotte, J. P. 2013. Phonon structure in dispersion curves and density of states of massive Dirac fermions. *Phys. Rev. B* 88:045417.

[67] Miller, D. L., Kubista, K. D., Rutter, G. M., Ruan, M., Heer, W. A. D., First, P. N., and Stroscio, J. A. 2009. Observing the quantization of zero mass carriers in graphene. *Science* 324:924–927.

[68] Pound, A., Carbotte, J. P., and Nicol, E. J. 2011. Phonon structures in the electronic density of states of graphene in magnetic field. *Europhys. Lett.* 94:57006.

[69] Gusynin, V. P., Sharapov, S. G., and Carbotte, J. P. 2006. Unusual microwave response of Dirac quasiparticles in graphene. *Phys. Rev. Lett.* 96:256802.

[70] Gusynin, V. P., Sharapov, S. G., and Carbotte, J. P. 2009. On the universal ac optical background in graphene. *New J. Phys.* 11:095103.

[71] Stille, L., Tabert, C. J., and Nicol, E. J. 2012. Optical signatures of the tunable band gap and valley-spin coupling in silicene. *Phys. Rev. B* 86:195405.

[72] Hiltunen, V.-M., Koskinen, P., Mentel, K. K., Manninen, J., Myllyperkiö, P., Pettersson, M., and Johansson, A. 2021. Ultrastiff graphene. *NPJ 2D Mater. Appl.* 5:49.

[73] Zhou, J., Xu, H., Shi, Y., and Li, J. 2021. Topological phase transition: Terahertz driven reversible topological phase transition of monolayer transition metal dichalcogenides. *Adv. Sci.* 8:2170072.

[74] Johannes, M. D., and Mazin, I. I. 2008. Fermi surface nesting and the origin of charge density waves in metals. *Phys. Rev. B* 77:165135.

[75] Varma, C. M., and Simons, A. L. 1983. Strong-coupling theory of charge-density-wave transitions. *Phys. Rev. Lett.* 51:138.

[76] Calandra, M., Mazin, I. I., and Mauri, F. 2009. Effect of dimensionality on the charge-density wave in few-layer 2H-NbSe$_2$. *Phys. Rev. B* 80:241108(R).

[77] Wilson, J. A., Di Salvo, F. J., and Mahajan, S. 1975. Charge-density waves and superlattices in the metallic layered transition metal dichalcogenides. *Adv. Phys.* 24:117–201.

[78] Liu, Y., Shao, D. F., Li, L. J., et al. 2016. Nature of charge density waves and superconductivity in 1T–TaSe$_{2-x}$Te$_x$. *Phys. Rev. B* 94:045131.

[79] Tanimura, K. 2018. Photoinduced discommensuration of the commensurate charge-density wave phase in 1T–TaS$_2$. *Phys. Rev. B* 97:245115.

[80] Marini, G., and Calandra, M. 2021. Light-tunable charge density wave orders in MoTe$_2$ and WTe$_2$ single layers. *Phys. Rev. Lett.* 127:257401.

[81] Krylow, S., Hernandez, F. V., Bauerhenne, B., and Garcia, M. E. 2020. Ultrafast structural relaxation dynamics of laser-excited graphene: Ab initio molecular dynamics simulations including electron-phonon interactions. *Phys. Rev. B* 101:205428.

[82] Hu, S.-Q., Zhao, H., Lian, C., Liu, X.-B., Guan, M.-X., and Meng, S. 2022. Tracking photocarrier-enhanced electron-phonon coupling in nonequilibrium. *NPJ Quantum Mater.* 7:14.

[83] Kamboj, V. S., Singh, A., Ferrus, T., Beere, H. E., Duffy, L. B., Hesjedal, T., Barnes, C. H. W., and Ritchie, D. A. 2017. Probing the topological surface state in Bi$_2$Se$_3$ thin films using temperature-dependent terahertz spectroscopy. *ACS Photon.* 4:2711.

[84] Chen, Y. L., Chu, J. H., Analytis, J. G., et al. 2010. Massive Dirac fermion on the surface of a magnetically doped topological insulator. *Science* 329:659–662.

[85] Jozwiak, C., Chen, Y. L., Fedorov, A. V., et al. 2011. Widespread spin polarization effects in photoemission from topological insulators. *Phys. Rev. B* 84:165113.

[86] Feng, W., Yao, Y., Zhu, W., Zhou, J., Yao, W., and Xiao, D. 2012. Intrinsic spin Hall effect in monolayers of group-VI dichalcogenides: A first-principles study. *Phys. Rev. B* 86:165108.

[87] Yin, Z., Li, H., Li, H., Jiang, L., Shi, Y. M., Sun, Y. H., Lu, G., Zhang, Q., Chen, X. D., and Zhang, H. 2012. Single-layer MoS$_2$ phototransistors. *ACS Nano.* 6:74–80.

[88] Siddique, S., Gowda, C. C., Tromer, R., et al. 2021. Scalable synthesis of atomically thin gallium telluride nanosheets for supercapacitor applications. *ACS Appl. Nano Mater.* 4:4829–4838.

[89] Huang, J., Dong, N., Zhang, S., Sun, Z., Zhang, W., and Wang, J. 2017. Nonlinear absorption induced transparency and optical limiting of black phosphorus nanosheets. *ACS Photon.* 4:3063.

[90] Zhang, G., Huang, S., Chaves, A., Song, C., Ozcelik, V. O., Low, T., and Yan, H. 2017. Infrared fingerprints of few-layer black phosphorus. *Nat. Commun.* 8:14071.

4 Plasmonic Materials and Their Applications

Jinfeng Zhu, Yinong Xie, and Yuan Gao

CONTENTS

4.1 Introduction..83
4.2 Fundamental Physics 84
    4.2.1 Surface Plasmon Polaritons 84
    4.2.2 Localized Surface Plasmon Resonance 87
    4.2.3 Spoof Surface Plasmon Polaritons...88
4.3 Plasmonic Materials..88
    4.3.1 Conventional Noble Metals 89
    4.3.2 Nonnoble Metals...89
    4.3.3 Semiconductors...92
    4.3.4 Two-Dimensional Materials 94
4.4 Fabrication Methods 95
    4.4.1 Electron Beam Lithography..95
    4.4.2 Focused Ion Beam Lithography..96
    4.4.3 Nanoimprint Lithography 96
    4.4.4 Laser Lithography...97
    4.4.5 Self-Assembly Technique 98
    4.4.6 Chemical Synthesis...98
4.5 Applications of Plasmonic Materials..99
    4.5.1 Refractive Index Sensing 100
    4.5.2 Label-Free Biosensing..101
    4.5.3 SEIRS 104
    4.5.4 Raman Spectroscopy 106
    4.5.5 Nonlinear Enhancement..107
    4.5.6 Absorbers...108
    4.5.7 Imaging..109
4.6 Summary and Outlook...110
Acknowledgments..110
Bibliography 111

4.1 INTRODUCTION

Plasmon is an important part of the profoundly influential field of modern photonics [1]. It mainly describes the interactions of electromagnetic radiation and conduction electrons in micro/nano metallic structures, which will result in a significant enhancement of the optical near-field in a sub-wavelength dimension. Due to its wide range of

DOI: 10.1201/9781003202608-4

applications, plasmons have attracted numerous investigations over the past 100 years. The earliest application of plasmons is for glass-stained metal nanoparticles, which can be traced back to the Roman era. In 1900, people could already describe the two important components of surface plasmons (surface plasmons polaritons (SPPs) and localized surface plasmons (LSPs)) and established a basic mathematical foundation.

At the beginning of the 20th century, mathematical descriptions about surface waves were established in the general context that radio waves can propagate along the surface of a conductor. Until the mid-20th century, this theory was verified in the visible region by observing anomalous light intensity reduction in the light reflectance spectrum of a metallic grating. Later, Sommerfeld realized the convenient excitation of surface waves from coupled prisms and made a unified definition of surface plasmon. Since then, research works in this field have increased explosively, initially focused on the visible spectral region, and then quickly extended to ultraviolet (UV), infrared (IR), microwave, and terahertz (THz) wave ranges.

There has been tremendous progress in the intrinsic physics of plasmons as well as their corresponding applications. In this chapter, we will briefly introduce the properties of plasmons and their research histories. In the following sections, we will describe the classification of plasmons and their intrinsic physics and excitation conditions. We will then discuss various plasmonic materials, fabrication methods, and their corresponding applications. Finally, a short future outlook will be provided.

4.2 FUNDAMENTAL PHYSICS

Plasmon is a kind of free electron oscillation quantum. When the conduction electrons in the conductor (such as metal) are disturbed from their equilibrium positions, plasmon can be produced by the collective oscillation of these electrons. Such interference can be triggered by electromagnetic waves, in which free electrons of metal are driven by an alternating electric field to coherently oscillate at a certain resonance frequency called surface plasmons resonance (SPR), accompanied with a significant localization and electric field enhancement. These characteristics, combined with the ability to change the local density of optical states (LDOS) and ultra-fast response to surrounding stimuli, are the core of various new technologies such as nanoelectronics, sensing, and imaging, and are widely used in different fields, from biomedicine and environmental monitoring to telecommunications and photovoltaics [2]. For infinite bulk metals, the bulk plasmon frequency ω_p can be described by $\omega_p = (Ne^2/\varepsilon_0 m_e)^{1/2}$, where N is the density of conduction electrons, ε_0 is the permittivity of vacuum, e and m_e are the charge and effective mass of electron, respectively [3]. Up to date, various metals, semiconductors and even two-dimensional (2D) materials are found to generate plasmon resonances from UV to far-IR. Here, we introduce three main types, namely, SPP, localized surface plasmon resonance (LSPR), and spoof SPP excited in the low-frequency range.

4.2.1 SURFACE PLASMON POLARITONS

SPPs are electromagnetic waves excited at the interface between a dielectric material and a metallic material due to the coupling of the electromagnetic fields to electron plasma oscillations in the metallic material. Because electromagnetic waves have a limited penetration depth into the metallic material (e.g., Ag and Au penetrate less than 50 nanometers), only plasmons caused by surface electrons are significant and are commonly referred as surface plasmons. If a surface plasmon is further associated with an extended metallic surface, it is called propagating surface plasmons [3]. The corresponding physical properties are governed by the Maxwell's equations at the interface between a semi-infinite metallic material and a semi-infinite dielectric material [1], each is characterized by a complex permittivity as:

$$| \varepsilon_m = \varepsilon'_m + i\varepsilon''_m$$
$$| \varepsilon_d = \varepsilon'_d + i\varepsilon''_d \tag{4.1}$$

where ε' and ε'' denote the real and imaginary parts of permittivity ε. Figure 4.1(a) illustrates that the generated SPPs are propagating along the metallic/dielectric interface (x-direction) and decay rapidly in the dielectric medium (y-direction), yielding an alternating positive and negative charge along the metal surface [4].

The dispersion relationship (the relationship between frequency and wave vector) of SPP is given by [4],

$$k_{SPP} = k_0 \sqrt{\frac{\varepsilon_m \varepsilon_d}{\varepsilon_m + \varepsilon_d}} \tag{4.2}$$

where $k_0 = \omega/c = 2\pi/\lambda$ is the vacuum wavevector. where ε' and ε'' denote the real and imaginary parts of permittivity ε. Note that the square root term in Eq. (2) is usually positive and greater than 1, which means that the wave vectors of the SPP modes are larger than the vacuum light frequency. As shown in Figure 4.1(b), a free-space photon has always less momentum ($\hbar k_0$) than a plasmon ($\hbar k_{SPP}$), and the propagating light in the free space cannot be directly coupled to the SPPs mode [5]. In this regard, the excitation of SPP can only arise if there is a momentum match between the plasmon and the incident light.

To achieve the momentum match condition, one needs to provide additional momentum by the evanescent coupling (prism coupling) of light into the metallic material. Two different configurations are commonly adopted. One is the Otto method in Figure 4.2(a), in which the prism is separated from the metal film by a thin dielectric gap (typically is air). Total internal reflection takes place at the interface between prism and dielectric, and the evanescent photons tunnel through the

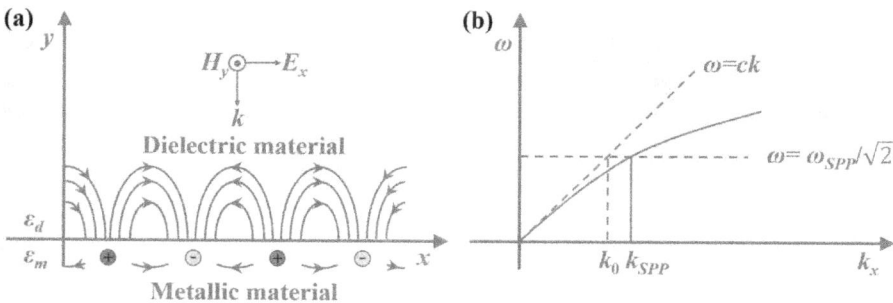

FIGURE 4.1 (a) Electric charge and electric field distribution at the interface of metal and dielectric. (b) Dispersion curve of SPPs.

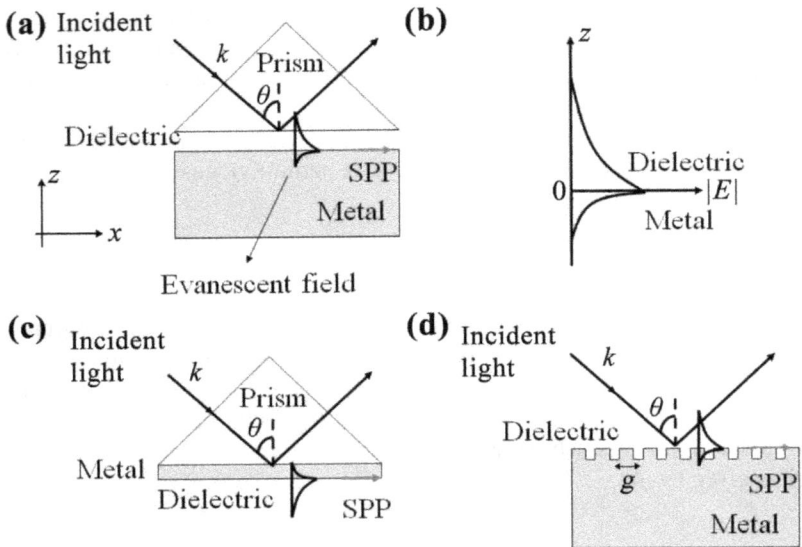

FIGURE 4.2 (a) Prism coupling of the Otto configuration. (b) Evanescent field at the interface. (c) Prism coupling of the Kretschmann configuration. (d) Phase-matching of incident light to SPP by grating coupling.

Source: Adopted from Zhang, D. et al. *Sensors.* 11, 5360-5382, 2011. No permission is required. Copyright 2011 MDPI.

dielectric/metal interface to excite SPPs. Figure 4.2(b) shows the evanescent field at the interface of dielectric and metal [6]. The other is the Kretschmann configuration in Figure 4.2(c), in which a thin metal film is evaporated on top of the prism, and the light is incident from the prism side in a direction greater than the critical angle of total internal reflection, and the evanescent photons tunnel through the metal film and excite SPPs at the metal-dielectric interface. For the prism coupling structure, the wave vector must satisfy Eq. (3) in the following,

$$k = k_0 \sqrt{\varepsilon_p} \sin \theta \qquad (4.3)$$

where the ε_p is the permittivity of the prism, θ is the incident angle. Clearly, the momentum of the original wavevector is enhanced, which makes it possible to match with k_{SPP}.

In addition to prism coupling, wave-vector mismatch can also be overcome by grating coupling. As shown in Figure 4.2(d), light propagates in the grating with a grating constant g, and the wavevector satisfies the following equation,

$$k = k_0 \sin \theta \pm \upsilon \frac{2\pi}{g}, \upsilon = \left(1, 2, 3, \cdots\right) \qquad (4.4)$$

Furthermore, SPPs can also be excited by waveguide coupling, charged particle impact, highly focused optical beam illumination and the near-field excitation. More details can be found in Ref. [1].

4.2.2 LOCALIZED SURFACE PLASMON RESONANCE

If the collective oscillation of free electrons does not occur at the interface but is confined to a certain volume, the resultant plasmon is called LSPs, where the electric field of incoming radiation can polarize the conduction electrons of metallic nanostructures [2]. For illustration, Figure 4.3 shows the interaction between the electric field of the incident light and the free electrons of a subwavelength metallic nanostructure [7]. The displacement of the electron cloud from the lattice creates a restoring force that attempts to pull the electrons back into the lattice. Therefore, the nanostructure behaves as an oscillator driven by the incident field and the recovered Coulomb force. When the light frequency matches the resonance frequency defined by the shape of the nanostructure, the LSPR occurs, and the electric near field of the nanostructure is significantly amplified. Remarkably, the LSPR does not require a momentum match and can be directly excited under arbitrary light incidence.

The properties of LSPR are closely related to the material, shape, and size of the nanostructure. Taking the metal nanoparticle as an example, its LSPR response can be divided into the absorption and scattering parts, and the sum of these two gives the total optical extinction. For a small particle, the absorption effect dominate and the incident light is mainly converted into heat due to the metal ohmic loss. As the particle size increases, so does the absorption and scattering, and the scattering increases much faster. For a sufficiently large nanoparticle, scattering replaces absorption as the main contributor, and the resonant position and width of LSPR also change, e.g., increasing the size of the Au sphere from 10 nm to 90 nm causes the formant to move from 400 nm to 800 nm. As discussed by Kreibig and Vollmer [8] and Bohren and Huffman [9], these changes can be explained by the polarization

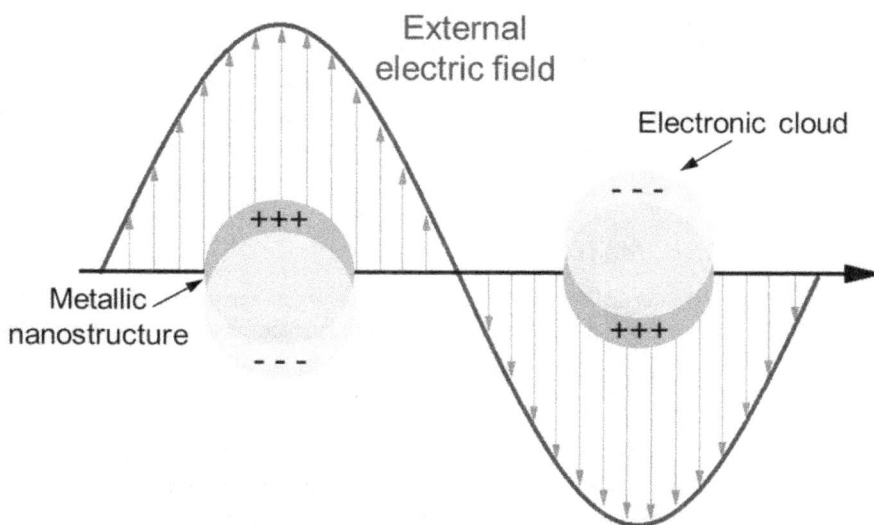

FIGURE 4.3 LSPRs of a metallic nanostructure.

Source: Reprinted with permission from Bozzola, A. et al. *Analyst.* 142, 883–898, 2017. Copyright 2017 Royal Society of Chemistry.

field induced by the surface charges that are affected by the amplitudes and relative phases of the scattering field and the incident field. Considering the optical dispersion of the plasmonic materials, the shift in the LSPR frequency also means that the resonances will experience varying degrees of damping.

4.2.3 SPOOF SURFACE PLASMON POLARITONS

At high frequency, surface plasmons can strongly confine electromagnetic fields on a length scale much smaller than the wavelength. At lower frequencies (such as microwave or THz frequencies), such field confinement is broken as the electromagnetic field can only penetrate a tiny depth into the metallic material due to its large permittivity that can be approximated as a perfect electric conductor. Since the inducible field amplitude inside the metallic material is essential to provide a non-zero component of the electric field parallel to the surface, which is necessary to establish an oscillatory space-charge distribution, the SPP vanishes for an ideal electrical conductor. Consequently, as the frequency decreases, the SPP gradually evolves into a grazing-incidence light field rather than the plasmon mode.

However, by fabricating an array of subwavelength holes or grating structures on the surface of plasmonic material, it is possible to create a layer on the surface that passes the evanescent field to enable effective field penetration. Pendry and co-workers have demonstrated that bound electromagnetic surface waves mimicking SPPs can be maintained even by a perfect electric conductor if its surface is periodically corrugated [10], which is named spoof SPPs and later confirmed experimentally by Hibbins *et al.* [11]. The discovery of spoof SPPs enables the excellent SPP-like characteristics in low-frequency ranges, resulting in a plethora of emerging applications [12]. Note that, materials with subwavelength structures and exhibiting such exotic photonic responses are also known as metamaterials or metasurfaces, which are discussed in Chapter 5.

Other studies have also revealed that localized plasmons can be excited at THz frequencies. For example, micrometer-sized silicon (Si) particles support dipole plasmonic resonances similar to Fröhlich modes [13]. Localized modes have also been observed in randomly distributed metallic particle clusters in the context of enhanced THz radiative transfer [14].

4.3 PLASMONIC MATERIALS

The fundamental material properties are important in determining the excitation and spectral feature of plasmon mode. Because the surface plasmon stems from the collective oscillation of the free charges, it features a negative real part permittivity. Metals are preferred plasmonic materials because of their large numbers of free electrons, large plasma frequencies and high electrical conductivities. However, they suffer from non-negligible losses, especially in the visible and UV spectrum, which are detrimental to the performance of plasmonic devices and severely limit the feasibilities of many applications [15,16]. Therefore, plasmonic materials with lower losses are developed, including metal alloys, metal compounds, semiconductors and even 2D materials, among others [17]. In this section, we will review these potential candidates.

4.3.1 Conventional Noble Metals

Noble metals are those with strong oxidation resistances. Among them, Ag and Au are the typical materials in plasmonic applications, because they have relatively low losses in the visible and near-IR ranges. Besides, many experimental studies prefer Ag or Au due to its excellent chemical stability and availability of various conjugation chemistries [17]. For example, Ag has been used to demonstrate superlenses, hyperlenses, negative refractive index materials, and extraordinary optical transmission. Au has been used to demonstrate negative refractive index materials for the first time in the near-IR spectrum, surface-enhanced Raman scattering (SERS), plasmonic waveguides, and LSPR sensors.

Ag has the lowest losses in the visible and near-IR range. However, in terms of fabrication, Ag degrades relatively quickly, with a thickness threshold of around 12 nm–23 nm for uniform continuous films, which makes Ag less suitable for use in transformation optics devices. Furthermore, the loss of Ag is strongly dependent on the surface roughness [18]. On the contrary, Au is chemically stable and can form continuous films even at thicknesses of 1.5 nm–7 nm [17,19,20]. In principle, Ag and Au thin films can be prepared by various physical vapor deposition techniques, and nanoparticles and metal-coated nanoparticles can be synthesized by liquid-phase chemical methods.

Although other noble metals have been used for plasmonic materials, they are not the primary options because of their high losses. For instance, platinum (Pt) and palladium (Pd) have been used as plasmonic catalysis systems [21,22]. In another study, Pt single crystalline surfaces have been used as a SERS substrate to obtain an enhancement of 10^5 [23].

4.3.2 Nonnoble Metals

The practical application of noble metal-based nanostructures is hindered by their high prices, significant optical losses, and limited LSPR wavelength ranges. In recent years, numerous alternative plasmonic materials are discovered, such as nonnoble metals, metal nitrides, metal chalcogenides, and graphene [24-27]. Compared with noble metal-based plasmonic nanostructures, their nonnoble analogs are relatively inexpensive and exhibit comparable or even better performance in specific applications. Though the development of nonnoble metal SPR is still in infancy, these materials show some exciting catalytic, magnetic, and optical features. Hereafter, we will introduce some typical nonnoble metal plasmonic materials, including copper (Cu), aluminum (Al), magnesium (Mg), indium (In), gallium (Ga), lead (Pb), nickel (Ni), cobalt (Co), and iron (Fe).

Cu nanostructures have received increasing attention due to their excellent optical, catalytic, and electronic properties. Because Cu is the second most conductive metal (after Ag), it is expected to exhibit good plasmonic properties. In fact, in the 600 nm – 750 nm range, Cu's ε'' is equivalent to that of Au. Considering the cost, Cu would be a good candidate to replace Ag and Au if its plasmonic performance is acceptable. Chan et al. demonstrated that oxide-free Cu nanospheres displayed sharper and narrower LSPR peaks comparable to Ag and Au [28]. Nevertheless, the fabrication of Cu-based structure is still challenging, due to the difficulty of Cu reduction in aqueous solution and the high sensitivity to oxidation. The current synthesis strategy of Cu nanostructure is mainly by polyol-based and micelle-based processes as well as thermal decomposition, which cannot provide as many structure morphologies as Au or Ag. There is still room for further improvement.

Al is the third most abundant element on earth, and its low cost and complementary metal-oxide-semiconductor (CMOS) compatibility make it ideal for large-scale synthesis and industrial device applications [29,30]. Al nanostructures generally support SPR in the UV range due to their high free electron densities, while Au/Ag-based nanostructures suffer from inherent limitations in this region owing to their interband transition thresholds. Al is easily oxidized in atmospheric conditions, and an aluminum oxide (Al_2O_3) layer of 2.5 nm–3 nm is naturally formed, resulting in a slight redshift of the LSPR peak. The presence of the surface alumina layer prohibits further oxidation of Al and maintains a high surface sensitivity. It has been demonstrated that large-area flexible passivated Al nanopillar metasurfaces in the visible light range, which has better biosensing performance than the Au counterparts [31]. So far, various Al plasmonic systems in the UV-blue spectral region are reported [17], such as LSPR, SPP propagation, surface-enhanced fluorescence, Raman spectroscopy, and biosensing.

Mg has favorable physicochemical properties, including low density, high hydrogen storage capacity (up to 7.6 wt.%), and availability (the seventh most abundant element). Moreover, Mg has been theoretically predicted to be a promising UV plasmonic material with even better far-field absorption and near-field enhancement than Al. Despite a few studies reported Mg as core materials for active plasmonics [32], chiral sensing [33] and color display [34], the application of Mg-based plasmonic nanostructures are not fully exploited, which demands further investigation.

In is widely used in optoelectronic devices, and In nanoparticles are promising candidates for short-wavelength plasmonic materials [35]. Schatz and coworkers showed that an In-based dimer configuration could yield an optimal SERS enhancement factor of 1.2×10^9 in the near-UV region [36]. Magnan and co-workers achieved plasmonic enhanced fluorescence in the UV region using In-Si core-shell nanoparticles, with a twofold to sevenfold enhancement factor [37]. In addition, the plasmonic properties of In nanoparticles can also enhance the performance of optoelectronic devices [38]. The photocurrent of a blue light photodetector using a single-layer graphene/single ZnSe nanoribbon Schottky junction (SLG/ZnSeNR) is greatly increased ($\approx$ 20 folds) after In nanoparticle array modification. This enhancement is attributed to the energy matching between the LSPR of the In nanoparticle array and the bandgap energy of the semiconductor nanoribbon, enabling incident light trapping and enhancing the electric field near the ZnSeNR. The low-cost and earth-abundant In nanoparticles can provide new opportunities for photocatalysts, photovoltaics, and optoelectronic applications.

Ga is an emerging plasmonic material with unique chemical and physical properties. Ga is in the same family as Al and has a high free-electron density that enables LSPR in the UV range [39,40]. SERS and fluorescence enhancement under UV laser (325 nm) have been demonstrated [40]. Everitt et al. investigated the fundamental optical properties of Ga nanostructures obtained by molecular beam epitaxy deposition (MBE), they found that the dielectric function of Ga is phase-dependent [40-42]. Furthermore, an inert self-terminating oxide layer (2 nm – 3 nm) has been formed on the surface of the synthesized Ga nanoparticles, protecting the metal core and endowing it with long-term (several months) stability. However, due to the low melting point (303 K) of Ga, the shape of the nanostructure is easily changed under ambient conditions, which might be detrimental to Ga plasmonic applications. This can be mitigated by alloying with other plasmonic metals.

Pb finds important applications due to its high-temperature superconductivity (transition temperature = 7.2 K). Although the plasmon dispersion of Pb nanowire arrays on Si has been investigated using electron energy loss spectroscopy (EELS) [43], the synthesis of Pb-based monodisperse nanoparticle generally demands strict controlled or ultrahigh vacuum environment to prevent surface oxidation. It also requires the use of strong reducing agents and anhydrous conditions during synthesis. In 2010, monodisperse colloidal Pb particles with sizes ranging from 4 nm to 20 nm have been synthesized by a bottom-up approach [44]. Meissner effect has been successfully observed, reflecting the superconducting properties of Pb nanoparticles. Recently, the growth mechanism of micron-scale Pb nanoparticles in solution has been investigated in real-time [45], and in situ electron microscopy imaging showed the sequential formation of nucleation, Ostwald ripening, and rapid aggregate growth.

The fundamental magnetic and optical properties of pure Ni nanostructures have been reported in 2011 [46]. Ni nanodiscs exhibit ferromagnetic responses with magneto-optical Kerr effect hysteresis loops, which allow the amplitude and phase of light to be controlled by an external magnetic field. A clear dipole plasmonic mode of Ni nanodisc has been observed by the near-field imaging, while a similar size dependent spectrum to that of noble metals has been confirmed by the far-field spectroscopy. Interestingly, as the size increases, the spectral redshift of Ni nanodiscs is significantly larger than that of Au counterpart. The high sensitivity of Ni nanoantenna to ambient refractive index changes has been used to fabricate ultrasensitive biosensors. The outstanding figure of merit (FOM, > 100 RIU^{-1}) is among the highest values of plasmon-based sensors [47].

Co produces a pronounced magnetic plasmon structure due to its greater damping. One strategy to circumvent the high ohmic loss and enhance the plasmonic properties is to form hybrid structures with noble metals. For example, Wang et al. demonstrated the enhanced magneto-optical Faraday rotation of Co-Ag core-shell nanoparticles [48]. The increased Co core diameter results in a blue-shifted Faraday rotation peak and absorption band. In addition, the large local electromagnetic field induced by the thicker Ag shell leads to stronger molybdenum activity enhancement. In addition, compared with the hollow Co nanotube arrays, the Co coated Au nanowire arrays also showed a largely improved molybdenum activity due to plasmonic enhancement [49]. It also shows that the remanent magnetization of ferromagnetic Co can be used to control the optical properties in the absence of an external magnetic field.

Fe has the highest saturation magnetization at room temperature (218 Am^2kg^{-1}). Fe-based super-paramagnetic nanomaterials have a wide range of applications in biosensors, drug delivery, magnetothermal therapy, and magnetic resonance contrast agents [50]. Due to the high reactivity, Fe nanoparticles are completely oxidized under ambient conditions, reducing the magnetic susceptibility [51]. Therefore, protective layers such as noble metal shells are introduced to enhance the stability and obtain higher magnetic responses in practical applications [51,52]. For example, monodisperse Fe-Ag core-shell nanoparticles have been synthesized in a non-aqueous solution by a two-step reduction process [53]. It is found that a larger Fe core leads to a better magnetic response because the reduced exchange coupling of Ag shell makes the critical dimension of super-paramagnetic smaller. In addition, another group synthesized Fe-Ag core-shell nanoparticles by sequential reduction using sodium borohydride in an aqueous solution [54], which exhibited super-paramagnetic hysteresis loops and absorption bands at 3 eV due to LSPR excitation. Overall, the introduction of magnetic elements enables further multifunctionality of plasmonic nanostructures.

In summary, plasmonic materials based on nonnoble metals significantly broaden the functional spectral range and allow to address many needs and challenges, including costs and bandwidths, that cannot be met by noble metals. Particularly, nonnoble metals such as Mg, Al, Ga, and In are suitable for UV plasmons which cannot be achieved by Ag or Au. However, low chemical stability is a major obstacle since the native oxide layers can easily form under ambient conditions, resulting in the broadening/disappearance of LSPR peaks and thus the low-quality factors (Q-factor) and weak field enhancements. Protective layers such as polymers and alloying can also be used to prevent nonnoble metals from oxidation [55]. Another challenge is size control, especially for the small UV plasmonic nanoparticles with quasi-static domain dimensions. For sizes greater than or equivalent to the resonant wavelength, LSPR linewidth broadening and redshift reduce the FOM [56,57]. There is an urgent need to develop facile routes to synthesize UV plasmonic nanoparticles with high-Q factors and various shapes and sizes for future applications.

On top of that, alloy nanostructures that benefit from synergistic effects between different metal materials offer more opportunities and better flexibilities to control plasmonic properties. Via varying the material composition, the permittivity of the alloy nanostructures can be tuned to engineer the plasmonic properties [58]. Furthermore, one can synthesize low-cost multifunctional plasmonic alloy nanoparticles with extra catalytic activities or magnetic properties [59,60]. Nevertheless, for noble and nonnoble metal materials, their plasmonic resonances are mainly within the UV-NIR region, and they behave almost like perfect electrical conductors in the long-wavelength region, hence cannot efficiently confine the electromagnetic field [61]. It is still challenging to extend the metal LSPR to the mid-IR (3 μm - 30 μm) range where most molecular vibrational and rotational absorption energies fall in [62].

4.3.3 SEMICONDUCTORS

Semiconductors are generally considered as dielectric materials with frequencies higher than hundreds of THz. However, in some cases, semiconductors can exhibit negative permittivity (real part) in this spectral region [17]. Plasmons based on semiconductors are promising due to the ease of fabrication and flexible tunability. To achieve low-loss plasmon, both the band gap and the plasmon frequency of the semiconductor should be larger than the target frequency range, which can be obtained via wide-bandgap, heavily doped semiconductors with high carrier mobilities.

Si is the most popular semiconductor material used in the current CMOS technology, offering unprecedented capabilities in mass production and ultrahigh integration capabilities. To this end, many technologies based on Si platforms, such as microelectromechanical systems and photonics, have been developed. Si photonics has moved toward commercialization [63], and the transition from photonics to plasmons can provide significant benefits in integration. Si can be doped with N-type group V elements such as phosphorus (P), arsenic (As), and stibium (Sb), or P-type group III elements such as boron (B), Al, and Ga. Heavily doped Si can have the properties of metals, which realize the Si plasmons [64-66].

Germanium (Ge) is another standard semiconductor commonly used in electronics along with Si platforms. Ge is attractive due to its higher electron mobility and smaller optical band gap than Si, which enables a strong optical response at telecommunication frequencies [25]. Compared to Si, Ge has a larger background permittivity, which requires higher doping to achieve metallic properties. At lower frequencies, Ge is transparent and can be highly doped to produce low loss plasmonic properties. However, at telecommunication frequencies, it is rather difficult to heavily dope Ge because the solid solution limit of dopants in Ge is much smaller than Si. In addition, large absorption is presented in Ge due to the intrinsic interband transitions. Currently, many Si and Ge compounds such as $Si_{1-x-y}Ge_xSn_y$ are being investigated for the possibility of plasmonic behavior at telecommunication wavelengths [27].

In recent decades, III–V semiconductors have provided alternative material platforms for many emerging technologies [25], such as high-speed switches, power electronics, and optoelectronics. These materials exhibit wide tunability in the optical bandgap, which can be controlled by changing the composition of their ternary and quaternary compounds. Since the integration of plasmonic devices on optoelectronic platforms will be a very important direction, investigating the plasmonic properties of these materials is a highly relevant step.

Group III-V semiconductor compounds, such as GaAs and InP, have intrinsic optical band gaps in the NIR region with comparable background permittivities to Si [67,68]. Furthermore, due to the small effective mass, the electron mobility in these materials is very high [69,70], which relaxes the carrier concentration requirement for achieving plasmonic properties. Typical carrier concentrations over 10 cm^3 - 20 cm^3 are required to observe plasmonic properties of these materials. P-type GaAs doped with beryllium or carbon can fulfill such carrier concentration [71]. However, holes have high effective mass and poor carrier mobility, which raises the minimum limit of carrier concentration to become plasmons. Furthermore, N-type InAs can have higher doping levels than GaAs. Law et $al.$ have reported a carrier concentration of about 7.5×10 cm^3 in InAs, exhibiting plasmonic behavior at wavelengths greater than 6 μm [72]. Further increasing the carrier concentration is proven to be difficult, so plasmonic applications of these materials are generally limited to the MIR range.

GaN is an emerging optoelectronic platform operating across the entire visible light range [73,74]. The large tunability of the band gap in the InGaN ternary system has been exploited for visible light optoelectronic applications. GaN is a wide bandgap semiconductor with a direct bandgap of about 3.3 eV. The wide band gap reduces its background permittivity to a moderate value, thus reducing the minimum carrier concentration required to obtain metallic properties. The carrier effective mass of GaN is larger than that of GaAs or InP [75], yielding a higher Drude loss. Nevertheless, this drawback can be easily compensated by a much higher N-type doping in GaN than GaAs [76]. Hageman et $al.$ demonstrated an ultrahigh N-type doping of Ge in GaN at about 3×10^{21} cm^3 [77]. With slightly higher doping, GaN can become metallic at telecommunication wavelengths, which holds promise as a low-loss plasmonic material in the near-IR spectrum.

4.3.4 Two-Dimensional Materials

2D materials, such as graphene, are unique plasmonic materials because of their dynamically tunable optical properties by electrical, chemical, electrochemical, and other means, triggering many paradigm-shift technologies in electronics and photonics as shown in Figure 4.4 [78-81]. As a typical example, graphene is a 2D system capable of exciting surface plasmons, similar to metal/dielectric interfaces. The optical properties of 2D materials are quite different from those of 3D bulk counterparts, which leads to completely different dispersion relations of 2D plasmons in graphene [82,83]. Note that graphene

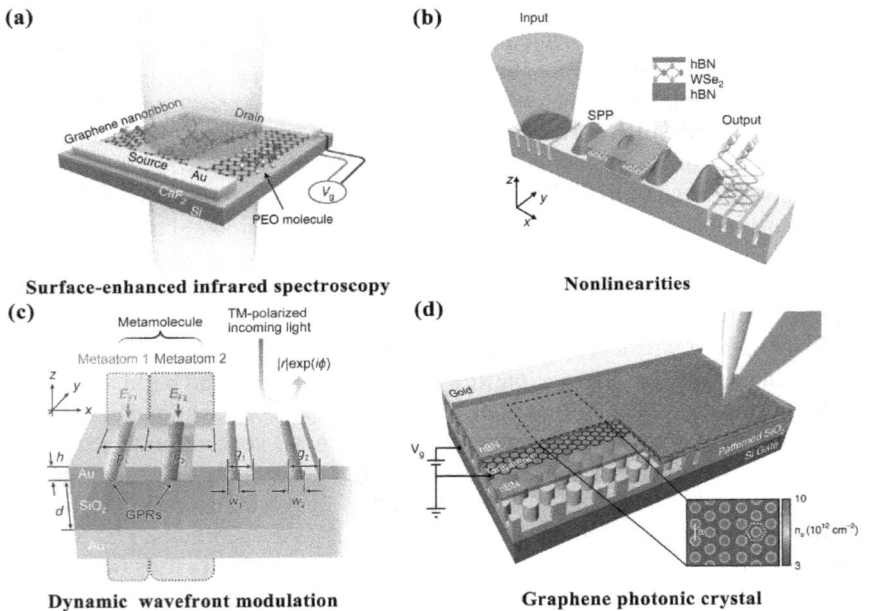

FIGURE 4.4 (a) Far-field nanoscale infrared spectroscopy of vibrational fingerprints of molecules with graphene plasmons.

Source: Adopted from Hu, H. et al. *Nature communications.* 7, 1–8, 2016. No permission is required. Copyright 2016 The authors.

(b) 2D semiconductor nonlinear plasmonic modulators.

Source: Adopted from Klein, M. et al. *Nature communications.* 10, 1–7, 2019. No permission is required. No permission is required. Copyright 2019 The authors.

(c) Complete Complex Amplitude Modulation with Electronically Tunable Graphene Plasmonic Metamolecules.

Source: Reprinted with permission from Han, S. et al. *ACS nano.* 14, 1166–1175, 2020. Copyright 2020 American Chemical Society.

(d) Photonic crystal for graphene plasmons.

Source: Adopted from Xiong, L. et al. *Nature communications.* 10, 1–6, 2019. No permission is required. Copyright 2019 The authors.

plasmons are only observed at low temperatures due to the high losses (high carrier mobilities) at room temperature. Apart from graphene, plasmons have been observed in other 2D materials such as molybdenum disulfide and some materials grown as 2D flakes, or more generally, in 2D electron gas systems such as semiconductor inversion layers [84], semiconductor heterostructures [85,86], and polar interfaces of oxides [87]. However, all these 2D plasmons are found in the MIR or longer wavelength range. Plasmons in the visible or near-IR range have not been reported in any of these 2D systems due to the insufficient carrier density. In addition, the synthesis of large-area single-crystal 2D materials presents another challenge that limits their practical applications [88,89].

4.4 FABRICATION METHODS

The shape and size of the plasmonic structure determine the intensity and frequency of plasmons [3]. Therefore, proper synthesis and manufacturing methods are the keys to realizing high performance plasmonic devices. In principle, an ideal fabrication method must meet the requirements of high resolution, productivity, reproducibility, and cost-efficiency. Currently, a variety of feasible fabrication methods have been developed, which can be roughly divided into top-down and bottom-up categories. For the top-down process, the most common method is lithography, which allows high-throughput reproduction of structures with well controlled shapes and sizes. Lithography uses exposure and development to depict geometric patterns on the photoresist, and then through etching and lift-off processes, the micro/nano patterns are transferred from the photomask to the substrate. The resolution of lithography is determined by the diffraction limit of the optical or electron sources used in the system. For the bottom-up approach, small assemblies of atomic or molecular dimensions self-assemble together (naturally or under external driving force) to generate larger and more ordered systems, empowering cost-efficient and large-scale fabrication. In this section, we will briefly review these techniques, including electron beam lithography (EBL), focused ion beam (FIB) lithography, nanoimprinting lithography (NIL), laser lithography, self-assembly technique and, chemical synthesis.

4.4.1 Electron Beam Lithography

EBL is a direct writing method that does not require a mask. It uses an electron beam to create patterns on the surface of materials, which is an extension of conventional photolithography. According to De Broglie's theory, electrons are waves with extremely short wavelengths, granting EBL with excellent flexibility and high resolution (about 10 nm) [90]. Hence, EBL is widely used to fabricate small and complex nanostructures even on unconventional substrates, such as the tips of optical fibers [91,92]. During the fabrication process, EBL first uses an improved scanning electron microscope to scan focused electron beams to write the desired pattern on electron resist. The electron beam changes the solubility of the resist, and by immersing it in a solvent (developing), the exposed areas (for positive resist) or unexposed areas (for negative resist) can be selectively removed as shown in Figure 4.5 [93]. Lastly, the formed nano/micro structures can be transferred to the substrate by etching. The old EBL systems typically use Gaussian-shaped beams and scan in a raster fashion, while the modern systems operate with shaped beams which can be deflected to various positions in the writing field (vector scan mode). Nevertheless, it generally takes a long time for EBL to realize the full structure pattern, because the electron beam can only focus on a single point of the pattern at a time. This drawback together with the high operating cost limits its commercial use for high throughput, large scale and large area device fabrication [94].

4.4.2 Focused Ion Beam Lithography

FIB was first discovered during the research of liquid metal ion sources in the early 1970s [95,96]. Later, the potential of this technology as a maskless manufacturing technology was realized [97]. Usually, FIB is combined with the scanning electron microscope (SEM) to ensure that vision and writing are conducted simultaneously. Unlike EBL, FIB is based on high energy focused ion beams that can directly write nanoscale patterns on various materials including metals without the use of resists, as shown in Figure 4.5, avoiding the pattern transfer step and significantly reducing the processing time. In addition, heavy ions (He^+ or Be^+ or Ga^+) are used in FIB to suppress scattering and ensure a stable and fine beam spot. The exposure sensitivity of FIB is higher than that of EBL [98], and the energy involved is usually around 100 to 200 kV [99]. Similar to EBL, the FIB method can also be used to make high resolution metasurfaces even at fiber tips [100]. The throughput and large area manufacturing of FIB is also a challenge for commercial production. Notably, FIB allows 3D structures to be manufactured in a single process run.

4.4.3 Nanoimprint Lithography

In order to solve the long processing time problem of EBL and FIB, NIL arises to establish a good compromise between resolution and cost in large-area patterning of nanostructures. NIL is a replication technique that utilizes conformal contact between a mold and a photoresist-coated substrate. The spatial resolution of NIL is determined by the mold pattern size, which is made via high-resolution tools such

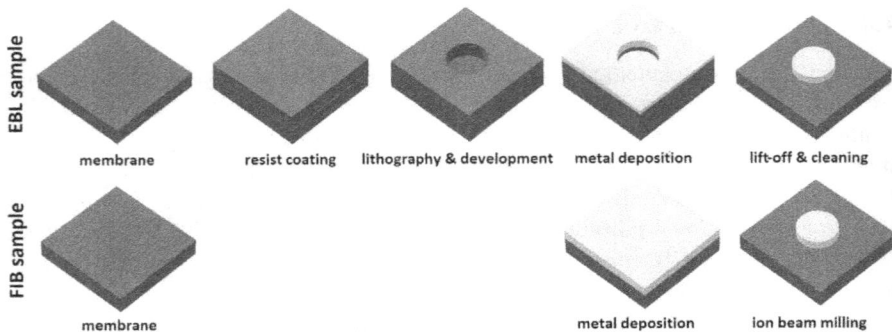

FIGURE 4.5   Schematic overview of the individual steps in the fabrication process of EBL and FIB.

Source: Adopted from Horák, M. et al. *Scientific Reports.* 8, 1–8, 2018. No permission is required Copyright 2018 The authors.

as EBL and FIB. In addition, imprint molds are usually made of inorganic materials with high Young's modulus, such as Si and quartz, to achieve high replication fidelity in conformal contact. Depending on the resist curing type, there are two NIL procedures: thermal NIL and UV-NIL [101]. The NIL process is mainly composed of two steps: (1) forming. Thermoplastic resist or photoresist flows into the cavity of the pattern imprinted mold. The resist is then cured by heat or UV; and (2) demolding. After the resist is cured, the mold is removed, leaving a nano pattern on the substrate. For illustration, Figure 4.6 shows the schematic drawings for the fabrications of the two complementary plasmonic nanopillar (PNP) and plasmonic nanoholes (PNHs) metasurfaces (Figure 4.6(a)), as well as the top-view SEM images for PNP metasurface (Figure 4.6(b)) and PNHs metasurface (Figure 4.6(c)) [102]. For most NIL processes, a thin residual layer remains, and an etching process is required to remove it. Lee and Jung have developed a residue-free NIL process that uses a benzyl methacrylate monomer solution as the resist to avoid the additional etching process [103].

4.4.4　Laser Lithography

Laser lithography is also a maskless technique that is suitable the for large-scale production of high-resolution plasmonic nanostructures. There are two cost-effective laser lithography methods: laser direct-write (LDW) lithography and laser interferometric lithography (LIL). LDW uses computer-controlled optics to achieve rapid maskless

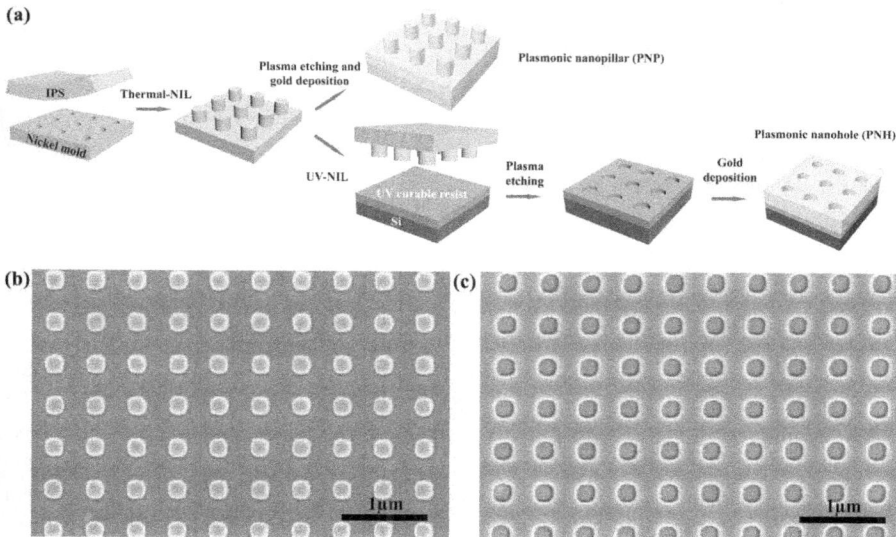

FIGURE 4.6 (a) The schematic drawing for the fabrication of the two complementary PNP and PNH metasurfaces. (b) and (c) Top-view SEM images for PNP metasurface and PNH metasurface, respectively.

Source: Adopted from Li, F. et al. *Biosensors and Bioelectronics.* 114038, 2022. No permission is required. Copyright 2022 Elsevier.

fabrication under non-vacuum conditions using continuous or pulsed lasers. LIL uses the interference and diffraction of light to regulate the distribution of light intensity through specific beam combinations, and the desired nano-pattern with different sizes and shapes can be easily obtained and directly projected onto the photoresist. Laser lithography allows working materials that are hard to be machined and generate 3D surface patterns by varying the energies of the employed laser. It does not require expensive projection optics, and the area of the exposure field is limited only by the light-transmitting aperture, and the resolution is limited only by the wavelength of light [104]. However, not all shapes can be patterned, and the minimum period of nanostructures is limited to half of the wavelength of light. Smaller features require the use of deep UV light [105], which makes LIL very expensive.

4.4.5   SELF-ASSEMBLY TECHNIQUE

Compared to these top-down lithograph techniques, bottom-up strategies such as the self-assembly technique are known for their low cost and high throughput advantages. Self-assembly is a technology in which basic structural units (molecules, nanomaterials, microns, or larger-scale substances) spontaneously form a large area ordered structure, including chains, sheets, and 3D structures [106]. The principle of this technique is to reduce the free energy required for each component to reach a local balance, which can be controlled from the outside using directional fields. During the self-assembly process, the basic structural units spontaneously organize or aggregate into a stable structure with a regular geometry under the interaction based on non-covalent bonds, thereby forming an etching template. In direct self-assembly technology, external fields and templates are used to support selective grouping and spatial sequence of self-assembly layers. This technique can obtain well-defined plasmonic nanoparticles in terms of shape and crystallinity, thereby ensuring the high optical quality of the individual nanoparticle. Zhu *et al.* reported a precisely controlled metallic nanomesh fabricated by this technique with excellent uniformity with hexagonally arrayed periodic circular holes, as shown in Figure 4.7 [107].

4.4.6   CHEMICAL SYNTHESIS

Direct chemical synthesis may be the most important bottom-up technology to grow various metal nanoparticle shapes, including spheres, stars, rods, boxes, and cages, because reaction conditions such as temperature, surfactants, and precursors can be independently controlled [108]. The diversified particle shapes and sizes provide flexible tunability on plasmonic resonances. Figure 4.8(a) shows the SEM image of the pentacle Au–Cu alloy nanocrystals obtained by chemical synthesis. The inset shows a SEM image of a typical individual nanocrystal. Figure 4.8(b) demonstrates the extinction spectra of aqueous suspensions of Au–Cu alloy nanocrystals with different shapes, revealing a well-controlled size dependent LSPR [109]. Furthermore, complex particle morphologies such as core-shell structures can also be obtained using chemical synthesis. Taking the Au-SiO$_2$ nanoshell as an example, the structure of the nanoshell particle is to first attach a small Au nanoparticle to the surface of the SiO$_2$ nanoparticle. Subsequently, the chemically adsorbed Au nanoparticles are used as nucleation points

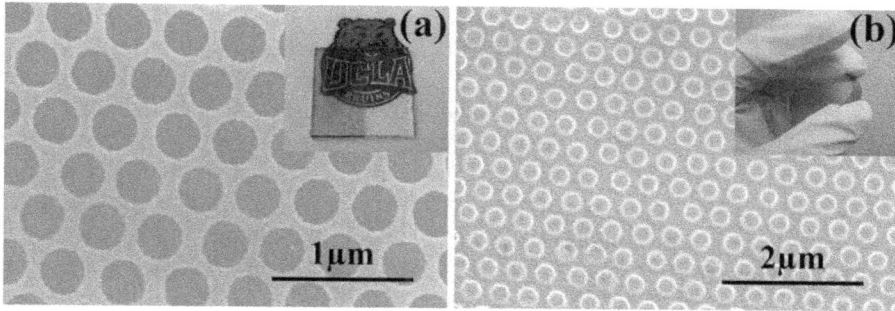

FIGURE 4.7 (a) SEM image of the Au nanomesh on a glass substrate. The inset is a photograph of the 1.5 × 1.5 cm² glass substrate with Au nanomesh on the left half. (b) is SEM image of the Au nanomesh on a polyethylene terephthalate substrate. The inset shows the photograph of the polyethylene terephthalate substrate with Au nanomesh.

Source: Adopted from Zhu, J. et al. *Applied Physics Letters.* 100, 143109, 2012. No permission is required. Copyright 2012 American Institute of Physics.

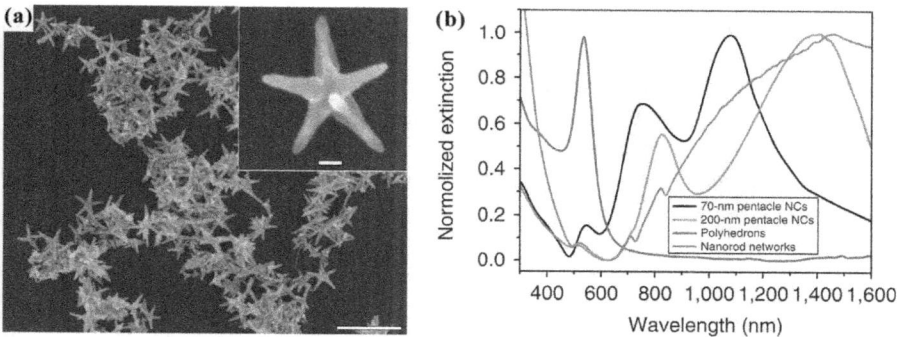

FIGURE 4.8 (a) SEM image of the pentacle Au-Cu alloy nanocrystals obtained by chemical synthesis. The scale bar of (a) is 500 nm and the scale bar of the inset is 50 nm. (b) extinction spectra of aqueous suspensions of Au-Cu alloy nanocrystals with different shapes.

Source: Adopted from He, R. et al. *Nature communications.* 5, 1–10, 2014. No permission is required. Copyright 2014 the authors.

to reduce the Au from the solution to the nucleus. Several other methods of synthesizing core-shell nanostructures have also been reported, including the growth of metal shells on core materials [110], chemical changes in reducing agents [111], and the synthesis of hollow crystalline shells by templating on block copolymers [112].

4.5 APPLICATIONS OF PLASMONIC MATERIALS

In past decades, photonic crystal with well-defined periodic structure is invented to tailor the electromagnetic wave propagation by creating an artificial band structure in the desired frequency range [113]. Plasmon-based metamaterials and metasurfaces share the same concept of photonic crystal, exploiting periodic lattice of metallic nanostructures to control the properties of electromagnetic wave. The major difference between plasmonic structure and photonic crystal is that subwavelength unit cells and periodicities are used in plasmonic technologies due to the strong optical confinement of surface plasmon, whereas unit cell sizes and periodicities on the wavelength scale are employed for the photonic crystal. As a result, the notorious diffraction limit is naturally overcome in the plasmonic metamaterial and metasurface, which is of paramount importance for high-resolution imaging and spectroscopy. In addition, the ability of plasmonic micro-nanostructures to confine and enhance the light field makes it useful in refractive index (RI) sensing, biosensing, surface-enhanced IR spectroscopy (SEIRS), SERS, nonlinear enhancement, absorbers, and many other fields, which will be briefly reviewed in this section.

4.5.1 REFRACTIVE INDEX SENSING

Because the plasmon resonant frequency is extremely sensitive to changes in the surrounding refractive index, it is ideally suited for ultra-sensitive label-free RI sensing, avoiding the expensive, time-consuming, and interfering labeling step to analytes. RI sensing is often performed by detecting the shift of resonant frequency, phase-shift, and the change of resonant intensity. In RI sensing, there are several important performance indicators, namely Q-factor, full width at half maximum (FWHM), RI sensitivity (S_{RI}), and FOM [114,115]. Among them, Q-factor reflects the resonance characteristics, the sharper the resonance, the larger the Q-factor and the higher the sensitivity. Larger Q-factor also means the higher resolution of the sensor. The Q factor is defined as,

$$Q = \frac{\lambda_0}{\text{FWHM}} \tag{4.5}$$

where λ_0 is the resonant wavelength. Next, the sensitivity S_{RI} is defined as,

$$S_{RI} = \frac{\Delta\lambda}{\Delta n} \tag{4.6}$$

where $\Delta\lambda$ is the change in the resonance wavelength and Δn is the change in refractive index. In addition to S_{RI}, the detection limit of sensor is also strongly restricted by the spectral width of the resonance. Therefore, FOM is proposed as a more reasonable parameter to evaluate the performance of a sensor, which is defined as,

$$\text{FOM} = \frac{S_{RI}}{\text{FWHM}} \tag{4.7}$$

Obviously, FOM can be improved by increasing S_{RI} or reducing FWHM. One way to reduce FWHM is to arrange metal nanoparticles into one-dimensional or 2D arrays [116,117]. Diffraction coupling between periodically arranged metal nanoparticles has been shown to produce lattice plasmonic resonances with FWHM below 10 nm [118-120]. Furthermore, by coupling it to the photonic microstructure, all-round improvement in FWHM, intensity, phase and sensing capabilities can be achieved [121]. Other than that, mode coupling provides another effective way to reduce FWHM. For example, Fano resonance based on the interference between a discrete state and a continuous state can yield a narrow resonance linewidth. Split ring resonator-bar structures, lifted cross-bar structures, closely packed nanodisc clusters and Ag nanocubes have been shown to possess high FOM values because of Fano resonances [47]. Liu *et al.* engineered periodic plasmonic structures with minimized feature sizes to suppress the radiation loss, as shown in Figure 4.9(a) [122]. As a result, a narrow line width is achieved in the wide wavelength range from 600 nm to 960 nm, with a minimum FWHM of 3 nm at 960 nm. A 3.5 nm shift in the resonance wavelength between distilled water (RI = 1.3330) and 0.5% glycerol (RI = 1.3336) was observed and a record FOM value of 730 was obtained, as shown in Figure 4.9(b). Such a sensor can detect a tiny RI change induced by the 10^{-10} Mol ultralow concentration of bovine serum albumin.

RI sensing is widely used in various fields, including qualitative and quantitative analysis of gaseous, liquid, and solid analytes. Chen *et al.* developed a plasmon nanoscale cylinder based on Au nanohole arrays to measure the nanoscale volume change, as shown in Figure 4.9(c) [123]. This study theoretically estimated the measurement function of plasmonic nanostructure and experimentally demonstrated the measurement sensitivity in the case of polystyrene nanospheres and paraffin wax, providing a promising method for non-destructive, low-cost, and rapid measurement of material volume changes at the nanoscale.

The high loss of the plasmonic structure due to the intrinsic absorption of the metallic material is a serious disadvantage for practical applications. To overcome this bottleneck while exploiting its high-field enhancement, hybrid metal-dielectric nanostructures that combine the advantages of plasmonic material and low-loss dielectric materials are emerging as a promising approach. Ray *et al.* proposed a hybrid nanoantenna made of Al, SiO_2 and Si in a layered stack (sandwich) geometry for bulk RI sensing with a sensitivity of 208 nm/RIU, as shown in Figures 4.9(d) and (e).

4.5.2 LABEL-FREE BIOSENSING

There is a special kind of application in plasmonic RI sensing, that is, label-free biosensing [124]. Plasmonic biosensing has a wide range of applications and can be used to detect a variety of biological analytes such as proteins, cells, and pathogens. Biological analytes are typically small compared to incident wavelengths or plasmonic structures, so more sophisticated ways to detect them are needed. In addition to the fabrication of plasmonic structures with high sensitivity, a series of bio-functionalization operations are often required in biosensing, enabling plasmonic structures to capture specific biomolecules for label-free specific recognition. The efficient ligand-receptor binding enables high sensitivity and specificity.

Carcinoembryonic antigen (CEA) is one of the most widely used bio-markers for cancer in the clinical analysis [125], and the detection of CEA has important implications for the cancer diagnosis and prognosis, especially in the early stage of the disease. Chemiluminescent immunoassay and enzyme-linked immunosorbent assay are commonly used in practice to detect CEA, which suffer from time-consuming sample labeling, large volume, high cost, and complex configuration. To resolve these problems, Zhu's

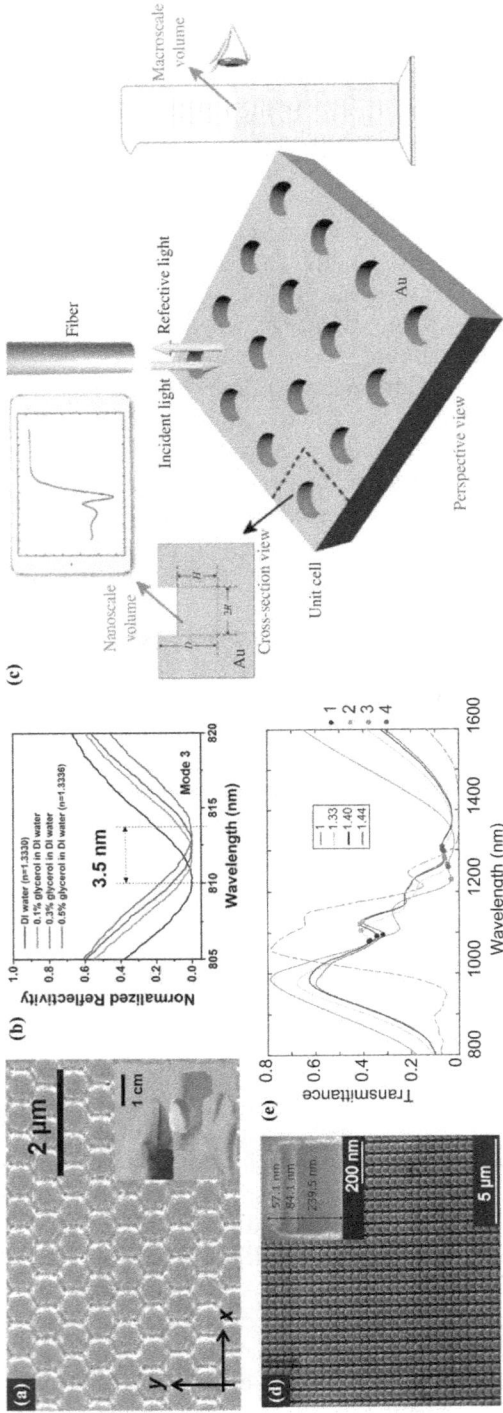

FIGURE 4.9 (a) and (b) The top-view SEM images of the plasmonic sensor array and the reflectance spectra of the sensor measured in solutions of different glycerol concentrations after normalization, respectively.

Source: Adopted from Liu, B. et al. *Advanced Materials*. 30, 1706031, 2018. No permission is required. Copyright 2018 Wiley.

(c) Measurement of nanoscale volumes by Au plasmonic nanocylinders using a fiber spectrometer, where the symbols D, R and H denote the depth and diameter of the nanocylinder, and the height of the specimen.

Source: Adopted from Zhu, J. et al. *Nanophotonics*. 9, 167–176, 2020. No permission is required. Copyright 2020 De Gruyter.

(d) and (e) SEM images of the nanoantenna array with a period of 670 nm and Experimental results for bulk refractive index sensing performed with the hybrid nanoantenna with four different glucose solutions as background; the inset in (d) shows a cross-section of a single nanoantenna, fabricated by FIB etching.

Source: Reprinted with permission from Ray, D. et al. *Nano Letters*. 20, 8752–8759, 2020. Copyright 2020 American Chemical Society.

team established a plasmonic immune chip platform for rapid and portable detection of CEA, as shown in the left part of Figure 4.10(a) [126]. Rapid quantitative CEA sensing has been achieved through a label-free scheme, and the ability to detect concentrations less than 5 ng/ml has been demonstrated. The clinical test on human serum shows that it has good consistency with routine medical examination, as illustrated in the right part of Figure 4.10(a). In addition, the team employed nanoimprinting and plasma etching on flexible polycarbonate substrates to realize low-cost flexible plasmonic metasurface sensors with Au nanobump arrays, as shown in Figure 4.10(b) [127]. This flexible plasmonic metasurface sensor achieves a high bulk refractive index sensitivity of 454.4 nm/RIU, and rapid quantitative sensing of less than 10 ng/mL CEA in human serum samples has been shown. This study elucidates the process of fabricating flexible lightweight plasmonic metasurface sensors with low cost and high throughput, which will facilitate the broad application of emerging flexible plasmonic biosensors.

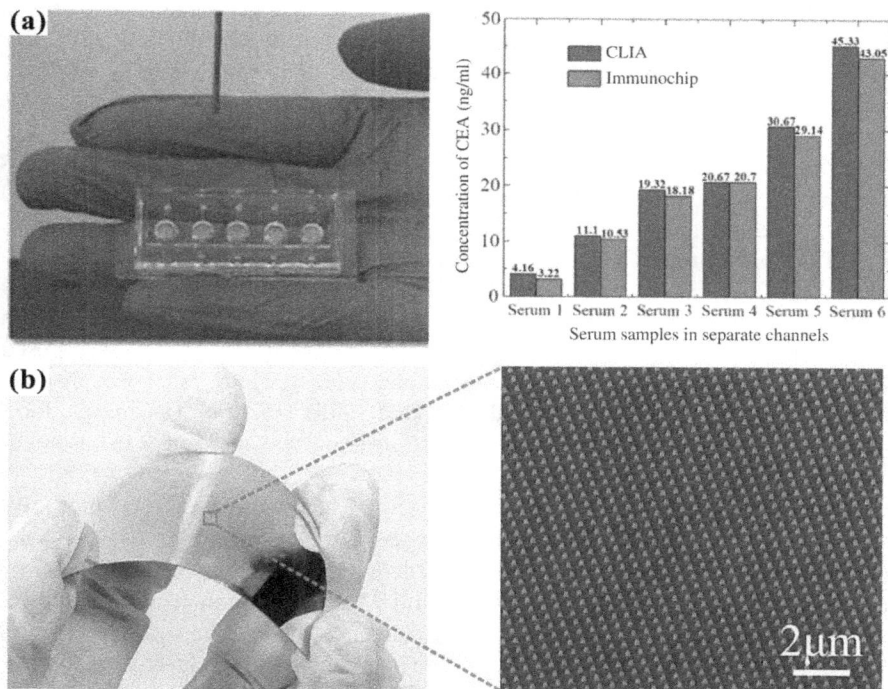

FIGURE 4.10   (a) Photograph of the immunochip and fiber probe (left) and the CEA detection results of serum samples from six individuals using the plasmonic immunochip, compared with using CLIA (right).

Source: Adopted from Zhou, J. et al. *Nanophotonics.* 8, 307–316, 2019. No permission is required. Copyright 2019 De Gruyter.

(b) Photograph of the large-area flexible metasurface and its corresponding SEM image.

Source: Adopted from Zhu, J. et al. *Biosensors and Bioelectronics.* 150, 111905, 2020. No permission is required. Copyright 2020 Elsevier.

Exosomes are cell-derived vesicles that show potential for cancer diagnosis because they are involved in transporting the molecular contents of cancer cells. Exosome detection and molecular analysis are technically challenging and often require numerous samples to be purified and labeled. Im et al. have demonstrated a detection of exosomes on periodic nanohole arrays sensor [128], which provides highly sensitive and label-free exosome analysis and the ability to monitor molecular binding in real-time. This work offers a label-free and high-throughput assay for clinical and functional proteomics applications.

In addition, several cell-to-cell interactions based on plasmonic nanopore arrays have been validated. For example, cell-cell interactions have been measured using a chemically modified nanopore array with spherical lipid bilayers [129]. The combination of spherical lipids and beads improves the signal-noise ratio and allows the construction of random color arrays with single image acquisition and eliminates sequential spotting. In another study, Tu et al. demonstrated label-free cell monitoring of C_3H_{10} and HeLa cells to understand their cell dynamics, which proved that HeLa cells grow more slowly than C_3H_{10} cells over the same period [130]. The retarded growth function has been fitted with mathematical expressions, and the difference in adhesion dynamics has been deduced. The entire cell attachment process is cross-confirmed by cell morphology.

For pathogen detection, Yanik et al. demonstrated the direct detection and identification of vesicular stomatitis virus (VSV), smallpox, and Ebola viruses in 2010 [131]. To identify a specific virus, antibodies to virus-specific membrane proteins (called glycoproteins or GPs) are immobilized on the surface of the plasmonic nanohole arrays. These envelope ribonucleic acid and envelope DNA viruses are detected based on the concept of exceptional optical transmission, resulting in a detection limit of 10^5 PFU/mL. Later, a new nanopore array design has been further proposed to produce overlapping transmission spectra and generate Fano resonances that can directly detect pathogens with naked eyes [132]. This label-free sensing platform can be used to detect pathogens and other bio-important molecules in resource-constrained environments.

The unique advantage of plasmonic biosensing has promoted its development for commercial applications. Currently, there are commercial plasmonic biosensors such as the BIAcore series based on the SPP effects of flat metal films, which are mainly used to analyze the binding kinetics and affinity testing of biomolecules. However, plasmonic biosensing technologies still face challenges in detecting small molecules at ultralow concentrations and producing compact devices for point-of-care applications. Better designs and integration techniques are in urgent demand to realize the mature plasmonic point-of-care product. It may be necessary to address the large-scale manufacturing of plasmonic chips while maintaining cost control to facilitate the commercialization process in the future.

4.5.3 SEIRS

Many kinds of molecules have unique vibration modes in the IR range (namely IR molecule fingerprints). SEIRS-based plasmonic sensors can significantly enhance the absorption spectral signal of molecules to detect the IR vibration modes of molecules. When the SPR resonance of the plasmonic sensor overlaps with the IR absorption fingerprint spectrum, the enhanced molecular-sensor coupling leads to a change in the resonance frequency or intensity, allowing the molecular fingerprint to be extracted. This concept, surface-enhanced IR absorption (SEIRA), has made significant progress both theoretically and experimentally and numerous applications have been demonstrated [133-136]. Chen *et al.* combined atomic layer lithography and template stripping to produce a new class of substrates for SEIRA [135]. As shown in Figure 4.11(a), the buried nanocavities are protected from contamination by the Si template until exposed by template stripping on demand. The exposed nanocavities produce strong IR resonances that tightly confine the infrared radiation within a 3 nm gap. The gap is filled with benzenethiol molecules, and a SEIRA enhancement factor of 10^5 is observed. In another work, Xu and co-workers developed a mid-IR hybrid nanofluidic-SEIRA liquid sensor based on Al_2O_3, as shown in Figure 4.11(b) [136]. They demonstrated the real-time dynamic monitoring of acetone molecular diffusion in deionized water using a single-port plasma system.

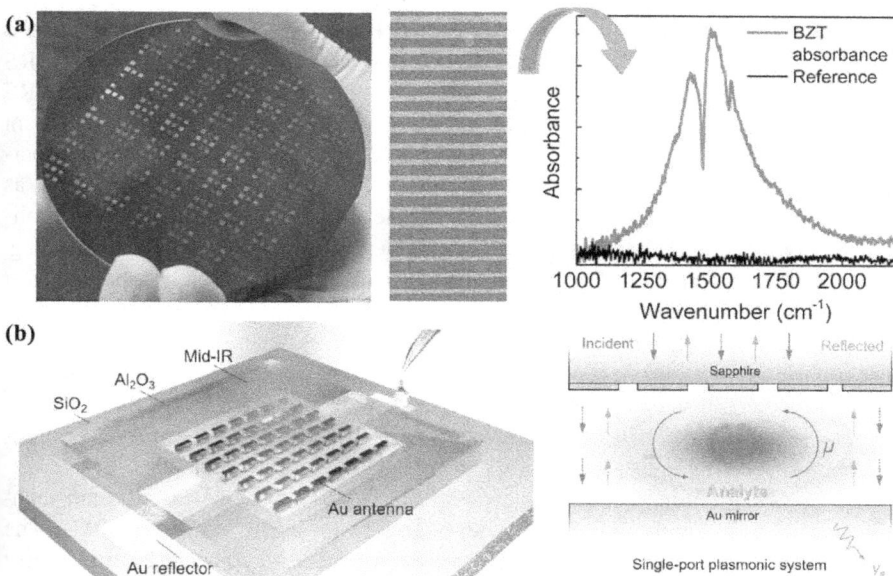

FIGURE 4.11 (a) Photograph of a 4-inch wafer containing metal stripes and the SEM image of an array of buried nanogaps on a chip as well as sensing of benzenethiol using a graded nanogap structure with a gap size of 3 nm.

Source: Reprinted with permission from Chen, X. et al. *Nano letters.* 15, 107–113, 2015. Copyright 2015 American Chemical Society.

(b) The device configuration of the self-driven 3D plasmonic Al_2O_3-based Mid-IR liquid sensing platform and the schematic diagram of the coupling principle for the single-port plasmonic system.

Source: Reprinted with permission from Xu, J. et al. *ACS nano.* 14, 12159–12172, 2020. Copyright 2020 American Chemical Society.

4.5.4 RAMAN SPECTROSCOPY

Raman spectroscopy has been known for over 80 years and has become a widely used technique for studying a wide variety of materials [137]. Raman spectrum is a kind of scattering spectrum, which can reflect the fingerprint vibration/rotation mode of the molecule. It can be used to obtain information on the chemical structure of molecules, identify compounds based on spectral fingerprints, or quantify a certain substance in a sample [138]. However, the main drawback of this method is the inherently small Raman cross-sections ($10^{-25} \sim 10^{-30}$ cm^2/molecule) of many analytes [139]. SERS technique provides an opportunity to overcome this limitation and examine weak Raman scatterings, even at low concentrations. It is generally accepted that the electromagnetic field enhancement mechanism dominates most SERS processes, which is caused by the optical amplification of the excitation of LSPRs between the analytes and the surfaces of plasmonic nanostructures [140]. In general, Au and Ag are frequently used as classical SERS substrates because they are air-stable, while Cu is more reactive. All three metals have LSPRs that cover most of the visible and near-IR regions, where most Raman measurements take place [141].

Over the past 30 years, tremendous efforts have been made to find new plasmonic materials [142,143], and develop novel configurations that support large SERS enhancements. Yao *et al.* report periodic bowtie SERS substrates with high SERS enhancement and large-area uniformity [144]. In this research, average enhancement factors up to 1×10^8 have been obtained, which are 50 times larger than Au nanoparticle assembly substrates and 140 times larger than commercial Klarite chips, as shown in Figures 4.12(a)–4.12(c). Along with the improved performance, the application scope of SERS has dramatically expanded, including sensing and imaging, single-molecule detection, and even extensions to ultrahigh vacuum and ultrafast science [145,146]. SERS biosensors are used to detect a variety of biological samples and diseases [141], including various cancers, Alzheimer's disease, and Parkinson's disease. Nearly 15 years ago, SERS began to evolve from large sample detection to single-molecule sensing [147,148]. Compared with single-molecule fluorescence, single-molecule SERS offers significant advantages, especially due to reduced fluorescence processing and richer fingerprint-like chemical information.

Tip-enhanced Raman spectroscopy (TERS) is one of the newest Raman techniques, typically combining SERS with scanning probe microscopy (SPM). The electromagnetic field enhancement is strongly confined at the sharp metal tip to boost Raman intensity, providing not only rich spectral and structural information of trace amount of molecules but also a high spatial resolution of 10 nm–20 nm [149,150]. The localized electromagnetic field in the tip-substrate cavity can be enhanced by two orders of magnitude, and the spatial resolution is determined by the radius of curvature of the SPM tip and the tip-substrate distance. Compared with SERS technique which requires the substrate surface to be roughened or nanostructured, TERS naturally bypasses this limitation and obtains a more versatile substrate [151,152]. In addition, TERS combined with electrochemical methods can be used to situ monitor the surface plasmons-driven decarboxylation and resolve the spatial distribution of hot carriers, as shown in Figure 4.12(d), where the electric field distribution shows a very strong field enhancement at the tip (Figure 4.12(e)).

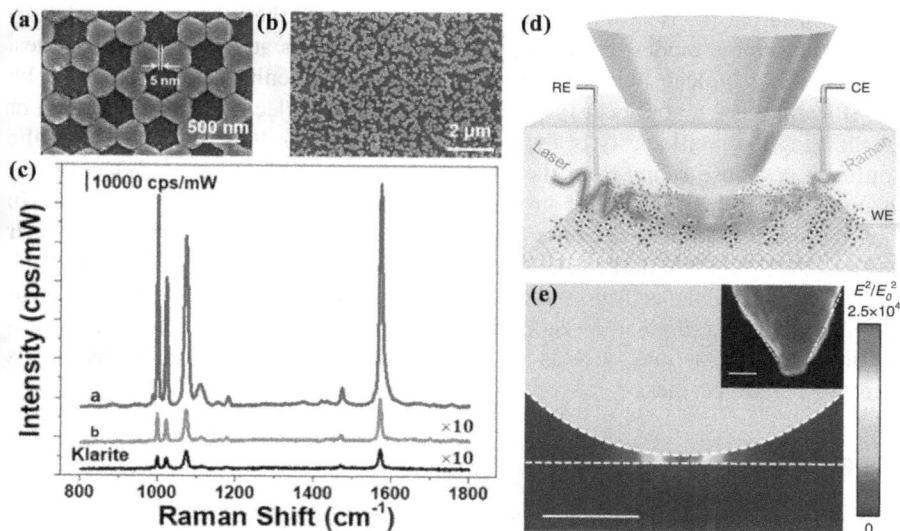

FIGURE 4.12 (a) SEM image of the bowtie array with a gap size of around 5 nm and (b) SEM image of the glass substrate assembly with 140 nm Au nanospheres anchored with the prefunctionalized 3-(Trimethoxysilyl)-1-propanamine. (c) SERS performance comparison of (a, b), and the commercial Klarite chip.

Source: Yao, X. et al. *ACS Applied Materials & Interfaces.* 12, 36505–36512, 2020. No permission is required. Copyright 2020 American Chemical Society.

(d) Schematic illustration of the electrochemical-TERS setup, where the decarboxylation reaction is induced by the strongly coupled SPR between the Ag tip and Au substrate. (e) Calculated plasmonic electric field (E^2) distribution when the Ag tip is positioned on the Au surface, the scale bar is 10 nm. Inset: scanning electron microscope image of the Ag tip, the scale bar is 50 nm.

Source: Adopted from Huang, S. et al. *Nature communications.* 11, 1–8, 2020. No permission is required. Copyright 2020 The authors.

4.5.5 Nonlinear Enhancement

Nonlinear enhancement enables many important applications in modern optical technology, including harmonic generation, mode-locking in ultrafast laser to holography, self-modulation and cross-modulation of optical signals, optical solitons, as well as quantum information [153]. However, nonlinear bulk materials exhibit observable nonlinear optical effects only at high light intensities, and typically require long propagation lengths to achieve significant nonlinear effects [154,155]. The strong near-field enhancement properties of SPR are widely used to improve the efficiency of light-matter interactions in linear domains, which also has profound consequences for nonlinear optical processes which depend on the local field intensity in a super-linear manner [156,157].

Nonlinear plasmons have been developed by exploiting the enhanced optical near-field of plasmonic modes, improving nonlinear responses at subwavelength scales. There are two typical methods: (1) using the near-field enhancement provided by the SPR to induce a nonlinear response in a nonlinear dielectric material nearby, or (2) utilizing the nonlinearity of the plasmonic material itself. In the latter case, the nonlinear response stems from the dynamics of non-equilibrium free electrons in the metal under the influence of strongly illuminated electromagnetic fields [153]. In fact, the metal nonlinear response is one of the strongest and fastest responses per unit of interaction length, and the femtosecond response time is determined by the relaxation of excited electrons to the equilibrium states, mainly affected by electron-electron and electron-phonon scatterings. Both methods are used to enhance coherent nonlinear interactions, such as harmonic generation and wave mixing, as well as Kerr-type nonlinearities.

4.5.6 ABSORBERS

The intrinsic loss of the plasmonic device is often considered to be harmful to many applications such as imaging, sensing, and stealth. However, it could become advantageous for electromagnetic absorbers, where incident radiation is first absorbed and then converted into ohmic heat or other forms of energy. Perfect absorption can be achieved by properly designing the geometries of the plasmonic unit cells [158,159]. In 2008, a perfect electromagnetic absorber based on plasmonic metamaterials was first reported in the microwave region [160]. Later, this concept has been extended to higher frequency regions at deep subwavelength scales [161-163]. Benefiting from advanced nanofabrication techniques, metamaterial-based plasmonic absorbers are experimentally possible with excellent adaptability and effectiveness.

In the past decades, electromagnetic absorbers based on metallic subwavelength plasmonic structures have been proposed, and the light trapping properties have been recognized. More importantly, the intensity and bandwidth of the plasmonic absorber can be flexibly tuned by changing the materials and geometries. Multiband absorption is also possible by adopting advanced techniques such as mixing multiple resonances, exciting phase resonances, and trapping light by anisotropic metamaterials. Therefore, various plasmonic absorbers with narrow or wide absorption bands and angle-insensitive or sensitive features have been realized, generating a plethora of applications. For instance, as shown in Figures 4.13(a)–4.13(c), the absorption efficiency and peak position of the plasmonic metamaterial consisting of Au nanohole array/SiO_2 spacer/Au film can be tuned by simply controlling the thicknesses of the hexagonal Au nanohole array and the spacer layer. Near-perfect absorption is achieved at the desired laser wavelength [163]. In other research, Zhu and co-workers combined the transmission line circuit theory with full-wave simulation to design plasmonic stack metamaterials in the near-infrared range [159], achieving broadband light trapping, as illustrated in Figures 4.13(d) and 4.13(e). In addition graphene-based absorbers have also been developed in recent years, providing new options for perfect electromagnetic absorption in different electromagnetic wave bands [164-167].

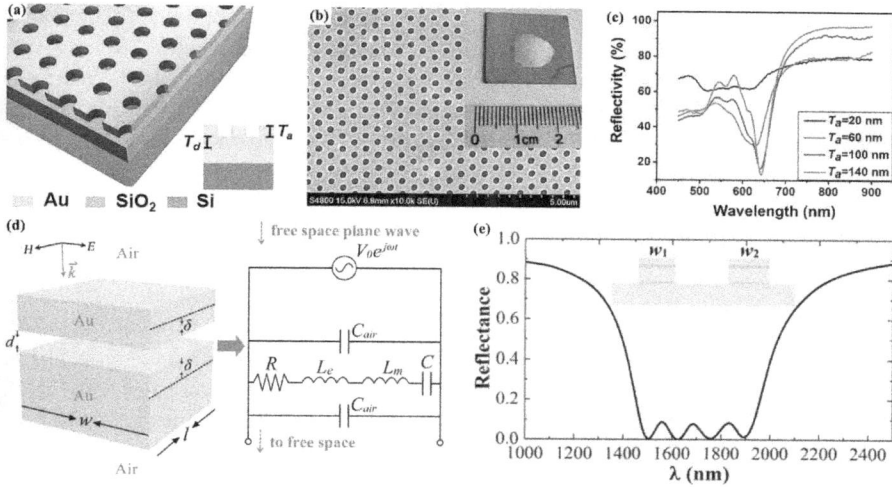

FIGURE 4.13 (a) The schematic view and (b) SEM image of the plasmonic metamaterial. (c) The measured reflectance spectra of the plasmonic metamaterial with different T_a; the inset in (b) shows the photo of the fabricated substrate with a size of as large as 10 mm in diameter.

Source: Adopted from Yang, K. et al. *Advanced Optical Materials.* 9, 2001375, 2021. No permission is required. Copyright 2021 Wiley.

(d) The metamaterial unit cell of the subwavelength component with two gold nanostructure blocks and the corresponding equivalent circuit with a series of RLC lumped elements. (e) The broadband absorption is achieved by using two subwavelength stacked nanostructures in a unit.

Source: Adopted from Zhu, J. et al. *Nanoscale.* 12, 2057-2062, 2020. Copyright 2020 Royal Society of Chemistry.

4.5.7 IMAGING

The discovery of SPR also paves the way for subwavelength resolution imaging, based on the extraordinary optical transmission properties of 2D plasmonic structures [168]. SPR microscopy has been demonstrated in a total internal reflection configuration in 2007 [169], providing an excellent spatial resolution that allows the study of protein binding kinetics [170]. Yang *et al.* demonstrated a plasmonic imaging technique for tracking the dynamics of individual organelles with 5 nm precision and 10 ms temporal resolution [171]. Compared to fluorescence tracking techniques, plasmonic imaging methods are label-free, fast, and provide precise positional information in all three dimensions. Plasmonic microscopy is also advantageous in exosome analysis because it only images samples within 200 nm of the surface, thereby suppressing medium noise. Recently, Syal *et al.* have reported a plasmonic imaging method to study the interaction dynamics between biomolecules and individual bacterial cells, demonstrating the ability to determine kinetic constants for ligand binding of individual live bacteria and to quantify heterogeneity in microbial populations [172].

In addition, the excitation of LSPR of plasmonic nanoparticles has also been applied in the field of microscopic imaging, where light scattering can be used in imaging techniques such as computed tomography and dark-field microscopy, and light absorption can be used in photoacoustic imaging tomography. Recently, Jing *et al.* report a time-resolved digital immunoassay based on plasmonic nanoparticle imaging for rapid detection of biomarkers with a wide dynamic range. Plasmonic imaging provides high-contrast and fast imaging of nanoparticles, allowing the detection of single-molecule binding on the sensor surface [173].

Notably, Section 4.5.1 to Section 4.5.7 cover just part of applications based on SPR, and there are more plasmonic applications such as sub-diffraction lensing [174], monochromatic [175] and color holography [176], polarization conversion [177], vortex plates [178], invisibility and cloaking [179], polarization-selection [180]. These applications are briefly captured in Chapter 5 when discussing metamaterials and metasurfaces. With the continuous development of plasmonic technologies, we do believe that more applications will be realized in the future.

4.6 SUMMARY AND OUTLOOK

In this chapter, we first reviewed the research and development of SPR, and introduce its classification and excitation conditions. Then, we reviewed various plasmonic materials and discussed their corresponding superiorities and inferiorities. Next, the fabrication methods of plasmonic micro/nanostructures were enumerated. Finally, some major applications of SPR and corresponding commercial products were introduced.

At the present stage, most plasmonic structures still suffer from high intrinsic losses. The emergence of low-loss dielectric materials has led to new options, and advanced hybrid structures can even combine strong near-field confinement of plasmonic materials and the low-loss high-Q properties of dielectric materials. More details can be found in Chapter 6. In the future, it is expected that more research efforts will be devoted to discovering more plasmonic materials and structures with excellent properties. In addition, with the continuous progress of manufacturing technology, large-scale integration will be a major development direction. Combining the CMOS technology, novel 2D materials with high spatial integration capability could further reduce the size of plasmonic devices and systems, providing an important foundation for more mature commercial equipment. There is little doubt that the field of plasmon will continue to boom, and will an important part in the next generation of nanophotonics and metamaterial/metasurface technologies.

ACKNOWLEDGMENTS

This work is supported by NSFC (62175205), NSAF (U1830116, U2130112), the Natural Science Foundation of Fujian Province (2020J06009).

BIBLIOGRAPHY

[1] S. A. Maier, *Plasmonics: fundamentals and applications*. Springer, 2007.

[2] E. D. Tommasi, E. Esposito, S. Romano *et al.*, "Frontiers of light manipulation in natural, metallic, and dielectric nanostructures," *La Rivista del Nuovo Cimento*, vol. 44, no. 1, pp. 1–68, 2021.

[3] B. Wiley, Y. Xia, S. Skrabalak *et al.*, "Chemical synthesis of novel plasmonic nanoparticles," *Annual Review of Physical Chemistry*, vol. 60, pp. 167–192, 2009.

[4] W. L. Barnes, A. Dereux, and T. W. Ebbesen, "Surface plasmon subwavelength optics," *Nature*, vol. 424, no. 6950, pp. 824–830, 2003.

[5] H. Reather, "Surface plasmons on smooth and rough surfaces and on gratings," *Springer Tracts in Modern Physics*, vol. 111, no. 71, pp. 345–398, 1988.

[6] D. Zhang, L. Men, and Q. Chen, "Microfabrication and applications of opto-microfluidic sensors," *Sensors*, vol. 11, no. 5, pp. 5360–5382, 2011.

[7] A. Bozzola, S. Perotto, and F. De Angelis, "Hybrid plasmonic–photonic whispering gallery mode resonators for sensing: a critical review," *Analyst*, vol. 142, no. 6, pp. 883–898, 2017.

[8] U. Kreibig, and M. Vollmer, *Optical properties of metal clusters*. Springer, 2013.

[9] C. F. Bohren, and D. Huffman, *Absorption and scattering of light by small particles*. Wiley, 1940.

[10] J. Pendry, L. Martin-Moreno, and F. Garcia-Vidal, "Mimicking surface plasmons with structured surfaces," *Science*, vol. 305, no. 5685, pp. 847–848, 2004.

[11] A. P. Hibbins, B. R. Evans, and J. R. Sambles, "Experimental verification of designer surface plasmons," *Science*, vol. 308, no. 5722, pp. 670–672, 2005.

[12] W. X. Tang, H. C. Zhang, H. F. Ma *et al.*, "Concept, theory, design, and applications of spoof surface plasmon polaritons at microwave frequencies," *Advanced Optical Materials*, vol. 7, no. 1, pp. 1800421, 2019.

[13] H. K. Nienhuys, and V. Sundström, "Influence of plasmons on terahertz conductivity measurements," *Applied Physics Letters*, vol. 87, no. 1, pp. 1759, 2005.

[14] K. J. Chau, G. D. Dice, and A. Y. Elezzabi, "Coherent plasmonic enhanced terahertz transmission through random metallic media," *Physical Review Letters*, vol. 94, no. 17, pp. 173904, 2005.

[15] J. Marton, and B. Jordan, "Optical properties of aggregated metal systems: Interband transitions," *Physical Review B*, vol. 15, no. 4, pp. 1719, 1977.

[16] P. B. Johnson *et al.*, "Optical constants of the noble metals," *Physical Review B*, vol. 6, pp. 4370, 1972.

[17] P. R. West, S. Ishii, G. V. Naik *et al.*, "Searching for better plasmonic materials," *Laser & Photonics Reviews*, vol. 4, no. 6, pp. 795–808, 2010.

[18] V. P. Drachev, U. K. Chettiar, A. V. Kildishev *et al.*, "The Ag dielectric function in plasmonic metamaterials," *Optics Express*, vol. 16, no. 2, pp. 1186–1195, 2008.

[19] Y. Yagil, P. Gadenne, C. Julien *et al.*, "Optical properties of thin semicontinuous gold films over a wavelength range of 2.5 to 500 μm," *Physical Review B*, vol. 46, no. 4, pp. 2503, 1992.

[20] D. D. Smith, Y. Yoon, R. W. Boyd *et al.*, "Z-scan measurement of the nonlinear absorption of a thin gold film," *Journal of Applied Physics*, vol. 86, no. 11, pp. 6200–6205, 1999.

[21] P. Tobiška, O. Hugon, A. Trouillet *et al.*, "An integrated optic hydrogen sensor based on SPR on palladium," *Sensors and Actuators B: Chemical*, vol. 74, no. 1–3, pp. 168–172, 2001.

[22] S. Baldelli, A. S. Eppler, E. Anderson *et al.*, "Surface enhanced sum frequency generation of carbon monoxide adsorbed on platinum nanoparticle arrays," *The Journal of
Chemical Physics*, vol. 113, no. 13, pp. 5432–5438, 2000.

[23] K. Ikeda, J. Sato, N. Fujimoto *et al.*, "Plasmonic enhancement of Raman scattering on
non-SERS-active platinum substrates," *The Journal of Physical Chemistry C*, vol. 113,
no. 27, pp. 11816–11821, 2009.

[24] A. N. Grigorenko, M. Polini, and K. Novoselov, "Graphene plasmonics," *Nature
Photonics*, vol. 6, no. 11, pp. 749–758, 2012.

[25] G. V. Naik, V. M. Shalaev, and A. Boltasseva, "Alternative plasmonic materials: Beyond
gold and silver," *Advanced Materials*, vol. 25, no. 24, pp. 3264–3294, 2013.

[26] U. Guler, V. M. Shalaev, and A. Boltasseva, "Nanoparticle plasmonics: Going practical
with transition metal nitrides," *Materials Today*, vol. 18, no. 4, pp. 227–237, 2015.

[27] S. Kim, J. M. Kim, J. E. Park *et al.*, "Nonnoble-metal-based plasmonic nanomaterials: recent advances and future perspectives," *Advanced Materials*, vol. 30, no. 42,
pp. 1704528, 2018.

[28] G. H. Chan, J. Zhao, E. M. Hicks *et al.*, "Plasmonic properties of copper nanoparticles
fabricated by nanosphere lithography," *Nano Letters*, vol. 7, no. 7, pp. 1947–1952, 2007.

[29] A. Moscatelli, "The aluminium rush," *Nature Nanotechnology*, vol. 7, no. 12, pp. 778–
778, 2012.

[30] D. Gérard, and S. K. Gray, "Aluminium plasmonics," *Journal of Physics D: Applied
Physics*, vol. 48, no. 18, pp. 184001, 2014.

[31] F. Jiao, F. Li, J. Shen *et al.*, "Wafer-scale flexible plasmonic metasurface with passivated
aluminum nanopillars for high-sensitivity immunosensors," *Sensors and Actuators B:
Chemical*, vol. 344, pp. 130170, 2021.

[32] F. Sterl, N. Strohfeldt, R. Walter *et al.*, "Magnesium as novel material for active plasmonics in the visible wavelength range," *Nano Letters*, vol. 15, no. 12, pp. 7949–7955,
2015.

[33] X. Duan, S. Kamin, F. Sterl *et al.*, "Hydrogen-regulated chiral nanoplasmonics," *Nano
Letters*, vol. 16, no. 2, pp. 1462–1466, 2016.

[34] X. Duan, S. Kamin, and N. Liu, "Dynamic plasmonic colour display," *Nature
Communications*, vol. 8, no. 1, pp. 1–9, 2017.

[35] J. M. McMahon, G. C. Schatz, and S. K. Gray, "Plasmonics in the ultraviolet with the
poor metals Al, Ga, In, Sn, Tl, Pb, and Bi," *Physical Chemistry Chemical Physics*, vol.
15, no. 15, pp. 5415–5423, 2013.

[36] M. B. Ross, and G. C. Schatz, "Aluminum and indium plasmonic nanoantennas in the
ultraviolet," *The Journal of Physical Chemistry C*, vol. 118, no. 23, pp. 12506–12514,
2014.

[37] F. Magnan, J. Gagnon, F.-G. Fontaine *et al.*, "Indium@ silica core–shell nanoparticles as plasmonic enhancers of molecular luminescence in the UV region," *Chemical
Communications*, vol. 49, no. 81, pp. 9299–9301, 2013.

[38] Y. Wang, C. W. Ge, Y. F. Zou *et al.*, "Plasmonic indium nanoparticle-induced high-
performance photoswitch for blue light detection," *Advanced Optical Materials*, vol. 4,
no. 2, pp. 291–296, 2016.

[39] P. C. Wu, C. G. Khoury, T.-H. Kim *et al.*, "Demonstration of surface-enhanced Raman
scattering by tunable, plasmonic gallium nanoparticles," *Journal of the American
Chemical Society*, vol. 131, no. 34, pp. 12032–12033, 2009.

[40] Y. Yang, J. M. Callahan, T.-H. Kim *et al.*, "Ultraviolet nanoplasmonics: a demonstration of surface-enhanced Raman spectroscopy, fluorescence, and photodegradation
using gallium nanoparticles," *Nano Letters*, vol. 13, no. 6, pp. 2837–2841, 2013.

[41] P. C. Wu, T. H. Kim, A. S. Brown *et al.*, "Real-time plasmon resonance tuning of liquid Ga nanoparticles by in situ spectroscopic ellipsometry," *Applied Physics Letters*, vol. 90, no. 10, pp. 197, 2007.

[42] Y. Yang, N. Akozbek, T.-H. Kim *et al.*, "Ultraviolet–visible plasmonic properties of gallium nanoparticles investigated by variable-angle spectroscopic and Mueller matrix ellipsometry," *ACS Photonics*, vol. 1, no. 7, pp. 582–589, 2014.

[43] T. Block, C. Tegenkamp, J. Baringhaus *et al.*, "Plasmons in Pb nanowire arrays on Si (557): between one and two dimensions," *Physical Review B*, vol. 84, no. 20, pp. 205402, 2011.

[44] P. Zolotavin, and P. Guyot-Sionnest, "Meissner effect in colloidal Pb nanoparticles," *Acs Nano*, vol. 4, no. 10, pp. 5599–5608, 2010.

[45] D. L. Delach, M. J. Dukes, A. C. Varano *et al.*, "Real-time imaging of lead nanoparticles in solution–determination of the growth mechanism," *RSC Advances*, vol. 5, no. 126, pp. 104193–104197, 2015.

[46] J. Chen, P. Albella, Z. Pirzadeh *et al.*, "Plasmonic nickel nanoantennas," *Small*, vol. 7, no. 16, pp. 2341–2347, 2011.

[47] Y. Shen, J. Zhou, T. Liu *et al.*, "Plasmonic gold mushroom arrays with refractive index sensing figures of merit approaching the theoretical limit," *Nature Communications*, vol. 4, no. 1, pp. 1–9, 2013.

[48] L. Wang, C. Clavero, Z. Huba *et al.*, "Plasmonics and enhanced magneto-optics in core–shell Co– Ag nanoparticles," *Nano Letters*, vol. 11, no. 3, pp. 1237–1240, 2011.

[49] B. Toal, M. McMillen, A. Murphy *et al.*, "Optical and magneto-optical properties of gold core cobalt shell magnetoplasmonic nanowire arrays," *Nanoscale*, vol. 6, no. 21, pp. 12905–12911, 2014.

[50] D. L. Huber, "Synthesis, properties, and applications of iron nanoparticles," *Small*, vol. 1, no. 5, pp. 482–501, 2005.

[51] S. Laurent, D. Forge, M. Port *et al.*, "Magnetic iron oxide nanoparticles: synthesis, stabilization, vectorization, physicochemical characterizations, and biological applications," *Chemical Reviews*, vol. 108, no. 6, pp. 2064–2110, 2008.

[52] S. Kayal, and R. V. Ramanujan, "Anti-cancer drug loaded iron–gold core–shell nanoparticles (Fe@ Au) for magnetic drug targeting," *Journal of Nanoscience and Nanotechnology*, vol. 10, no. 9, pp. 5527–5539, 2010.

[53] L. Lu, W. Zhang, D. Wang *et al.*, "Fe@ Ag core–shell nanoparticles with both sensitive plasmonic properties and tunable magnetism," *Materials Letters*, vol. 64, no. 15, pp. 1732–1734, 2010.

[54] L. Wang, K. Yang, C. Clavero *et al.*, "Localized surface plasmon resonance enhanced magneto-optical activity in core-shell Fe–Ag nanoparticles," *Journal of Applied Physics*, vol. 107, no. 9, pp. 09B303, 2010.

[55] M. Lee, J. U. Kim, K. J. Lee *et al.*, "Aluminum nanoarrays for plasmon-enhanced light harvesting," *ACS Nano*, vol. 9, no. 6, pp. 6206–6213, 2015.

[56] M. B. Ross, and G. C. Schatz, "Radiative effects in plasmonic aluminum and silver nanospheres and nanorods," *Journal of Physics D: Applied Physics*, vol. 48, no. 18, pp. 184004, 2014.

[57] G. V. Hartland, "Optical studies of dynamics in noble metal nanostructures," *Chemical Reviews*, vol. 111, no. 6, pp. 3858–3887, 2011.

[58] M. B. Cortie, and A. M. McDonagh, "Synthesis and optical properties of hybrid and alloy plasmonic nanoparticles," *Chemical Reviews*, vol. 111, no. 6, pp. 3713–3735, 2011.

[59] D. Wang, H. Xin, R. Hovden *et al.*, "Structurally ordered intermetallic platinum–cobalt core–shell nanoparticles with enhanced activity and stability as oxygen reduction electrocatalysts," *Nature Materials*, vol. 12, pp. 81–87, 2012.

[60] C. Chen, Y. Kang, Z. Huo *et al.*, "Highly crystalline multimetallic nanoframes with three-dimensional electrocatalytic surfaces," *Science*, vol. 343, no. 6177, pp. 1339–1343, 2014.

[61] S. Law, L. Yu, A. Rosenberg *et al.*, "All-semiconductor plasmonic nanoantennas for infrared sensing," *Nano Letters*, vol. 13, no. 9, pp. 4569–4574, 2013.

[62] Y. Zhong, S. D. Malagari, T. Hamilton *et al.*, "Review of mid-infrared plasmonic materials," *Journal of Nanophotonics*, vol. 9, no. 1, pp. 093791, 2015.

[63] R. Soref, "The past, present, and future of silicon photonics," *IEEE Journal of Selected Topics in Quantum Electronics*, vol. 12, no. 6, pp. 1678–1687, 2007.

[64] J. A. Dionne, L. A. Sweatlock, M. T. Sheldon *et al.*, "Silicon-based plasmonics for on-chip photonics," *IEEE Journal of Selected Topics in Quantum Electronics*, vol. 16, no. 1, pp. 295–306, 2010.

[65] A. Hryciw, Y. C. Jun, M. L. Brongersma, "Electrifying plasmonics on silicon," *Nature Mater*, vol. 9, pp. 3–4, 2010.

[66] R. Soref, "Mid-infrared photonics in silicon and germanium," *Nature Photonics*, vol. 4, no. 8, pp. 495–497, 2010.

[67] M. A. Afromowitz, "Refractive index of Ga1− xAlxAs," *Solid State Communications*, vol. 15, no. 1, pp. 59–63, 1974.

[68] B. Broberg, and S. Lindgren, "Refractive index of In1− x Ga x As y P1− y layers and InP in the transparent wavelength region," *Journal of Applied Physics*, vol. 55, no. 9, pp. 3376–3381, 1984.

[69] W. Walukiewicz, J. Lagowski, L. Jastrzebski *et al.*, "Electron mobility and free-carrier absorption in InP; determination of the compensation ratio," *Journal of Applied Physics*, vol. 51, no. 5, pp. 2659–2668, 1980.

[70] W. Walukiewicz, L. Lagowski, L. Jastrzebski *et al.*, "Electron mobility and free-carrier absorption in GaAs: Determination of the compensation ratio," *Journal of Applied Physics*, vol. 50, no. 2, pp. 899–908, 1979.

[71] T. Yamada, E. Tokumitsu, K. Saito *et al.*, "Heavily carbon doped p-type GaAs and GaAlAs grown by metalorganic molecular beam epitaxy," *Journal of Crystal Growth*, vol. 95, no. 1–4, pp. 145–149, 1989.

[72] D. Law, D. C. Adams, A. M. Taylor *et al.*, "Mid-infrared designer metals," *Optics Express*, vol. 20, pp. 12155, 2012.

[73] B. Monemar, "III-V nitrides—important future electronic materials," *Journal of Materials Science: Materials in Electronics*, vol. 10, no. 4, pp. 227–254, 1999.

[74] A. Hangleiter, "III–V nitrides: A new age for optoelectronics," *MRS Bulletin*, vol. 28, no. 5, pp. 350–353, 2003.

[75] A. Kasic, M. Schubert, S. Einfeldt *et al.*, "Free-carrier and phonon properties of n-and p-type hexagonal GaN films measured by infrared ellipsometry," *Physical Review B*, vol. 62, no. 11, pp. 7365, 2000.

[76] J.-K. Sheu, and G. Chi, "The doping process and dopant characteristics of GaN," *Journal of Physics: Condensed Matter*, vol. 14, no. 22, pp. R657, 2002.

[77] P. Hageman, W. Schaff, J. Janinski *et al.*, "N-type doping of wurtzite GaN with germanium grown with plasma-assisted molecular beam epitaxy," *Journal of Crystal Growth*, vol. 267, no. 1–2, pp. 123–128, 2004.

[78] H. Hu, X. Yang, F. Zhai *et al.*, "Far-field nanoscale infrared spectroscopy of vibrational fingerprints of molecules with graphene plasmons," *Nature Communications*, vol. 7 no. 1, pp. 1–8, 2016.

[79] M. Klein, B. H. Badada, R. Binder *et al.*, "2D semiconductor nonlinear plasmonic modulators," *Nature Communications*, vol. 10, no. 1, pp. 1–7, 2019.

[80] S. Han, S. Kim, S. Kim *et al.*, "Complete complex amplitude modulation with electronically tunable graphene plasmonic metamolecules," *ACS Nano*, vol. 14, no. 1, pp. 1166–1175, 2020.

[81] L. Xiong, C. Forsythe, M. Jung *et al.*, "Photonic crystal for graphene plasmons," *Nature Communications*, vol. 10, no. 1, pp. 1–6, 2019.

[82] F. Stern, "Polarizability of a two-dimensional electron gas," *Physical Review Letters*, vol. 18, no. 14, pp. 546, 1967.

[83] E. Hwang, and S. D. Sarma, "Dielectric function, screening, and plasmons in two-dimensional graphene," *Physical Review B*, vol. 75, no. 20, pp. 205418, 2007.

[84] T. N. Theis, "Plasmons in inversion layers," *Surface Science*, vol. 98, no. 1–3, pp. 515–532, 1980.

[85] B. Van Wees, H. Van Houten, C. Beenakker *et al.*, "Quantized conductance of point contacts in a two-dimensional electron gas," *Physical Review Letters*, vol. 60, no. 9, pp. 848, 1988.

[86] O. Ambacher, J. Smart, J. Shealy *et al.*, "Two-dimensional electron gases induced by spontaneous and piezoelectric polarization charges in N-and Ga-face AlGaN/GaN heterostructures," *Journal of Applied Physics*, vol. 85, no. 6, pp. 3222–3233, 1999.

[87] H. Y. Hwang, Y. Iwasa, M. Kawasaki *et al.*, "Emergent phenomena at oxide interfaces," *Nature Materials*, vol. 11, no. 2, pp. 103–113, 2012.

[88] S. Bae, H. Kim, Y. Lee *et al.*, "Roll-to-roll production of 30-inch graphene films for transparent electrodes," *Nature Nanotechnology*, vol. 5, no. 8, pp. 574–578, 2010.

[89] C. Zhi, Y. Bando, C. Tang *et al.*, "Large-scale fabrication of boron nitride nanosheets and their utilization in polymeric composites with improved thermal and mechanical properties," *Advanced Materials*, vol. 21, no. 28, pp. 2889–2893, 2009.

[90] A. Broers, W. Molzen, J. Cuomo *et al.*, "Electron-beam fabrication of 80-Å metal structures," *Applied Physics Letters*, vol. 29, no. 9, pp. 596–598, 1976.

[91] G. Kostovski, P. R. Stoddart, and A. Mitchell, "The optical fiber tip: An inherently light-coupled microscopic platform for micro-and nanotechnologies," *Advanced Materials*, vol. 26, no. 23, pp. 3798–3820, 2014.

[92] P. Malara, A. Crescitelli, V. Di Meo *et al.*, "Resonant enhancement of plasmonic nanostructured fiber optic sensors," *Sensors and Actuators B: Chemical*, vol. 273, pp. 1587–1592, 2018.

[93] M. Horák, K. Bukvišová, V. Švarc *et al.*, "Comparative study of plasmonic antennas fabricated by electron beam and focused ion beam lithography," *Scientific Reports*, vol. 8, no. 1, pp. 1–8, 2018.

[94] A. N. Broers, "Fabrication limits of electron beam lithography and of UV, X-ray and ion-beam lithographies," *Philosophical Transactions of the Royal Society A: Mathematical, Physical and Engineering Sciences*, vol. 353, no. 1703, 1995.

[95] V. Krohn, "Electrohydrodynamic capillary source of ions and charged droplets," *Journal of Applied Physics*, vol. 45, no. 3, pp. 1144–1146, 1974.

[96] V. Krohn, and G. Ringo, "Ion source of high brightness using liquid metal," *Applied Physics Letters*, vol. 27, no. 9, pp. 479–481, 1975.

[97] R. L. Seliger, and W. P. Fleming, "Focused ion beams in microfabrication," *Journal of Applied Physics*, vol. 45, no. 3, pp. 1416–1422, 1974.

[98] K. Gamo, "Nanofabrication by FIB," *Microelectronic Engineering*, vol. 32, no. 1–4, pp. 159–171, 1996.

[99] S. Matsui, K. Mori, K. Saigo *et al.*, "Lithographic approach for 100 nm fabrication by focused ion beam," *Journal of Vacuum Science & Technology B: Microelectronics Processing and Phenomena*, vol. 4, no. 4, pp. 845–849, 1986.

[100] M. Principe, M. Consales, A. Micco *et al.*, "Optical fiber meta-tips," *Light: Science & Applications*, vol. 6, no. 3, pp. e16226–e16226, 2017.

[101] M. Colburn, S. C. Johnson, M. D. Stewart *et al.*, "Step and flash imprint lithography: a new approach to high-resolution patterning," *Proceedings of SPIE – The International Society for Optical Engineering*, vol. 3676, pp. 379–389.

[102] F. Li, J. Shen, C. Guan *et al.*, "Exploring near-field sensing efficiency of complementary plasmonic metasurfaces for immunodetection of tumor markers," *Biosensors and Bioelectronics*, pp. 114038, 2022.

[103] H. Lee, and G.-Y. Jung, "Full wafer scale near zero residual nano-imprinting lithography using UV curable monomer solution," *Microelectronic Engineering*, vol. 77, no. 1, pp. 42–47, 2005.

[104] J. H. Moon, S.-M. Yang, D. J. Pine *et al.*, "Multiple-exposure holographic lithography with phase shift," *Applied Physics Letters*, vol. 85, no. 18, pp. 4184–4186, 2004.

[105] M. Helgert, M. Burkhardt, K. Rudolf *et al.*, "High-frequent structures generated by interference lithography in the DUV," *Diffractive Optics and Micro-Optics*, p. DTuC3.

[106] M. Grzelczak, J. Vermant, E. M. Furst *et al.*, "Directed self-assembly of nanoparticles," *ACS Nano*, vol. 4, no. 7, pp. 3591–3605, 2010.

[107] J. Zhu, X. Zhu, R. Hoekstra *et al.*, "Metallic nanomesh electrodes with controllable optical properties for organic solar cells," *Applied Physics Letters*, vol. 100, no. 14, pp. 143109, 2012.

[108] B. Wiley, Y. Sun, J. Chen *et al.*, "Shape-controlled synthesis of silver and gold nanostructures," *Mrs Bulletin*, vol. 30, no. 5, pp. 356–361, 2005.

[109] R. He, Y.-C. Wang, X. Wang *et al.*, "Facile synthesis of pentacle gold–copper alloy nanocrystals and their plasmonic and catalytic properties," *Nature Communications*, vol. 5, no. 1, pp. 1–10, 2014.

[110] D. I. Gittins, A. S. Susha, B. Schoeler *et al.*, "Dense nanoparticulate thin films via gold nanoparticle self-assembly," *Advanced Materials*, vol. 14, no. 7, pp. 508–512, 2002.

[111] C. Graf, and A. van Blaaderen, "Metallodielectric colloidal core– shell particles for photonic applications," *Langmuir*, vol. 18, no. 2, pp. 524–534, 2002.

[112] Y. Sun, B. T. Mayers, and Y. Xia, "Template-engaged replacement reaction: a one-step approach to the large-scale synthesis of metal nanostructures with hollow interiors," *Nano Letters*, vol. 2, no. 5, pp. 481–485, 2002.

[113] S. Mair, *Plasmonics-Fundamentals and Applications*. Springer, 2007.

[114] K. M. Mayer, and J. H. Hafner, "Localized surface plasmon resonance sensors," *Chemical Reviews*, vol. 111, no. 6, pp. 3828–3857, 2011.

[115] S. Wang, L. Xia, H. Mao *et al.*, "Terahertz biosensing based on a polarization-insensitive metamaterial," *IEEE Photonics Technology Letters*, vol. 28, no. 9, pp. 986–989, 2016.

[116] B. Auguié, and W. L. Barnes, "Collective resonances in gold nanoparticle arrays," *Physical Review Letters*, vol. 101, no. 14, pp. 143902, 2008.

[117] E. M. Hicks, S. Zou, G. C. Schatz *et al.*, "Controlling plasmon line shapes through diffractive coupling in linear arrays of cylindrical nanoparticles fabricated by electron beam lithography," *Nano Letters*, vol. 5, no. 6, pp. 1065–1070, 2005.

[118] G. Vecchi, V. Giannini, and J. G. Rivas, "Surface modes in plasmonic crystals induced by diffractive coupling of nanoantennas," *Physical Review B*, vol. 80, no. 20, pp. 201401, 2009.

[119] V. Kravets, F. Schedin, and A. Grigorenko, "Extremely narrow plasmon resonances based on diffraction coupling of localized plasmons in arrays of metallic nanoparticles," *Physical Review Letters*, vol. 101, no. 8, pp. 087403, 2008.

[120] W. Zhou, and T. W. Odom, "Tunable subradiant lattice plasmons by out-of-plane dipolar interactions," *Nature Nanotechnology*, vol. 6, no. 7, pp. 423–427, 2011.

[121] D. Chanda, K. Shigeta, T. Truong *et al.*, "Coupling of plasmonic and optical cavity modes in quasi-three-dimensional plasmonic crystals," *Nature Communications*, vol. 2, pp. 479, 2011.

[122] B. Liu, S. Chen, J. Zhang *et al.*, "A plasmonic sensor array with ultrahigh figures of merit and resonance linewidths down to 3 nm," *Advanced Materials*, pp. 1706031, 2018.

[123] J. Zhu, X. Chen, Y. Xie *et al.*, "Imprinted plasmonic measuring nanocylinders for nanoscale volumes of materials," *Nanophotonics*, vol. 9, no. 1, pp. 167–176, 2020.

[124] J. Mejía-Salazar, and O. N. Oliveira Jr, "Plasmonic biosensing: Focus review," *Chemical Reviews*, vol. 118, no. 20, pp. 10617–10625, 2018.

[125] N. Wada, Y. Kurokawa, Y. Miyazaki *et al.*, "The characteristics of the serum carcinoembryonic antigen and carbohydrate antigen 19–9 levels in gastric cancer cases," *Surgery Today*, vol. 47, no. 2, pp. 227–232, 2017.

[126] J. Zhou, F. Tao, J. Zhu *et al.*, "Portable tumor biosensing of serum by plasmonic biochips in combination with nanoimprint and microfluidics," *Nanophotonics*, vol. 8, no. 2, pp. 307–316, 2019.

[127] J. Zhu, Z. Wang, S. Lin *et al.*, "Low-cost flexible plasmonic nanobump metasurfaces for label-free sensing of serum tumor marker," *Biosensors and Bioelectronics*, vol. 150, pp. 111905, 2020.

[128] H. Im, H. Shao, Y. I. Park *et al.*, "Label-free detection and molecular profiling of exosomes with a nano-plasmonic sensor," *Nature Biotechnology*, vol. 32, no. 5, pp. 490–495, 2014.

[129] N. J. Wittenberg, T. W. Johnson, and S.-H. Oh, "High-density arrays of submicron spherical supported lipid bilayers," *Analytical Chemistry*, vol. 84, no. 19, pp. 8207–8213, 2012.

[130] X. Li, M. Soler, C. I. Özdemir *et al.*, "Plasmonic nanohole array biosensor for label-free and real-time analysis of live cell secretion," *Lab on a Chip*, vol. 17, no. 13, pp. 2208–2217, 2017.

[131] A. A. Yanik, M. Huang, O. Kamohara *et al.*, "An optofluidic nanoplasmonic biosensor for direct detection of live viruses from biological media," *Nano Letters*, vol. 10, no. 12, pp. 4962–4969, 2010.

[132] A. A. Yanik, A. E. Cetin, M. Huang *et al.*, "Seeing protein monolayers with naked eye through plasmonic Fano resonances," *Proceedings of the National Academy of Sciences*, vol. 108, no. 29, pp. 11784–11789, 2011.

[133] R. Adato, and H. Altug, "In-situ ultra-sensitive infrared absorption spectroscopy of biomolecule interactions in real time with plasmonic nanoantennas," *Nature communications*, vol. 4, no. 1, pp. 1–10, 2013.

[134] D. Dregely, F. Neubrech, H. Duan *et al.*, "Vibrational near-field mapping of planar and buried three-dimensional plasmonic nanostructures," *Nature Communications*, vol. 4, no. 1, pp. 1–9, 2013.

[135] X. Chen, C. Ciracì, D. R. Smith *et al.*, "Nanogap-enhanced infrared spectroscopy with template-stripped wafer-scale arrays of buried plasmonic cavities," *Nano Letters*, vol. 15, no. 1, pp. 107–113, 2015.

[136] J. Xu, Z. Ren, B. Dong *et al.*, "Nanometer-scale heterogeneous interfacial sapphire wafer bonding for enabling plasmonic-enhanced nanofluidic mid-infrared spectroscopy," *ACS Nano*, vol. 14, no. 9, pp. 12159–12172, 2020.

[137] C. V. Raman, "A new radiation," *Indian Journal of Physics*, vol. 2, pp. 387–398, 1928.

[138] E. Smith, and G. Dent, *Modern Raman Spectroscopy: A Practical Approach.* John Wiley & Sons, 2019.

[139] W. Demtröder, *Laser Spectroscopy: Basic Concepts and Instrumentation.* Springer Science & Business Media, 2013.

[140] P. L. Stiles, J. A. Dieringer, N. C. Shah *et al.*, "Surface-enhanced Raman spectroscopy," *Annual Review of Analytical Chemistry*, vol. 1, pp. 601–626, 2008.

[141] B. Sharma, R. R. Frontiera, A.-I. Henry *et al.*, "SERS: Materials, applications, and the future," *Materials Today*, vol. 15, no. 1–2, pp. 16–25, 2012.

[142] A. Boltasseva, and H. A. Atwater, "Low-loss plasmonic metamaterials," *Science*, vol. 331, no. 6015, pp. 290–291, 2011.

[143] K. Kosuda, J. Bingham, K. Wustholz *et al.*, "Nanostructures and surfaceenhanced Raman spectroscopy," *Handbook of Nanoscale Optics and Electronics*, vol. 309, 2010.

[144] X. Yao, S. Jiang, S. Luo *et al.*, "Uniform periodic bowtie SERS substrate with narrow nanogaps obtained by monitored pulsed electrodeposition," *ACS Applied Materials & Interfaces*, vol. 12, no. 32, pp. 36505–36512, 2020.

[145] W. E. Doering, and S. Nie, "Single-molecule and single-nanoparticle SERS: examining the roles of surface active sites and chemical enhancement," *The Journal of Physical Chemistry B*, vol. 106, no. 2, pp. 311–317, 2002.

[146] P. G. Etchegoin, and E. Le Ru, "A perspective on single molecule SERS: current status and future challenges," *Physical Chemistry Chemical Physics*, vol. 10, no. 40, pp. 6079–6089, 2008.

[147] K. Kneipp, Y. Wang, H. Kneipp *et al.*, "Single molecule detection using surface-enhanced Raman scattering (SERS)," *Physical Review Letters*, vol. 78, no. 9, pp. 1667, 1997.

[148] S. Nie, and S. R. Emory, "Probing single molecules and single nanoparticles by surface-enhanced Raman scattering," *Science*, vol. 275, no. 5303, pp. 1102–1106, 1997.

[149] R. M. Stöckle, Y. D. Suh, V. Deckert *et al.*, "Nanoscale chemical analysis by tip-enhanced Raman spectroscopy," *Chemical Physics Letters*, vol. 318, no. 1–3, pp. 131–136, 2000.

[150] B. Pettinger, B. Ren, G. Picardi *et al.*, "Nanoscale probing of adsorbed species by tip-enhanced Raman spectroscopy," *Physical Review Letters*, vol. 92, no. 9, pp. 096101, 2004.

[151] V. Shalaev, and S. Kawata, *Advances in Nano-optics and Nano-photonics.* Elsevier, 2007.

[152] Z.-Q. Tian, B. Ren, J.-F. Li *et al.*, "Expanding generality of surface-enhanced Raman spectroscopy with borrowing SERS activity strategy," *Chemical Communications*, no. 34, pp. 3514–3534, 2007.

[153] A. V. Krasavin, P. Ginzburg, and A. V. Zayats, "Free-electron optical nonlinearities in plasmonic nanostructures: a review of the hydrodynamic description," *Laser & Photonics Reviews*, vol. 12, no. 1, pp. 1700082, 2018.

[154] R. Boyd, and B. Masters, *Nonlinear Optics*, 3rd ed. Academic, 2008.

[155] Y.-R. Shen, *Principles of Nonlinear Optics.* Wiley-Interscience, 1984.

[156] M. Kauranen, and A. V. Zayats, "Nonlinear plasmonics," *Nature Photonics*, vol. 6, no. 11, pp. 737–748, 2012.

[157] A. A. Maradudin, J. R. Sambles, and W. L. Barnes, *Modern Plasmonics.* Elsevier, 2014.

[158] C. M. Watts, X. Liu, and W. J. Padilla, "Metamaterial electromagnetic wave absorbers," *Advanced Materials*, vol. 24, no. 23, pp. OP98–OP120, 2012.

[159] J. Zhu, L. Zhang, S. Jiang *et al.*, "Selective light trapping of plasmonic stack metamaterials by circuit design," *Nanoscale*, vol. 12, no. 3, pp. 2057–2062, 2020.

[160] N. I. Landy, S. Sajuyigbe, J. J. Mock *et al.*, "Perfect metamaterial absorber," *Physical Review Letters*, vol. 100, no. 20, pp. 207402, 2008.

[161] Z. H. Jiang, S. Yun, F. Toor *et al.*, "Conformal dual-band near-perfectly absorbing mid-infrared metamaterial coating," *ACS Nano*, vol. 5, no. 6, pp. 4641–4647, 2011.

[162] Q.-Y. Wen, H.-W. Zhang, Y.-S. Xie *et al.*, "Dual band terahertz metamaterial absorber: Design, fabrication, and characterization," *Applied Physics Letters*, vol. 95, no. 24, pp. 241111, 2009.

[163] K. Yang, J. Wang, X. Yao *et al.*, "Large-area plasmonic metamaterial with thickness-dependent absorption," *Advanced Optical Materials*, vol. 9, no. 1, pp. 2001375, 2021.

[164] J. Zhu, Q. H. Liu, and T. Lin, "Manipulating light absorption of graphene using plasmonic nanoparticles," *Nanoscale*, vol. 5, no. 17, pp. 7785–7789, 2013.

[165] L. Ye, Y. Chen, G. Cai *et al.*, "Broadband absorber with periodically sinusoidally-patterned graphene layer in terahertz range," *Optics Express*, vol. 25, no. 10, pp. 11223–11232, 2017.

[166] Y. Cai, J. Zhu, and Q. H. Liu, "Tunable enhanced optical absorption of graphene using plasmonic perfect absorbers," *Applied Physics Letters*, vol. 106, no. 4, pp. 043105, 2015.

[167] Y. Cai, J. Zhu, Q. H. Liu *et al.*, "Enhanced spatial near-infrared modulation of graphene-loaded perfect absorbers using plasmonic nanoslits," *Optics Express*, vol. 23, no. 25, pp. 32318–32328, 2015.

[168] T. W. Ebbesen, H. J. Lezec, H. Ghaemi *et al.*, "Extraordinary optical transmission through sub-wavelength hole arrays," *Nature*, vol. 391, no. 6668, pp. 667–669, 1998.

[169] B. Huang, F. Yu, and R. N. Zare, "Surface plasmon resonance imaging using a high numerical aperture microscope objective," *Analytical Chemistry*, vol. 79, no. 7, pp. 2979–2983, 2007.

[170] W. Wang, Y. Yang, S. Wang *et al.*, "Label-free measuring and mapping of binding kinetics of membrane proteins in single living cells," *Nature Chemistry*, vol. 4, no. 10, pp. 846–853, 2012.

[171] Y. Yang, H. Yu, X. Shan *et al.*, "Label-free tracking of single organelle transportation in cells with nanometer precision using a plasmonic imaging technique," *Small*, vol. 11, no. 24, pp. 2878–2884, 2015.

[172] K. Syal, W. Wang, X. Shan *et al.*, "Plasmonic imaging of protein interactions with single bacterial cells," *Biosensors and Bioelectronics*, vol. 63, pp. 131–137, 2015.

[173] W. Jing, Y. Wang, Y. Yang *et al.*, "Time-resolved digital immunoassay for rapid and sensitive quantitation of procalcitonin with plasmonic imaging," *Acs Nano*, vol. 13, no. 8, pp. 8609–8617, 2019.

[174] S. Kawata, Y. Inouye, and P. Verma, "Plasmonics for near-field nano-imaging and superlensing," *Nature Photonics*, vol. 3, no. 7, pp. 388–394, 2009.

[175] X. Ni, A. V. Kildishev, and V. M. Shalaev, "Metasurface holograms for visible light," *Nature Communications*, vol. 4, no. 1, pp. 1–6, 2013.

[176] S. Choudhury, U. Guler, A. Shaltout *et al.*, "Pancharatnam–Berry phase manipulating metasurface for visible color hologram based on low loss silver thin film," *Advanced Optical Materials*, vol. 5, no. 10, pp. 1700196, 2017.

[177] F. Ding, Z. Wang, S. He *et al.*, "Broadband high-efficiency half-wave plate: a supercell-based plasmonic metasurface approach," *ACS Nano*, vol. 9, no. 4, pp. 4111–4119, 2015.

[178] F. Yue, D. Wen, J. Xin *et al.*, "Vector vortex beam generation with a single plasmonic metasurface," *ACS Photonics*, vol. 3, no. 9, pp. 1558–1563, 2016.

[179] X. Ni, Z. J. Wong, M. Mrejen *et al.*, "An ultrathin invisibility skin cloak for visible light," *Science*, vol. 349, no. 6254, pp. 1310–1314, 2015.

[180] S. Ishii, A. V. Kildishev, V. M. Shalaev *et al.*, "Metal nanoslit lenses with polarization-selective design," *Optics Letters*, vol. 36, no. 4, pp. 451–453, 2011.

5 Artificial Metamaterials, Metasurfaces, and Their Applications

Dacheng Wang

CONTENTS

5.1 Introduction...121
5.2 Fundamental Physics 123
    5.2.1 Principle of Metamaterials..123
    5.2.2 Principles of Metasurfaces..126
5.3 Applications of Metamaterials and Metasurfaces....................................129
    5.3.1 Wavefront Shaping..130
    5.3.2 Polarization Conversion..134
    5.3.3 Metasurface Holography...137
    5.3.4 Active Metasurfaces..139
5.4 Summary and Outlook..143
Acknowledgements...145
Bibliography 145

5.1 INTRODUCTION

Modern advanced optical devices and systems operating from visible to infrared, terahertz, and millimeter wave, are derived from naturally available bulk material, such as metals, dielectrics, plastics, and ceramic. Despite the excellent performances of these bulk materials in electromagnetic wave control, they are either obtained from natural environment or synthesized through excessive experiments with tedious trial-and-error process. More importantly, the inherent properties of these bulk materials (*e.g.*, positive refractive index) constrain the scientists and technicians to design and realize exotic electromagnetic responses (*e.g.*, invisible cloaking). Hence, a long-standing dream for researchers is to develop an artificial means to define the properties of materials at will and control the electromagnetic waves flexibly and precisely. Such fabulously scientific fantasy motivates the birth of metamaterials. The prefix *meta* is a Greek word with the meaning of beyond, indicating metamaterials with exotic characteristics that have not been seen in natural materials. Metamaterials are normally composed by unique subwavelength micro- and nanostructures, in which the interactions between light and these man-made structures, rather than the intrinsically chemical composition, give rise to the desired properties. As a result, metamaterials shed light on artificial control of electromagnetic

DOI: 10.1201/9781003202608-5

waves in a pre-designed manner, which can hardly be achieved with conventional bulk materials.

The earliest history of metamaterials can be traced back to the end of 19th century. In 1898, Jagadish Chandra Bose studied twisted structures in microwave regime with chiral properties [1]. These twisted designs are similar to those recent concepts of artificial chiral metamaterials. In the early 20th century, Karl Ferdinand Lindman investigated metallic wire helices with the chirality properties [2]. Afterwards, most metamaterials related works were focusing on artificial dielectrics for microwave beam shaping at around 1950s under the physics of antenna theory [3]. In 1968, a pioneering work was performed theoretically by Victor Veselago with the concept of negative-index materials [4]. He proposed a left-handed medium, which simultaneously satisfied negative permittivity and permeability. When light transmitted through this medium, the wave vector, electric field vector and magnetic field vector followed the left-handed law. The phase velocity of the propagated light is antiparallel to the direction of Poynting vector, which is contrary to conventional materials. Such negative-index material concept existed for years in theory without experimental realization until 1990s. Sir John Pendry proposed revolutionary structures made of periodic metallic split ring resonators and thin wires, which simultaneously generated negative permittivity and permeability [5]. In 2000, David R. Smith's group experimentally demonstrated negative index metamaterials in microwave by stacking periodical split ring resonators and thin wires [6]. Since then, metamaterials have attracted widespread attentions with diversified functionalities. Various exotic applications, such as invisibility cloaking, super-resolution imaging and near-zero refractive index, have been investigated based on the effective medium model, covering a broad spectral range from visible to infrared, terahertz, and millimeter waves [7-9]. Although metamaterials promise tremendous fascinating characteristics and applications, fabricating three dimensional metamaterials is still challenging with current fabrication techniques, especially in optical frequency. Other inevitable constrains of metamaterials are the high loss and strong dispersion accompanied by the resonators, limiting the efficiency and performance of meta-devices. To resolve these restrictions, metasurfaces have promptly emerged as an alternative design paradigm in controlling electromagnetic waves.

The metasurfaces can be defined as an artificially structured interface with sub-wavelength thickness. Metasurfaces manipulate electromagnetic wave through spatially arranged meta-atoms, which can be regarded as the two-dimensional metamaterials. However, different from metamaterials with designed constitutive effective optical parameters in three-dimensional space, metasurfaces control electromagnetic wave by specific boundary conditions. Such planar interface design approaches show great potentials to overcome the dilemma that volumetric metamaterials confronted with. The history of metasurfaces can be traced back to the early investigation of Wood's anomaly in metallic gratings and the discovery of surfaces plasmon polariton at metal surfaces (see Chapter 4) [10]. In microwave technology, some conventional designs, such as frequency selective surfaces, can also be categorized as metasurfaces. In 2011, a milestone work was proposed by Capasso's group with the first demonstration of generalized Snell's law based on phase discontinuities created by V-shape plasmonic metasurfaces [11]. After that, various plasmonic metasurfaces have been proposed to realize exotic functionalities, including polarization convertors, metalenses, holograms, vortex generators, among others [12,13]. However, plasmonic metasurfaces encounter high intrinsic ohmic losses in the optical ranges, which limit the efficiencies of meta-devices and hinder their practical applications. Thus, dielectric metasurfaces, composed by high index all-dielectric antennas, are proposed to overcome the loss issue [14]. In addition, dielectric metasurfaces show great potential for fabrication using the complementary metal oxide semiconductor (CMOS) techniques. Another unique advantage of dielectric metasurfaces is their abilities to support magnetic type resonances, which enable versatile means to control light-matter interaction and mold electromagnetic waves. More recently, active metasurfaces have drawn broad research interests for the pursuit of compact and multifunctional meta-devices with high performances.

5.2 FUNDAMENTAL PHYSICS

5.2.1 PRINCIPLE OF METAMATERIALS

In the electromagnetic theory, the response of an optical element under the excitation of electromagnetic waves is mainly defined by its macroscopic parameters of electric permittivity ε and magnetic permeability μ (See Chapter 2) [15]. Hence, materials can be classified into different categories with respect to different values of permittivities and permeabilities, as shown in Figure 5.1. When both permittivity and permeability are larger than zero ($\varepsilon > 0$, $\mu > 0$), the material is noted as a double positive (DPS) material. Most naturally available materials (*e.g.*, dielectrics) are filled into this category. When a material possesses negative permittivity and positive permeability ($\varepsilon < 0$, $\mu > 0$), this material is defined as an epsilon-negative (ENG) material. Various plasmas show such properties in certain wavelength ranges. For instance, many noble metals, such as gold and silver, are ENG materials in visible and infrared ranges.

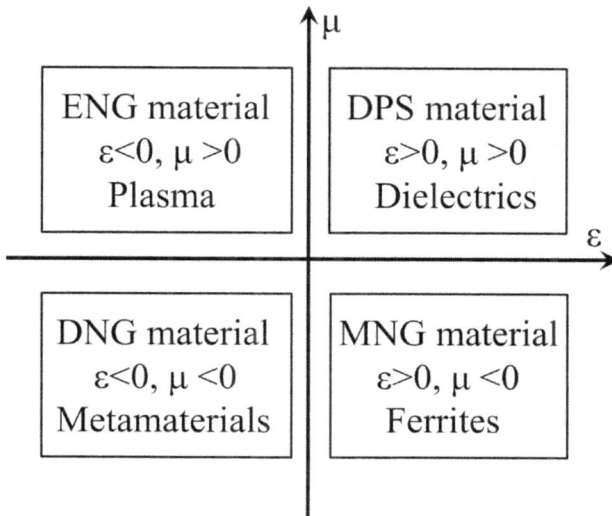

FIGURE 5.1   Classification of materials by their permittivity and permeability.

When a material exhibits positive permittivity and negative permeability ($\varepsilon > 0$, $\mu < 0$), it is named as a mu-negative (MNG) material with typical examples of some ferrites. When a material presents negative permittivity and negative permeability ($\varepsilon < 0$, $\mu < 0$), it is designated as a double-negative (DNG) material. DNG materials were only a theoretical proposal until the discovery and realization of metamaterials. As mentioned before, Sir John Pendry proposed periodic thin wires and periodic split ring resonators, which independently achieved negative effective electric permittivity and magnetic permeability at low frequencies. Next, we will discuss the operating principle of DNG metamaterials.

Negative electric permittivity Permittivity smaller than zero is not rare in natural materials. A typical example is plasma with a dispersion expressed as $\varepsilon(\omega) = 1 - \left(\omega_p^2 / \omega^2 \right)$, where ω_p is the plasma frequency. It is observed that when $\omega < \omega_p$, the permittivity of plasma medium could be negative. In addition, some negative metals behave like plasma under the excitation of electromagnetic waves with the plasma frequency as [5]:

$$\omega_p^2 = \frac{n_e e^2}{\varepsilon_0 m_{eff}}, \tag{5.1}$$

where n_e is the density of electrons, e is the electron charge, m_{eff} is the effective mass of electrons and ε_0 is the vacuum dielectric constant. The plasma frequencies of most metals are within UV and visible frequencies. When the operating frequency is lower than the plasma frequency, these metals show negative permittivities. However, when the frequency decreases, the permittivity of bulk metal decreases dramatically accompanied with a significant increase in loss. How to effectively control the negative permittivity is critical, which could be resolved by the metallic thin wire array model proposed by John Pendry. Based on Eq. (5.1), one could engineer the plasma frequency by controlling the density and the effective mass of electrons in a plasma system. For metallic wires with the radius of r and the periodicity of a, the density of electrons can be described as:

$$n_{eff} = \frac{\pi r^2}{a^2} n_e. \tag{5.2}$$

Due to the self-inductance effect, the effective mass of electrons in the metallic wire arrays can be expressed as:

$$m_{eff} = \frac{\mu_0 \pi r^2 e^2 n_e}{2\pi} \ln(a/r) \tag{5.3}$$

By changing the structure parameters of r and a, the density and the effective mass of electrons can be tuned, which further modify the plasma frequency. It was predicted that the plasma frequency of aluminum wire arrays can reach 8.2 GHz with the radius at 1 μm and the periodicity of 5 mm. Pendry's work provides the theoretical basis for the realization of negative permittivity in DNG materials in microwave ranges, which has been verified later in experiments. The effective permittivity of metallic wire arrays is derived as [16]:

$$\varepsilon_{eff} = 1 - \frac{\omega_p^2 - \omega_0^2}{\omega^2 - \omega_0^2 + i\omega\Gamma}, \tag{5.4}$$

where ω_0 is the resonant frequency defined by the structural parameters, Γ is the loss rate. When the frequency satisfies the relation of $\omega_0 < \omega < \omega_p$, the effective permittivity is negative. It should be noted that the structural parameters are much smaller than the wavelength of the incident electromagnetic wave, it is reasonable to describe the optical properties of the metallic wire arrays based on the effective permittivity.

Negative magnetic permeability Compared with negative permittivity, there are only a few materials that exhibit negative permeability. For instance, some ferrites show negative permeability when ferromagnetic resonances are excited under external magnetic field and illumination of electromagnetic wave [17]. However, such phenomena mostly occur at low frequency ranges (*e.g.* microwave or radio frequency). Beyond terahertz and at even higher frequency, natural materials generally do not present negative permeability. In principle, materials with magnetic responses should have magnetic dipoles, which are generated by loop current caused by electron spin and/or orbital motions. To overcome the limitation of nature substance, John Pendry proposed split ring resonators to produce artificial loop current (or pseudo loop current) [18]. As shown in Figure 5.2, when the split ring resonators are excited by electromagnetic waves with the magnetic field perpendicular to the plane of the resonators, induced loop current can be generated due to the electromagnetic inductance effect. This induced current would further create induced magnetic field mimicking Ampere's law, leading to strong magnetic responses and potentially negative magnetic permeabilities. The split in the resonator can generate transient accumulated charges, which acts as a capacitor. The ring in the resonator acts as an inductor. Therefore, the electromagnetic responses of the split ring resonators can be treated as

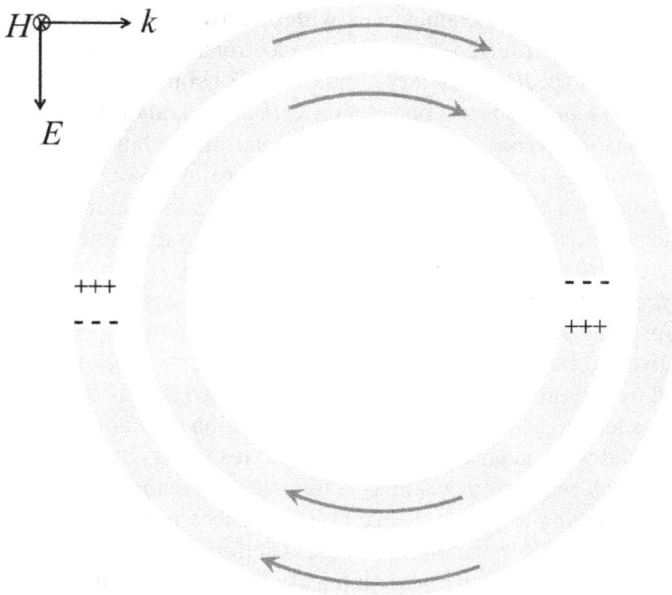

FIGURE 5.2 Schematic of split ring resonators.

equivalent LC circuit, which can be tuned by the geometries and dimensions of the resonators. It should be emphasized that the gap in the resonator not only enhances the resonance strength, but also decreases the resonance frequency remarkably. As a result, the operating wavelength of the split ring resonator is much smaller than the dimensions of the resonators, making effective magnetic permeability applicable in describing the overall magnetic response of these resonator arrays. The effective magnetic permeability of periodic split ring resonator arrays can be derived as:

$$\mu_{eff} = 1 - \frac{F\omega_0^2}{\omega^2 - \omega_0^2 + i\omega\Gamma}, \tag{5.5}$$

where F is the filling factor of the split ring resonator. Assuming that the magnetic plasma frequency of the split ring resonator is $\omega_{mp} = \omega_0/\sqrt{1-F}$, when the frequency of electromagnetic wave is $\omega_0 < \omega < \omega_{mp}$, the metamaterials composed by split ring resonator arrays would exhibit an effective negative permeability. Based on Eq. (5.5), David R. Smith's group utilized conventional circuit processing techniques to fabricate the wire arrays and split ring resonators arrays and measured the transmission spectra, which verified the existence of DNG materials [6]. Afterwards, a variety of novel DNG metamaterials have been proposed and demonstrated, including Ω-shape, U-shape, double-rod, and fishnet structures [19-22]. The operating frequencies are extended from microwave to terahertz, infrared and visible ranges, offering a versatile means to control electromagnetic waves with exotic functionalities.

5.2.2 Principles of Metasurfaces

Unlike three dimensional metamaterials with effective macroscopic permittivities and permeabilities, metasurfaces manipulate electromagnetic waves through artificial interfaces with special boundary conditions. Generally, metasurfaces are tackling with optical responses based on antenna resonances and/or Pancharatnam-Berry phase (i.e., geometric phase). The antenna resonance utilized in metasurfaces mainly focuses on controlling linearly polarized electromagnetic waves, while the geometric phase is associated with circularly polarized waves. In this section, we discuss the basic principle of metasurfaces involved with the antenna resonance and the geometric phase for electromagnetic wave control.

Metasurfaces with antenna resonances Conventional optical elements for wavefronts control rely on the phase gradually accumulated along the optical path. At the interface between different bulky materials, the propagations of electromagnetic waves are governed by Fresnel equations and Snell's law, which strictly determine the transmission and reflection coefficients and their propagation directions [15]. If this interface is embedded with an array of subwavelength resonators of deep-subwavelength thicknesses, which refers to metasurfaces, the transmission and reflection properties could be altered owing to phase change at the interface induced by the variation of boundary conditions. When the phase change is uniformly distributed along the interface, it would not modify the transmission and reflection properties. When the phase change is not uniformly distributed at the interface (e.g., gradient phase change), the metasurfaces modify the propagation properties and the wavefront shapes. According to Fermat's principle that light travels along an extremum path, generalized laws of refraction and reflection with respect to metasurfaces can be derived as:

$$\begin{cases} n_t \sin(\theta_t) - n_i \sin(\theta_i) = \dfrac{1}{k_0} \dfrac{d\Phi}{dx} \\[2ex] \cos(\theta_t)\sin(\varphi_t) = \dfrac{1}{n_t k_0} \dfrac{d\Phi}{dy} \end{cases} \tag{5.6}$$

$$\begin{cases} \sin(\theta_r) - \sin(\theta_i) = \dfrac{1}{n_i k_0} \dfrac{d\Phi}{dx} \\[2ex] \cos(\theta_r)\sin(\varphi_r) = \dfrac{1}{n_r k_0} \dfrac{d\Phi}{dy} \end{cases} \tag{5.7}$$

where the angles are defined in Figure 5.3. $d\Phi/dx$ and $d\Phi/dy$ are the phase gradients that are parallel and perpendicular to the plane of the incidence, respectively. The generalized Snell's laws indicate that by tuning the interfacial phase gradient of the metasurfaces, the transmitted and reflected light can be guided into arbitrary directions in their corresponding half spaces. To realize generalized Snell's laws, a

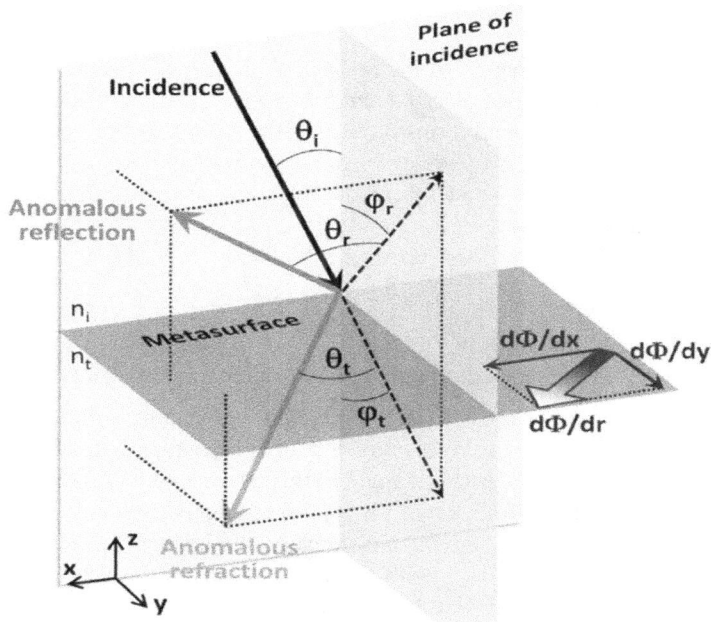

FIGURE 5.3 Schematic of anomalous reflection and refraction in the metasurfaces with gradient phase shifts.

Source: Reprinted with permission from Chen, H. T. et al. *Rep. Prog. Phys.* 79, 076401 2016. Copyright 2016 IOP Publishing.

pioneer work was performed by Capasso's group, in which they introduced gradient phase discontinuity at the interface with V-shape antennas [11]. These anisotropic antennas support two different plasmonic eigenmodes with completely different optical properties. By modifying the geometries and dimensions of the V-shape antennas (*e.g.*, different opening angles, orientations, and arm lengths), a phase gradient with a step of $\pi/4$ for the cross polarized incidence can be achieved, while the scattering amplitude of each antenna maintains uniform. Considering that the periodicity of the antennas is much smaller than the wavelength in free space, such metasurfaces satisfy the generalized Snell's laws and demonstrate anomalous refraction and reflection. Besides, other types of antennas with different geometries, such as C-shape, cross-shape and U-shape split ring resonators, have also been proposed to manipulate linearly polarized waves [23,24].

Metasurfaces with geometric phase Different from the control of linearly polarized light with the resonance behaviors of the gradient metasurfaces, the manipulation of circularly polarized light in metasurfaces is mainly based on the geometric phase. The geometric metasurfaces are composed by identical and anisotropic subwavelength antennas, but with spatially varying in-plane orientations. When circularly polarized light (*e.g.*, right-handed circular polarization (RCP)) is illuminated on the antenna, the electric field of the transmitted light can be expressed as:

$$E_t \propto J(\varphi)E_{inc}^{RCP} = \cos\frac{\varphi}{2}\begin{bmatrix}1\\i\end{bmatrix} + i\sin\frac{\varphi}{2}\begin{bmatrix}1\\-i\end{bmatrix}, \tag{5.8}$$

where $J(\varphi)$ is the Jones matrix of the anisotropic antenna. It is observed that the transmitted light contains co-polarized and cross-polarized components, which correspond to the first and second terms of Eq. (5.8), respectively. The cross-polarized component is related to the phase delay between the short and long axes in the anisotropic antenna. When the antenna is rotated in-plane by an angle of θ, the transmitted electric field can be expressed as:

$$E_t \propto J_{(\varphi)}^{(\theta)}\begin{bmatrix}1\\i\end{bmatrix} = \cos\frac{\varphi}{2}\begin{bmatrix}1\\i\end{bmatrix} + i\sin\frac{\varphi}{2}e^{i2\theta}\begin{bmatrix}1\\-i\end{bmatrix}, \tag{5.9}$$

where $J_{(\varphi)}^{(\theta)}$ is the Jones matrix of the rotated antenna. It can be seen from the second term in Eq. (5.9) that the cross polarized component obtains an additional phase shift of 2θ with respect to the rotation of the antenna, which is the origin of geometric phase [25]. When the rotation angle of the antenna varies from 0 to 180°, the geometric phase shift can fully cover the $[0, 2\pi]$ range. This phenomenon can also be considered as that the rotated antenna converts the incident circularly polarized light into its opposite handedness with an extra phase change of 2θ. Based on this principle, by properly control the rotation angle of the anisotropic antenna in different spatial positions, geometric phase based gradient metasurfaces can be designed to convert incident circular polarized light into its opposite handedness and guide the cross polarized light into different directions. Compare to metasurfaces with particular spectral resonances, a major advantage of geometric phase metasurface is that the phase shift is dispersionless as it only depends on the rotation angle of the antenna, which naturally gives a broadband operation. Nevertheless, the main limitation of geometric phase based metasurfaces is the low cross polarization conversion efficiency. Despite the versatile functionalities realized in various designs, a single layer plasmonic metasurfaces encountered physical limits for the conversion efficiency. Bi-layer and multilayer configurations were proposed to enhance the efficiency [26,27]. Recently, dielectric geometric metasurfaces enabled an alternative approach to accomplish high conversion efficiencies[28]. The dielectric metasurfaces support both electric and magnetic Mie resonances with low intrinsic material absorption. Based on Huygens principle which results from the coherent interference between electric and magnetic Mie resonances (see Chapter 6), a variety of dielectric functional metasurfaces have been demonstrated to control circular polarized light with high efficiencies. It should be noted that besides most metasurfaces with periodic arrangements of the meta-atoms, disorder metasurfaces with randomly positioned meta-atoms, promise another exotic approach to control electromagnetic waves. For instance, disorder metasurfaces can achieve wavefront shaping with aberration-free focusing and subwavelength focusing [29]. A perfect optical diffuser can be realized in dielectric disorder metasurfaces [30]. Such disorder metasurfaces might continue to contribute the design of novel meta-devices.

5.3 APPLICATIONS OF METAMATERIALS AND METASURFACES

The last two decades have witnessed the rapid development of metamaterials and metasurfaces with many intriguing applications. Particularly, by controlling the effective permittivity and permeability, three dimensional metamaterials reveal bizarre optical behaviours, such as negative refraction, perfect lens imaging, anomalous Cherenkov radiation and anomalous Doppler effect [4,8,19,31]. Based on these optical properties of metamaterials, a variety of applications can be realized, among which invisible cloaking and super-resolution imaging are two most typical and eye-catching examples. Thus, we mainly focus on reviewing the applications of three dimensional metamaterials in these two examples.

The concept of invisible cloaking using metamaterials was first proposed by John Pendry in 2006, which was experimentally verified by David R. Smith's group soon after in the microwave ranges [7]. The invisible cloaking was built on the capability of artificial manipulation of the effective electromagnetic parameters and their spatial distributions under the theory of transformation optics. However, these three dimensional metamaterial based cloaking devices encounter narrow band operation, high loss and complex fabrication process, which limit their practical applications. Such limitations become even more serious for cloaking devices operating at optical frequency ranges. In 2009, a two dimensional ground-plane cloaking device was proposed, which simultaneously achieved low-loss and broadband operation with non-resonant structures [32]. In 2010, a three dimensional ground-plane cloaking device was realized by engineering the distribution of the refractive index with respect to different designs of the dielectric unit cells [33]. By modifying the dimensions of the dielectric resonators, gradient refractive index distributions can be achieved. The measured and simulated electric field distributions indicated the cloaking device can effectively conceals the object inside. On the road of extending the operating frequencies to visible ranges, multilayered cloaking designs using dielectric structures have been proposed with the first experimental verification demonstration in 2009 [34]. However, such cloaking devices can only cover micrometer-scale objects. Despite that numerous cloaking concepts have been proposed, obvious constrains still exist for three dimensional metamaterials, such as limited operation bandwidth, complex fabrication process and bulky designs. In recent years, two dimensional metasurface based invisibility was realized with advantages of compact device design and easy for fabrication, which endows an alternative and effective means for invisible cloaking.

Another fascinating application in metamaterials is super-resolution imaging. Conventionally, it is not possible to achieve sub-wavelength resolution in optical lenses due to the notorious Abbe's diffraction limit. However, metamaterials assisted perfect lens can break such fundamental limit because the negative index can enhance the evanescent waves and contribute to the super-resolution imaging. The earliest demonstration of perfect lens is made of metal materials, which could enhance the evanescent waves via surface plasmon resonances and subsequently converted to propagating waves. In 2005, superlens was proposed by Xiang Zhang's group for super-resolution imaging, in which the resolution was decreased to one sixth of the wavelength [8]. Many other designs, such as hyperlens, have also been widely investigated for super-resolution imaging [35]. However, metal based superlens and hyperlens inevitably suffer high losses [8,35], rendering it inapplicable for practical applications. Besides, the near-field operation principle of surface plasmon also brings considerable challenges in reality. In order to circumvent the limitations of three dimensional metamaterials, two dimensional metasurfaces are proposed and attract tremendous research interests for meta-devices and applications. In the following, we will review recent progress of flat metasurfaces-based applications in detail.

5.3.1 Wavefront Shaping

Conventional lenses, which are indispensable tools among different scientific applications, shape the wavefronts of electromagnetic waves with curved surfaces and gradually accumulated phase retardation. Such lenses are inevitably large and impose restrictions on the compact system integration. Metalenses, which are metasurface-based lenses, enable an efficient approach to shape the wavefront of electromagnetic wave at will. Beyond the previous discussed anomalous reflection and refraction under the concept of generalized Snell's law, metalenses are demonstrated with much more possibilities. Compared with conventional lenses, metalenses are flat and can mold electromagnetic wave locally and artificially at subwavelength scale, providing unprecedented capabilities to control the functionalities of devices.

To realize a typical planar metalens with focusing functionality, the metasurface should follow a defined phase profile as follows:

$$\varphi(x, y) = \frac{2\pi}{\lambda}(\sqrt{x^2 + y^2 + f^2} - f) \qquad (5.10)$$

where λ is the operating wavelength and f is focal length. This phase profile function can be replaced with others to realize different functionalities. The first flat metalens has been experimentally realized at telecom wavelength in 2012 by Capasso's group [36]. As shown in Figure 5.4(a), V-shaped plasmonic metal antennas were adopted to induce phase discontinuities that follow well-defined phase profile required in lenses and axicons. Based on these precisely defined antennas, spherical wavefronts and nondiffracting Bessel beams were generated. This metalens also presents high numerical aperture (NA) without spherical aberration, which is highly desirable for practical applications. Besides the V-shaped antennas, many other designs, including C-shaped antennas and rectangular antennas, have been experimentally demonstrated for planar metalens [37,38]. According to Babinet's principle, complementary designs, such as V-shaped apertures and U-shaped aperture, can also focus electromagnetic waves and suppress the background light [39,40]. One of the main limitations of these metalenses is the low focusing efficiency, because they utilized only single layer resonators with small filling factors. Meanwhile, their focusing performance relies on cross-polarization conversion with respect to the incident light, which yields a theoretical maximum efficiency of 25% for the single layer design. In recent years, flat lenses composed by catenary resonators have been proposed with continuous geometric phases, which achieve the efficiency close to this theoretical ceiling boundary [41]. To further improve the efficiency, multilayered metalenses and dielectric metalenses have been demonstrated recently. As shown in Figure 5.4(b), a sandwiched design, consisted of three layers of metallic resonators separated by dielectric spacers, has been proposed to achieve a transmission efficiency beyond 45% [42]. Another tri-layered metalens has exhibited a high focusing efficiency of 68% at 400 GHz [43]. In these multilayer designs, the Fabry-Perot-like resonances among different layers make a significant contribution to the overall efficiency. Despite the progress of plasmonic metalenses in recent years, their performance still remains unsatisfactory due to the inherent high energy dissipations of metal materials.

Lossless dielectric metalenses are introduced based on the concept of Huygens' principle, in which the spectral overlapping of electric and magnetic dipole resonances can accomplish high efficiency and 2π phase coverage. For instance, Huygens metalenses were proposed by Zhang et al. and achieved diffraction-limited focusing and imaging in mid-infrared ranges, as shown in Figure 5.4(c) [44]. The focusing efficiency reaches 75% with a deep subwavelength device thickness. The drawback of this metalens is the narrow operation bandwidth accompanied by the electric and magnetic modes overlapping. Another significant problem is the mutual coupling effect between adjacent dielectric resonators, which deteriorates the performance of the metalenses, especially for high NA with steep phase gradient. To circumvent these disadvantages, high contrast dielectric metasurfaces are developed, in which high refractive index resonators are designed to maintain a device thickness comparable to the wavelength. These high index dielectric resonators can be regarded as truncated waveguides to confine light with the waveguide resonances. The propagation phase of light in the waveguide can be defined as:

$$\varphi = \frac{2\pi H}{\lambda} n_{eff} \tag{5.11}$$

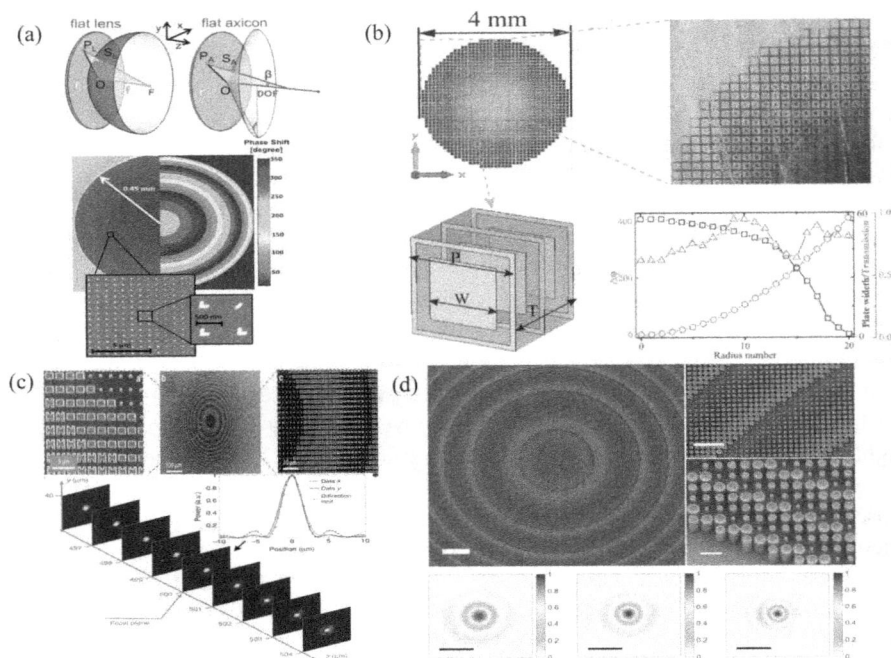

FIGURE 5.4 (a) Schematic of the design of flat lenses and axicons and the phase shift profile with the insets of patterned antennas.

Source: Reprinted with permission from Aieta, F. et al. *Nano Lett.* 12, 4932–4936, 2012. Copyright 2012 American Chemical Society.

(b) Design of the multilayered metalens.

Source: Reprinted with permission from Yang, Q. et al. *Adv. Opt. Mater.* 5, 1601084, 2017. Copyright 2017 John Wiley and Sons.

(c) Design of the dielectric metalens with the focusing performance.

Source: Adopted from Zhang, L. et al. *Nat. Commun.* 9, 1481, 2018. No permission is required. Copyright 2018 the authors.

(d) The fabricated metalens with the focal spots at the designed wavelengths.

Source: Reprinted with permission from Khorasaninejad, M. et al. *Nano Lett.* 16, 7229–7234, 2016. Copyright 2016 American Chemical Society.

where λ is the operating wavelength, H is height of the resonator and n_{eff} is the effective index determined by the structural dimensions of the resonators. Thus, by tailoring the dimensions of dielectric resonators, one can design highly efficient metalenses with large phase coverage. In the visible frequency ranges, many dielectric materials can be adopted to design high contrast metalenses, such as titanium dioxide (TiO_2) and gallium nitride (GaN). The refractive indices of these materials are around 2 to 2.5 which require high aspect ratio to obtain full wavefront control. A TiO_2 metalens has been demonstrated at the wavelength of 405 nm, 532 nm, and 660 nm with the efficiencies of 86%, 73% and 66%, respectively [45]. GaN-based metalenses have also been realized in the visible frequency ranges with high efficiencies [46]. In the infrared frequency ranges, silicon is commonly utilized for designing dielectric metalenses owing to its advantages of high refractive index ~3.5, low loss and mature fabrication techniques. Arbabi et al. proposed silicon-based metalens in infrared with high efficiency and diffraction limited focusing [47]. Many other metalenses operating at mid- and far-infrared regions have also been demonstrated with good performance. At lower frequency ranges, such as terahertz and millimeter waves, dielectric metalenses exhibited similar performance for focusing light into a diffraction limited spot.

To characterize the performance of metalens, the primary key indicator is the efficiency. As shown in Figure 5.4(d), circular nano-pillar based dielectric metalens has been demonstrated at the wavelength of 600 nm, in which the focusing efficiency is up to 90% with the focus spot of around 0.64λ [48]. Beside efficiency, the numerical aperture is another important parameter. Generally, the numerical aperture can be manipulated by the refractive index of the environment, the phase distribution of the lens and the distribution of the diffracted energy. A numerical aperture of 1.1 at 532 nm has been achieved by Chen et al. in metalens immersed in oil [49]. Similarly, a silicon-based metalens has been demonstrated with the numerical aperture of 1.48 when it is immersed in oil [50]. Another strategy to increase the numerical aperture is to design a steep phase shift at the edge. In conventional lenses, this requires an extremely sharp change of the thickness of the lenses, which is challenging in the grinding and polishing fabrication process. In metalenses, such steep phase shift can be easily achieved by arranging particular resonators to the proper positions [50]. On top of that, controlling the diffracted light and concentrating it into a particular position is another useful technique to realize metalenses with high numerical aperture. For instance, asymmetric dielectric resonators can diffract light and control the diffraction angles, which have been applied to realize a metalens with a near-unity numerical aperture [51].

In certain applications, especially high-quality imaging, aberration correction is a crucial problem of metalenses. Monochromatic and chromatic aberration corrections in metalenses have been widely investigated in recent years. Monochromatic aberration stems from the non-paraxial effect of the incidence and the chromatic aberration originates from the structural and material dispersions, which would deteriorate the image quality. To resolve the monochromatic aberration, multiple metalenses are stacked together. Typical examples are the doublet meta-lenses, which have been demonstrated with good monochromatic aberration correction over 30° to 50° field of view at near infrared and visible ranges [52,53]. Chromatic aberration correction has been studied and extended to multi-wavelengths and broadband regions. A multi-wavelength infrared metalens has been realized for chromatic aberration correction, which operates at the wavelengths of 1300 nm, 1500 nm, and 1800 nm, respectively [54]. Another metalens proposed by Zhao et al. presents chromatic aberration corrections at the wavelengths of 532 nm, 632 nm, and 785 nm, respectively [55]. To further extend multi-wavelength operation to broadband operation, Tsai et al. proposed to utilize resonant phase and geometric phase simultaneously and achieved broadband achromatic metalenses in visible and infrared ranges [56,57]. Capasso et al. proposed to correct chromatic aberration by precisely tune the phase, group delay and group delay dispersion. Based on this principle, an achromatic metalens composed by coupled nanopillars has been demonstrated with the working wavelength from 470 nm to 670 nm and an average efficiency of 20% [58]. The limitation of such achromatic metalens is the small device size restricted by the limited group delay. This limitation can be mitigated by merging metalenses with conventional optical lenses, which has been demonstrated as the metacorrector [59].

The wavefront shaping by metalenses can be applied to many optical systems, including imaging, sensing, endoscopy, virtual and augmented reality. Other application scenes, such as mini-drones, unmanned aerial vehicle, autonomous driving, also demand high performance and compact meta-components for radar detection and high-speed communication. These emerging applications may drive metalenses into a new era of meta-optics.

5.3.2 POLARIZATION CONVERSION

Polarization state represents one of the most fundamental properties of electromagnetic waves, which conveys valuable information for many photonic applications, such as communication, sensing and imaging. Conventional methods to control polarization rely on birefringent materials with bulky sizes and limited performance. Metasurfaces have garnered intense research attentions in recent years for polarization control owing to their great capabilities and flexibilities in electromagnetic wave control. In this section, we mainly focus on the progress of metasurface-based polarization conversion, including linear-to-circular polarization conversion and polarization rotation.

Linear-to-circular polarization conversion can be realized using anisotropic subwavelength scale resonators. As shown in Figure 5.5(a), when a resonator can support two orthogonal resonances modes with $\pi/2$ phase delay, it can convert a linear polarized light into a circular polarized light [60]. Based on this principle, an ultrathin terahertz quarter-wave plate was experimentally demonstrated using complementary asymmetric cross-shape metasurfaces [61]. By controlling the two dipole modes along the two orthogonal directions, the transmission amplitudes along these two directions can be the same, while their relative phase delay is 90°, leading to a linear-to-circular polarization conversion. Another approach to realize this functionality is to arrange the spatial positions of different anisotropic resonators. As shown in Figure 5.5(b), a broadband quarter-wave plate metasurface was demonstrated in the infrared region with the ellipticity above 0.97. Two sub-units containing various V-shaped antennas are adopted in this design, where the gradient phase variation is utilized to form the anomalously transmitted beam with linear-to-circular polarization conversion [60]. To further enhance the efficiency of polarization conversion, multilayer metasurfaces are proposed. Similar as the multilayer metalens discussed in Section 5.3.1, the Fabry-Perot-like coupling effect among different layers can greatly enhance the polarization conversion efficiency and extend the operating bandwidth. Figure 5.5(c) shows a bilayer design, which achieves a broadband linear-to-circular polarization conversion in the terahertz region [62]. This bilayer design consists of two layers of twisted metallic wire gratings with a relative rotation angle of 45°. The phase dispersions of two gratings can compensate with each other, forming a phase delay of 90° within a relative bandwidth of 40%. The metallic wire gratings are inserted into polyimide, in which the multiple interference effect among different interfaces contributes to the high transmission efficiency. Anisotropic tri-layer metasurfaces have also been investigated for high efficiency polarization conversion. Circular-to-circular polarization has been studied using tri-layer metasurfaces, which present high transmissions in a broadband infrared region [63]. When the anisotropic metasurfaces are backed with a ground metal planar reflector, such designs can enhance the polarization conversion in a broadband manner in the reflection mode. In addition to metallic metasurfaces, dielectric metasurfaces can also be applied to realize highly efficiency polarization conversion. As shown in Figure 5.5(d), silicon

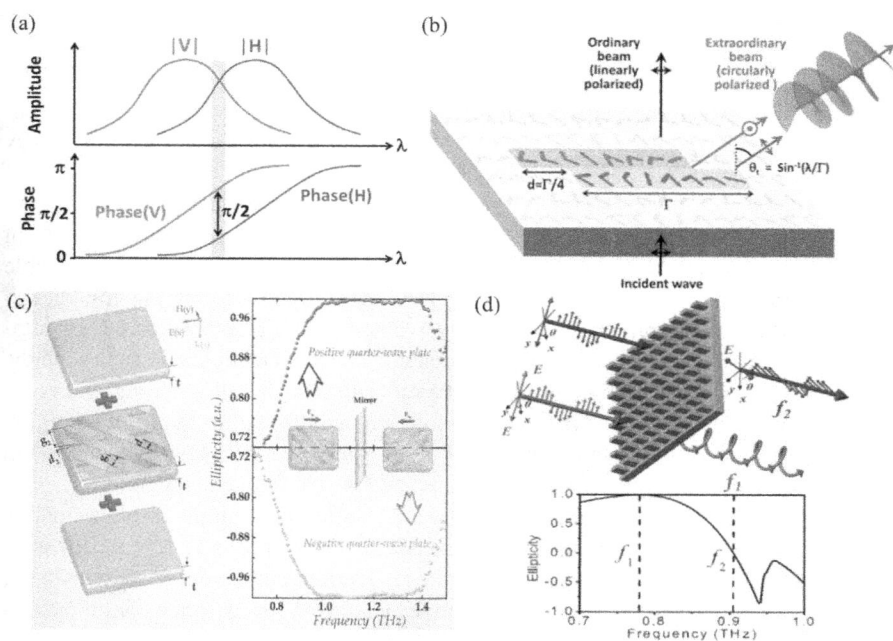

FIGURE 5.5 (a) Schematic of two orthogonal resonance modes with similar amplitude and phase responses at different frequency points. In their spectral overlapped region, the amplitudes are the same with π/2 phase delay, making the device operate as a quarter-wave plate. (b) Schematic of the plasmonic metasurface based broadband quarter-wave plate.

Source: Reprinted with permission from Yu, N. et al. *Nano Lett.* 12, 6328–6333, 2012. Copyright 2012 American Chemical Society.

(c) The design and performance of broadband terahertz metamaterial quarter-wave plate.

Source: Reprinted with permission from Cong, L. et al. *Laser Photonics Rev.* 8, 626–632, 2014. Copyright 2014 John Wiley and Sons.

(d) Schematic of dielectric metasurfaces for dual-functional polarization conversion. *Source:* Reprinted with permission from Wang, D. C. et al. *Appl. Phys. Lett.* 113, 201103, 2018. Copyright 2018 AIP Publishing LLC.

pillar based metasurfaces can realize linear-to-circular polarization conversion and polarization rotation simultaneously at different frequencies. The silicon pillars can support different dipole and quadrupole Mie modes, in which the interference among these modes can suppress backward scattering and expand the phase coverage range, leading to the dual-function polarization conversion with near unity transmission [64]. In addition, reflective type dielectric linear-to-circular polarization conversion metasurfaces have also been demonstrated, which operates over a broadband regime (relative bandwidth up to 80%) with close to unit efficiency [65]. Other dielectric resonators have been studied for broadband terahertz mirrors with a 0.63 THz bandwidth and 92% conversion efficiency [66].

Metasurface-based polarization rotation has been widely investigated in recent years. Anisotropic resonators in metasurfaces can manipulate both the amplitudes and phases efficiently and give rise to a variety of polarization rotators. A simple design of metasurface-based half wave plate was reported by arranging identical cur-wire arrays along the diagonal direction and backing with a ground metal reflector. Based on the multiple interferences between the cut-wire and the ground plane, such metasurfaces can convert linearly polarized light into cross polarized light with a high conversion efficiency beyond 80% over a broad bandwidth from 0.73 to 1.80 THz [67]. The key mechanism of this broadband performance is the destructive interference of the co-polarized reflected components at different frequencies. This concept has also been applied to a variety of multiband and broadband polarization rotation meta-devices from microwave to visible frequency regions in the reflection mode [68,69]. Again, implementing low loss dielectric metasurfaces is a straightforward strategy to obtain high efficiency polarization rotation. Metasurfaces combining rectangle silicon and a ground metal plane can rotate the polarization of the incident light by 90° in the reflection mode with the conversion efficiency above 98% and the operating bandwidth beyond 200 nm in the infrared region [70]. These dielectric resonators can greatly reduce the inevitable loss encountered in metallic metasurfaces, especially in the infrared and visible frequency regions.

Besides the reflection operation, polarization rotation in the transmission mode is highly desirable in many photonic applications. Bilayer and tri-layer metasurfaces have been reported to realize flexible cross polarization conversion at narrow band, multiband and broadband. A typical example of bilayer metasurfaces is composed by one layer of asymmetric split-ring resonator arrays and one layer of S-shaped resonator arrays with a low loss dielectric spacer between them [71]. Under the excitation of linearly polarized light, the split-ring resonator can induce an effective electric dipole, which produces scattered fields along two orthogonal directions. The S-shaped resonators are designed to have the same resonant frequency as the split-ring resonator for propagating co-polarized light, but are transparent for cross-polarized light. By properly tuning the geometric parameters and controlling the Fabry-Perot resonances between the two layers, polarization rotation can be achieved with high efficiency. Another interesting design is proposed by Grady et al. with tri-layer metasurfaces. Two layers of metallic gratings are positioned orthogonal to each other and 45° tilted cur-wire arrays covered by a polyimide spacer are sandwiched between these two layers. Due to a multi-reflection process among different layers, the transmitted co-polarized light is suppressed with the destructive interference and the cross-polarized light is enhanced with the constructive interference. The cross-polarization conversion efficiency is up to 80% with the bandwidth beyond two octaves [67]. Similar design strategies have been applied to other metasurface-based polarization rotators with superior performances. Some chiral designs have been reported in polarization control and asymmetric transmission with the capability of tuning the handedness of circularly polarized light [72]. Active polarization control is another intriguing research area, which will be discussed in the latter section.

5.3.3 METASURFACE HOLOGRAPHY

Holography is an imaging technique that generates images by recording and reconstructing the wave information of the target. Traditional approaches to generate holograms are based on the interference effects between a reference wave and the scattering wave from the target, in which the amplitude and phase information can be fully recorded. This is completely different from other lens-based imaging techniques, where only the intensity of reflected light is captured. The milestone work of holography is the concept of computer-generated hologram (CGH), which produces a hologram image by encoding the required phase information onto a spatial light modulator. This technique provides an accurate means to generate hologram images with high qualities. To further pursue holographic imaging with high resolution, high efficiency, broadband responses and wide field of view, metasurface-based holography has been proposed in recent years. The basic procedure to realize a metasurface-based hologram includes the following five steps. First, a mathematical model to describe the object and the hologram is required. Second, numerical calculation is performed to give the amplitude and phase information in the hologram plane and digitize the information into an array. Third, the digitized amplitude and phase information in the hologram plane are encoded into physical media of metasurfaces, in which the properties of the hologram are transferred into the building blocks. The fourth step is to fabricate the metasurfaces by modern nanofabrication techniques. The last step is to reconstruct the hologram image through a traditional optical process. During the reconstruction process, no reference light is required, making metasurface-based hologram completely different from those conventional optical holograms.

In general, metasurface holography can be categorized into three kinds: phase-only holography, amplitude-only holography, and amplitude-phase holography. For a phase-only hologram, the Gerchberg-Saxton iterative algorithm is commonly adopted to calculate the phase profile, which is then encoded onto a metasurface without any amplitude information. Geometric phase (see Section 5.2.2) based metasurfaces have exhibited excellent control of the phase profile by simply varying the spatial orientation of the building block. Based on this principle, metasurface hologram, composed by nanorod arrays, has shown three dimensional reconstructed images. Due to the subwavelength scale of the nanorod, the reconstructed hologram image presents high resolution and wide field of view. High order diffraction can also be eliminated in such holograms. The nature of geometric phase ensures that the hologram is operating in a broadband manner with helicity dependent properties. The efficiency of geometric metasurface based hologram can be greatly enhanced by adding a ground metal reflector and a dielectric spacer with the metasurfaces. As shown in Figure 5.6(a), a metasurface hologram with the efficiency of 80% was achieved in

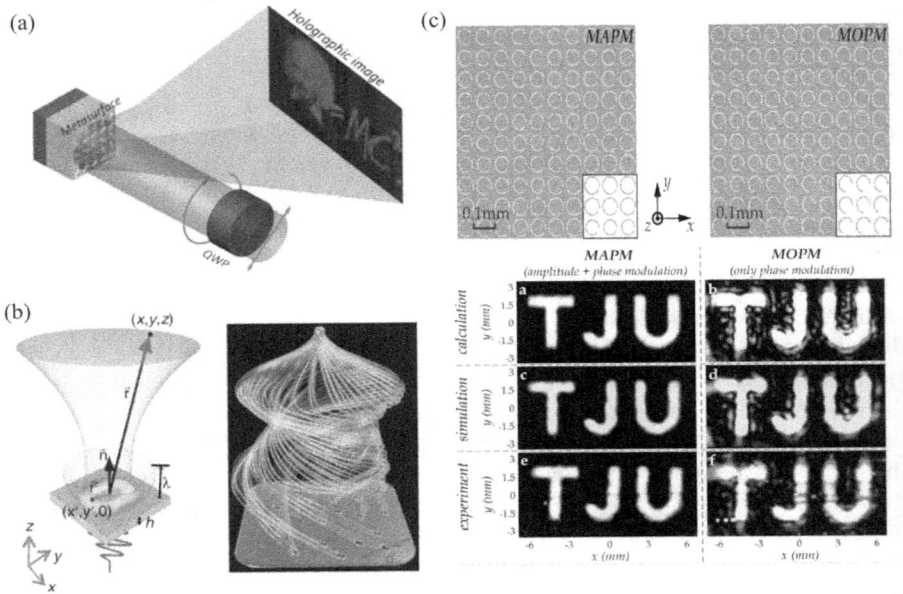

FIGURE 5.6: (a) Illustration of the reflective nanorod-based computer-generated hologram under a circularly polarized incident beam.

Source: Reprinted with permission from Zheng, G. *Nat. Nanotechnol.* 10, 308–312, 2015. Copyright 2015 Springer Nature.

(b) Far-field diffraction of light from a unit hole perforated on a perfect-electric-conductor film and the sketch of controlling light beyond the evanescent region by a photon sieve.

Source: Adopted from Huang, K. et al. *Nat. Commun.* 6, 7059, 2015. No permission is required. Copyright 2015 the authors.

(c) The optical images of fabricated hologram metasurfaces with the performance of holographic imaging.

Source: Adopted from Wang, Q. et al. *Sci. Rep.* 6, 32867, 2016. No permission is required. Copyright 2016 the authors.

the reflection mode [73]. Other complementary designs have also been demonstrated in phase-only holography, in which nanoapertures are adopted. In addition to those metallic designs, all-dielectric metasurfaces have been investigated for efficient hologram. Anisotropic dielectric resonators can be engineered to operate as a half-wave plate and convert the incident circularly polarized into its opposite handedness. The phase profile can be tuned by simply rotating the resonator, which can fully cover the $[0, 2\pi]$ range. Based on Huygens principle, silicon nanodiscs have been demonstrated in phase-only hologram. The transmission efficiency reaches 86% within a broad bandwidth [74]. In the infrared region, dielectric metasurface hologram has shown a high resolution with the transmission beyond 90% [75]. Metasurfaces made of titanium dioxide, which possesses a large bandgap to prevent absorption at high frequency, have extended the operating wavelength of holograms into visible frequency region.

Amplitude-only holography follows similar design principle of phase-only counterpart, but the control parameters become the local amplitude in either transmission or reflection mode. A typical binary amplitude-only hologram relies on the mutual interference among different resonators. As shown in Figure 5.6(b), a high efficiency hologram was realized by a random photon sieve, in which the subwavelength nanoholes were optimized by the genetic search algorithm under the consideration of diffraction problems [76]. However, both phase-only holography and amplitude-only holography suffer from poor image quality and low resolution because part of information is not encoded in metasurface.

To achieve perfect holographic images, the capability of manipulating both the amplitude and phase information is highly desirable. Once the complex wavefronts can be artificially controlled by metasurfaces, the reconstructed hologram images would contain all the information of the target in principle. As shown in Figure 5.6(c), C-shaped split-ring resonators were proposed as the basic building blocks for complex amplitude-phase holograms [77]. By tuning the opening and orientation angles of the C-shaped resonators, the amplitude and phase of cross-polarized light can be controlled. By properly choosing geometric parameters, five-level amplitude control and eight-level phase control were achieved. Arranging the resonators at pre-designed spatial positions, a complex amplitude-phase hologram is realized. Compared with those phase-only or amplitude-only holograms, the amplitude-phase hologram presents much higher resolution.

Metasurface holography can be applied to a variety of functionalities, such as holographic multiplexing, surface wave holography, nonlinear holography, and active holography. All these fascinating metasurface holograms promise various practical applications, including optical information processing, chemical sensing, optical security, and counterfeit.

5.3.4 ACTIVE METASURFACES

Despite the excellent performance of metasurfaces in manipulating electromagnetic waves, the functionalities of most metasurfaces are static owing to their fixed geometric parameters and material properties, which trigger the research of active metasurfaces with dynamical control capabilities. Active meta-devices play a vital role in various photonic systems, ranging from signal processing to communication and imaging. In this section, we will discuss typical active metasurfaces with different active media, which include graphene, semiconductors, phase change materials, liquid crystals, and microelectromechanical systems (MEMS).

Graphene is a two-dimensional carbon-based material, which exhibits tunable Fermi levels under external bias. By hybridizing graphene with metasurfaces, active voltage-controlled metasurfaces are achievable. The resonance modes of metasurfaces can enhance the local electric fields due to the strong light-mater interactions. When graphene is integrated into metasurfaces, these enhanced electric fields can be modified by graphene. Thus, the resonance behaviors of metasurfaces can be actively controlled by graphene with different applied voltages. A terahertz modulator has been reported by hybridizing graphene with metallic metasurfaces [78]. In the infrared region, graphene hybrid metasurfaces have been demonstrated with tunable absorption [79]. As shown in Figure 5.7(a), the metallic resonators are backed with a ground metal plane and separated by a dielectric spacer. The metasurfaces present almost zero reflection and nearly perfect absorption when a graphene film is positioned beneath the resonators. By tuning the applied voltage, the reflection in such hybrid metasurfaces can be modified at a MHz scale tuning speed and almost 100% modulation depth. Similarly, metasurfaces with high Q-factor resonances, such as Fano resonances and electromagnetic induced transparency, can be actively controlled when hybridized with graphene. Another interesting design is graphene patterned metasurfaces, in which the properties originate from the graphene resonators [80]. The optical performance and resonant behavior of metasurface can be directly controlled by the carrier density of graphene. The key advantages of graphene metasurfaces are the voltage tunable characteristics with ultra-compact device design.

Semiconductors have been widely applied in active metasurfaces owing to the variable conductivity through optical or electrical doping. A pioneer work has been reported to achieve active control of terahertz wave by semiconductor-based metasurfaces [81]. As shown in Figure 5.7(b), gold resonator arrays were designed on a semiconductor substrate. By optical pumping and applying a gate voltage, the carrier injection and depletion process modified the conductivity of the semiconductor, which would further affect the resonant response of the gold resonators. This hybrid design achieves dynamical modulation of terahertz wave with the modulation depth of 50%. In addition, semiconductor materials can be patterned and embedded into resonators to locally tune the optical response. Silicon bars have been integrated into split-ring resonators for switchable properties. When silicon bars are illuminated with near-infrared light, the conductivity of silicon increases dramatically, making silicon a metallic-like material. This would alter the effective capacitance of the split-ring resonators and tune the resonance frequency [82]. By embedding silicon into chiral metamaterials and applying the optical pumping, an active chiral metadevice was realized, in which the circular dichroism can be dynamically controlled [83]. By utilizing semiconductor materials with double-channel heterostructures, hybrid metasurfaces can realize GHz modulation speed, which have been applied to a high-performance wireless communication system [84]. By locally controlling the transmission and reflection of each unit cell, semiconductor-based metasurfaces can also realize spatial light modulator, which has been implemented in the terahertz region for imaging applications [85].

Phase change materials are promising candidates to actively control electromagnetic waves when integrated with metasurfaces. Through the phase transition, the dielectric constant of phase change materials can be modified, which shows a powerful means to alter the resonance behaviors of metasurfaces. For instance, vanadium dioxide (VO_2) is a common phase change material, which exhibits four orders of conductivity variation through the phase transition. This phase transition process can be activated by temperature changing, electrical gating or optical pumping. VO_2 behaves as dielectric before phase transition and metal after phase transition. Based on these properties, VO_2-integrated metasurfaces can realize active amplitude modulation with the modulation depth of 20% in the terahertz region. As shown in

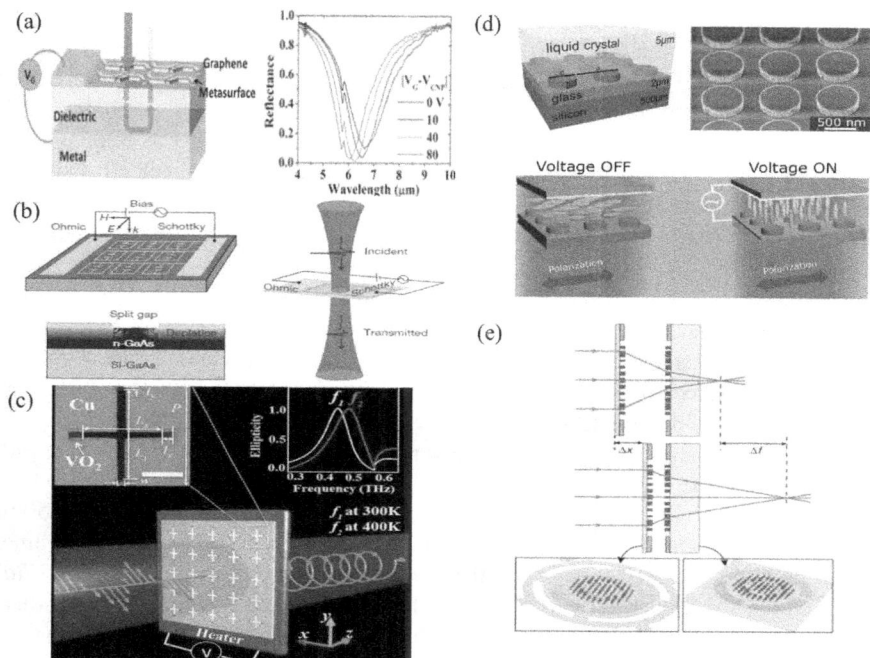

FIGURE 5.7 (a) The design of graphene hybrid metasurfaces and the reflectance spectrum.

Source: Reprinted with permission from Yao Y. et al. *Nano Lett.* 14, 6526–6532, 2014. Copyright 2014 American Chemical Society.

(b) The design of semiconductor hybrid active metasurfaces.

Source: Reprinted with permission from Chen, H. T. *Nature* 444, 597–600, 2006. Copyright 2006 Springer Nature.

(c) Design and fabrication of phase change material hybrid switchable terahertz metasurfaces.

Source: Adopted from Wang, D. C. et al. *Sci. Rep.* 5, 15020, 2015. No permission is required. Copyright 2015 the authors.

(d) Realization of liquid crystal hybrid active metasurfaces.

Source: Reprinted with permission from Komar, A. *Appl. Phys. Lett.* 110, 071109, 2017. Copyright 2017 AIP Publishing LLC.

(e) Schematic illustration of the MEMS-tunable dielectric metasurface lens.

Source: Adopted from Arbabi, E. et al. *Nat. Commun.* 9, 812, 2018. No permission is required. Copyright 2018 the authors.

Figure 5.7(c), by inserting VO_2 pads into asymmetric cross-shaped metallic apertures, a thermally switchable terahertz quarter-wave plate was demonstrated [86]. Active metalenses were realized by designing the resonators of the metalenses on a VO_2 substrate [87]. Dynamic holograms were achieved in VO_2 hybrid metasurfaces with thermally controlled images [88]. Besides VO_2, the phase change chalcogenide glass has also been applied to active metasurfaces. The most well-known chalcogenide glass is germanium-antimony-tellurium (GST), which has been widely used in optical disks and DVDs. GST presents a reversible and rewritable functionality by phase transition between its amorphous and crystalline states. By combining GST films and metasurfaces, active infrared transmission modulation was realized [89]. Phase change materials offer a non-volatile means to actively control electromagnetic waves with compact device design, fast response time and versatile control schemes. In Chapter 9, phase change materials and their applications will be discussed in more details.

Liquid crystals are affordable active materials that have been widely used in many photonic systems. By applying a gate voltage onto liquid crystals, the effective refractive index can be electrically manipulated, which provides a robust method for designing active photonic devices. For instance, nematic liquid crystals have been inserted into metamaterials composed by metallic split-ring resonators, which achieves active manipulation of negative permeability in the microwave region [90]. In the infrared and visible regions, liquid crystal based meta-devices have also been demonstrated with an amplitude modulation up to 30% at the wavelength of 1550nm [91]. Dielectric metasurfaces, composed by silicon nanodisks, have been combined with liquid crystals for active spectral response tuning. As shown in Figure 5.7(d), by switching the gate voltage on and off, the Mie resonances of silicon nanodisks can be effectively modified, which can achieve the modulation depth of transmission beyond 75% [92]. To further extend the functionality, nematic liquid crystals were infiltrated into a silica metalens, which realize a dynamical control of focusing performance [93]. The main bottlenecks of the liquid crystal based meta-devices are the small frequency tuning range and limited modulation speed. Nevertheless, the mature industry technology of liquid crystals enables an efficient approach to design active meta-devices, which may push active meta-devices into commercial applications in dynamic displays, imaging, and holograms.

MEMS are miniaturized mechanical and electro-mechanical devices that capable of moving certain tiny components to realize tunable characteristics, which have been widely applied to various functional devices, such as sensors, actuators, and microelectronics systems. The most significant feature of MEMS meta-devices that distinguishing from other active media is that MEME technology can locally modify the geometry and relative distance of the resonators, Utilizing MEMS technology for active meta-devices is of great significance for compact and functional meta-systems. MEMS metamaterials were initially proposed to tune the transmission lines, which function as MEMS filters [94]. Similar works have been reported to use microelectromechanical actuators for locally controlling of meta-atoms and realizing active negative refractive index. In the terahertz region, active MEMS metasurfaces were demonstrated by arranging the metallic split-ring resonators onto the bimaterial cantilevers. Under a thermal excitation the cantilevers would bend the resonators, which result in modified optical responses. The same work mechanism was adopted in MEMS metasurfaces for dynamical control the phase response and the dispersion of the meta-cavity [95]. As shown in Figure 5.7(e), a MEMS-based tunable dielectric metalens was demonstrated with varifocal lengths [96]. Such a metalens achieved the optical power variation over 60 diopters with respect to one micrometer movement and the scanning speed can reach a few kHz. A compact microscope with a large field of view was realized by further adding another metasurface, which is favorable in three-dimensional imaging. By decreasing the size into tens of nanometers, nano-electromechanical systems (NEMS) can be realized and integrated into metasurfaces for visible light manipulation. It is foreseeable that MEMS and NEMS technology will have significant impact on the tunable photonic systems.

In addition to those active media discussed above, there are many other active materials that make a significant contribution to active metasurfaces. Superconducting materials have been designed to replace the convention metallic resonators and form superconducting metasurfaces [97]. The resonance behaviors can be dynamically tuned by temperature and external magnetic field. Particular quantum interference effects can be studied in superconducting metasurfaces with Josephson junction rings. The low loss and high sensitivity of superconductor can potentially enhance the performance of meta-devices in a new fashion. Nonlinear materials exhibit nonlinear optical responses, which provide a viable means to design nonlinear metasurfaces. In the microwave region, the varactor diodes were integrated into metasurfaces, which achieved dynamic tuning of the transmission [98]. Furthermore, metasurfaces can enhance light-matter interaction and the local field confinement, which in turn can enhance the nonlinear effect. VO_2-based metasurfaces have been proposed for nonlinear responses when illuminated with intensity terahertz fields [99]. Second harmonic generation is another area that has been widely studied by integrating nonlinear media into metasurfaces, in which the nonlinear effect can be greatly enhanced. Last but not least, transparent conducting oxide materials have been applied to active metasurfaces in visible and infrared regions by optically or electrically tuning the dielectric properties [100]. All these active metasurfaces might play a vital role in developing compact and multi-functional photonic systems.

5.4 SUMMARY AND OUTLOOK

Metamaterials and metasurfaces have led to a fundamental revolution of electromagnetic wave control in an artificial means with a variety of potential applications. This chapter discussed the historical development of metamaterials and metasurfaces, illustrated the basic operation principle and reviewed various typical applications. The pursuit of novel functional meta-devices and meta-systems will continue to inspire the researchers to obtain better performance and push them into practical applications. A rapid growing field of research is digital and programmable metasurfaces, which can achieve digitalized optical information control through different coding schemes and real-time information communication. For instance, Cui et al. demonstrated programmable metasurfaces by integrating diodes with metallic resonators [101]. The optical properties can be electrically modified by the field-programmable gate array with different loaded voltage onto the diodes. This field is a fast-growing research area with intriguing applications in high speed communication, imaging and detection. A typical example is the concept of reconfigurable intelligent surface, which exhibits superior performance in active beamforming for 5G/6G communication. In 2021, China Mobile Group Co., Ltd and the team of Professor Cui Tiejun of Southeast University have successfully verified such intelligent metasurfaces with reconfigurable properties for active beamforming on the Nanjing live network [102]. In the residential outdoor test, the average signal at the edge coverage can be enhance by 3 ~ 4 dB with the user throughout enhancement by 10 times. It is expected that reconfigurable intelligent surfaces would be a significant infrastructure in future 6G communication networks.

Inverse design of metasurfaces attracts intense research attentions for convenient and rapid design of high-performance meta-devices. Conventional methods to design metamaterials and metasurfaces adopt forward design strategies, including physical design modeling, trial-and-error testing, parameter sweep and optimization, experimental testing, and physical model re-optimization. The numerical simulation of a full-scale meta-device demands heavy computing resources with a long computing time, which is expensive and inefficient. The tedious parameter sweep and optimization also limit the design efficiency. If the needs of the meta-device are changed, the whole design process would have to be repeated again. How to realize an efficient and automated design is of significance to meet the practical requirements of complex photonic systems. Inverse design enlightens us a promising means via advanced algorithms. For instance, machine learning, or deep learning, can automatically learn from examples obtained in the past and build connections between the input and output data. Based on the learned experience, machine learning can make logical decisions automatically and find the optimum data for the target results. The whole design process requires less computing resources, less design time with high accuracy and flexibility. A deep learning-based metasurface has been investigated for detecting the inner rules between the unit cell and its optical properties with an accuracy of 76.5% [103]. An inverse designed metalens has been reported with the numerical aperture of 0.9 by combing adjoint optimization with coupled mode theory [104]. Such high numerical aperture metalenses exhibited much higher efficiencies than those obtained in traditional designed metalenses, where brute force parameter sweep is used. The inverse design will continue to involve and benefit many meta-devices.

Device miniaturization and system integration are of great values to drive meta-devices into practical applications. One research direction is to cascade multiple metasurfaces in one system for multiple functionalities. Another approach is to integrate different functional devices in a single device. For instance, spintronic emitters and metasurfaces are co-designed together to realize polarization tunable emitter, which inherit the advantages of spintronic materials for terahertz wave emission and metasurfaces for polarization control [105]. Additionally, wafer-scale functional metasurfaces are enabling technology for practical applications. By combining metasurfaces with complementary metal-oxide-semiconductor (CMOS) compatible manufacturing techniques, mass production with low cost is achievable. CMOS compatible amorphous silicon metalenses have been demonstrated on a 12-inch glass wafer with the numerical aperture of 0.496 at the wavelength of 940 nm [106]. These highly miniaturized meta-devices show superior performance in reducing system sizes, which will continue to benefit for compact photonics and consumer electronics. In recent years, we can observe obvious potential of meta-devices entering into industry and consumer markets. Samsung and Sony are developing advanced meta-devices in their optical systems [107,108]. Many startup companies have emerged to push meta-devices into practical applications. Metalenz Inc. and Tunoptix Inc. in USA, Metalenx Inc. in China, NIL Technology Inc. in Denmark have developed high performance metalenses for consumer electronics [109–111]. Lumotive Inc. has demonstrated metasurface-based Lidar for automotive vehicles [112]. DARPA has established a five-year program of "Extreme Optics" for revolutionary optical components. We can envision that in the near future meta-devices will appear in our daily life.

ACKNOWLEDGEMENTS

This work is supported by the National Natural Science Foundation of China (NSFC) (61905225) and Science Challenge Project (TZ2018003).

BIBLIOGRAPHY

[1] Emerson, D. T. 1997. The work of jagadis chandra bose: 100 years of mm-wave research. *IEEE Trans. Microw. Theory Tech.* 45: 2267.

[2] Lindell, I., Sihvola, A., Kurkijarvi, J. 1992. Karl f. Lindman: The last hertzian, and a harbinger of electromagnetic chirality. *IEEE Antennas Propag. Mag.* 34: 24–30.

[3] Kock, W. 1949. Path-length microwave lenses. *Proc. Inst. Radio Eng.* 37: 852–855.

[4] Veselago, V. G. 1968. The electrodynamics of substances with simultaneously negative values of ε and μ. *Sov. Phys. Uspekhi.* 10: 509–514.

[5] Pendry, J. B., Holden, A. J., Stewart, W. J., Youngs, I. 1996. Extremely low frequency plasmons in metallic mesostructures. *Phys. Rev. Lett.* 76: 4773–4776.

[6] Smith, D. R., Padilla, W. J., Vier, D. C., Nemat-Nasser, S. C., Schultz, S. 2000. Composite medium with simultaneously negative permeability and permittivity. *Phys. Rev. Lett.* 84: 4184–4187.

[7] Schurig, D., Mock, J. J., Justice, B. J., Cummer, S. A., Pendry, J. B., Starr, A. F., Smith, D. R. 2006. Metamaterial electromagnetic cloak at microwave frequencies. *Science* 314: 977–980.

[8] Fang, N., Lee, H., Sun, C., Zhang, X. 2005. Sub–diffraction-limited optical imaging with a silver superlens. *Science* 308: 534–537.

[9] Moitra, P., Yang, Y., Anderson, Z., Kravchenko, I. I., Briggs, D. P., Valentine, J. 2013. Realization of an all-dielectric zero-index optical metamaterial. *Nat. Photonics* 7: 791–795.

[10] Wood, R. W. 1902. On a remarkable case of uneven distribution of light in a diffraction grating spectrum. *Proc. Phys. Soc. Lond.* 18: 269–275.

[11] Yu, N., Genevet, P., Kats, M. A., Aieta, F., Tetienne, J.-P., Capasso, F., Gaburro, Z. 2011. Light propagation with phase discontinuities: Generalized laws of reflection and refraction. *Science* 334: 333–337.

[12] Chen, H. T., Taylor, A. J., Yu, N. 2016. A review of metasurfaces: Physics and applications. *Rep. Prog. Phys.* 79: 076401.

[13] Zheludev, N. I., Kivshar, Y. S. 2012. From metamaterials to metadevices. *Nat. Mater.* 11: 917–924.

[14] Arbabi, A., Horie, Y., Bagheri, M., Faraon, A. 2015. Dielectric metasurfaces for complete control of phase and polarization with subwavelength spatial resolution and high transmission. *Nat. Nanotechnol.* 10: 937–943.

[15] Kong, J. A. 1975. *Theory of electromagnetic waves.* John Wiley: New York.

[16] Gay-Balmaz, P., Maccio, C., Martin, O. J. F. 2002. Microwire arrays with plasmonic response at microwave frequencies. *Appl. Phys. Lett.* 81: 2896–2898.

[17] Kang, L., Zhao, Q., Zhao, H., Zhou, J. 2008. Magnetically tunable negative permeability metamaterial composed by split ring resonators and ferrite rods. *Opt. Express* 16: 8825–8834.

[18] Pendry, J. B., Holden, A., Robbins, D. J., Stewart, W. J. 1999. Magnetism from conductors, and enhanced non-linear phenomena. *IEEE Trans. Microw. Theory Tech.* 47: 2075–2084.

[19] Huangfu, J., Ran, L., Chen, H., Zhang, X.-M., Chen, K., Grzegorczyk, T. M., Kong, J. A. 2004. Experimental confirmation of negative refractive index of a metamaterial composed of ω-like metallic patterns. *Appl. Phys. Lett.* 84: 1537–1539.

[20] Xiong, X., Sun, W.-H., Bao, Y.-J., Peng, R.-W., Wang, M., Sun, C., Lu, X., Shao, J., Li, Z., Ming, N.-B. 2009. Construction of chiral metamaterial with u-shaped resonator assembly. *Phys. Rev. B* 81.

[21] Zhou, J., Zhang, L., Tuttle, G., Koschny, T., Soukoulis, C. M. 2006. Negative index materials using simple short wire pairs. *Phys. Rev. B* 73: 041101.

[22] Zhang, S., Fan, W., Malloy, K. J., Brueck, S. R. J., Panoiu, N. C., Osgood, R. M. 2005. Near-infrared double negative metamaterials. *Opt. Express* 13: 4922–4930.

[23] Zhang, X., Tian, Z., Yue, W., Gu, J., Zhang, S., Han, J., Zhang, W. 2013. Broadband terahertz wave deflection based on c-shape complex metamaterials with phase discontinuities. *Adv. Mater.* 25: 4567–4572.

[24] Zeng, H., Zhang, Y., Lan, F., Liang, S., Wang, L., Song, T., Zhang, T., Shi, Z., Yang, Z., Kang, X., Zhang, X., Mazumder, P., Mittleman, D. M. 2019. Terahertz dual-polarization beam splitter via an anisotropic matrix metasurface. *IEEE Trans. Terahertz Sci. Technol.* 9: 491–497.

[25] Huang, L., Chen, X., Mühlenbernd, H., Li, G., Bai, B., Tan, Q., Jin, G., Zentgraf, T., Zhang, S. 2012. Dispersionless phase discontinuities for controlling light propagation. *Nano Lett.* 12: 5750–5755.

[26] Qin, F., Ding, L., Zhang, L., Monticone, F., Chum Chan, C., Deng, J., Mei, S., Li, Y., Teng, J., Hong, M., Zhang, S., Alù, A., Qiu, C.-W. 2016. Hybrid bilayer plasmonic metasurface efficiently manipulates visible light. *Sci. Adv.* 2: e1501168.

[27] Luo, J., Yu, H., Song, M., Zhang, Z. 2014. Highly efficient wavefront manipulation in terahertz based on plasmonic gradient metasurfaces. *Opt. Lett.* 39: 2229–2231.

[28] Jiang, X., Chen, H., Li, Z., Yuan, H., Cao, L., Luo, Z., Zhang, K., Zhang, Z., Wen, Z., Zhu, L.-G., Zhou, X., Liang, G., Ruan, D., Du, L., Wang, L., Chen, G. 2018. All-dielectric metalens for terahertz wave imaging. *Opt. Express* 26: 14132–14142.

[29] Jang, M., Horie, Y., Shibukawa, A., Brake, J., Liu, Y., Kamali, S. M., Arbabi, A., Ruan, H., Faraon, A., Yang, C. 2018. Wavefront shaping with disorder-engineered metasurfaces. *Nat. Photonics.* 12: 84–90.

[30] Arslan, D., Rahimzadegan, A., Fasold, S., Falkner, M., Zhou, W., Kroychuk, M., Rockstuhl, C., Pertsch, T., Staude, I. 2022. Toward perfect optical diffusers: Dielectric huygens' metasurfaces with critical positional disorder. *Adv. Mater.* 34: 2105868.

[31] Seddon, N., Bearpark, T. 2003. Observation of the inverse doppler effect. *Science* 302: 1537–1540.

[32] Liu, R., Ji, C., Mock, J. J., Chin, J. Y., Cui, T. J., Smith, D. R. 2009. Broadband ground-plane cloak. *Science* 323: 366–369.

[33] Ma, H. F., Cui, T. J. 2010. Three-dimensional broadband ground-plane cloak made of metamaterials. *Nat. Commun.* 1: 21.

[34] Qiu, C.-W., Hu, L., Xu, X., Feng, Y. 2009. Spherical cloaking with homogeneous isotropic multilayered structures. *Phys. Rev. E* 79: 047602.

[35] Jacob, Z., Alekseyev, L. V., Narimanov, E. 2006. Optical hyperlens: Far-field imaging beyond the diffraction limit. *Opt. Express* 14: 8247–8256.

[36] Aieta, F., Genevet, P., Kats, M. A., Yu, N., Blanchard, R., Gaburro, Z., Capasso, F. 2012. Aberration-free ultrathin flat lenses and axicons at telecom wavelengths based on plasmonic metasurfaces. *Nano Lett.* 12: 4932–4936.

[37] Wang, Q., Zhang, X., Xu, Y., Tian, Z., Gu, J., Yue, W., Zhang, S., Han, J., Zhang, W. 2015. A broadband metasurface-based terahertz flat-lens array. *Adv. Opt. Mater.* 3: 779–785.

[38] Chen, X., Huang, L., Mühlenbernd, H., Li, G., Bai, B., Tan, Q., Jin, G., Qiu, C.-W., Zhang, S., Zentgraf, T. 2012. Dual-polarity plasmonic metalens for visible light. *Nat. Commun.* 3: 1198.

[39] Ni, X., Ishii, S., Kildishev, A. V., Shalaev, V. M. 2013. Ultra-thin, planar, babinet-inverted plasmonic metalenses. *Light Sci. Appl.* 2: e72–e72.

[40] Kang, M., Feng, T., Wang, H.-T., Li, J. 2012. Wave front engineering from an array of thin aperture antennas. *Opt. Express* 20: 15882–15890.

[41] Guo, Y., Ma, X., Pu, M., Li, X., Zhao, Z., Luo, X. 2018. High-efficiency and wide-angle beam steering based on catenary optical fields in ultrathin metalens. *Adv. Opt. Mater.* 6: 1800592.

[42] Yang, Q., Gu, J., Xu, Y., Zhang, X., Li, Y., Ouyang, C., Tian, Z., Han, J., Zhang, W. 2017. Broadband and robust metalens with nonlinear phase profiles for efficient terahertz wave control. *Adv. Opt. Mater.* 5: 1601084.

[43] Chang, C.-C., Headland, D., Abbott, D., Withayachumnankul, W., Chen, H.-T. 2017. Demonstration of a highly efficient terahertz flat lens employing tri-layer metasurfaces. *Opt. Lett.* 42: 1867–1870.

[44] Zhang, L., Ding, J., Zheng, H., An, S., Lin, H., Zheng, B., Du, Q., Yin, G., Michon, J., Zhang, Y., Fang, Z., Shalaginov, M. Y., Deng, L., Gu, T., Zhang, H., Hu, J. 2018. Ultra-thin high-efficiency mid-infrared transmissive huygens meta-optics. *Nat. Commun.* 9: 1481.

[45] Khorasaninejad, M., Chen Wei, T., Devlin Robert, C., Oh, J., Zhu Alexander, Y., Capasso, F. 2016. Metalenses at visible wavelengths: Diffraction-limited focusing and subwavelength resolution imaging. *Science* 352: 1190–1194.

[46] Chen, B. H., Wu, P. C., Su, V.-C., Lai, Y.-C., Chu, C. H., Lee, I. C., Chen, J.-W., Chen, Y. H., Lan, Y.-C., Kuan, C.-H., Tsai, D. P. 2017. Gan metalens for pixel-level full-color routing at visible light. *Nano Lett.* 17: 6345–6352.

[47] Arbabi, A., Horie, Y., Ball, A. J., Bagheri, M., Faraon, A. 2015. Subwavelength-thick lenses with high numerical apertures and large efficiency based on high-contrast transmitarrays. *Nat. Commun.* 6: 7069.

[48] Khorasaninejad, M., Zhu, A. Y., Roques-Carmes, C., Chen, W. T., Oh, J., Mishra, I., Devlin, R. C., Capasso, F. 2016. Polarization-insensitive metalenses at visible wavelengths. *Nano Lett.* 16: 7229–7234.

[49] Chen, W. T., Zhu, A. Y., Khorasaninejad, M., Shi, Z., Sanjeev, V., Capasso, F. 2017. Immersion meta-lenses at visible wavelengths for nanoscale imaging. *Nano Lett.* 17: 3188–3194.

[50] Liang, H., Lin, Q., Xie, X., Sun, Q., Wang, Y., Zhou, L., Liu, L., Yu, X., Zhou, J., Krauss, T. F., Li, J. 2018. Ultrahigh numerical aperture metalens at visible wavelengths. *Nano Lett.* 18: 4460–4466.

[51] Paniagua-Domínguez, R., Yu, Y. F., Khaidarov, E., Choi, S., Leong, V., Bakker, R. M., Liang, X., Fu, Y. H., Valuckas, V., Krivitsky, L. A., Kuznetsov, A. I. 2018. A metalens with a near-unity numerical aperture. *Nano Lett.* 18: 2124–2132.

[52] Arbabi, A., Arbabi, E., Kamali, S. M., Horie, Y., Han, S., Faraon, A. 2016. Miniature optical planar camera based on a wide-angle metasurface doublet corrected for monochromatic aberrations. *Nat. Commun.* 7: 13682.

[53] Groever, B., Chen, W. T., Capasso, F. 2017. Meta-lens doublet in the visible region. *Nano Lett.* 17: 4902–4907.

[54] Aieta, F., Kats Mikhail, A., Genevet, P., Capasso, F. 2015. Multiwavelength achromatic metasurfaces by dispersive phase compensation. *Science* 347: 1342–1345.

[55] Zhao, Z., Pu, M., Gao, H., Jin, J., Li, X., Ma, X., Wang, Y., Gao, P., Luo, X. 2015. Multispectral optical metasurfaces enabled by achromatic phase transition. *Sci. Rep.* 5: 15781.

[56] Hsiao, H.-H., Chen, Y. H., Lin, R. J., Wu, P. C., Wang, S., Chen, B. H., Tsai, D. P. 2018. Integrated resonant unit of metasurfaces for broadband efficiency and phase manipulation. *Adv. Opt. Mater.* 6: 1800031.

[57] Wang, S., Wu, P. C., Su, V.-C., Lai, Y.-C., Hung Chu, C., Chen, J.-W., Lu, S.-H., Chen, J., Xu, B., Kuan, C.-H., Li, T., Zhu, S., Tsai, D. P. 2017. Broadband achromatic optical metasurface devices. *Nat. Commun.* 8: 187.

[58] Chen, W. T., Zhu, A. Y., Sanjeev, V., Khorasaninejad, M., Shi, Z., Lee, E., Capasso, F. 2018. A broadband achromatic metalens for focusing and imaging in the visible. *Nat. Nanotechnol.* 13: 220–226.

[59] Chen, W. T., Zhu, A. Y., Sisler, J., Huang, Y.-W., Yousef, K. M. A., Lee, E., Qiu, C.-W., Capasso, F. 2018. Broadband achromatic metasurface-refractive optics. *Nano Lett.* 18: 7801–7808.

[60] Yu, N., Aieta, F., Genevet, P., Kats, M. A., Gaburro, Z., Capasso, F. 2012. A broadband, background-free quarter-wave plate based on plasmonic metasurfaces. *Nano Lett.* 12: 6328–6333.

[61] Wang, D., Gu, Y., Gong, Y., Qiu, C.-W., Hong, M. 2015. An ultrathin terahertz quarter-wave plate using planar babinet-inverted metasurface. *Opt. Express* 23: 11114–11122.

[62] Cong, L., Xu, N., Gu, J., Singh, R., Han, J., Zhang, W. 2014. Highly flexible broadband terahertz metamaterial quarter-wave plate. *Laser Photonics Rev.* 8: 626–632.

[63] Pfeiffer, C., Grbic, A. 2014. Bianisotropic metasurfaces for optimal polarization control: Analysis and synthesis. *Phys. Rev. Appl.* 2: 044011.

[64] Wang, D.-C., Sun, S., Feng, Z., Tan, W., Qiu, C.-W. 2018. Multipolar-interference-assisted terahertz waveplates via all-dielectric metamaterials. *Appl. Phys. Lett.* 113: 201103.

[65] Chang, C.-C., Zhao, Z., Li, D., Taylor, A. J., Fan, S., Chen, H.-T. 2019. Broadband linear-to-circular polarization conversion enabled by birefringent off-resonance reflective metasurfaces. *Phys. Rev. Lett.* 123: 237401.

[66] Lee, W. S. L., Ako, R. T., Low, M. X., Bhaskaran, M., Sriram, S., Fumeaux, C., Withayachumnankul, W. 2018. Dielectric-resonator metasurfaces for broadband terahertz quarter- and half-wave mirrors. *Opt. Express* 26: 14392–14406.

[67] Grady, N. K., Heyes, J. E., Chowdhury, D. R., Zeng, Y., Reiten, M. T., Azad, A. K., Taylor, A. J., Dalvit, D. A., Chen, H. T. 2013. Terahertz metamaterials for linear polarization conversion and anomalous refraction. *Science* 340: 1304–1307.

[68] Wu, J.-L., Lin, B.-Q., Da, X.-Y. 2016. Ultra-wideband reflective polarization converter based on anisotropic metasurface. *Chinese Physics B* 25: 088101.

[69] Xia, R., Jing, X., Gui, X., Tian, Y., Hong, Z. 2017. Broadband terahertz half-wave plate based on anisotropic polarization conversion metamaterials. *Opt. Mater. Express* 7: 977.

[70] Yang, Y., Wang, W., Moitra, P., Kravchenko, I. I., Briggs, D. P., Valentine, J. 2014. Dielectric meta-reflectarray for broadband linear polarization conversion and optical vortex generation. *Nano Lett.* 14: 1394–1399.

[71] Cong, L., Cao, W., Zhang, X., Tian, Z., Gu, J., Singh, R., Han, J., Zhang, W. 2013. A perfect metamaterial polarization rotator. *Appl. Phys. Lett.* 103: 171107.

[72] Ma, Z., Li, Y., Li, Y., Gong, Y., Maier, S. A., Hong, M. 2018. All-dielectric planar chiral metasurface with gradient geometric phase. *Opt. Express* 26: 6067–6078.

[73] Zheng, G., Mühlenbernd, H., Kenney, M., Li, G., Zentgraf, T., Zhang, S. 2015. Metasurface holograms reaching 80% efficiency. *Nature Nanotechnology* 10: 308–312.

[74] Zhao, W., Jiang, H., Liu, B., Song, J., Jiang, Y., Tang, C., Li, J. 2016. Dielectric huygens' metasurface for high-efficiency hologram operating in transmission mode. *Sci. Rep.* 6: 30613.

[75] Wang, L., Kruk, S., Tang, H., Li, T., Kravchenko, I., Neshev, D. N., Kivshar, Y. S. 2016. Grayscale transparent metasurface holograms. *Optica* 3: 1504–1505.

[76] Huang, K., Liu, H., Garcia-Vidal, F. J., Hong, M., Luk'yanchuk, B., Teng, J., Qiu, C.-W. 2015. Ultrahigh-capacity non-periodic photon sieves operating in visible light. *Nat. Commun.* 6: 7059.

[77] Wang, Q., Zhang, X., Xu, Y., Gu, J., Li, Y., Tian, Z., Singh, R., Zhang, S., Han, J., Zhang, W. 2016. Broadband metasurface holograms: Toward complete phase and amplitude engineering. *Sci. Rep.* 6: 32867.

[78] Lee, S. H., Choi, M., Kim, T.-T., Lee, S., Liu, M., Yin, X., Choi, H. K., Lee, S. S., Choi, C.-G., Choi, S.-Y., Zhang, X., Min, B. 2012. Switching terahertz waves with gate-controlled active graphene metamaterials. *Nat. Mater.* 11: 936–941.

[79] Yao, Y., Shankar, R., Kats, M. A., Song, Y., Kong, J., Loncar, M., Capasso, F. 2014. Electrically tunable metasurface perfect absorbers for ultrathin mid-infrared optical modulators. *Nano Lett.* 14: 6526–6532.

[80] Mousavi, S. H., Kholmanov, I., Alici, K. B., Purtseladze, D., Arju, N., Tatar, K., Fozdar, D. Y., Suk, J. W., Hao, Y., Khanikaev, A. B., Ruoff, R. S., Shvets, G. 2013. Inductive tuning of fano-resonant metasurfaces using plasmonic response of graphene in the mid-infrared. *Nano Lett.* 13: 1111–1117.

[81] Chen, H.-T., Padilla, W. J., Zide, J. M. O., Gossard, A. C., Taylor, A. J., Averitt, R. D. 2006. Active terahertz metamaterial devices. *Nature* 444: 597–600.

[82] Chen, H.-T., O'Hara, J. F., Azad, A. K., Taylor, A. J., Averitt, R. D., Shrekenhamer, D. B., Padilla, W. J. 2008. Experimental demonstration of frequency-agile terahertz metamaterials. *Nat. Photonics* 2: 295–298.

[83] Zhang, S., Zhou, J., Park, Y. S., Rho, J., Singh, R., Nam, S., Azad, A. K., Chen, H. T., Yin, X., Taylor, A. J., Zhang, X. 2012. Photoinduced handedness switching in terahertz chiral metamolecules. *Nat. Commun.* 3: 942.

[84] Zhang, Y., Qiao, S., Liang, S., Wu, Z., Yang, Z., Feng, Z., Sun, H., Zhou, Y., Sun, L., Chen, Z., Zou, X., Zhang, B., Hu, J., Li, S., Chen, Q., Li, L., Xu, G., Zhao, Y., Liu, S. 2015. Gbps terahertz external modulator based on a composite metamaterial with a double-channel heterostructure. *Nano Lett.* 15: 3501–3506.

[85] Chan, W. L., Chen, H.-T., Taylor, A. J., Brener, I., Cich, M. J., Mittleman, D. M. 2009. A spatial light modulator for terahertz beams. *Appl. Phys. Lett.* 94: 213511.

[86] Wang, D., Zhang, L., Gu, Y., Mehmood, M., Gong, Y., Srivastava, A., Jian, L., Venkatesan, T., Qiu, C.-W., Hong, M. 2015. Switchable ultrathin quarter-wave plate in terahertz using active phase-change metasurface. *Sci. Rep.* 5: 15020.

[87] Wang, T., He, J., Guo, J., Wang, X., Feng, S., Kuhl, F., Becker, M., Polity, A., Klar, P. J., Zhang, Y. 2019. Thermally switchable terahertz wavefront metasurface modulators based on the insulator-to-metal transition of vanadium dioxide. *Opt. Express* 27: 20347–20357.

[88] Liu, X., Wang, Q., Zhang, X., Li, H., Xu, Q., Xu, Y., Chen, X., Li, S., Liu, M., Tian, Z., Zhang, C., Zou, C., Han, J., Zhang, W. 2019. Thermally dependent dynamic meta-holography using a vanadium dioxide integrated metasurface. *Adv. Opt. Mater.* 7: 1900175.

[89] Sámson, Z. L., MacDonald, K. F., De Angelis, F., Gholipour, B., Knight, K., Huang, C. C., Di Fabrizio, E., Hewak, D. W., Zheludev, N. I. 2010. Metamaterial electro-optic switch of nanoscale thickness. *Appl. Phys. Lett.* 96: 143105.

[90] Zhao, Q., Kang, L., Du, B., Li, B., Zhou, J., Tang, H., Liang, X., Zhang, B. 2007. Electrically tunable negative permeability metamaterials based on nematic liquid crystals. *Appl. Phys. Lett.* 90: 011112.

[91] Minovich, A., Farnell, J., Neshev, D. N., McKerracher, I., Karouta, F., Tian, J., Powell, D. A., Shadrivov, I. V., Hoe Tan, H., Jagadish, C., Kivshar, Y. S. 2012. Liquid crystal based nonlinear fishnet metamaterials. *Appl. Phys. Lett.* 100: 121113.

[92] Komar, A., Fang, Z., Bohn, J., Sautter, J., Decker, M., Miroshnichenko, A., Pertsch, T., Brener, I., Kivshar, Y. S., Staude, I., Neshev, D. N. 2017. Electrically tunable all-dielectric optical metasurfaces based on liquid crystals. *Appl. Phys. Lett.* 110: 071109.

[93] Lininger, A., Zhu, A. Y., Park, J.-S., Palermo, G., Chatterjee, S., Boyd, J., Capasso, F., Strangi, G. 2020. Optical properties of metasurfaces infiltrated with liquid crystals. *PNAS* 117: 20390.

[94] Gil, I., Martın, F., Rottenberg, X., Raedt, W. D. 2007. Tunable stop-band filter at q-band based on rf-mems metamaterials. *Electron. Lett.* 43: 1153–1154.

[95] Cong, L., Pitchappa, P., Lee, C., Singh, R. 2017. Active phase transition via loss engineering in a terahertz mems metamaterial. *Adv. Mater.* 29(26): 1700733.

[96] Arbabi, E., Arbabi, A., Kamali, S. M., Horie, Y., Faraji-Dana, M., Faraon, A. 2018. Mems-tunable dielectric metasurface lens. *Nat. Commun.* 9: 812.

[97] Ricci, M., Orloff, N., Anlage, S. M. 2005. Superconducting metamaterials. *Appl. Phys. Lett.* 87: 034102.

[98] Powell, D. A., Shadrivov, I. V., Kivshar, Y. S., Gorkunov, M. V. 2007. Self-tuning mechanisms of nonlinear split-ring resonators. *Appl. Phys. Lett.* 91: 144107.

[99] Liu, M., Hwang, H. Y., Tao, H., Strikwerda, A. C., Fan, K., Keiser, G. R., Sternbach, A. J., West, K. G., Kittiwatanakul, S., Lu, J., Wolf, S. A., Omenetto, F. G., Zhang, X., Nelson, K. A., Averitt, R. D. 2012. Terahertz-field-induced insulator-to-metal transition in vanadium dioxide metamaterial. *Nature* 487: 345–348.

[100] Forouzmand, A., Salary, M. M., Kafaie Shirmanesh, G., Sokhoyan, R., Atwater, H. A., Mosallaei, H. 2019. Tunable all-dielectric metasurface for phase modulation of the reflected and transmitted light via permittivity tuning of indium tin oxide. *Nanophotonics* 8: 415–427.

[101] Zhang, L., Cui, T. J. 2021. Space-time-coding digital metasurfaces: Principles and applications. *Research* 2021: 9802673.

[102] See https://min.News/en/tech/93fce0916854f11efe07a3cb5fb64c48.Html (last accessed January 21, 2022).

[103] Qiu, T., Shi, X., Wang, J., Li, Y., Qu, S., Cheng, Q., Cui, T., Sui, S. 2019. Deep learning: A rapid and efficient route to automatic metasurface design. *Adv. Sci.* 6: 1900128.

[104] Mansouree, M., McClung, A., Samudrala, S., Arbabi, A. 2021. Large-scale parametrized metasurface design using adjoint optimization. *ACS Photonics* 8: 455–463.

[105] Changqin, L., Shunjia, W., Sheng, Z., Qingnan, C., Peng, W., Chuanshan, T., Lei, Z., Yizheng, W., Zhensheng, T. 2021. Active spintronic-metasurface terahertz emitters with tunable chirality. *Adv. Photonics* 3: 1–10.

[106] Hu, T., Zhong, Q., Li, N., Dong, Y., Xu, Z., Fu, Y. H., Li, D., Bliznetsov, V., Zhou, Y. Lai, K. H., Lin, Q., Zhu, S., Singh, N. 2020. Cmos-compatible a-si metalenses on a 12-inch glass wafer for fingerprint imaging. *Nanophotonics* 9: 823–830.

[107] See https://www.Sait.Samsung.Co.Kr/saithome/about/collabo_apply.Do for "samsung global collaboration" (last accessed January 21, 2022).

[108] See https://www.Sony.Com/research-award-program#focusedresearchaward for "sony research award program" (last accessed January 21, 2022).
[109] See https://www.Metalenz.Com/ for "metalez inc."
[110] See https://www.Tunoptix.Com/ for tunoptix inc.
[111] See http://www.Metalenx.Com/ for metalenx inc.
[112] See https://www.Laserfocusworld.Com/optics/article/14036818/metasurface-beamsteering-enables-solidstate-highperformance-lidar for lumotive inc. (last accessed January 21, 2022).

6 Low Loss Dielectric Materials and Their Applications

Song Sun

CONTENTS

6.1 Introduction.. 153
6.2 Fundamental Physics and Fabrication Methods 154
    6.2.1 Low Loss Properties of Dielectric Materials 154
    6.2.2 Electric and Magnetic Multipole Mie Resonances of
        Dielectric Structure.. 155
    6.2.3 Fabrication Methods for Dielectric Structures................................ 157
6.3 Applications of Dielectric Materials.. 159
    6.3.1 Electric and Magnetic Hotspots... 159
    6.3.2 Enhanced Raman Scattering 161
    6.3.3 Enhanced Fluorescence Spontaneous Emission 163
    6.3.4 Optical Force.. 164
    6.3.5 Engineering Far Field Scattering 167
    6.3.6 Non-Radiating Mode and Enhanced Nonlinearity 168
    6.3.7 Optoelectronic and On-Chip Photonic Devices............................. 170
    6.3.8 Metamaterial and Metasurface... 172
6.4 Summary and Outlook.. 172
Acknowledgements... 174
Bibliography 174

6.1 INTRODUCTION

While plasmonic materials based on noble metals have provided a versatile platform to build functional devices for various optical, infrared, and terahertz applications (see Chapter 4), they inevitably suffer from high dissipation losses that originate from the intrinsically non-zero imaginary parts of permittivities, which are undesired for applications that demand high energy efficiencies [1]. To overcome this limitation, dielectric materials with high refractive indices are adopted and undergo rapid development not only because of their low loss properties but also the potential compatibility with complementary metal oxide semiconductor (CMOS) fabrication process. On top of that, dielectric materials facilitate both electric and magnetic types of resonance modes, which are difficult to be achieved via the plasmonic counterparts. These distinct resonance modes grant dielectric structure a strong capability to tailor the electromagnetic fields at sub-wavelength scale,

DOI: 10.1201/9781003202608-6

obtaining comparable functionalities as those of plasmonic structures. In addition, the interplay between electric and magnetic resonance modes of a delicately designed dielectric structure offers an additional degree of freedom to manipulate the near- and far-field characteristics of the electromagnetic fields, which triggers a variety of exotic phenomena such as unidirectional scattering, non-radiating anopole mode, optical magnetism, among others, greatly expanding the application scope [2]. Besides, the intrinsic material properties of dielectric can be modified by altering the carrier density via the common doping technique, further increasing the flexibility. All these make dielectric structure a promising build-block for optical, infrared, and terahertz applications, which could add much value to plasmonic counterparts. Hereafter, we shall first elaborate some key fundamental physics of dielectric materials, followed by reviewing their development and applications. Finally, a brief outlook will be provided.

6.2 FUNDAMENTAL PHYSICS AND FABRICATION METHODS

6.2.1 LOW LOSS PROPERTIES OF DIELECTRIC MATERIALS

The loss of dielectric materials is determined by its band gap. For photon energy smaller than the band gap, the dielectric material will not absorb the photon and appear to be transparent and lossless. Whereas for photon energy larger than the band gap, it will absorb the photon by pumping the carrier from valence band to conduction band, which might eventually convert into heat due to electron-phonon coupling and induce energy loss. In electromagnetism, it is captured in the complex value of material's permittivity $\varepsilon = \varepsilon_r + i\cdot\varepsilon_i$, since the absorption loss can be expressed as a volume integration of electric field inside the dielectric structure as $P_{abs} = \sigma\int|E|^2\cdot dV$, where the conductivity is directly related to the imaginary part of permittivity as $\sigma = \omega\varepsilon_0\varepsilon_i$. Figure 6.1 compares the (a) real and (b) imaginary parts of permittivities between some commonly used high refractive index dielectric (GaAs,

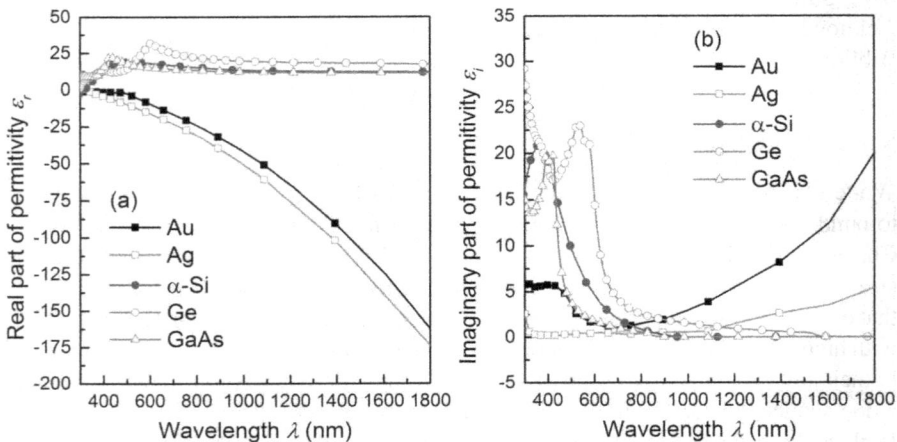

FIGURE 6.1 (a) Real and (b) imaginary parts of permittivities for some commonly used metal (Au and Ag) and high refractive index dielectric (α-Si, Ge, GaAs) materials.

Ge and α-Si) and metal (Au and Ag) materials in the optical to near infrared regime (300 nm–1800 nm). The material properties are taken from an online database [3].

It can be seen that that plasmonic metal materials have negative real part permittivities ε_r whereas dielectric materials have positive ones. More importantly, the dielectric materials have pronounced imaginary part permittivities ε_i in particular at short wavelengths, indicating strong absorptions that are even larger than the metal materials. As the wavelength goes beyond some cut-off value, ε_i diminishes completely and dielectric becomes lossless. This cut-off wavelength is directly related to the dielectric bandgap, for example, α-Si has a cut-off wavelength around 1000 nm which corresponds to its bandgap of 1.1 eV.

There are a few things to take note: 1) the results in Figure 6.1 considers only the linear response of material to the incident electromagnetic wave. Any nonlinear interaction such as two-photon absorption is not included; 2) the permittivities are mainly used to describe the bulk materials. For ultrathin film or nanoparticle, there could be some deviations on the permittivities; 3) for some materials, the permittivities could be different depending on their crystalline structures, e.g., single crystal or amorphous [4]. Besides, external environmental conditions (e.g., temperature, pressure) could also induce changes on the material properties [5]. Therefore, caution must be taken when using the online database.

6.2.2 Electric and Magnetic Multipole Mie Resonances of Dielectric Structure

For any arbitrary micro-nano structure, its electromagnetic spectral response can be rigorously decomposed into a series of electric and magnetic current multipoles (also referred to multipole Mie resonances [6]. For plasmonic metal materials, it is generally dominated by the electric type resonance modes since the plasmonic resonance essentially results from collective oscillation of electron at metal surface. Only a specially designed geometry (e.g., split ring, diabolo, etc.) could grant the capability to generate magnetic type resonance mode by constructing a virtual current loop, which gives a local magnetic field enhancement based on Ampere's law [7,8]. Whereas for the dielectric materials, the incident light could induce a circular displacement current of electric field, owing to the field penetration and phase retardation effects inside the structure, which could easily produce a significant magnetic type resonance mode as long as the wavelength inside the dielectric structure is comparable to the structure size [9]. For illustration purpose, Figure 6.2(a)-(c) show the multipole spectral responses within the optical frequency for a 140 nm Ag nanosphere, a 160 nm high refractive index Si nanosphere ($n = 3.5$) and a 200 nm middle refractive index TiO_2 nanosphere ($n = 2.5$), respectively, under a plane wave incidence. The electric dipole mode (ED), magnetic dipole mode (MD), electric quadrupole mode (EQ) and magnetic quadrupole (MQ) are demonstrated here, while other higher order modes are omitted due to their insignificant contributions. The environment is assumed to be air ($n = 1$). The corresponding scattering patterns of these multipoles are schematically depicted in Figure 6.2(d). Note that the overlapping between electric and magnetic type resonance modes can be easily manipulated by adopting anisotropic geometries [10].

Clearly, the Ag nanosphere only possesses pronounced ED and EQ responses in the extinction cross-section (CS), whereas for the Si and TiO_2 nanosphere, significant

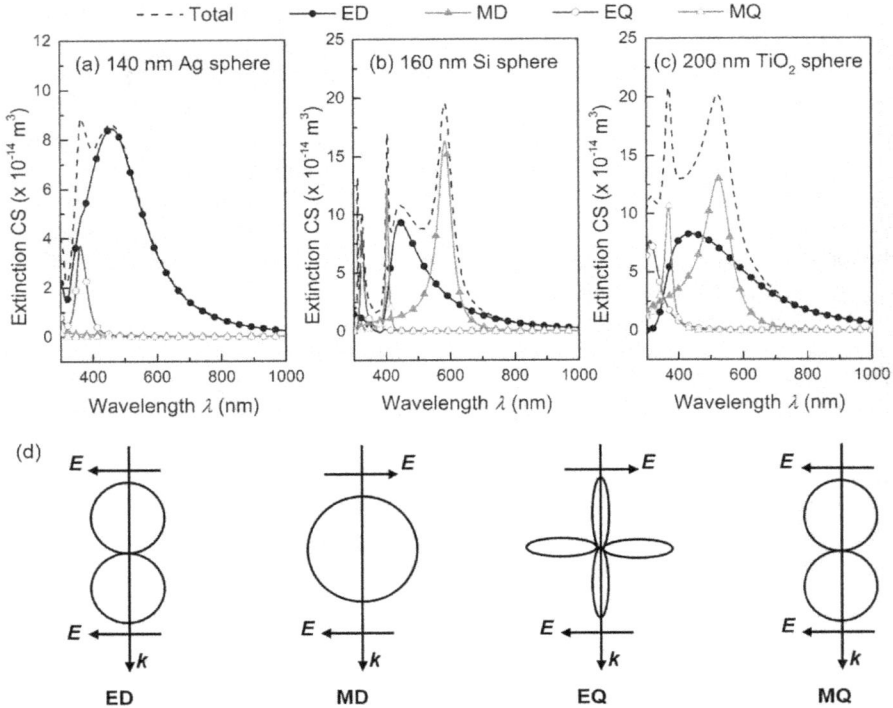

FIGURE 6.2 Multipole decomposition results for (a) a 140 nm Ag nanosphere, (b) a 160 nm Si nanosphere and (c) a 200 nm TiO_2 nanosphere, respectively, under a plane wave illumination. Electric/magnetic dipole modes (ED/MD) and quadrupole modes (EQ/MQ) are presented. (d) Schematic scattering patterns of these multipole modes.

electric and magnetic type resonance modes can be observed. Furthermore, the quality factor of the resonance mode is larger for a high refractive index Si nanosphere than that of a lower refractive index TiO_2 nanosphere, since the electromagnetic energy is better confined in the high refractive index material. That's why in the design of artificial dielectric micro-nanostructures for a particular frequency band, people always prefer to choose the material with a refractive index as high as possible to harvest a strong and efficient resonance mode, in addition to the lossless characteristic. Ref. [11] summaries some of the known dielectric materials with high refractive indices in the visible and infrared regimes. It is found that in the mid-infrared regime, narrow bandgap semiconductors (e.g., GeTe, PbTe, Te) and polar crystals (e.g., SiC) are attractive, while in the visible and near-infrared regime, wide bandgap semiconductor materials (e.g., Si, Ge, GaAs) can be exploited. However, caution must be taken when dealing with these materials because a high refractive index is generally accompanied with a cost of increased absorption due to the Kramers-Kronig relation [12]. In reality, the resonance performance of dielectric material is ultimately limited by the trade-off between absorption and high refractive index.

The abundance of multipole resonance modes in the dielectric materials is considered to be attractive in generating electric and magnetic hotspots with enhanced local electric or magnetic fields, which is directly useful in enhanced Raman spectroscopy, fluorescence spectroscopy as well as optical force. On top of that, the interference between various electric and magnetic resonance modes via intra-particle or inter-particle approaches could enable many exciting phenomena in tailoring the near and far field characteristics at subwavelength scale, allowing unique applications that are generally difficult to be conceived in plasmonic materials. Just to name a few, the coherent coupling between electric dipole mode and magnetic dipole mode could resemble the Kerker condition, creating a unidirectional scattering pattern. The interference between low order electric dipole mode and high order toroidal mode could generate a nonradiative anapole mode, significantly boosting the local fields that is beneficial for nonlinear application and lasing. No need to mention the metamaterial and metasurfaces with various functionalities. These details will be elaborated in Section 6.3.

6.2.3 FABRICATION METHODS FOR DIELECTRIC STRUCTURES

To fully utilize the advantages of high refractive index dielectric materials, appropriate cutting-edge fabrication methods need to be adopted [11,13]. For the applications in optical, infrared, and terahertz regimes, high standards on fabrication are needed not only for the fine control of dielectric structure geometry and location, but also for the repeatability and throughput. Besides, simplicity and cost also need to be considered in the perspective of mass production. In general, there are four types of fabrication methods that are commonly used: lithography, chemical method, dewetting, and laser-assisted method. Table 6.1 compares these four methods in terms of five representative parameters: repeatability, productivity, resolution, position control, and complexity. It is worthy mention that there is no much difference in fabrication methods between dielectric and plasmonic materials. Readers can also refer to Chapter 4 for discussion on various fabrication techniques.

Lithography. Perhaps the most straightforward way to fabrication dielectric structure is the lithography, which could provide high repeatability and the best resolution. Depending on the required resolution, electron-beam-lithograph (EBL), focused-ion-beam lithography (FIB) or optical lithography could be used. The general procedure

---

TABLE 6.1
Comparison of Various Fabrication Techniques

	Repeatability	Productivity	Resolution	Position Control	Complexity
Lithography	High	Medium	High	Medium	High
Chemical method	Medium	High	Medium	Low	Medium
Dewetting	Medium	High	Medium	Medium	Medium
Laser-assisted method	High	Medium	Medium	High	Low

Other important properties: cost, area of fabrication, flexibility, precision

Source: Materials are adapted from Ref. [11]

of lithograph starts with a projection of designed geometry pattern into the photoresist on a substrate, followed by an etching process. Multistep lithograph process could be further implemented to fabricate complicate geometries, accompanied with increased cost and complexity. Up to present, lithograph technique is still considered to be the most popular and reliable method for fabrication of functional dielectric structures, and has been successfully implemented in various dielectric materials such as Si, GaAs, Te, Si_3N_4, and TiO_2, among others. A detail review on the lithograph approach can be found in Reference [13]. Nevertheless, the current lithography process is difficult to realize spherical or spheroidal shape structures. In addition, for EBL and FIB, wafer-scale fabrication remains difficult due to the high expense and low productivity, in particular for the sub-micrometer-scale fabrication. Nanoimprint could provide a better yield, but the cost of template is still high compared to that of conventional optical lithography.

Chemical method. The key benefit of chemical methods relies on the high productivity and low cost, which is considered to be a good complementary to lithograph. With the common chemical vapor deposition technique (CVD), high refractive index dielectric nanoparticles (e.g., Si, GaAs, Ge) with various shapes can be easily obtained to cover a large area. In addition, middle refractive index dielectric materials (e.g., TiO_2, ZnO) can be synthesized using sol-Gel synthesis method or microemulsion technique. Nevertheless, the main drawback of chemical method is the poor controllability and repeatability on the structure's shape, size, and location, thereby hindering its application in constructing functional meta-device that demands a large number of repeated unit cells. This problem can be partially resolved by combining the chemical methods with the lithography techniques, at the cost of additional fabrication step. Besides, chemical methods generally involve toxic or corrosive substances which might cause environmental contamination and damage to the fabricated sample. Bio-assisted method using biological substances (e.g., microorganism and plant extracts) as template or reactor is developed to synthesize metal oxide nanoparticles, which is considered to be environment friendly and low-toxic. More details on chemical method can be found in Ref. [14,15].

Dewetting. Dewetting process is essentially due to the agglomeration of nanoparticles when heating a thin film in order to minimize the total energy of film surface. The temperature during heat and the properties of thin film (e.g., thickness, density of defect, initial pattern, etc.) are the main controlling parameters for the dewetting process. In particular, the thickness of the film has a significant effect. The smaller the thickness, the higher the driving force for dewetting. Therefore, dewetting could be conducted at a temperature less than the melting point of the thin film. The advantage of dewetting method is the high productivity and simplicity, and can be implemented in any materials. So far, this method has been successfully used to form nanoparticles of different sizes on various dielectric materials such as Si and Ge over a whole wafer scale. Nevertheless, similar to the chemical method, it is difficult to flexible control the shapes of the particles as well as their spatial arrangements. More details on dewetting method can be found in Ref [16].

Laser-assisted method. The idea of laser-assisted method is to exploit the advantage of laser techniques (e.g., material selectivity, submicron resolution, high energy density, etc.) to improve the precision of fabrication. This could be done via the following two ways. First, direct laser ablation using an ultrashort laser pulse has been successfully implemented to fabricate single crystalline sub-micron or micron sized microspheres for various dielectric materials (e.g., ZnO, CdSe, PdTe, etc.). The resultant particles are distributed closely to the laser focal point. However, this approach cannot control the size and location of the fabricated particle, since it is different to control the interaction between an ultrashort laser pulse and the targeted material. The second way is the laser printing, where the laser spot is focused on the interface between the printed materials and donor substrate, followed by a transfer process to another receiver substrate. More importantly, laser annealing technique can be subsequently applied for post processing of laser printing, which could manipulate the phase of fabricated dielectric particle (e.g., from amorphous to crystal) to tune its electromagnetic properties. It has been demonstrated that Si dimer with well-controlled gap distance can be achieved with this technique [17]. Besides, by combining the laser printing and dewetting techniques, high productivity, high resolution, and high position control can be simultaneously achieved, which is promising for large-scale fabrication of dielectric particles. Nevertheless, the main constraint about this method is the requirement of the high-quality laser source and high precision of the position system [18].

3D printing. Besides the four main stream fabrication methods mentioned earlier, we would like to briefly discuss another potential method—3D printing. Also known as addictive manufacturing or rapid prototyping, the key advantage of 3D printing is the ability to produce very complex three-dimensional shapes or geometries from a CAD model that would be otherwise impossible to construct by other fabrication techniques, which is very attractive for metamaterials and metasurface applications. So far, 3D printing has made a great commercial success in various industry sectors including aerospace, automotive, and medical industries, and continues to undergo a rapid expansion. However, there are two limitations for this method. First, the current 3D printing technology only works for a few particular types of materials (e.g., polymers, metals, ceramics), and is not applicable for high refractive index semiconductor materials (e.g., Si, Ge, GaAs). Second, the resolution of 3D printing is generally around a few micrometers scale, which might be capable of constructing meta-structures in the far infrared or terahertz regime since the size of the unit meta-atom is at a few hundred micrometers scale. However, for optical or near infrared regime, the size of the unit cell is typically at nanometer scale, therefore requiring a much higher resolution of just a few nanometers and is beyond the capability of the current 3D printing technology, even with the sophisticate two-photon lithography (~ 100 nm resolution) [19,20]. Future research should focus on overcoming these two obstacles to enable a full functionality of 3D printing in dielectric materials for optical, infrared, and terahertz applications.

6.3 APPLICATIONS OF DIELECTRIC MATERIALS

6.3.1 Electric and Magnetic Hotspots

As shown in Section 6.2.2, the key advantage of dielectric micro-nanostructure is the capability to support both electric and magnetic types resonance modes, which could effective manipulate the local electric and magnetic fields at a deep subwavelength scale. One of the direct consequences of these multipole resonances is the generation of electric and magnetic hotspots, producing an enhanced local electromagnetic field that is orders of magnitudes larger than the incident one. Taken dielectric nanosphere as an example, Figure 6.3(a) illustrates the electric and magnetic field distributions

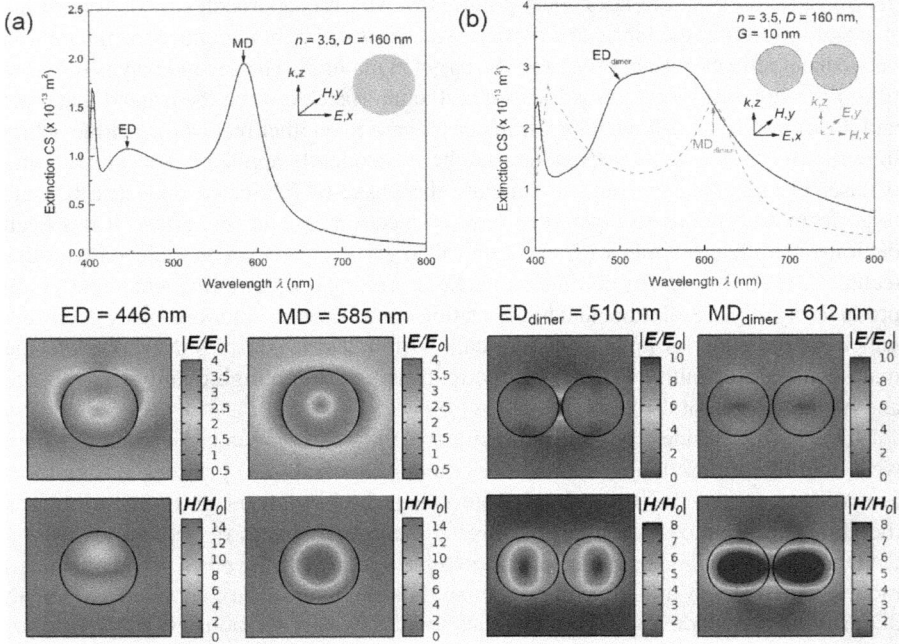

FIGURE 6.3 (a) Top: extinction cross-section (CS) of a dielectric spherical particle, with a diameter $D = 160$ nm and refractive index $n = 3.5$. Bottom: electric and magnetic field distributions at electric dipole mode (ED) and magnetic dipole mode (MD). (b) Top: extinction cross-section (CS) of a dielectric dimer consisting of two identical spherical particles in (a), with a gap distance $G = 10$ nm. Depending on whether the electric field (solid line) or magnetic field (dashed line) is parallel to the dimer central axis, the near field coupling between the two constituents could produce a hybrid ED_{dimer} or MD_{dimer} mode. Bottom: electric and magnetic field distributions at ED_{dimer} and MD_{dimer}.

at its ED and MD resonances, respectively. Clearly, both ED and MD can produce electric field enhancements $|E/E_0|$ and magnetic field enhancements $|H/H_0|$ inside and outside the nanoparticle. While $|E/E_0|$ are comparable for both resonances, the $|H/H_0|$ of MD is much higher than that of ED. Nevertheless, the $|H/H_0|$ inside the dielectric particle is significantly larger than the outside one, which makes it difficult to be exploited. To create a magnetic hotspot outside the structure, dielectric dimer is proposed by utilizing the near field coupling between two adjacent dielectric particles, similar to the plasmonic dimer. Depending on the incident light polarization, electric or magnetic hotspot could be created within the dimer gap, as shown in Figure 6.3(b). Obviously, the coupling between the two constituents could induce a much larger electric or magnetic field in the gap region compared with that of the individual nanosphere, accompanied with a shift in the resonant wavelength. Other structures such as doughnut-shape dielectric disk and dielectric oligomer have also been demonstrated to support electric and magnetic hotspots outside the structures [21].

Note that although dielectric structure could also provide an electric hotspot, its electric field enhancement capability is generally weaker compared with that of plasmonic metal counterpart, especially when the environment medium is water ($n = 1.33$) or oil ($n = 1.45$) [22]. This can be qualitatively understood from the material permittivity profiles in Figure 6.1. Intuitively, the near field enhancement is stronger for a larger permittivity contrast $|\varepsilon - \varepsilon_0|$ between the resonant structure $\varepsilon = \varepsilon_r + i \cdot \varepsilon_i$ and the surrounding environment ε_0. Assuming a lossless material $\varepsilon_i = 0$, for the dielectric material, $\varepsilon_r > 0$ and $|\varepsilon - \varepsilon_0|$ decreases as the medium permittivity increases, which causes diminished electric and magnetic field enhancements. On the contrary, $\varepsilon_r < 0$ for plasmonic metal and $|\varepsilon - \varepsilon_0|$ increases as the medium permittivity increases, which results in even greater electric and magnetic field enhancements. Therefore, appropriated materials should be chosen depending on the practical situation to get an optimized performance.

It is also worthy to emphasize that the control of magnetic field is not trivial in the optical, infrared, and terahertz regimes since most natural substances at these frequency bands are not endowed with microscopic magnetization (e.g., relative permeability $\mu = 1$), and intuitively should have only weak interactions with the magnetic components of the electromagnetic waves. The existence of magnetic resonance modes provides an effective way to exploit the previously unusable energy and information of optical magnetic field based on an intrinsically non-magnetic dielectric material. Though plasmonic metal materials could also demonstrate magnetic properties, they generally depend on the design and fabrication of complex structures (e.g., split ring, oligomer, diabolo) to produce a pseudo-current loop so as to generate magnetic type resonance, which is practically difficult at optical frequency. In addition, the inherent loss of plasmonic metal in the optical frequency severely prohibits its applications from efficient optical metadevices. In the contrary, the high refractive dielectric material only has a negligible loss as long as the incident photon energy is smaller than its bandgap (see Section 6.2.1), which could easily overcome some detrimental issues such as heat dissipation and radiation loss. More important, the magnetic type resonance modes are universally existed in the dielectric structures, even for a simple spherical particle, thereby significantly relaxing the requirement on the device design and fabrication. All these make the dielectric material a promising platform in studying optical magnetic phenomena.

6.3.2 Enhanced Raman Scattering

The hotspots demonstrated in Section 6.3.1 build a solid foundation to enhance the Raman scattering. The Raman scattering accounts the inelastic scattering of photons due to vibrations in molecules or phonon modes in crystalline materials, whose signal strength scales dramatically to the electric near field intensity. Since the electric and magnetic field enhancements are presented both inside and outside the dielectric structure, they can be exploited to enhance either the intrinsic Raman scattering supported by the optical phonon mode of dielectric structure, or alternatively the Raman signal from external molecules. In particular, the former case is a unique property for the dielectric structure, since the plasmonic metal counterpart in general does not support hotspots inside the structure.

As shown in Figure 6.4(a), the relationship between the intrinsic Raman scattering signal and the multipole electric and magnetic resonance modes is unambiguously

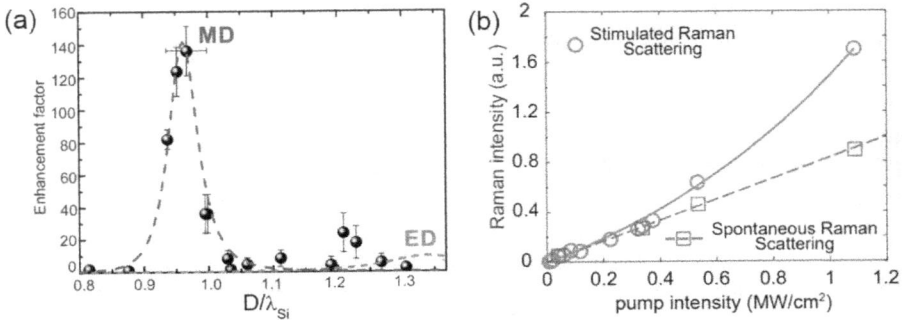

FIGURE 6.4   (a) Theoretical (dashed curves) and experimental (black dots) dependencies of Raman scattering enhancement for a spherical Si nanoparticle on their diameter D normalized to the incident wavelength.

Source: Reprinted with permission from Dmitriev, P. A. et al. *Nanoscale* 8, 9721–9726, 2016. Copyright 2016 Royal Society of Chemistry.

(b) Nonlinear enhancement on Raman intensity.

Source: Reprinted with permission from Zograf, G. P., et al. *Nano Lett.* 20, 5786–5791, 2020. Copyright 2020 American Chemical Society.

demonstrated by Demitriev et al. in Ref. [23], based on a spherical Si nanoparticle with well-separated ED and MD responses. It is found that a 140-fold enhancement on the Raman intensity can be achieved due to MD since a decent electric field enhancement is generated inside the particle. Whereas for ED, only a modest Raman enhancement is observed since the electric hotspot is mainly generated outside the particle (see also Figure 6.3(a)). Later, more complex mechanisms are introduced to obtain a strong Raman signal, by further increasing the electric field strength. For example, it is shown that a nonradiative anapole mode could significantly enhance the Raman scattering of optical phonon mode of a Si nanodisc [24]. This is practically useful to probe those low radiative resonance modes (e.g., magnetic quadrupole mode, anapole mode) of dielectric structure that are difficult to be observed in the conventional far field scattering spectrum, since now they can be clearly reflected as a peak in the Raman spectrum. More importantly, since the enhanced Raman scattering of dielectric crystal is directly related to the internal electric field enhancement inside the structure, it can be used to measure the near field distribution inside the dielectric structure. Compared to other near-field instruments, such as near-field scanning optical microscopy (NSOM) or cathodoluminescence (CL), the far field Raman spectroscopy could provide quantitative information of near field based on a noninvasive approach, albeit its spatial resolution is relatively lower [25]. Furthermore, since the optical phonon modes of many crystals have high thermal sensitivities, the enhanced intrinsic Raman scattering could function as a nanothermometry, which could be utilized in various applications including tracking temperature during laser-reshaping and crystallization process, phototherapy, and drug release

The dielectric resonances is also applied to the surface-enhanced Raman spectroscopy (SERS), which could be useful for fingerprint-like sensing of traced molecules. While the multipole resonance mode of an individual dielectric nanoparticle already gives a Raman enhancement of 10^2, an even large enhancement up to 10^3 could be achieved using a silicon nanodisc dimer configuration [26]. On top of that, enhanced stimulated Raman scattering (SRS) has also been demonstrated with dielectric nanostructures, which are potentially beneficial for the miniaturized Raman lasing devices. The SRS is a third order nonlinear light amplification process that involves two optical fields, one at excitation frequency and the other at Stokes-shifted emission frequency. Since Si has a relatively large Raman gain, SRS in nanoscale has been recognized as a promising solution for compact light source in integrated silicon photonic circuit. For illustration, a simple Si nanodisc supporting higher-order multipole resonances is sufficient to achieve SRS, resulting in a nonlinear enhancement on the Raman intensity as shown in Figure 6.4(b) [27]. Besides the conventional multipole Mie resonator, other judiciously designed dielectric nano-structures possessing Fano resonances, bounded-state-in-continuums (BIC), or whisper-gallery-mode (WGM) can also be exploited to produce large Raman enhancements, due to the strong electric field enhancements associated with high quality factor resonance modes. All these details can be found in recent review articles [25,28].

6.3.3 ENHANCED FLUORESCENCE SPONTANEOUS EMISSION

The rich variety of electric and magnetic multipole Mie modes of dielectric structure is promising for flexibly engineering the fluorescence spontaneous emission of quantum emitter. The name quantum emitter is used because it is usually described theoretically as a two-level quantum system. Since the size of quantum emitter is typically much smaller than the emission wavelength, microscopically it is also treated as a point dipole source for convenience. From the perspective of fundamental physical mechanism, there are generally two categories of quantum emitters—electric dipole emitters and magnetic dipole emitters. The electric dipole emitter is represented by a pair of electric charges of equal magnitudes but opposite signs separated by a small distance, whose strength is characterized by the electric dipole momentum. There exist many forms of electric dipole emitter in nature, such as organic dye molecule, quantum dot, nitrogen-vacancy center (NV), and most of the atomic emission sources, whose emission wavelengths span over the optical and infrared regions. Even for some optical and terahertz devices (e.g., light emitting diode (LED), spintronic tera-hertz thin film), the emission properties of their active regimes (e.g., multi-quantum well) can be well approximated as an assembly of electric dipole emitters [29].

In contrast, the magnetic dipole emitter is comparatively rare, because the intrinsic magnetic dipole emissions of most known materials in optical, infrared, and tera-hertz regions are typically four to five orders magnitudes smaller than those of electric dipole emission. In addition, there is no "magnetic charge" in nature, and the magnetic dipole is typically produced by a closed electric current loop. However, in the optical, infrared, and terahertz region, it is rather difficult to construct such a small electric current loop at such a high frequency. It was believed that there is no access to the opto-magnetism till the discovery of trivalent lanthanide rare-earth ion [30]. Due to the unique atomic structures, the intra-$4f^n$ optical transitions of trivalent lanthanide rare-earth ions could induce pronounced magnetic dipole emissions at some particular wavelengths (e.g., Eu^{3+}: $^5D_0 \rightarrow {}^7F_1$ at 588 nm and Er^{3+}: $^4I_{13/2} \rightarrow {}^4I_{15/2}$ at 1500 nm). These magnetic dipole emitters could interact with the magnetic components of the electro-magnetic waves, which are valuable complements to its electric dipole counterparts.

Regardless of electric or magnetic dipole emitter, the intrinsic spontaneous emission intensity from a single emitter is rather weak, thereby hindering its applications in many emergent fields such as single molecule fingerprint-like sensing, single photon source, high resolution imaging, quantum memory, among other. Similar like plasmonic materials, the dielectric materials could enhance the fluorescence emission of quantum emitter based on three mechanisms [31,32]. First, the electric and magnetic field enhancements of the multipole resonances could boost the excitation rate (e.g., pumping the carriers from ground state to high energy state) of electric and magnetic dipole emitter, respectively. Second, the excited carrier subsequently returns to the ground state and release a photon, which could again interact with the dielectric structure. The multipole resonances could magnify the emission intensity by enhancing the local density of optical states (LDOS), which is also referred as Purcell effect. Depending on the orientation, the electric or magnetic dipole emitter selectively couples to certain multipole Mie resonances. Lastly, the presence of dielectric structure could engineer the local environment, altering the directivity of fluorescence emission to improve the signal collection efficiency. Compare to the plasmonic metal counterpart, the distinctive advantage of dielectric material relies on its low loss property, which could effective suppress the non-radiative loss to avoid the notorious fluorescence quenching and enhance the quantum efficiency of dipole emitter. Sun et al. systematically compared the fluorescence enhancements between a plasmonic and dielectric nanosphere for an electric dipole emitter [22]. It is shown that in a low refractive index medium (e.g., air), dielectric could outperform metal due to its much lower loss and decent electric field enhancement. Whereas in a high refractive index medium (e.g., water, oil), dielectric is no longer the best option due to its reduced electric field enhancement. Besides, the interference of electric and magnetic multipole modes could enable a highly directional scattering by realizing Kerker condition (see Section 6.3.5), which is also applicable for point dipole source.

Many works have shown significant fluorescence enhancements of quantum emitters with dielectric structures, here we just give two examples for the sake of illustration. Regmi et al. demonstrated single molecule level fluorescence enhancement with a pure Si dimer configuration [33]. A 270 times fluorescence enhancement has been achieved for Alexa Fluor 647 dye molecule, which is larger than that of the Au dimer, as shown in Figure 6.5(a). On the other hand, Vaskin et al. has demonstrated an enhanced the magnetic dipole transition of a Eu^{3+} doped thin polymer film with a Mie-resonant Si nanocylinders [34], as shown in Figure 6.5(b). The averaged enhancement factor is approximately four times. In general, the enhancement of magnetic dipole emitter is comparatively unexplored than that of electric dipole emitter, and there still remains a large space to improve.

6.3.4 Optical Force

The subwavelength confinement capability of dielectric structure could naturally create an optical gradient force or extinction force. The optical gradient force heavily depends on the gradient of light intensity, which can be enhanced

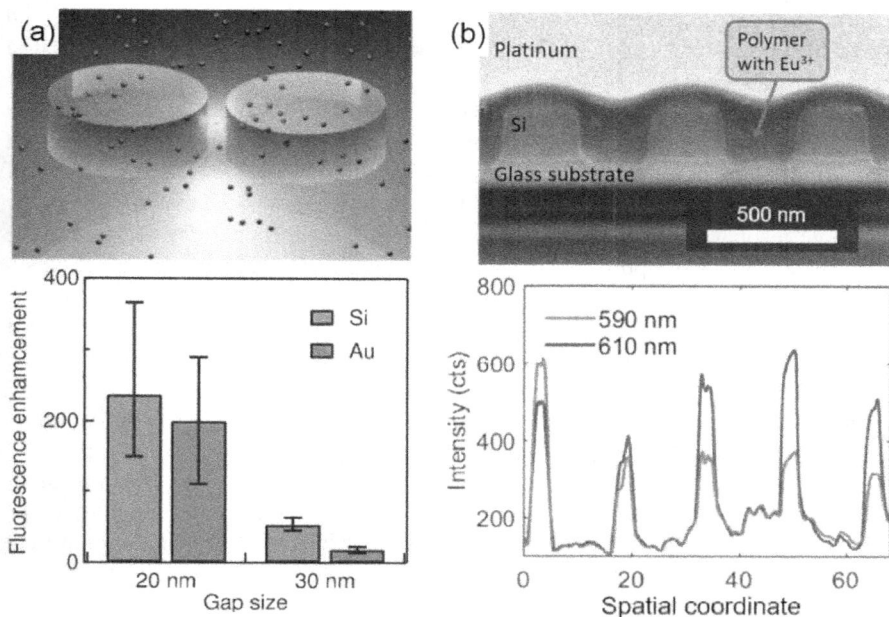

FIGURE 6.5 (a) Si dimer enhanced electric dipole emission of dye molecule, and comparison on the fluorescence enhancement factors between Si and Au dimers.

Source: Reprinted with permission from Regmi, R. et al. *Nano Lett.* 16, 5143–5151, 2016. Copyright 2016 American Chemical Society.

(b) Si nanodisk enhanced magnetic dipole emission of Eu^{3+} layer.

Source: Reprinted with permission from Vaskin, A. et al. *Nano Lett.* 19, 1015–1022, 2019. Copyright 2019 American Chemical Society.

by reducing the size of hotspot. While the optical extinction force, which is the combination of scattering and absorption forces, is linearly proportional to the intensity of electromagnetic wave. These optomechanical properties make dielectric resonator capable of manipulating the movement of the small particle, which could be used in a plethora of applications such as biomedicine, nanoparticle sorting, microfluidic, optical trapping, particle rotation etc. When a small particle is placed near the hotspot regime, it could exchange the momentum with the evanescent electromagnetic field, being pushing away or pulling towards the dielectric structure. So far, various biomolecules (e.g., DNA, bacteria, and virus) or atoms have been demonstrated to be effectively confined within a 60 nm trap size, with a nearly perfect trapping efficiency [35]. On top of that, the optical force is also presented in a system of coupled dielectric structures, which is important for another set of applications such as integrated photonic circuit, optical signal processing, optomechanical cooling, sensor and actuator, tunable laser etc. For instance, an optomechanical actuator made of two coupled waveguides was demonstrated with a maximum displacement of 67 nm and a response time of 94.5 nm, under an attractive optical force of 1 pN/μm/mW [36]. More examples can be found in a recent review paper [37].

Compared to the metal counterpart whose plasmonic resonance could generate a large amount of heat that might damage the structure of biomolecule, the low heat generation of dielectric resonances could safely preserve the intrinsic property of biological substance. In addition, the interplay between the electric and magnetic multipole resonances of dielectric structure offers an additional degree freedom to manipulate the sign of optical force (attractive or repulsive), which is potentially useful for the control of nanoparticle assembling in aqueous medium. For example, the Si nanodimer could possess either repulsive or attractive binding forces in water for an incident wave polarized along the dimer axis as shown in Figure 6.6, whereas it only exhibits attractive force for a gold nanodimer [38]. The exotic repulsive effect is a consequence of the Lorentz force induced by the magnetic dipole mode of dimer, which is considered to be unique for the dielectric structure.

Note that there exists a special type of force called Casimir force, which is mediated by virtual photon, for a coupled nanostructure system. The Casimir force can only be described with the quantum electrodynamic framework, and does not present in classic electromagnetics. In general, the Casimir force for dielectric structure is much weaker than the plasmonic counterpart due to the suppressed thermal effect, but it is still significant when two structure are in close proximity (e.g., gap distance

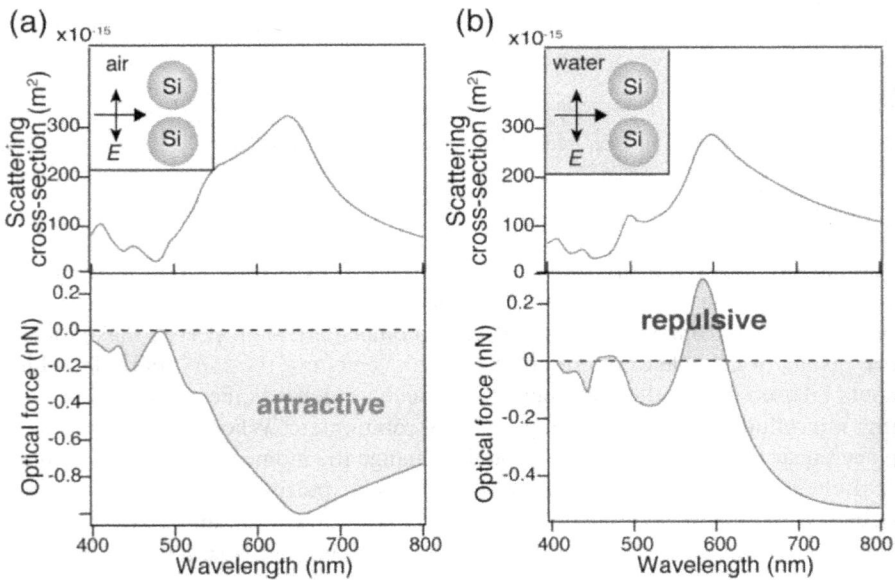

FIGURE 6.6   Scattering cross-sections (top) and optical binding forces (bottom) of a Si nanodimer in (a) air and (b) water. Incident light is polarized along the dimer axis.

Source: Reprinted with permission from Yano, T. et al. *Opt. Express* 25, 431–439, 2017. Copyright 2017 The Optical Society of America.

< 100 nm). We will not elaborate this particular force type in this book, interested readers can refer to a recent review [39].

6.3.5 ENGINEERING FAR FIELD SCATTERING

Early in 1983, it was first proved theoretically by Kerker et al. that the backward or forward scattering of a hypothetic magnetic sphere can be completely eliminated, if the electric permittivity and magnetic permeability are equal ($\varepsilon = \mu$), which is essentially attributed to the in-phase or out-of-phase oscillations of electric and magnetic dipoles [40]. While it is an interesting phenomenon, it is rather difficult to be realized in reality due to the lack of natural magnetic material, especially in the high frequency regime. As explained in Section 6.2.2, the dielectric structure universally possesses a full family of electric and magnetic multipole Mie resonances regardless of its shape or material, thus providing a viable platform to not just realize the originally proposed Kerker condition but also largely expand the scope. The rich variety of multipole resonances have different phase symmetries of far field radiations (e.g., see Figure 6.2(d)), and the constructive or destructive interference between them could flexibly engineer the scattering powers as well as the angular scattering patterns. A rule of thumb is that the parities are opposite for multipoles of the same nature and adjacent orders (e.g., even parity for ED, while odd parity for EQ), and for multipole of the same order but different natures (e.g., even parity for ED, while odd parity for MD). Since the relative contribution of each Mie mode as well as their spectral overlapping can be easily controlled by altering the morphology of dielectric structure, it is rather convenient to design various functional dielectric structures with desired scattering patterns. Further considering its low loss property, the exotic scattering generated by dielectric structure has a great potential in many flourishing fields such as cloaking, nanoantenna, optical communications, sensing and imaging, among others.

Based on the earlier-mentioned mechanism, the Kerker condition using high refractive index dielectric material was first experimentally demonstrated with a single Si spherical microparticle in the microwave regime and later extended to optical and infrared regime, using only dipolar multipole contributions. Such simplest scenario with overlapped ED and MD resonances resembles the concept of Huygens's source in the antenna engineering. After that, it is found that by involving higher order multipole resonances (e.g., ED + MD + EQ + MQ), unidirectional forward or backward scattering can be achieved with enhanced scattering directionality and efficiency, which is particularly useful to obtain collimated electromagnetic beam. The contribution of higher order mode also enables exotic transverse scattering governed by the interference between dipolar and quadrupolar Mie modes (ED + MD = - (EQ + MQ)). In this case, both forward and backward scattering are simultaneously suppressed, which might be beneficial for on-chip electromagnetic wave transmission or anti-reflective coating. In addition, the Kerker effect is presented not only for the conventional linear polarized incident electromagnetic wave, but also applicable for other structured waves (e.g., radically polarized incident) or even electron beam, thereby greatly expanding the application scope. In particular, it holds valid for a quantum emitter source (e.g., electric or magnetic dipole emitter as introduced in Section 6.3.3), which is promising for fluorescence related applications [41]. These generalized Kerker effect is summarized in Ref. [42]

Other than isolated scattering body such as nanoparticle or particle clusters, the Kerker effects also hold valid for periodic or aperiodic particle lattices, as well as their corresponding complementary structures (e.g., holes or slits) based on Babinet's principle. In a lattice structure, the multipole Mie resonances of the individual unit cell is modified by the lattice boundary condition, endowing various functionalities such as perfect transmission, perfect reflection, perfect absorption, and higher order diffraction control [42]. For perfect reflection, transmission and absorption, periodic lattice structure with subwavelength periodicity is sufficient to realize the functionalities (for perfect absorption, lossy material is usually required). Whereas for the higher order diffraction control, the lattice periodicity is generally larger than the incident wavelength. The unit cell of lattice can be deliberately designed to eliminate certain diffraction orders, leading electromagnetic energy to a desired direction, which can be used for wave-routing or beam steering. For example, Khaidarov et al. demonstrated that a great portion could be channeled into the -1 diffraction order using an asymmetrical TiO_2 dimer lattice as shown in Figure 6.7, while all other possible diffraction channels are effective suppressed [43].

6.3.6 NON-RADIATING MODE AND ENHANCED NONLINEARITY

The interference between multipolar resonances could create a special type of resonance mode called *anapole*, which is a consequence of destructive interference between the fundamental dipolar mode and toroidal dipole moments, as shown in Figure 6.8 [44]. The toroidal dipole moment is a result of poloidal current excitation, which has identical far field radiation pattern as that of dipolar mode but is opposite

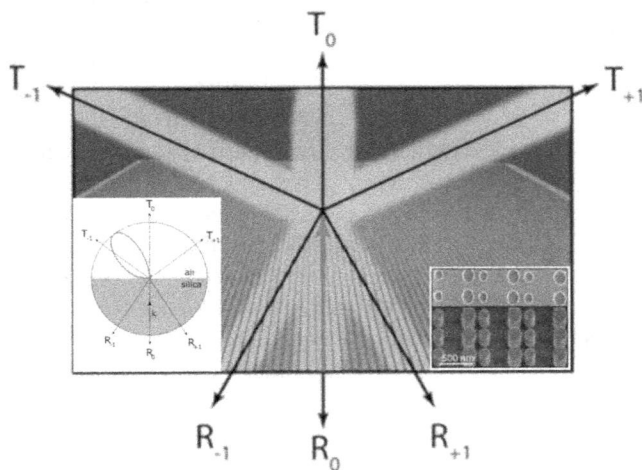

FIGURE 6.7 Light transmission to -1 diffraction order using asymmetrical TiO_2 dimer lattice.

Source: Reprinted with permission from Khaidarov, E. et al. *Nano Lett.* 17, 6267–6272, 2017. Copyright 2017 American Chemical Society.

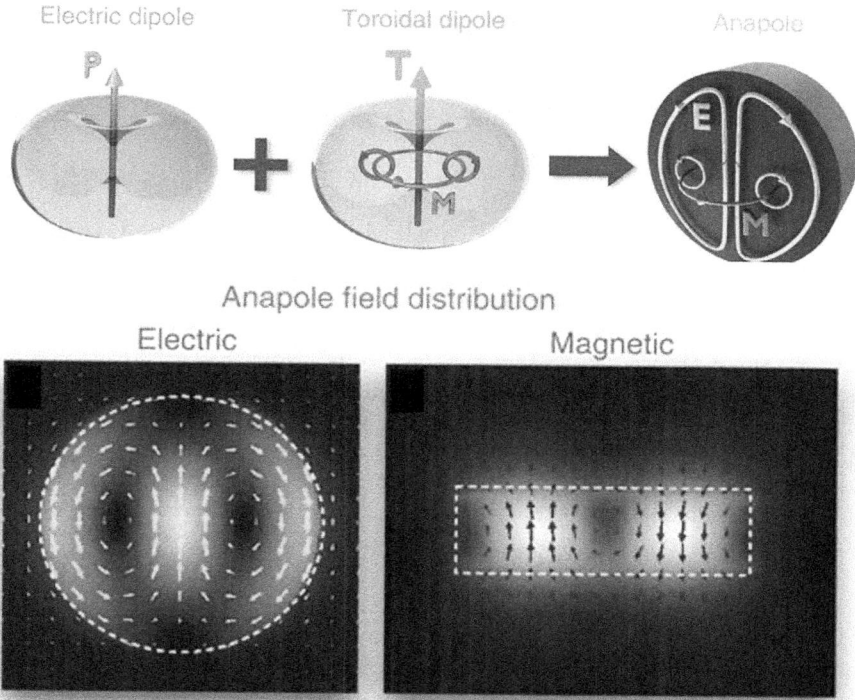

Anapole field distribution

FIGURE 6.8 Top: Schematic illustration of anapole mode, which results from the destructive interference of electric dipole mode and toroidal dipole mode. Bottom: electric and magnetic field distribution of anapole mode of a dielectric disk.

Source: Adopted from Miroshnichenko, A. E. et al. *Nat. Commun.* 6, 8069, 2015. No permission is required. Copyright 2015 The authors.

in parity. Rigorously govern by the Cartesian multipole decomposition method, a full set of toroidal dipole moments with different natures (electric or magnetic) is given in Ref. [45] for convenience. Initially, the concept of toroidal mode was only conceived theoretically, because it is generally weak and often hidden by the fundamental ED and MD, making it difficult to be measured directly. It took more than 30 years to experimentally realize these peculiar resonances modes, first at microwave frequencies, and later at terahertz and optical frequencies [45,46], which is endowed by the rapid development of fabrication and measurement techniques.

The unique property of anapole is that it has a non-radiating far field, which usually exhibits a pronounced dip in the scattering spectrum, and might be useful for electromagnetic invisibility and cloaking. More importantly, it allows an extremely high optical confinement, enabling a giant local electromagnetic enhancement which is very suitable for nonlinear applications. Compared to conventional bulk nonlinear crystals (e.g., $LiNbO_3$), the anapole mode based on subwavelength micro-nanostructure could induce a large nonlinear conversion efficiency without the need of phase-matching, which is beneficial for the device miniaturization and compact system integration. For any material, its nonlinear electromagnetic response is characterized by the nonlinear polarization P as,

$$P = \varepsilon_0[\chi^{(1)}E + \chi^{(2)}E^2 + \chi^{(3)}E^3 \ldots] \tag{6.1}$$

where $\chi^{(n)}$ is the nth-order susceptibility which is an intrinsic property of material, E is the electric field and ε_0 is the vacuum permittivity. For conventional linear electromagnetics, only the first term is considered because the nonlinear effect is generally weak. Since the electric field could be significantly boosted by the anopole mode, a series of nonlinear effects would become easier to achieve, such as nonlinear lasing, second harmonic generation (SHG), third harmonic generation (THG), four-wave mixing, frequency up-conversion etc. For example, Grinblat et al. demonstrated a 0.0001% THG conversion efficiency with the anapole mode of Ge nanodisc [47], which is four orders of magnitudes larger than the unstructured Ge film, and is at least one order of magnitude larger than the fundamental ED or MD resonances.

Besides anapole mode, there is another type of non-radiating state named *the bounded state in continuum* (BIC) (sometimes also refer to trapped mode, embedded eigenstate, dark state) [48], exhibiting a very high-quality factor resonance mode and a large associated field enhancement. The concept of BIC was originally introduced in quantum mechanics, corresponding to those peculiar resonances that are located within the radiation continuum spectrum with no energy decay. The universal nature of the Schrodinger equation implies the ubiquitous presence of BIC in all domains of wave physics, ranging from microwave to optics, acoustics, and water waves. The ideal BIC, which is strictly non-radiating and has an infinitely large quality factor, remains a pure mathematically expression for a lossless material with extreme (zero or infinite) dielectric function. In practice, it is often coupled to a certain radiation channel and becomes a leaky mode with a finite quality factor, which is experimentally measurable and termed as *quasi-BIC*. Koshelev et al. demonstrated enhanced SHG with quasi-BIC in an individual AlGaAs cylindrical resonator placed on a three-layer substrate ($SiO_2/ITO/SiO_2$). The quasi-BIC was obtained by overlapping several Mie modes, which resulted in a nonlinear conversion coefficient of $\sim 10^6$ W^{-1} with a maximum sustainable pump power of 10 W. Other structures and strategies to yield quasi-BIC effect can be found in a recent review [49].

6.3.7 OPTOELECTRONIC AND ON-CHIP PHOTONIC DEVICES

The dielectric micro-nanostructures could be engineered to integrate with the commercially available optoelectronic devices to improve their performance. Since many electromagnetic wave sources can be physically treated as an assembly of coherent (e.g., laser) or incoherent (e.g., LED) dipole emitters, it would be rather convenient to exploit the transparent dielectric resonator to enhance the emission efficiency of electromagnetic source device based on the Purcell effect, or to improve the extraction efficiency by manipulating the directivity of electromagnetic wave. For illustration, Kuo et al. achieved functional regulation by preparing SiO_2/GaN nanoantenna arrays embedded inside the active layer of the LED, and greatly improved the color conversion efficiency of the LED to 32.4% as shown in Figure 6.9, which is attributed to the enhanced extraction efficiency and suppression of nonradiative energy transfer [50]. The full description on dielectric structure enhanced LED performance, including various mechanisms and plenty of examples, are summarized in our recent review paper [51]. Another straight forward application is to use dielectric structure for photovoltaic to enhance the light absorption or reduce reflection. Zhong et al. designed a dielectric array composed of Si Mie resonator buried in a low refractive index SiN layer as a transparent antireflection coating. The reflection could be effectively suppressed within a broadband wavelength (400 nm–1000 nm), and eventually lead to a certified conversion efficiency of 18.47% for the nanostructured silicon solar cells on a 156 mm by 156 mm wafer [52].

Beside the optoelectronic applications, dielectric subwavelength structure is also promising for on-chip photonic device and quantum device, aiming for ultrafast and low-power nanoscale devices that are urgently demanded for fast information processing, spectroscopy, and others in a multiplicity of applications from cyber security, banking, healthcare, communications, defense, and non-destructive detection. A properly designed dielectric micro-nanostructure could be beneficial in both passive (e.g., waveguide, power splitter, coupler, Bragg grating) and active (e.g., modulator, source) components of on-chip photonic and quantum devices. An inspiring example is to incorporate the dielectric antireflective structure into on-chip waveguide to reduce the facet reflection loss, or to utilize dielectric metasurfaces to transform the eigenmode of waveguide from one to another. In quantum photonics, embedding quantum emitter (e.g., quantum dot or NV center) into a special design dielectric structure is common practice to achieve an excellent source efficiency. Interested reader may refer to Ref. [53] for more details.

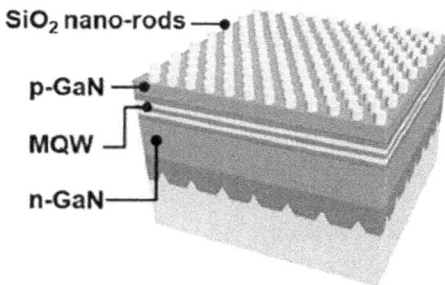

FIGURE 6.9   Schematic of GaN LED color-conversion with a layer of SiO_2 nanorod array.

6.3.8 METAMATERIAL AND METASURFACE

Similar like plasmonic metal, the dielectric micro-nanostructure could be used a building block for constructing novel metamaterials and metasurfaces, to take a full control on the phases, polarizations, and amplitudes of electromagnetic waves in optical, infrared, and terahertz regime [54–56]. In terms of functionalities, there is no much difference between metal and dielectric. All the meta-functions reported for plasmonic metal materials could also be achieved with the dielectric counterparts, which have been systematically discussed in Chapter 5. Here, we just highlight two advantages of dielectric material: first, the low loss feature of dielectric is highly demanded for those devices requires high energy efficiencies, e.g., lens, polarizer, coloring, hologram, structured light source, etc.; second, the presence of both electric and magnetic multipole resonances endows dielectric more flexibility in tailoring the electromagnetic wavefront (e.g., Huygens metasurface), spectral line shape (e.g., fano resonance, electromagnetic induced transparency (EIT)) and near field distribution (e.g., BIC, anapole). In particular, the contributions from higher order mode provides a convenient approach to achieve broadband functional meta-devices.

6.4 SUMMARY AND OUTLOOK

Table 6.2 briefly summaries the differences between dielectric and its primary competitor—plasmonic metal, and briefly discusses the applicability of these two types of materials. It is clear that though there is a large overlap in terms of the application scope, plasmonic metal and dielectric materials have their own advantages and preferred practical scenarios. This inspires the researcher to investigate the performance of metal-dielectric hybrid structure, which may leverage the advantages of both types of materials. For example, Sun et al. demonstrated that a hybrid nanoantenna could simultaneous possess the high electric field enhancement from metal and low loss property from dielectric, achieving all round enhancements in dipole excitation rate, quantum yield and fluorescence emission intensity [57–60]. Using a hybrid heterodimer structure (see Figure 6.10), they also showed that the plasmonic resonance could also interfere with dielectric multipole Mie resonances, which resembled a nearly ideal first Kerker condition close to the resonance peak of dimer, creating a highly efficient unidirectional forward scattering with a large forward-to-backward scattering ratio (F/B ratio) of ~ 50 dB. The scattering intensity of hybrid dimer is much higher than that of isolated dielectric nanoparticle at its Kerker condition wavelength [61]. We expect that the hybrid platform could potentially outperform bare metal or dielectric structure in many applications, which is worthy of study in the future.

Another rapidly growing field of research is to use active materials to gain dynamic control on the functionality of dielectric structures. This could be achieved by either integrating the passive dielectric structure with some active media (e.g., VO_2, graphene, etc.) [62] or directly implementing the active materials as the main building block [63]. In view of this, Berestennikov et al. pointed out that halide perovskite is a promising candidate because it has a decent refractive index (2.0 ~ 2.5) and exceptional electric property, which could easily obtain tunable structures while benefit from various multipole Mie resonances [63]. Active dielectric structure is essential to realize switchable and tunable multifunctional meta-device, which is highly desired

TABLE 6.2

Comparison between Plasmonic Metal and Dielectric Materials

	Dielectric	Plasmonic Metal
Resonant nature	1) Resonance generated due to displacement current distribution inside structure. 2) Co-existence of both electric and magnetic multipole modes. 3) Large resonance quality factor	1) Resonance generated due to collective oscillation of surface electrons. 2) In general dominated by electric type resonances, except some special configurations. 3) Small resonance mode volume.
Electric and magnetic field enhancements	1) Capable of generate both electric and magnetic hotspots. 2) Local field enhancement is presented both inside and outside the structure. 3) Field enhancement decreases as the environmental refractive index increases.	1) Mainly yield electric field enhancement, except some special configurations. The enhancement factor is generally larger than the dielectric structure. 2) Local field enhancement is concentrated only at the surface of plasmonic structure. 3) Field enhancement increases as the environmental refractive index increases.
Loss	Lossless if incident photon energy is less than the bandgap.	High loss is always presented.
Applicability and advantage	1) Applications in air or vacuum. 2) Applications requires high energy efficiency, e.g. polarizer. 3) Applications requires surface and/or volumetric enhancement technique. 5) Applications involving magnetic component of electromagnetic waves, e.g. optomagnetism. 6) Exotic phenomenon generated by the interference between electric and magnetic resonance, e.g. anapole mode, Kerker condition, etc.	1) Applications in air, water, or oil. 2) Applications requires ohmic loss, e.g. plasmonic heating, thermoelectric. 3) Applications requires mainly surface enhancement technique. 4) Applications demand high electric field enhancement, e.g. Raman scattering, nonlinear optics 5) Applications requires small footprint and mode volume, e.g. single molecule Rabi-splitting.

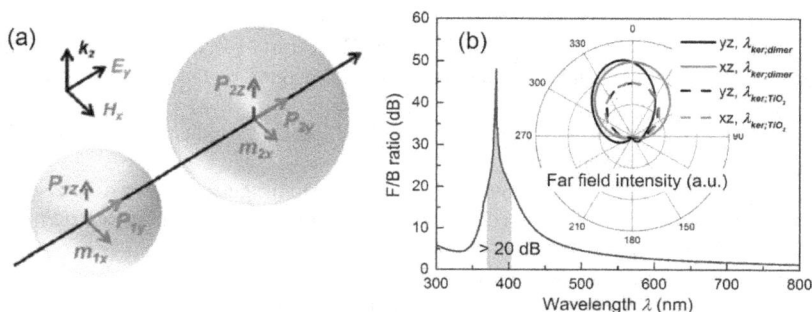

FIGURE 6.10 (a) Schematic of metal-dielectric heterodimer. (b) Highly efficient unidirectional forward scattering pattern with large F/B ratio.

Source: Adopted from Sun, S. et al. *Nanoscale* 12, 22289–22297, 2020. No permission is required. Copyright 2020 Royal Society of Chemistry.

in the practical applications. Moving along the line, we foresee more research efforts should be devoted into this direction.

Last but not least, the 3D fabrication technique offers a vital tool to satisfy the everlasting demand on improving the performances of dielectric structures and creating more complex functionalities. Most of current dielectric meta-devices are based on planar geometries, which does not utilize the potential in the vertical space direction. Developing novel 3D fabrication techniques allow more compact dielectric devices while maintaining high spatial resolutions, which are along the mainstream of integrated photonic systems [64]. Industrialization of these techniques requires high compatibility with the CMOS production line, which is another important research direction in the future.

ACKNOWLEDGEMENTS

This work is supported by the National Natural Science Foundation of China (NSFC) (62005256) and China Academy of Engineering Physics Innovation and Development Fund (CX20200011).

BIBLIOGRAPHY

[1] Kuznetsov, A. I., Miroshnichenko, A. E., Brongersma, M. L., Kivshar, Y. S., and Luk'yanchuk, B. 2016. Optically resonant dielectric nanostructures. *Science* 354:aag2472.

[2] Kivshar, Y., and Mieoshnichenko, A. 2017. Meta-optics with Mie resonances. *Opt. Photon. News* 28:24–31.

[3] Materials' refractive index database. https://refractiveindex.info/ (accessed April 30, 2021).

[4] Thutupalli, G. K. M., and Tomlin, S. G. 2001. The optical properties of amorphous and crystalline silicon. *J. Phys. C-Solid State Phys.* 10:467.

[5] Goldschmidt, D. 1982. Optical coefficients of single-crystal and amorphous germanium at elevated temperatures and at 6328 Å. *J. Opt. Soc. Am.* 72:1692–1697.

[6] Grahn, P., Shevchenko, A., and Kaivola, M. 2012. Electromagnetic multipole theory for optical nanomaterials. *New J. Phys.* 14:093033.

[7] Monticone, F., and Alu, A. 2014. The quest for optical magnetism: from split-ring resonators to plasmonic nanoparticles and nanoclusters. *J. Mater. Chem. C* 2:9059–9072.

[8] Grosjean, T., Mivelle, M., Baida, F. I., Burr, G. W., and Fischer, U. C. 2011. Diabolo nanoantenna for enhancing and confining the magnetic optical field. *Nano Lett.* 11:1009–1013.

[9] Kuznetsov, A. I., Miroshnichenko, A. E., Fu, Y. H., Zhang, J., and Luk'yanchuk, B. 2012. Magnetic light. *Sci. Rep.* 2:492.

[10] Luk'yanchuk, B., Voshchinnikov, N. V., Paniagua-Dominguez, R., and Kuznetsov, A. I. 2015. Optimum forward light scattering by spherical and spheroidal dielectric nanoparticles with high refractive index. *ACS Photon.* 2:993–999.

[11] Baranov, D. G., Zuev, D. A., Lepeshov, S. I., et al. 2017. All-dielectric nanophotonics: the quest for better materials and fabrication techniques. *Optica* 4:814–825.

[12] Nussenzveig, H. M. 1972. *Causality and dispersion relations.* New York: Academic Press.

[13] Decker, M., and Staude, I. 2016. Resonant dielectric nanostructures: a low loss platform for functional nanophotonics. *J. Opt.* 18:103001.

[14] Dhand, C., Dwivedi, N., Loh, X. J., et al. 2015. Methods and strategies for the synthesis of diverse nanoparticles and their applications: a comprehensive review. *RSC Adv.* 5:105003–105037.

[15] Kolahalam, L. A., Viswanath, I. V. K., Diwakar, B. S., Govindh, B., Reddy, V., and Murthy, Y. L. N. 2019. Review on nanomaterials: synthesis and applications. *Mater. Today* 18:2182–2190.

[16] Thompson, C. V. 2012. Solid-state dewetting on thin films. *Annu. Rev. Mater. Res.* 42:339–434.

[17] Zywietz, U., Schmidt, M., Evlyukhin, A., Reinhardt, C., Aizpurua, J., and Chichkov, B. 2015. Electromagnetic resonances of silicon nanoparticle dimers in the visible. *ACS Photon.* 2:913–920.

[18] Schaaf, P. 2010. *Laser Processing of Materials.* Berlin and Heidelberg: Springer-Verlag.

[19] Ngo, T. D., Kashani, A., Imbalzano, G., Nguyen, K. T. Q., and Hui, D. 2018. Additive manufacturing (3D printing): a review of materials, methods, applications and challenges. *Compos. Part B-Eng.* 143:172–196.

[20] Vyatskikh, A., Delalande, S., Kudo, A., Zhang, X., Portela, C. M., and Greer, J. R. 2018. Additive manufacturing of 3D nano-architected metals. *Nat. Commun.* 9:593.

[21] Calandrini, E., Cerea, A., Angelis, F. D., Zaccaria, R. P., and Toma, A. 2019. Magnetic hot-spot generation at optical frequencies: from plasmonic metamolecules to all-dielectric nanoclusters. *Nanophotonics.* 8:45–62.

[22] Sun, S., Wu, L., Bai, P., and Png, C. E. 2016. Fluorescence enhancement in visible light: dielectric or noble metal? *Phys. Chem. Chem. Phys.* 18:19324–19335.

[23] Dmitriev, P. A., Baranov, D. G., Milichko, V. A., et al. 2016. Resonant Raman scattering from silicon nanoparticles enhanced magnetic response. *Nanoscale* 8:9721–9726.

[24] Baranov, D. G., Verre, R., Karpinski, P., and Kall, M. 2018. Anapole-enhanced intrinsic Raman scattering from silicon nanodisks. *ACS Photon.* 5:2730–2736.

[25] Raza, S., and Kristensen, A. 2021. Raman scattering in high-refractive-index nanostructures. *Nanophotonics* 10:1197–1209.

[26] Caldarola, M., Albella, P., Cortes, E., et al. 2015. Non-plasmonic nanoantennas for surface enhanced spectroscopies with ultralow heat conversion. *Nat. Commun.* 6:7915.

[27] Zograf, G. P., Dyabov, D., Rutckaia, V., et al. 2020. Stimulated Raman scattering from Mie-resonant subwavelength nanoparticles. *Nano Lett.* 20:5786–5791.

[28] Alessandri, I., and Lombardi, J. R. 2016. Enhanced Raman scattering with dielectrics. *Chem. Rev.* 116:14921–14981.

[29] Ma, L., Yu, P., Wang, W. H., et al. 2021. Nanoantenna-enhanced light-emitting diodes: fundamental and recent progress. *Laser Photon. Rev.* 15:2000367.

[30] Dodson, C. M., and Zia, R. 2012. Magnetic dipole and electric quadrupole transitions in the trivalent lanthanide series: calculated emission rates and oscillator strengths. *Phys. Rev. B* 86:125102.

[31] Baranov, D. G., Savelev, R. S., Li, S. V., Krasnok, A. E., and Alu, A. 2017. Modifying magnetic dipole spontaneous emission with nanophotonic structures. *Laser Photon. Rev.* 11:1600268.

[32] Novotny, L., and Hecht, B. 2012. *Principle of nano-optics.* Cambridge: Cambridge University Press.

[33] Regmi, R., Berthelot, J., Winkler, P. M., et al. 2016. All-dielectric silicon nanogap antennas to enhance the fluorescence of single molecules. *Nano Lett.* 16:143–151.

[34] Vaskin, A., Mashhadi, S., Steinert, M., et al. 2019. Manipulation of magnetic dipole emission from Eu^{3+} with Mie-resonant dielectric metasurfaces. *Nano Lett.* 19:1015–1022.

[35] Shi, Y., Zhao, H., Nguyen, K. T., et al. 2019. Nanophotonics array induced dynamic behavior for label-free shape-selective bacteria sieving. *ACS Nano.* 13:12070–12080.

[36] Cai, H., Xu, K. J., Liu, A. Q., et al. 2012. Nano-opto-mechanical actuator driven by gradient optical force. *Appl. Phys. Lett.* 100:013108.

[37] Chin, L. K., Shi, Y. Z., and Liu, A. Q. 2020. Optical forces in silicon nanophotonics and optomechanical systems: science and applications. *Adv. Dev. Instru.* 2020:1964015.

[38] Yano, T., Tsuchimoto, Y., Zaccaria, R. P., Toma, A., Portela, A., and Hara, M. 2017. Enhanced optical magnetism for reversed optical binding forces between silicon nanoparticles in the visible region. *Opt. Express* 25:431–439.

[39] Gong, T., Corrado, M. R., Mahbub, A. R., Shelden, C., and Munday, J. N. 2021. Recent progress in engineering the Casimir effect—applications to nanophotonics, nanomechanics, and chemistry. *Nanophotonics* 10:523–536.

[40] Kerker, M., Wang, D. S., and Giles, C. L. 1983. Electromagnetic scattering by magnetic spheres. *J. Opt. Soc. Am.* 73:765–767.

[41] Vercruysse, D., Zheng, X., Sonnefraud, Y., et al. 2014. Directional fluorescence emission by individual V-antennas explained by mode expansion. *ACS Nano* 8:8232–8241.

[42] Liu, W., and Kivshar, Y. S. 2018. Generalized Kerker effects in nanophotonics and metaoptics. *Opt. Express.* 26:13085–13105.

[43] Khaidarov, E., Hao, H., Paniagua-Dominguez, R., et al. 2017. Asymmetric nanoantennas for ultrahigh angle broadband visible light bending. *Nano Lett.* 17:6267–6272.

[44] Miroshnichenko, A. E., Evlyukhin, A. B., Yu, Y. F., et al. 2015. Nonradiating anapole modes in dielectric nanoparticles. *Nat. Commun.* 6:8069.

[45] Gurvitz, E. A., Ladutenko, K. S., Dergachev, P. A., Evlyukhin, A. B., Miroshnichenko, A. E., and Shalin, A. S. 2019. The high-order toroidal moments and anapole states in all-dielectric photonics. *Laser Photon. Rev.* 1800266.

[46] Baryshnikova, K., Smirnova, D., Lukyanchuk, B., and Kivshar, Y. 2019. Optical anapoles: concepts and applications. *Adv. Opt. Mater.* 7:1801350.

[47] Grinblat, G., Li, Y., Nielsen, M. P., et al. 2016. Enhanced third-harmonic generation in single germanium nanodisks excited at the anapole mode. *Nano Lett.* 16:4635–4640.

[48] Koshelev, K., Kruk, S., Melik-Gaykazyan, E. 2020. Subwavelength dielectric resonators for nonlinear nanophotonics. *Science* 367:288–292.

[49] Koshelev, K., Favraud, G., Bogdanov, A., Kivshar, Y., and Fratalocchi, A. 2019. Nonradiating photonics with resonant dielectric structures. *Nanophotonics* 8:725–745.

[50] Liu, C. Y., Chen, T. P., Kao, T. S., Huang, J. K., Kuo, H. C., Chen, Y. F., and Chang, C. Y. 2016. Color-conversion efficiency enhancement of quantum dots via selective area nanorods light emitting diodes. *Opt. Express* 24:19978–19987.

[51] Ma, L., Yu, P., Wang, W. H., et al. 2021. Nanoantenna-enhanced light-emitting diodes: fundamental and recent progress. *Laser Photon. Rev.* 2000367.

[52] Zhong, S. H., Zeng, Y., Huang, Z. G., and Shen, W. Z. 2015. Superior broadband antireflection from buried Mie resonator arrays for high-efficiency photovoltaics. *Sci. Rep.* 5:8915.

[53] Karabchevsky, A., Katiyi, A., Ang, A. S., and Hazan, A. 2020. On-chip nanophotonics and future challenges. *Nanophotonics* 9:3733–3753.

[54] Kamali, S. M., Arbabi, E., Arbabi, A., and Faraon, A. 2018. A review of dielectric optical metasurfaces for wavefront control. *Nanophotonic* 7:1041–1068.

[55] Zhang, X. F., Yao, B. S., Chen, Li, et al. 2021. Metasurfaces for manipulating terahertz waves. *Light Adv. Manuf.* 2:10.

[56] Krishnamoorthy, H. N. S., Adamo, G., Yin, J., Savinov, V., Zheludev, N. I., and Soci, C. 2020. Infrared dielectric metamaterials from high refractive index chalcogenides. *Nat. Commun.* 11:1692.

[57] Sun, S., Li, M., Du, Q. G., Png, C. E., and Bai, P. 2017. Metal-dielectric hybrid dimer nanoantenna: coupling between surface plasmons and dielectric resonances for fluorescence enhancement. *J. Phys. Chem. C* 121:12871–12884.

[58] Sun, S., Li, R., Li, M., Du, Q. G., Png, C. E., and Bai, P. 2018. Hybrid mushroom nanoantenna for fluorescence enhancement by matching the Stokes shift of the emitter. *J. Phys. Chem. C* 122:14771–14780.

[59] Sun, S., Zhang, T. P., Liu, Q., Ma, L., Du, Q. G., and Hui, G. D. 2019. Enhanced directional fluorescence emission of randomly oriented emitters via a metal-dielectric hybrid nanoantenna. *J. Phys. Chem. C* 123:21150–21160.

[60] Sun, S., Rasskazov, I. L., Carney, P. S., Zhang, T. P., and Moroz, A. 2020. Critical role of shell in enhanced fluorescence of metal-dielectric core-shell nanoparticles. *J. Phys. Chem. C* 124:13365–13373.

[61] Sun, S., Wang, D. C., Feng, Z., and Tan, W. 2020. Highly efficient unidirectional forward scattering induced by resonant interference in a metal-dielectric heterodimer. *Nanoscale.* 12:22289–22297.

[62] Shaltout, A. M., Shalaev, V. M., and Brongersma, M. L. 2019. Spatiotemporal light control with active metasurfaces. *Science* 364:eaat3100.

[63] Berestennikov, A. S., Voroshilov, P. M., Makarov, S. V., and Kivshar, Y. S. 2019. Active meta-optics and nanophotonics with halide perovskites. *Appl. Phys. Rev.* 6:031307.

[64] Harinarayana, V., and Shin, Y. C. 2021. Two-photon lithography for three-dimensional fabrication in micro/nanoscale regime: a comprehensive review. *Opt. Laser Technol.* 142:107180.

7 Chiral Metamaterials and Their Applications

Yidong Hou, Xuannan Wu, and Xiu Yang

CONTENTS

7.1   Introduction..179
7.2   Intriguing Effects and Fundamental Physics 180
    7.2.1   Intrinsic and Extrinsic Chiroptical Effect 180
    7.2.2   Nonlinear Chiroptical Effect...182
    7.2.3   Circular Conversion Dichroism and Asymmetric Transmission
          Effect..183
    7.2.4   Chirality Transfer..184
    7.2.5   Trochoidal Spin (TS) and Trochoidal Dichroism (TD) 186
7.3   Design and Fabrication Method and Fundamental Physics.........................188
    7.3.1   Design Principle...188
    7.3.2   Fabrication Method..190
7.4   Applications...193
    7.4.1   Nanophotonic Devices 193
    7.4.2   Chiral Biosensor 197
    7.4.3   Asymmetric Excitation and Emission 198
    7.4.4   Others..199
7.5   Conclusion and Outlook 200
Acknowledgements...201
Bibliography 201

7.1 INTRODUCTION

"I call any geometrical figure, or group of points, chiral, and say it has chirality, if its image in a plane mirror, ideally realized, cannot be brought to coincide with itself."

Lord Kelvin (1904) [1]

Chirality widely exists in nature life, and almost all of the living substances, such as the essential amino acids, polysaccharides, proteins, and DNAs, are chiral and possess even the same chirality, *i.e.*, left- or right-handedness. This happens despite the fact that both of the left- and right-hand chiral substances own the same physical and chemical properties in an achiral environment. Substances of the same chirality form a chiral biochemical environment in living lives and maintain the life activities, whereas substances of opposite chirality may destroy this chiral environment, and

DOI: 10.1201/9781003202608-7

bring sicknesses or deaths to living lives. Due to the utter importance, determining the chirality of living substances is urgently demanded in the fields of bioscience, medicine, clinical science, and food security.

Chirality is essential a Boolean data type and can't be measured quantitatively. However, the chiral geometry will greatly affect the cross coupling between the electric and magnetic dipoles in chiral medium, and break the degeneracy between the helicity eigenmodes of lights, leading to different extinction (circular dichroism, CD) or transmission (optical activity, OA) for left- and right-circularly polarized (LCP and RCP) waves. Nevertheless, the chiroptical effects of nature substances are very weak due to the size mismatches between the chiral molecules of sub-nanometers and the spiral optical wavelengths of sub-microns. The weak chiroptical effect prohibits nature substance from effectively controlling the phase and polarization state of light, and eventually limits the detection sensitivity. Recently, CMs made by artificially engineered chiral resonators have attracted a lot of research interest, because they demonstrate large chiroptical enhancements (> 1000) than that of nature chiral molecules, empowering many fascinating phenomena and applications, such as the nonlinear chiroptical effect, negative refractive index, Casimir effect, circular conversion dichroism (CCD), trochoidal dichroism (TD), and the high-sensitive chiral sensor [2,3]. Hereafter, we will review the fundamentals and application of CMs.

7.2 INTRIGUING EFFECTS AND FUNDAMENTAL PHYSICS

7.2.1 INTRINSIC AND EXTRINSIC CHIROPTICAL EFFECT

For any chiral substance, it can't be superposed on its mirror object through the translation, rotation or flipping operations. If this phenomenon only exists when the operations are confined in a two-dimensional (2D) space, the object is called 2D chiral object, and the chirality is 2D intrinsic chirality [4,5], for example, the twisted nebula and the "卐" shape (Figure 7.1(a)) [5,6]. Otherwise the object is a 3D chiral object and the chirality is the 3D intrinsic chirality, for example, the spiral-like geometry of shells [7,8] or springs [9,10] (Figure 7.1(b)). The intrinsic chirality can also be achieved by properly arranging achiral objects. As shown in Figure 7.1(d), both of the unit cell object and lattice are rectangular and achiral. However, the intrinsic 2D lattice chirality emerges if there is a small angle between the longitudinal direction of the rectangle object and the rectangle lattice; and the 3D lattice chirality will emerge if we further give a spiral-like rotation between the planar lattices or the objects in the neighbor planar layers (as shown in Figure 7.1(e)).

Other than the intrinsic chiroptical effect originated from structure itself, the illumination configuration will bring additional degree of freedom to make the whole system chiral, which is the so-called extrinsic chirality [11,12]. The extrinsic chirality includes the extrinsic chirality of single object and the extrinsic lattice chirality [13]. As shown in Figure 7.1(c) and 7.1(f), the illumination plane breaks the mirror symmetry of the achiral object when the illumination plane is not parallel to the symmetrical axis, rendering the whole system to be chiral. Notably, due to the 3D illumination geometry, all the extrinsic chirality is naturally 3D.

In the context of plasmonics (see Chapter 4), strong intrinsic and extrinsic chiroptical effects have also been demonstrated. In general, the linear chiroptical effect

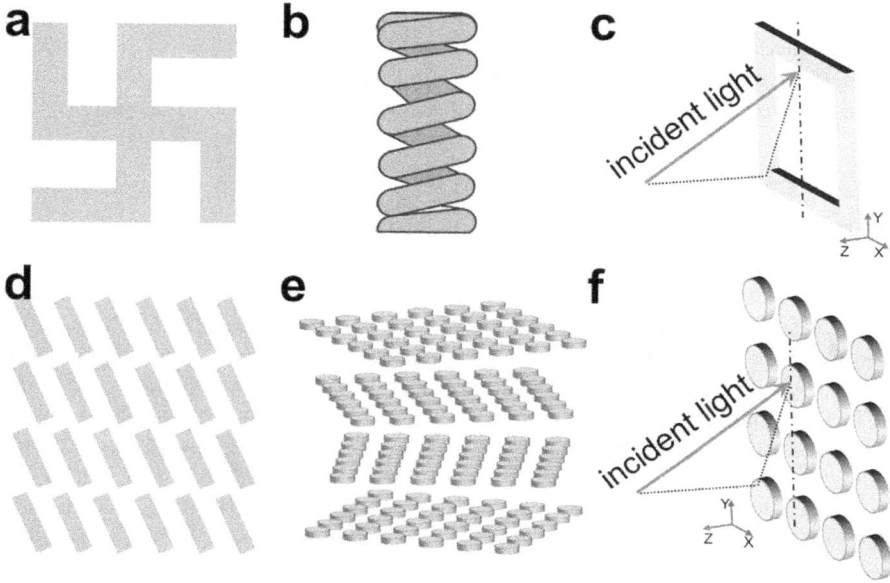

FIGURE 7.1 (a) and (b) Intrinsic 2D and 3D chiral geometries, (e) and (d) Intrinsic 2D and 3D chiral lattice, (c) and (f) Extrinsic chirality from the specific illumination geometry on the achiral geometry and the achiral lattice.

includes CD and OA in transmission, reflection or scattering. CD is defined by the ellipticity angle η of the polarized electromagnetic wave,

$$\eta = \frac{1}{2} arcsin \left(\frac{\left(|E_+|^2 - |E_-|^2 \right)}{\left(|E_+|^2 + |E_-|^2 \right)} \right) \tag{7.1}$$

and OA is defined by the azimuth angle Φ,

$$\Phi = -\frac{1}{2} \left[arg(E_+) - arg(E_-) \right] \tag{7.2}$$

Thus, the maximum values of η and Φ in theory are 45° and 90°, respectively. The CD values of nature chiral substances are usually on the order of mere 10^{-3} degree; while the CD values of delicately designed CMs can reach to the theoretical limit [14], which is about 1000 times larger than that of nature chiral substances. The typical intrinsic 3D and 2D CMs are the spiral-like CMs and the planar CMs with twisted arms, such as the helix, the twisted nano-rods, and the 卐-shape CMs. The corresponding chiroptical resonance intensity is generally dominated by the matching-degree of the twisted geometries between the incident light and the chiral structures.

While for the extrinsic chiroptical effect, the typical CMs are the fish-scale achiral planar structures and the achiral nano-hole or nano-particle arrays. The illumination geometry dominates the extrinsic chiroptical resonance intensity, and a larger illuminated angle usually results in stronger chiroptical effect.

7.2.2 Nonlinear Chiroptical Effect

Nonlinear chiroptical effect is the normal nonlinear optical effect combing with the influence of the polarization states of the excitation and emission lights. Due to the existence of the extrinsic or intrinsic chirality, the light of different spin states can result in nonlinear signals of different intensities. This is the excitation-based nonlinear chiroptical effect. The emission-based nonlinear chiroptical effect is focused on the high harmonic lights of difference spin states, where the emitted LCP and RCP lights may own different intensities or phases between each other, referring to the nonlinear CD or OA respectively. Certainly, both the excited and emission differences can be combined to form a more complex version of nonlinear chiroptical effect, which, however, has not been systematically investigated to date. In theory, all of these nonlinear chiroptical effects come from the introduced chiral medium, which changes the nonlinear susceptibility. Taking the second harmonic (SHG), for example, the nonlinear susceptibility $\chi^{(2)}$ for an in-plane isotropic achiral sample can be written as [2]:

$$\chi^{(2)} = \begin{pmatrix} 0 & 0 & 0 & 0 & \chi_{xxz} & 0 \\ 0 & 0 & 0 & \chi_{yyz} & 0 & 0 \\ \chi_{zxx} & \chi_{zyy} & \chi_{zzz} & 0 & 0 & 0 \end{pmatrix} \qquad (7.3)$$

where $\chi_{zxx} = \chi_{zyy}$ and $\chi_{xxz} = \chi_{yyz}$, and the z-axis is along the sample normal. If the sample surface exhibits both in-plane isotropic symmetry and chirality, e.g., by random distribution of chiral molecules or nanoparticles (NP) in the x-y plane, the additional tensor components appear, i.e., $\chi_{xyz} = -\chi_{yxz}$, and the nonlinear susceptibility becomes:

$$\chi^{(2)} = \begin{pmatrix} 0 & 0 & 0 & \chi_{xyz} & \chi_{xxz} & 0 \\ 0 & 0 & 0 & \chi_{yyz} & \chi_{yxz} & 0 \\ \chi_{zxx} & \chi_{zyy} & \chi_{zzz} & 0 & 0 & 0 \end{pmatrix} \qquad (7.4)$$

The additional components lead to the appearance of nonlinear chiroptical effect, such as second harmonic CD and OA in both of the excitation and emission processes. While the non-zero tensor components in Eq. (7.3) refer to the achiral ones, which remain the same for both chiral and achiral cases. The circular polarized components in SHG signals can be extracted as,

$$E(2\omega) = E(2\omega)^{LCP} + E(2\omega)^{RCP} \propto P(2\omega) = \chi^{(2)}E^2 = \chi^{(2)}\left(E_{LCP} + E_{RCP}\right)^2 \qquad (7.5)$$

where $E(2\omega)^{LCP}$ and $E(2\omega)^{RCP}$ are the LCP and RCP components of the emitted electromagnetic field, respectively; and E_{LCP} and E_{RCP} are the LCP and RCP components of the excited electromagnetic field, respectively. Notably, the complex tensor computation can be simplified by changing the coordinate system as discussed in Ref. [15].

To date, lots of chiral CMs with different geometries have been demonstrated to show strong nonlinear chiroptical effect, including the well-arranged G-shape structure [16,17], the synthesized chiral nanoparticles [18] and the two-arm chiral structure [19]. Comparing with normal nonlinear effect, the nonlinear chiroptical effect requires both the electromagnetic "hotspots" and the chiral arrangement, where the former gives a sufficiently large nonlinear signal for detecting, and the latter provides a sensitivity to the circularly polarized states of the incident and emission lights. Note that, although arbitrary chiral plasmonic structures seem to meet these two requirements, the anisotropic arrangement of G-shape structures doesn't show different SHG intensities under LCP and RCP excitations [17].

7.2.3 CIRCULAR CONVERSION DICHROISM AND ASYMMETRIC TRANSMISSION EFFECT

Considering a simple optical system with two ports, the LCP (or RCP) lights go inside the system through one port, and the out-going lights though the other port can be either the LCP (or RCP) or elliptically polarized light. In fact, the polarized state of out-going lights depends on the chosen ports and the symmetry of chiral media. For the widely applied cases of normal incidence and normal transmission, the eigenstate of isotropic chiral medium is circularly polarized lights (LCP and RCP), such as the nature chiral medium and the Faraday rotational medium with the external magnetic field that parallels to the incident lights. Their chiroptical effect intuitively comes from the non-conversion parts. However, this case completely changes in the anisotropic CMs [4,20]. The eigenstates of planar (2D) anisotropic CMs are the elliptically polarized states, where both of the LCP and RCP lights exist in the transmission upon the injection of LCP or RCP lights. Their chiroptical effect only comes from the difference between the circular conversion parts, and the resultant CD and OA can be calculated by

$$\eta_{ani-CM} = \frac{1}{2} arcsin\left(\frac{\left(|E_{-+}|^2 - |E_{+-}|^2 \right)}{|E_+|^2 + |E_-|^2} \right) \quad (7.6)$$

$$\Phi = -\frac{1}{2}\left[arg\left(E_{+-} + E_{++}\right) - arg\left(E_{--} + E_{-+}\right) \right] \quad (7.7)$$

where $E_{++} = E_{--}$, $E_+ = E_{-+} + E_{++}$ and $E_- = E_{+-} + E_{--}$. E_{ij} is the j-polarized field in transmission under the illumination of i-polarized lights, and i and j refer to the LCP (-) or RCP (+) lights. For the planar or 3D isotropic CMs, $E_{+-} = E_{-+} = 0$, and the CD and OA can be directly calculated by Eqs. (7.1) and (7.2), which are similar to those of nature chiral medium. Whereas for the normal 3D anisotropic CMs, we have $E_{++} \neq E_{--} \neq E_{+-} \neq E_{-+} \neq 0$, and the CD and OA should be calculated by Eqs. (7.1) and (7.7), respectively.

The chiroptical effect in planar CMs is asymmetric when light illuminates from opposite directions. In fact, the planar CMs own an interested geometry that shows an opposite twisted sense when reversing the observing direction. This phenomenon is similar to that of the Faraday rotational medium, where the OA effect depends on the intensity and direction of the extra-magnetic field, and the OA value changes its sign (+ or -) when reversing the illumination directions. Thus, we can get the following relationship between the forward and backward transmission components, i.e. $\vec{E}$ and $\bar{E}$:

$$\vec{E}_{ii} = \bar{E}_{jj}$$

for the planar isotropic CMs, and

$$\begin{cases} \vec{E}_{ii} = \bar{E}_{ii} \\ \vec{E}_{ij} = \bar{E}_{ji} \end{cases}$$

for the planar anisotropic CMs. In short, the light transmission in the planar CMs is asymmetric for circularly polarized lights when reversing the wave-vector direction. The asymmetric transmission difference comes from the non-conversion/conversion parts for the planar isotropic/anisotropic CMs. While for the anisotropic 3D CMs, all of the forward and backward components are different with each other, and an asymmetric transmission phenomenon can be observed in both the circularly and linearly polarized lights.

It should be noted that the asymmetric transmission phenomenon in CMs is fundamentally different from that of the Faraday rotational medium. The asymmetric transmission effect in the Faraday rotational medium is nonreciprocal, and works for all of the possible ports; whereas the asymmetric transmission effect in plasmonic CMs is reciprocal, and only works for some specific polarized ports, which means that the forward and backward transmittances are equal under un-polarized lights, i.e., $\sum_{ij} \vec{E}_{ij} \equiv \sum_{ij} \bar{E}_{ij}$. The CCD and the asymmetric transmission effect widely exists in the natural chiral medium and Faraday medium, plasmonic CMs and their hybrid systems.

7.2.4   CHIRALITY TRANSFER

Chirality transfer plays a crucial role in the self-assembling process of life due to the inherent chiral property of living bodies. The research of chirality transfer effect may reveal the life mechanism and provide new nanotechnology in future. In general, the chirality transfer effect includes the chiroptical transfer effect and the chiral geometry transfer effect [3]. The chiroptical transfer effect is also called the induced chiroptical effect, and usually happens in a system combining both of chiral and achiral substances. The underlying physics for the induced chiroptical effect can be briefly attributed to the following three mechanisms: i) the orbital coupling of electronic wave functions in the system combining the inorganic nanocrystals and the organic chiral molecules [21–24]. In fact, new chemical bonds usually formed between the inorganic nanocrystals and the organic molecules, where the chiral molecular structure greatly modifies the inter-band transitions of the inorganic nanocrystals. The resultant chiroptical effect happens near the inter-band transition of the inorganic nanocrystals, rather than the inter-band transition of the chiral molecules; ii) the near-field coupling between the inorganic nanocrystals and the organic chiral molecules, *i.e.*, the electric charge coupling [25–28]. The near-field coupling refers to the interaction between the excitons in chiral molecules and the excitons formed in the inter- or intra-band transition processes of the inorganic nanocrystals. In principle, the excitons formed in the inorganic nanocrystals can change the electromagnetic field distributions inside the molecules, and thus their electric and magnetic dipole moments, leading to the induced chiroptical effect. In return, the dipole moments of molecules also can induce chiral current in the inorganic nanocrystals, and result in the induced chiroptical effect. This near-field coupling effect is very sensitive to the distance d between the chiral molecules and the inorganic nanocrystals, and the coupling intensity decays rapidly as $\sim 1/\left(d^{n}\right)$ [28], where n is larger than 3 due to the multipole interactions; iii) the far-field coupling effect [29–31]. Similar to the dipole-dipole interactions, the electromagnetic field radiated from the chiral molecules or metal nanostructures will in turn change the electric and magnetic dipole moments of molecules or induce chiral current in metal nanostructures, leading to strong induced chiroptical effect. The electromagnetic radiative coupling is less sensitive to d and decays as $1/d$ [32]. Comparing to the near-field coupling effect and the orbital coupling effect, the electromagnetic radiative coupling distance can reach the scale of wavelength, and provides a more convenient method for achieving induced chiroptical effect.

Chiral geometry transfer widely happens all the times in all of the living bodies. The newly generated biomacromolecules are all chiral due to the inherent chiral feature of living body. Thus, the chiral geometry transfer effect is a universal phenomenon. However, repeating this enantioselective synthesis in lab didn't succeed until the beginning of the 20th century. One example was the monodecarboxylation of the monobrucine salt of methylethyl malonic acid, and the resulted methylethylacetic acid showed a very small optical rotation. In the following 100 years, the enantioselective synthesis has been rapidly developed in the medicine and food areas, where chiral geometry transfer was demonstrated between molecules and molecules, or molecules and inorganic nanocrystals [3]. In 2018, Prof. Ki Tae Nam and his collaborators demonstrated the chiral geometry transfer between the chiral cysteines and the metal nanoparticles for the first time, and the key requirements for this asymmetric synthesis are the formation of high-Miller-Index surfaces ({hkl}, h ≠ k ≠ l ≠ 0), which are intrinsically chiral. This is the first time that human achieves chirality transfer from chiral molecules to plasmonic metamaterials, and inspires lots of novel plasmonic CM synthesis [33]. In 2020, Prof. Zhifeng Huang and his collaborators demonstrated that the chiral geometry also can transfer from plasmonic CMs to biomolecules, *i.e.*, from the physical deposited metal helixes to the cyclodimers, during their photo-induced cyclodimerization [34]. It should be noted that the chiral geometry transfer can happen between the submicron-scale plasmonic metamaterials and the angstrom-scale biomolecules, where the size difference is on the scale of about 1000 times. These chiral geometry transfer effects provide an effective method for obtain CMs and chiral molecules, and also give deep insight into the chiral transfer mechanism.

7.2.5 TROCHOIDAL SPIN (TS) AND TROCHOIDAL DICHROISM (TD)

Photons possess two kinds of momentums, *i.e.*, spin momentum and orbital momentum. It is generally accepted that the spin momentum is directly related to the circular polarization and the chiroptical effects discussed earlier, and it is parallel to the main propagation direction. However, recent researches demonstrate that photons can also possess TS in the non-paraxial case, and the spin direction is orthogonal to the main propagation direction. A typical wave that carrying TS is the evanescent wave excited at the interface between two optical media [35]. As shown in Figure 7.2(a), the polarized light illuminates on the *yz*-interface through the *xz*-plane

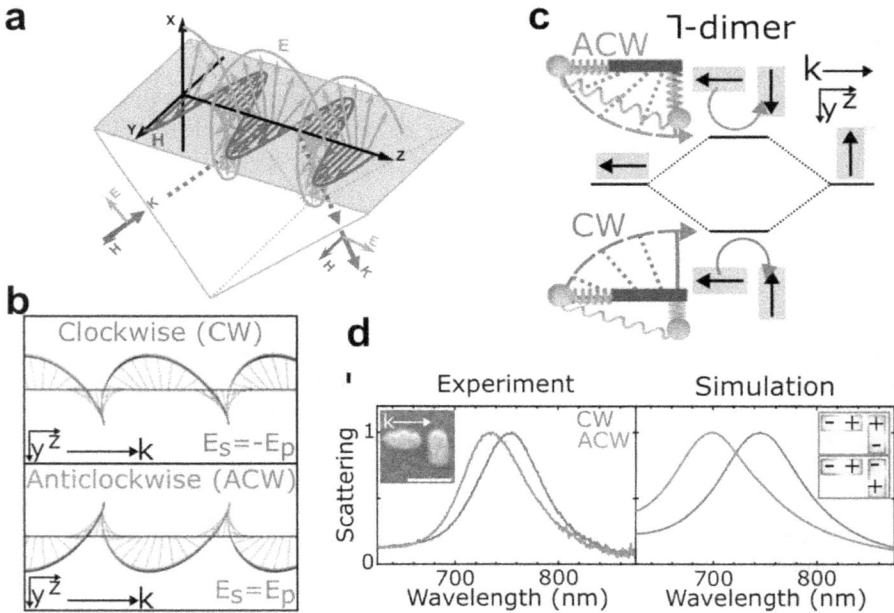

FIGURE 7.2 (a) Evanescent wave generated by the total internal reflection of p-polarized light, and the inset image shows the TS AM of the electric-field component. (b) The clockwise (CW) and anticlockwise (ACW) TS wave in the YZ plane generated by the plane wave with $E_s = -E_p$ and $E_s = E_p$, respectively, under the condition of total internal reflection. (c) The CW (ACW) wave interacts with the plasmonic dimer, giving a low (high) energy mode. (d) The measured (left) and simulated (right) TD spectra. The inset images show the dimer SEM image and the simulated charge distribution, and the scale bar is 100 nm.

with an incident angle larger than the total internal reflection angle, the formed evanescent wave can be described as:

$$k = k_z z_0 + i\kappa x_0, \ k^2 = k_z^2 - k^2.$$

where k is the wave-vector, and z_0 and x_0 are the unit vectors of z- and x- axis, respectively.

The electric and magnetic fields of the evanescent wave can be written as [36]:

$$E = \frac{A_0}{\sqrt{1+|m|^2}} \left(x_0 + m\frac{k}{k_z} y_0 - i\frac{\kappa}{k_z} z_0 \right) exp\left(ik_z z - \kappa x \right) k_z z_0 + ik x_0,$$

$$H = \frac{k}{k} \times E, \tag{7.8}$$

where A_0 is a constant field amplitude, and m is the complex polarization parameter. The real component E_x and the "imaginary" longitudinal component $E_z \propto -i\kappa / k_z$ produce a rotation electric field in the xz-plane. Since the wave propagates along the z direction, its electric field follows a cycloidal trajectory, which is similar to a point on a moving and spinning wheel. The elliptical polarization in the xz-plane implies a spin angular momentum (AM) along y-axis, i.e., orthogonal to the wave-vector of evanescent wave. These are inherently different with the usual longitudinal spins with the helical trajectories for circularly polarized lights. The spin AM S of the evanescent wave can be calculated by:

$$S = \frac{W}{\omega} \frac{Re(k) \times Im(k)}{Re(k)^2} \tag{7.9}$$

Thus, the spin AM of TS is solely determined by the real and imagine parts of the wave-vector, rather than the polarization state and helical trajectory for the longitudinal spin. In addition, the transverse spin owns asymmetric electric and magnetic properties. As shown in the inset image of Figure 7.2(a), the transverse spin excited by a p-polarized light shows a rotated electric field but a linear magnetic field along y-axis, which only emerge in some special cases, for example, the evanescent wave excited by circularly polarized lights. While the longitudinal spin owns the same rotation of electric and magnetic fields.

To date, TS has been demonstrated in both the localized and propagated waves, such as the earlier-mentioned evanescent wave [35–38], the two-wave interference [39], and the focused Gaussian beam [40]. The TS-related physical effect and applications also have been developed and discussed. Similar to the usual CD in the longitudinal spin, the degeneration of the LH- and RH-TS states also can be broken by specific molecular or artificial structures, leading to an absorption or scattering difference between the LH- and RH-TS lights. This difference is defined as TD [41]. The traditional Born-Kuhn model for CD includes two charged masses attached to three springs, where the wave-propagation direction is perpendicular to the direction of two charged masses; while in the modified Born-Kuhn model for TD, the TS wave and two charged masses are located in the same plane, and one mass moves vertically to the wave propagating direction and the other moves parallelly to the wave propagating direction (Figure 7.2(b)). Figure 7.2(c) shows that the clockwise and anticlockwise TS waves in the yz-plane can be excited by the polarized light with $E_s = -E_p$ and $E_s = E_p$, respectively. The clockwise and anticlockwise TS waves excite different electromagnetic modes in the L- or J-shaped planar chiral structures, and result in strong scattering difference in the far-field. The researcher has experimentally demonstrated that the formed scattering difference doesn't come from different linear polarized lights, and should be explained by the TD effect (Figure 7.2(d)). However, it should be noted that the TD effect relies on both of the TS wave-vector and the structure geometry. The TD signal can be obtained from achiral structures through properly aligning the propagating direction of the TS wave. Thus, the TD effect is a kind of extrinsic chiroptical effect.

7.3 DESIGN AND FABRICATION METHOD AND FUNDAMENTAL PHYSICS

7.3.1 DESIGN PRINCIPLE

Strong chiroptical resonance, low loss and simple geometry are the three basic indicators for high-performance CMs. However, a strong chiroptical resonance will in general bring a high loss in CMs, especially in the ultraviolet-visible waveband. Therefore, a balance between the chiroptical resonance and loss is required to match the specific applications. The fabrication of CMs working in the ultraviolet-visible waveband is still a challenge due to the complex chiral geometry and the sub-wavelength size. To obtain high-performance CMs, lots of theories or methods have been proposed and experimentally demonstrated, for examples, the Metric-Torsion theory based a generalized framework of transformation optics [42], the precise shaping of bound states in the continuum (BIC) [43], and the phase-transition evoking in the 2D CMs [44]. Strong chiroptical effects approaching the theoretical limit has been achieved in experiment. However, the proposed methods are usually elusive and often restricted to specific geometries, and the designed CMs cannot simultaneously fulfill the three requirements.

Hereafter, we introduce a general designed principle based symmetric consideration. In case for the 2D CMs, a strongly twisted geometry is usually expected, such as the 'L'-shaped, six-armed, 冊-shaped, and cataphracted structures. The in-plane rotational symmetry should be considered carefully, where a threefold or higher rotation symmetry will lead to the optical eigenstates of circular polarizations, while a lower rotation symmetry will lead to the optical eigenstates of elliptical polarizations. The 2D chiral geometry also can be obtained by destroying the symmetry of achiral structure with specific optical resonance, such as the phase-transition method, and the resultant chiroptical effect can inherit the advantages of such optical resonance such as high-quality factor and low loss. In case for the 3D CMs, the CMs should match one of the helical electromagnetic fields to give strong chiroptical effect. For simplicity, we consider a double-layer system as shown in Figure 7.3, where the twisting angle between layers is fixed at 90° [7]. If the constituting elements possess

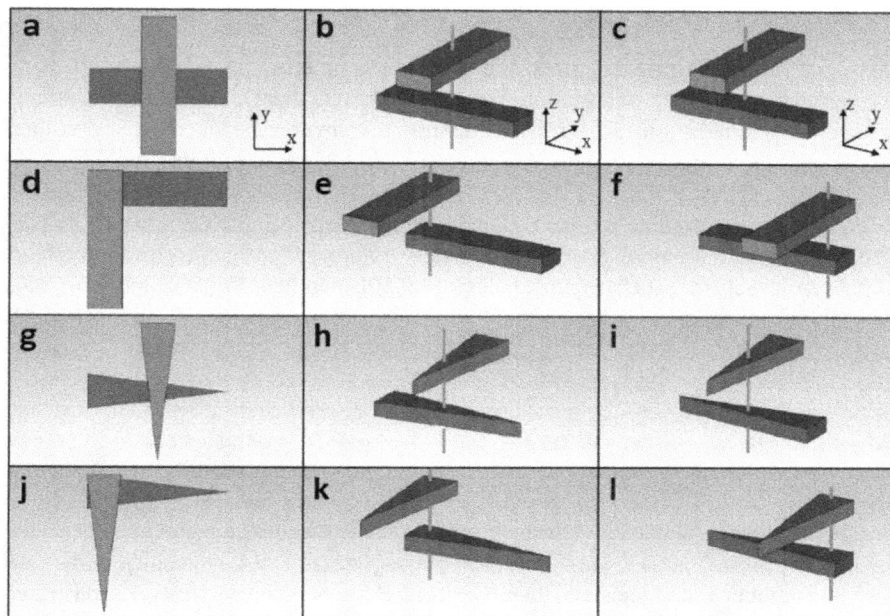

FIGURE 7.3 Chiral structure design based simple symmetry consideration of a double-layer system. (a–f) The constituting elements possess C2 symmetry, and displacement is required to form chirality. (g–i) The constituting elements lack C2 symmetry, and chirality can exist without displacement.

Source: Reprinted with permission from Hou, Y. et al., *Adv. Funct. Mater.* 26, 7807–7816, 2016. Copyright 2016 John Wiley and Sons.

two-fold rotational symmetry, chirality exists only when both layers are aligned off the center; if the constituting elements don't possess any rotational symmetry, the system is always chiral, and large chirality is expected when the layers are aligned at the end. It should be noted that the same principle demonstrated for the double layer system holds valid for other multilayer systems.

Both metal and dielectric materials are commonly adopted to create chiral structures. The discrete and continuous metal structures can support the localized surface plasmonic resonance (LSPR) [45] and the propagating plasmonic polariton (SPP) [46], respectively, which are highly sensitive to the structure size and the constituent material. More information about the plasmonic materials can be found in Chapter 4. Alternatively, the dielectric materials can support high-quality surface phonon polaritons and extremely low loss in the formed structures, and have attracted a great amount of interest in the field of chiral metamaterial and metasurface. More details on dielectric nanophotonics can be found in Chapter 6. Other than the fundamental materials, various exotic optical resonances, such lattice resonance [47] and Fano resonance [48], can be implemented to improve the chiroptical resonance, which may find potential applications in the design of high-performance CMs.

7.3.2 Fabrication Method

Compared with achiral structures, the complex geometries of CMs usually bring greater challenge in the fabrication process. However, there is no much difference between the fabrication of chiral and achiral metamaterials via the conventional lithography techniques (*e.g.*, photon-, electron- and ion-beam lithography). Thus, we will focus on the two synthesis strategies in this section: the wet chemical assembled techniques, and template-assisted glancing angle∙deposition techniques (GLAD). The wet chemical assembled technique usually employs the organic chiral templates to pattern structures, such as chiral fibers [49,50], peptides [51,52], liquid crystals [53], chiral mesoporous silicon [54,55] and organogelators [56,57]. In principle, metal nanoparticles can attach to chiral templates through noncovalent or electrostatic interactions, or grow in situ on chiral templates. These nanoparticles form a chiral geometry, and their collective interaction between each other leads to strong chiroptical effect. One typical case is that Liz-Marzán's group succeeded in assembling gold nanorods in the helical arrangement by using the twisted supramolecular fibers, and a large CD signal of about 190 mdeg was observed [49].

DNA molecules is one class of the most promising assembled technique due to the unique programmable Watson-Crick base pairing. Short DNA molecules with well-designed number and sequence of the basic groups have been successfully demonstrated in the assembly of metal nanoparticles, quantum dots, or up-conversion nanoparticles in chiral configurations, include the triangular and pyramidal structures. A strong chiroptical effect emerges in the transmission, scattering or fluorescence spectra. In addition, one single long DNA strands can be folded into nearly arbitrary 2D and 3D shaped origami template with the help of multiple smaller "staple" strands. The specific ligands can be precisely knitted on the DNA template, and the nanoparticles captured by these ligands form the designed chiral arrangement [58–61]. As shown in Figures 7.4(a) and (b), a chiral dimer is created by arranging the Au nanorods on the DNA origami bundles [60]. In particular, the twisted angle between the gold nanorods can be actively controlled by adding suitable molecules to selectively active fixed strands on the DNA plate and bundles. For example, to form a left-hand chiral center, the right-hand blocking molecules and the left-hand activating molecules are added, and *vice versa*. Thus, active CMs also can be achieved by applying DNA origami template with bundles that can response to external stimuli., such as the new molecules, the light or PH values. The chirality of the helical chiral structures can be reversed by adding new blocking and releasing strands, leading to an active chiroptical effect. DNA origami technique has been regarded as a breakthrough of DNA nanotechnology, and provide a new platform for engineering structures in nanoscale.

GLAD is a kind of large scale, materials-independent, low-cost, and high-efficient methods. Combining with the templates, GLAD shows powerful fabrication capability for arbitrary planar structures and various specific 3D structures with working wavelength down to ultraviolet waveband. One typical method for planar structures is the micro-sphere lithography technique, where the self-assembled microsphere monolayer is used as template, and the material vapor beam transmitting through the hole between micro-spheres performs like a pencil that can write any desired pattern on substrates [62–66]. As shown in Figure 7.4(c), the material vapor falls from the hole on the substrate, and the hole size and the deposition angle determine the feature size of the deposited structures. Continuously changing the deposition angle and rate, a typical helical chiral structure is formed on substrates, and the measured transmittance difference between LCP and RCP can reach about 0.13, as show in Figure 7.4(d) [65]. In fact, one can apply multiple deposition, combing with different materials and deposition parameters, to form more complex planar structures, for example, the triangle nanoparticle array, the chiral oligomers, the umbrella-shaped structures and the alloy structures [67,68], which are usually beyond the capability of other fabrication techniques.

Besides, GLAD is also capable for forming 3D structures with specific geometries, such as the spiral and the shell-like structures, which are highly depended on the templates that used in deposition process. For CMs, the typical spiral structures

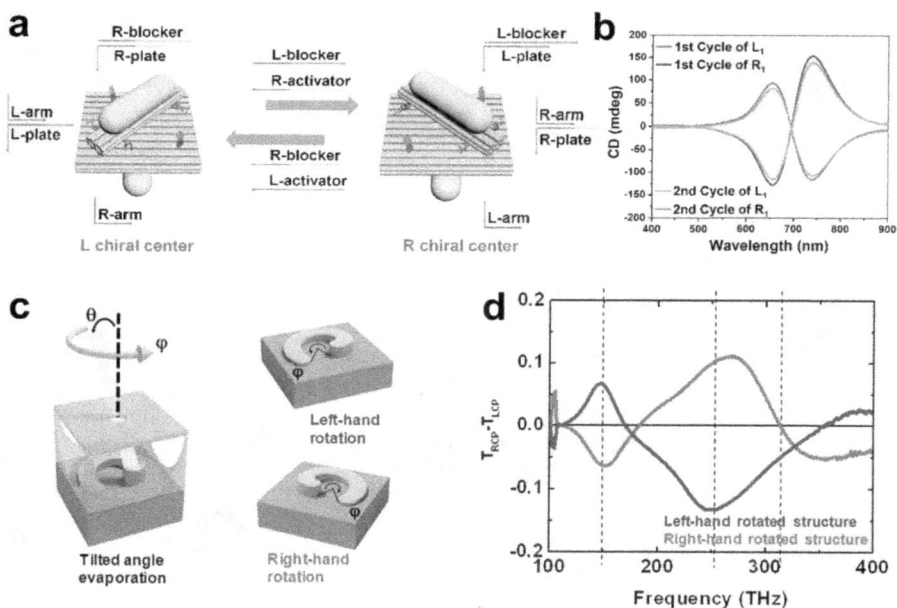

FIGURE 7.4 (a) Schematic illustration of the self-assembly and reconfiguration of chiral plasmonic structures based on DNA origami technique. Two gold nanorods are fixed on the DNA plate and bundle, respectively, and the twisted angle between the nano-rods can be reversed by adding related blocking and activating strands, leading a conversion between the formed LH and RH structures. (b) The measured CD spectra of the LH- and RH-structures designed in (a). (Reprinted with permission from Wang, M. et al. *ACS nano* 13, 13702–13708, 2019. Copyright 2017 American Chemical Society.) (c) Angle-dependent vapor deposition through a hole-mask. Precisely controlling the deposition angle and speed results in LH- and RH- CMs on substrates. (d) The measured transmittance difference between LCP and RCP lights. The red and blue curves refer to the formed RH- and LH-CMs, respectively.

Source: Reprinted with permission from Frank, B. et al. *Acs Nano* 7, 6321–6329, 2013. Copyright 2013 American Chemical Society.

with feature sizes down to 20 nm and pitches of 34 nm are successfully achieved in experiment by the low-temperature GLAD [9]. The achieved spiral size is 40 times smaller than that fabricated by the two-photon lithography, and is also smaller than the stereometamaterials fabricated by the electron-beam lithography. The measured CD signal can reach to about 1 deg around 600 nm when the spirals are dispersed in solution, and the CD intensity can be further improved by increasing the pitch number or employing other materials, such as the Ag:Cu (65:35) alloy.

Another typical chiral structure fabricated by GLAD is the chiral shell-like structures formed on the surface of microsphere monolayer [69,70]. As shown in Figure 7.5,

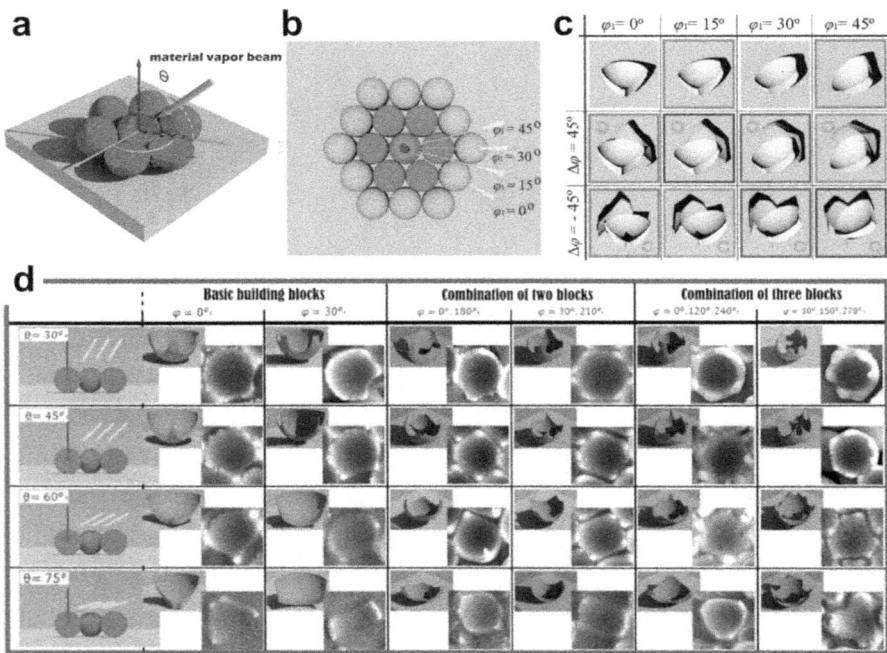

FIGURE 7.5   (a) Schematic illustration of angle-dependent material deposition on the microsphere monolayer. The deposition angle θ refers to the angle between the substrate normal and the vapor direction, and the azimuthal angle φ refers to the angle between the reference line (yellow line) and the projection of the vapor in the substrate surface. (b) Schematic illustration of micro-sphere monolayer with azimuthal angles as defined. The gray lines with yellow arrows are the projections of material vapor deposition directions in the substrate plane. (c) Simulated shells under the single-step (top panel) and two-steps (middle and bottom panels) deposition processes. The frames of same color denote the structural enantiomers. (d) The 3D geometrical model and the SEM images of the shell-like structures formed on the micro-sphere surface.

Source: (a) and (d) Reprinted with permission from Su, Y. et al. *Microelectron. Eng.* 115, 6–12, 2014. Copyright 2014 Elsevier. (b) and (c) Reprinted with permission from Hou, Y. et al. *Adv. Funct. Mater.* 26 7807–7816, 2016. Copyright 2016 John Wiley and Sons.

the geometry of the shell-like structures formed on the surface of the center sphere depends on the shadow effect from the neighboring spheres, *i.e.*, the deposition angle θ and the azimuthal angle φ. Due to the hexagonal lattice of PS sphere monolayer, the azimuthal angle φ owns a periodic influence on the shadow effect, and the period is 60 deg. When $\varphi \neq 0$ and 30 deg, the chiral shell-like structures can be formed. Otherwise, the structures behave as a pair of enantiomers, which will bring the cancellation effect for the entire sample due to the disorder property of the assembled PS sphere monolayer, where both of the left- and right-hand CMs are formed in one sample [70,71]. To overcome such cancellation effect, one can use the multiple screwed deposition method to enhance the chiroptical effect of one particular enantiomer [71], or use the two-step deposition method to fabricate pure left- or right-hand shell-like structures in one sample [7]. The latter results a broadband chiroptical effect with a CD value of up to 10 deg in experiment. This chiroptical effect can be further enhanced by using the anisotropy of CMs to form chiral cavity, where the anisotropy will lead to the circular conversion parts in transmission and the circular un-conversion parts in reflection [72]. Ideally for GLAD, a microsphere monolayer with uniform lattice is highly demanded. The recent research show that the uniform micro-sphere monolayer can be achieved by the Langmuir-Blodgett deposition method [47], and the development of chiral shell-like structures based on this method are in progress. It should be noted that the microsphere size, deposition materials, or scheme can be conveniently changed to adjust the chiroptical resonance intensity or tune the chiroptical resonance from ultraviolet waveband to infrared waveband. These advantages make the shell-like CMs promising for practical applications.

7.4 APPLICATIONS

7.4.1 Nanophotonic Devices

Broadband wave plate or chiral beam splitter. Wave-plate is widely used to control the polarization states of electromagnetic waves. The traditional wave-plates are usually made of bulky birefringent crystals, where the phase retardation difference between the lights along the fast and slow axis leads to the new polarization state. However, the traditional wave-plate suffers from low polarization rotation sensitivity unless a thick crystal is used, or a narrow bandwidth due to the dispersion characteristic of nature medium. For broadband wave-plate, both the phase velocities and the axis rations of the two eigenstates must not change in a wide spectral range. The novel CMs with strong chiroptical effects provide a viable approach. For the working wavelength of several GHz, the helices have been demonstrated as a high-performance broadband wave-plates [10]. The two eigenstates (*i.e.*, left- and right-elliptical polarization—LEP and REP) on the transverse plane of metal helix array can be effectively controlled by the Bragg scattering and the electromagnetic coupling derived from the continuous helical symmetry. The ellipticity and difference between the two eigenstates can be fixed in a wide waveband, *e.g.*, from 3.5 GHz to 10 GHz. Similar to normal quarter wave-plates, the incident linearly polarized light can be converted to linearly or circularly polarized lights with a nearly 100% efficiency.

For the near infrared waveband, the broadband CM wave-plates usually rely on the selectively consumption of LCP (or RCP) lights and only the RCP (or LCP) lights can pass through the device, where the Ohmic loss and resonance scattering makes the energy utilization efficiency below 50% [14]. This consumption becomes extremely strong in the visible waveband, and the energy utilization efficiency will become smaller than 5%. Although the dielectric metamaterials seem to have very low loss in the visible waveband, an effective design has not been demonstrated yet. In contrast to CD, OA of high value seems easier to be realized in the visible waveband. A near-complete cross-polarization conversions (90 deg optical rotation) have been demonstrated in various structures, for example, the S-shaped holes that support both LSP and SPP [73], and the L-shaped holes that support tunable coupling strength between meta-atoms [44]. However, the strong dispersions of materials in the visible waveband limit the bandwidths of these OA devices.

Chiral mirror. Mirrors are the most important optical components to redirect electromagnetic waves and are widely used in telescopes, lasers, microscopes, satellite dishes, among others. For conventional metal mirrors, the handedness of incident circularly polarized lights usually reverses after reflecting from the mirrors; whereas the chiral mirror provides a handedness-preserving reflection scheme [74]. In fact, the conventional metal mirrors are a class of electric field response mirror, where the electric field component reverses direction when reflecting from the mirrors; whereas the CM counterparts can also provide magnetic field response, where magnetic field component reverses direction when reflecting from the mirrors. Combining the concept of perfect absorber, researchers have further developed the LCP and RCP mirrors, which will reflect one handedness polarized light, but absorb the other. The general design of LCP or RCP mirror is the three-layer structure consisted of one layer of anisotropic planar CMs, one layer of metal film and one dielectric layer sandwiched between them. The CCD effect from the anisotropic planar CMs can enable a differential micro-cavity interference effect between LCP and RCP lights, and leads to a strong absorbance of either LCP or RCP light depending on the equivalent optical length between the CMs and the metal layer [75]. It is believed that the chiral mirror will find many potential applications, such as the chiral cavity formed by chiral mirrors.

Active CMs. Active optical devices are highly demanded in practical applications, such as all-optical nanophotonic circuits, sensors, displays, imaging, and data storage. However, all of the CMs mentioned earlier are passive, whose functionalities are fixed once the fabrication are complete. To obtain active CMs, lots of dynamically controlled mechanisms have been introduced, including graphene injection of free carriers in semiconductors, the phase change materials of GeSbTe (GST) and VO_2, mechanical deformations, and the chemisorption of gas molecules. In 2015, Xinghui Yin and his collaborators reported an active chiral plasmonic device enabled by GST [76], which consisted by 50 nm GST-326 layer and two gold nanorods of strong chiral geometry placed on the top and bottom sides. Heating the device to 160°C or above, the amorphous GST (refractive index $n \approx 3.5 + 0.01i$) would change to the crystalline GST ($n \approx 6.5 + 0.06i$), resulting a large redshift of the chiroptical resonance from about 4.3 μm to 5.3 μm. More information on phase-change materials can be found in Chapter 9. Meanwhile, Liu Na and her group demonstrated a new class of hybrid plasmonic CMs composed of magnesium and gold nanoparticles, where continuously change of chiroptical resonance can be obtained by loading/unloading hydrogen since the magnesium parts are sensitive to the hydrogen [77]. In addition, the active metamaterials based on shape change also has been demonstrated by employing vertically deformable MEMS spirals [78]. The vertical deformation by a pneumatic force can create a 3D spiral from a planar spiral, and the handedness and chiroptical resonance intensity depend on the direction and magnitude of the pneumatic force. A polarization rotation of up to ±28 deg was experimentally demonstrated in the terahertz range. Besides, other control parameters can be involved to actively change the chiroptical resonance of CM, such as the humidity, stress and voltage, which may find potential applications in real time polarization state control, or sensing.

Stokes parameter detection or imaging. Stokes parameter detection and imaging techniques show wide applications in bio-imaging, remote sensing, artificial object detection, etc. The conventional detection and imaging method usually relies on rotating the quarter wave-plate and the polarization analyzer to measure the intensities of the linearly and circularly polarized components, and the stokes parameters can be extracted from these measured data. The rapid development of micro/nanophotonics provide an integration scheme for the stokes parameter detection and imaging. In general, the full stokes parameter detection requires four grating-like structures of different orientations and two chiral structures of opposite handedness, where the former refers to the linearly polarized components with polarization angles of 0 deg, 45 deg, 90 deg and 135 deg, and the latter refers to the LCP and RCP components [79]. All six devices function together to form one super-pixel for the full stokes parameter imaging. In 2019, Prof. Federico Capasso and his group demonstrated a compact full-stokes polarization camera by applying four different dielectric metasurfaces with the elliptically polarized eigenstates [80]. Their research further demonstrates that the polarization imaging of S_3 is very useful for the detection of the shape change induced by stress. In addition, one also can combine chiral structures and 2D materials to form novel integrated photo-detectors.

Chiral motors. The invention of micro-nanomachine has inspired infinite imaginations in bioscience, clinic, medicine and so on, for example, the micro-biorobots can travel in the blood vessels to transport drugs or catch diseased cells. Motors as the basic power unit has attracted a great amount of research interest. To date, various nano-motors based on the supplied chemical fuels or externally physical stimuli have been developed. However, the chemical fuel driven nano-motors are usually toxic and are limited for practical applications, especially in biological tissues [81]. From this point of view, the fuel-free nano-motors are highly desired. Optical force based on the momentum transfer of photon when interacting with object provides a promising approach. In 2010, Prof. Xiang Zhang and his group developed a nanoscale plasmonic motor with a speed of about 0.5 r.p.m by combining the chiral geometry and linear polarized light [82]. It demonstrated that the light-matter interaction can lead to a rotational force on the plasmonic structures due to the specific chiral geometry, and drive a silica micro-disk of 4000 times larger in volume to rotate, where the linear momentum of photons transfers to the rotational angular momentum of the plasmonic structure and the silica micro-disk. The rotational rates can be further increased to 300 rpm by changing the plasmonic chiral structures to the carbon-coated SU-8 micro-gears and placing it at the liquid-air interface, where the thermocapillary propulsion dominates the light actuations [83]. In addition to photons, the magnetic and acoustic fields can also be employed as the external stimuli to drive the motors. In 2015, Prof. Joseph Wang and his group reported the magneto-acoustic hybrid nanomotor composed by a Ni coated Pd nano-helix and a Au concave nanorod end [84], as shown in Figure 7.6(a). The movement direction and speed under different propulsion can be precisely controlled. These hybrid nanomotors also showed diverse biomimetic collective behavior, such as the swarm motion, stable aggregation and swam vortex. More examples can be found in Ref. [85].

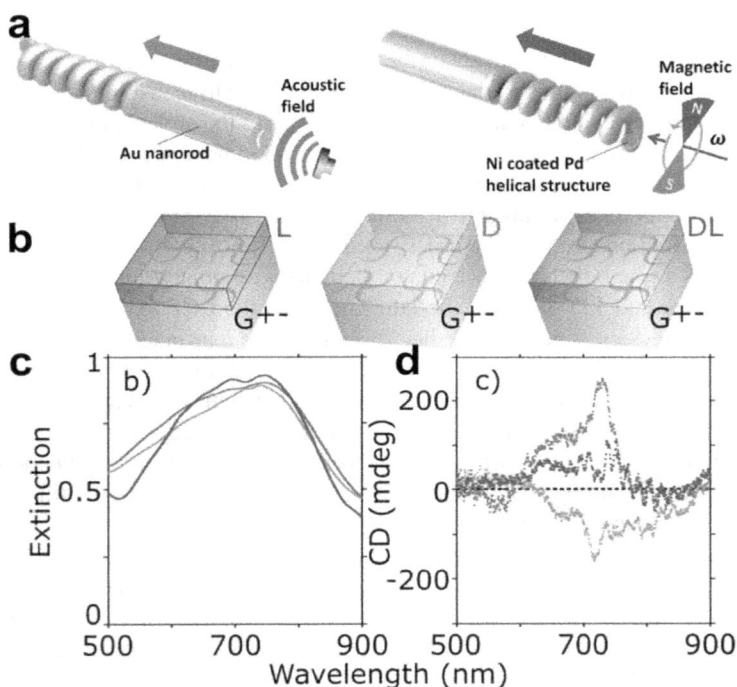

FIGURE 7.6 (a) Schematic illustration of the magneto-acoustic hybrid nanomotor. The nanomotor is comprised by a gold nanorod and a Ni-coated Pd helix, which are response to the acoustic field and magnetic field, respectively.

Source: Reprinted with permission from Li, J. et al. Nano Letters 15, 4814–4821, 2015. Copyright 2015 American Chemical Society.

(b) Schematic illustration of L-, D- and LD-molecular layer covered racemic chiral structure arrays. And the relatively measured CD spectra are shown in the following (c). A strong CD signal is observed at about 700 nm, and the sign of CD value depends on the molecular chirality.

Source: Reprinted with permission from García-Guirado, J. et al. *Nano Lett* 18, 6279–6285, 2018. Copyright 2018 American Chemical Society.

7.4.2 Chiral Biosensor

Detection of molecular chirality is of great significance in the fields of bioscience, drug, and food safety, among others. However, the extremely small size of molecule can't effectively feel the twist field of lights, which greatly limits the detection sensitivity. To date, researchers have spent lots of efforts to improve the sensor sensitivity, and two main methods are established: one is based on the chirality-transfer mechanism and the other is the chiral SPR biosensor. The chirality-transfer method is trying to make the molecular-chirality more visible to human through copying or moving the molecular chirality to an optical oscillator [3]. The chiroptical resonance of the optical-oscillator can be tuned to an easily detected waveband (e.g., visible region), and the resonance intensity is usually several orders of magnitude higher than the molecular intrinsic chiral intensity. Alternatively, the chiral SPR biosensor usually employs the chiral plasmonic structure to perceive and enhance the changes induced by the molecular chirality [86]. In fact, the chiral plasmonic resonance is very sensitive to the environment refractive index, which, together with the enhanced chiral electromagnetic field, can greatly enhance the sensitivity of molecular chirality.

Chirality Transfer. As mentioned earlier, the chirality transfer effect includes the induced chiroptical effect and the chiral geometry transfer effect. For the induced chiroptical effect, three main mechanisms are identified: the orbit coupling effect, the near- and far-field coupling effects. Among them, only the orbit coupling effect has not been reported for chiral biosensor because it can generate new chiral center to dilute the molecular chirality, whereas the other two have been widely investigated for chiral biosensors. Although the near- and far-field couplings are fundamentally different, the resultant induced chiroptical effects show lots of similar properties, for example, their induced chiroptical resonance intensity is proportional to the near-field intensity and the signal sign is directly related to the molecular chirality. Thus, a strong electromagnetic field, or hot-pot, is highly desired for highly performance sensor, which can be well engineered by applying proper plasmonic or dielectric structures. In general, the investigated system is composed of chiral medium and achiral structures, where the measured chiroptical signals are only related to the molecular chirality. However, several systems composed of chiral molecules and chiral structures have also been investigated. To eliminate the influence from the artificial chirality, the racemic chiral structures are used, where the chiroptical signal intensity should be zero in theory due to the cancellation effect [87], as shown in Figure 7.6(b-d). In addition, one also can deposit chiral medium on both of the left- and right-hand chiral structures, and then make a sum on the chiroptical signal from these two systems to remove the chiroptical signals from chiral structures [88]. Comparing with the systems with achiral structures, the asymmetric interactions between chiral molecules and chiral structures could bring stronger chiroptical signals, and an extremely high sensitivity down to zeptomole levels has been demonstrated.

Geometrical chirality transfer also can be used for chiral sensor. The molecular chirality can be transferred to chiral metal nanoparticles in the wet-chemical synthesis process, or the assembled nanoparticles in the wet-assembled process. To date, only the geometrical chirality transfer in the assembled process has been demonstrated for high performance chiral biosensors [89,90]. It has been shown that the two metallic NPs bridged by chiral biomolecules possess a scissor-like chiral geometry, and the twisting angle between the long axes of NPs is about 9 deg [89]. This small dihedral angle results in a chiral geometry and hence strong chiroptical effect in the plasmonic dimer, while the sign or the handedness of the measured chiroptical signal is dominated by the specific conformation of the bridging biomacromolecules. Because the chiroptical resonance from two coupled NPs is far stronger than that of biomolecules themselves, this method provides an effective method for sensing. In addition, the so-called "majority rules" (MR) principle in the supramolecular assemblies also works in the self-assembled organic-inorganic nanocomposite system, where a slight excess of one enantiomer can result in a strong bias toward the supramolecular helicity of that enantiomeric monomers [91]. A nonlinear amplification of chiroptical effect has been demonstrated in the multiple NPs assembled with chiral biomolecules, which can significantly improve the sensor sensitivity. When further combining with single-particle circular differential scattering spectroscopy, one can even achieve the detection of single chiral molecules [92].

Chiral SPR biosensor. In principle, the chiral SPR biosensor relies on the super-chiral near-field of chiral structures to reduce the space mismatch between the twisted circularly polarized electromagnetic field and the chiral molecules. The shift of chiroptical resonance peak is often measured to indicate the molecular chirality. To obtain a high-performance chiral biosensor, the employed chiral structures should simultaneously own the super-chiral near-field, high refractive index sensitivity, and high-quality chiroptical resonance. Among which, the super-chiral near-field is considered to be the most important factor for chiral biosensor. Researchers have proposed and investigated lot of structures that can support the super-chiral near-field, including the nano-helix, two- or four-armed chiral structures, chiral plasmonic oligomer, the offset nano-slit, and the diagonal slit on mirror. In fact, the chiral near-field depends on both electric and magnetic components of light, but the metallic materials can only response to the electric component. Thus, a specific plasmonic geometry should be included to improve the response to the magnetic field, and hence the chiral field. One of the most interested designs is the nano-cub dimer on mirror, which theoretically shows a super-chiral near-field enhancement of up to 3000-times. In 2010, M. Kadodwala and his groups demonstrated the ultrasensitive detection and characterization of supramolecular chirality by using superchiral field excited near the planar chiral nanostructures, yielding a high detection sensitivity that is 10^6 times of that without chiral nanostructures [86]. To date, researchers have successfully detected the chirality of both the micromolecules, such as the L- and D-cysteine [93], and the macromolecules, such as the proteins. Various types of chiral structures have been proposed and investigated, such as six-armed chiral structure [94,95] and 卍-shaped chiral structure [86]. We believe that the continuous efforts will be devoted on this direction to provide an effective method for detecting molecular chirality.

7.4.3 ASYMMETRIC EXCITATION AND EMISSION

Spontaneous emission widely exists in nature, and most of the currently known light emission phenomenon can be explained by the spontaneous emission model, such as the fluorescence, luminescence, Raman scattering and the up-conversion emission

The polarization state of the emitted light contains the inherent physical properties of emitters, which is critical for various applications, such as information processing [96,97] and transillumination under heavy fog [98]. CMs, together with the excited superchiral field, can effectively modify this polarization-dependent emission to enhance the differential response to the excited lights of opposite spins [99], or the differential spontaneous decay rate of quantum emitters coupled to left- and right-hand resonators [100].

There are usually two processes for the light emissions: one is the excited process to lift the system from ground state to excited state, and the other is the emission process, where the system returns to the ground state by releasing a photon. When considering the influence of optical chirality, the asymmetric excitation theory [99] and chiral Purcell effect [100] have been proposed, indicating both the exciting process and emission process from a chiral system is asymmetric for lights of opposite spins. To experimentally verify the asymmetric excitation theory, the chiral field at the nodes of the standing wave was used to excite the fluorescence from the p- and m-enantiomers of a binaphthyleneperylenebiscarboxydiimide dimer, and 11-fold asymmetric enhancement was observed [101]. For the chiral Purcell effect, one type of chiral cavity model has been proposed, but has not yet been demonstrated in experiments [100].

In addition, the chiral nanoparticles or chiral plasmonic nano-cavity can selectively interact with the valley spin states of 2D materials, such as MoS_2 and WSe_2, due to the degeneracy-lifted circularly polarized local density of states in the nano-cavity, and a high degree of circular polarization > 45% of photoluminescence is demonstrated [102]. The controllable valley-encoded signals at room temperature indicate great potentials for valleytronic devices. For the system combing CMs and chiral molecules, a selectively enhanced Raman scattering of L/D-cysteine has been demonstrated [103], providing a viable method for label-free chiral recognition, asymmetric catalysis, chiral sensor, and so on. Nevertheless, the current works have not clearly distinguished where the asymmetrically enhanced phenomenon come from or identified the separate contributions of asymmetric excitation, chiral Percell factor and filter effect. More work should be done on this direction.

7.4.4 OTHERS

Besides these applications, CMs also show lots of advantages in achieving some special physical effects, such as negative refractive index [104,105] and repulsive Casimir force [106]. The chiral self-assembled structures with the reversible self-assembly and dissociation features have been used in the real-time bio-imaging of high signal-to-noise ratio and high spatial resolution [107]. The hot-electrons generated from CMs is also asymmetric under the excitation lights of opposite spin, which can enhance the difference in photochemical processes [108,109]. CMs also can be used in the circularly polarized photodynamic therapy [110,111], where the initial non-toxic drug (photosensitizer) can induce cell death when activated by the circularly polarized lights. The energy utilization efficacy is highly related to the optical properties of CMs, since the plasmonic resonance can greatly modify the recombination rates of electrons and holes, and hence the generation of singlet oxygen species. Last but not least, the loss of plasmonic structures can convert the absorbed light energy into heat to increase the temperature [112]. The distinct response to light of opposite spins can results in a chiral thermal effect, which may find potential applications in bioscience.

7.5 CONCLUSION AND OUTLOOK

This chapter briefly summarizes the fundamental physics, fabrication techniques and applications of CMs, which is a rising star in the field of metamaterials. Despite remarkable success, there still remains some key challenges that are discussed in the following paragraphs.

The design and fabrication of high performance CMs in the ultraviolet waveband. The artificial subwavelength resonators highly rely on the structure sizes, and smaller structures usually work in short waveband. This rises the fabrication difficulty of CMs in the ultraviolet waveband, whose feature sizes are usually smaller than 100 nm. In fact, there are only several successes in the development of ultra-violent CMs to date, such as the nano-helix made by TiO_2. In addition, inevitable high intrinsic dissipation, including the Ohmic loss and the resonance scattering, is presented in the ultra-violent waveband, usually resulting in weak chiroptical resonance or small energy utilization efficiency. The Ohmic loss in plasmonic structures can be overcome by applying dielectric materials. However, there are very few reports of dielectric CMs working in the ultraviolent waveband. The dissipation from the resonance scattering can't be completely eliminated, but can be optimized to be very small. More efforts should be paid in this area to develop the ultraviolent CMs and their applications.

The design of uniform and exposed super-chiral field. CMs have been widely used to selectively enhance the interaction between chiral substances and circularly polarized lights, where the super-chiral field play a crucial role in the interaction process. The super-chiral field depends on both of electric and magnetic components of electromagnetic waves. Nevertheless, the current plasmonic materials can mainly response to the electric field components, leaving the magnetic field component largely overlooked. Although the magnetic hot-spots also have been realized, its utilization in super-chiral field is still limited. In addition, the idea super-chiral field should be uniform to give a consistent interaction between the CMs and the nearby chiral substance. However, both the negative and positive chiral fields are simultaneously formed in the adjacent spaces, and this will greatly weaken the selective interaction between the super-chiral field and the chiral substances. This issue is very difficult to overcome due to the strong perturbation of electromagnetic field near CMs. Certainly, the superchiral field also can be generated by exciting achiral structures with the linearly polarized or elliptically polarized lights. In this case, the selective-interaction between the super-chiral field and CMs will only happen in the excitation process. Lastly, the super-chiral field should be exposed outside CMs, and hence the chiral substance can be easily placed in the super-chiral field.

The spin-orbit interaction in the system containing CMs. The traditional research of CMs is focused on the interaction between the spin of lights and CMs. Until the recent years, researchers find that the plasmonic structures can greatly enhance the spin-orbit interactions, and the conversion between the orbit and spin momentum is very drastic near plasmonic structures. In fact, the momentum is nearly conservative near plasmonic structures, where the increment of spin momentum will be compensated by a reduce of the orbit momentum. In particular, this spin-orbit interaction can be easily detected in the far field, for example, the recently observed photon-spin Hall effect in CMs, where one linearly polarized beam splits into two beam lights of opposite spins. This breakthrough paves the way for manipulating the spin and orbit interaction, controlling the orbit and spin state of lights, and also finding new applications. For example, combing CMs and a single emitter can result in a single-photon source with controllable orbit and spin states. The spin-orbit interaction also may inspire new theory for controlling spontaneous emission and new design method of CMs.

Other interesting research directions in near future may include the new applications in bioscience, and new nanophotonic devices. For example, the wet-synthesized chiral nanoparticles show a distinct property in regulating the maturation of immune cells, and the efficiency from LH-chiral nanoparticle is about 1258 times larger than that of its enantiomer, opening a new path for the immunology by using nanoscale chirality. The CMs also can combine with the artificial intelligence technique to form new devices, such as the polarization-detected devices.

ACKNOWLEDGEMENTS

This work was supported by the International/Hong Kong, Macao and Taiwan Cooperation Fund of Science and Technology Innovation of Sichuan Provincial Department of Science and Technology (Grant No. 2021YFH0137), the International Visiting Program for Excellent Young Scholars of SCU (Grant No. 20181504), and the National Natural Science Foundation of China (NSFC) (Grant No. 11604227).

BIBLIOGRAPHY

[1] Kelvin, L. W. T. 1904. *Baltimore lectures on molecular dynamics and the wave theory of light.* CUP Archive.

[2] Valev, V. K., et al. 2013. Chirality and chiroptical effects in plasmonic nanostructures: fundamentals, recent progress, and outlook. *Adv. Mater.* 25(18): 2517–2534.

[3] Cao, Z., et al. 2020. Chirality transfer from sub-nanometer biochemical molecules to sub-micrometer plasmonic metastructures: physiochemical mechanisms, biosensing, and bioimaging opportunities. *Adv. Mater.* 32(41): 1907151.

[4] Fedotov, V., et al. 2006. Asymmetric propagation of electromagnetic waves through a planar chiral structure. *Phys. Rev. Lett.* 97(16): 167401.

[5] Papakostas, A., et al. 2003. Optical manifestations of planar chirality. *Phys. Rev. Lett.* 90(10): 107404.

[6] Kuwata-Gonokami, M., et al. 2005. Giant optical activity in quasi-two-dimensional planar nanostructures. *Phys. Rev. Lett.* 95(22): 227401.

[7] Hou, Y., et al. 2016. Ultrabroadband optical superchirality in a 3D stacked-patch plasmonic metamaterial designed by two-step glancing angle deposition. *Adv. Func. Mater.* 26(43): 7807–7816.

[8] He, Y., et al. 2014. Tunable three-dimensional helically stacked plasmonic layers on nanosphere monolayers. *Nano Lett.* 14(4): 1976–1981.

[9] Mark, A. G., et al. 2013. Hybrid nanocolloids with programmed three-dimensional shape and material composition. *Nat. Mater.* 12(9): 802–807.

[10] Wu, C., et al. 2011. Metallic helix array as a broadband wave plate. *Phys. Rev. Lett.* 107(17): 177401.

[11] Plum, E., et al. 2009. Extrinsic electromagnetic chirality in metamaterials. Journal of Optics A: Pure and *Appl. Opt.* 11(7): 074009.

[12] Plum, E., Fedotov, V. A., and Zheludev, N. I. 2010. Asymmetric transmission: a generic property of two-dimensional periodic patterns. *J. Opt.* 13(2): 024006.

[13] Maoz, B. M., et al. 2012. Chiroptical effects in planar achiral plasmonic oriented nano-hole arrays. *Nano Lett.* 12(5): 2357–2361.

[14] Gansel, J. K., et al. 2009. Gold helix photonic metamaterial as broadband circular polarizer. *Science.* 325(5947): 1513–1515.

[15] Kauranen, M., et al. 1994. Second-harmonic generation from chiral surfaces. *J. Chem. Phys.* 101(9): 8193–8199.

[16] Valev, V., et al. 2010. Asymmetric optical second-harmonic generation from chiral G-shaped gold nanostructures. *Phys. Rev. Lett.* 104(12): 127401.

[17] Valev, V.K., et al. 2009. Plasmonic ratchet wheels: switching circular dichroism by arranging chiral nanostructures. *Nano Lett.* 9(11): 3945–3948.

[18] Ohnoutek, L., et al. 2020. Single nanoparticle chiroptics in a liquid: optical activity in hyper-Rayleigh scattering from Au helicoids. *Nano Lett.* 20(8): 5792–5798.

[19] Rodrigues, S.P., et al. 2014. Nonlinear imaging and spectroscopy of chiral metamaterials. *Adv. Mater.* 26(35): 6157–6162.

[20] Fedotov, V., et al. 2007. Asymmetric transmission of light and enantiomerically sensitive plasmon resonance in planar chiral nanostructures. *Nano Lett.* 7(7): 1996–1999.

[21] Ben-Moshe, A., et al. 2016. Probing the interaction of quantum dots with chiral capping molecules using circular dichroism spectroscopy. *Nano Lett.* 16(12): 7467–7473.

[22] Tsay, J.M., et al. 2005. Enhancing the photoluminescence of peptide-coated nanocrystals with shell composition and UV irradiation. *J. Phys. Chem B.* 109(5): 1669–1674.

[23] Ben Moshe, A., Szwarcman, D., and Markovich, G. 2011. Size dependence of chiroptical activity in colloidal quantum dots. *ACS Nano.* 5(11): 9034–9043.

[24] Frederick, M. T., Amin, V. A., and Weiss, E. A. 2013. Optical properties of strongly coupled quantum dot–ligand systems. *J. Phys. Chem. Lett.* 4(4): 634–640.

[25] Zhou, Y., et al. 2011. Optical coupling between chiral biomolecules and semiconductor nanoparticles: size-dependent circular dichroism absorption. *Angew. Chem.* 123(48): 11658–11661.

[26] Lu, F., et al. 2013. Discrete nanocubes as plasmonic reporters of molecular chirality. *Nano Lett.* 13(7): 3145–3151.

[27] Bao, Z. Y., et al. 2017. Interband absorption enhanced optical activity in discrete Au@Ag core–shell nanocuboids: probing extended helical conformation of chemisorbed cysteine molecules. *Angew. Chem.* 129(5): 1303–1308.

[28] Maoz, B. M., et al. 2013. Amplification of chiroptical activity of chiral biomolecules by surface plasmons. *Nano Lett.* 13(3): 1203–1209.

[29] Govorov, A. O., and Fan, Z. 2012. Theory of chiral plasmonic nanostructures comprising metal nanocrystals and chiral molecular media. *Chem. Phys. Chem.* 13(10): 2551–2560.

[30] Zhang, W., et al. 2017. Surface-enhanced circular dichroism of oriented chiral molecules by plasmonic nanostructures. *J. Phys. Chem. C.* 121(1): 666–675.

[31] Nesterov, M. L., et al. 2016. The role of plasmon-generated near fields for enhanced circular dichroism spectroscopy. *ACS Photon.* 3(4): 578–583.

[32] Abdulrahman, N. A., et al. 2012. Induced chirality through electromagnetic coupling between chiral molecular layers and plasmonic nanostructures. *Nano Lett.* 12(2): 977–983.

[33] Lee, H.-E., et al. 2018. Amino-acid-and peptide-directed synthesis of chiral plasmonic gold nanoparticles. *Nature.* 556(7701): 360–365.

[34] Wei, X., et al. 2020. Enantioselective photoinduced cyclodimerization of a prochiral anthracene derivative adsorbed on helical metal nanostructures. *Nat. Chem.* 12(6): 551–559.

[35] Bliokh, K. Y., and Nori, F. 2012. Transverse spin of a surface polariton. *Phys. Rev. A.* 85(6): 1577–1581.

[36] Bliokh, K. Y., Bekshaev, A. Y., and Nori, F. 2014. Extraordinary momentum and spin in evanescent waves. *Nat. Commun.* 5(1): 1–8.

[37] Kim, K. Y., et al. 2012. Time reversal and the spin angular momentum of transverse-electric and transverse-magnetic surface modes. *Phys. Rev. A.* 86(6): 11987–11987.

[38] Kim, K.-Y., and Wang, A. X. 2015. Spin angular momentum of surface modes from the perspective of optical power flow. *Opt. Lett.* 40(12): 2929–2932.

[39] Banzer, P., et al. 2012. The photonic wheel: demonstration of a state of light with purely transverse angular momentum. *J. Europ. Opt. Soc. Rap. Public.* 8:13032.

[40] Neugebauer, M., et al. 2015. Measuring the transverse spin density of light. *Phys. Rev. Lett.* 114(6): 63901–63901.

[41] McCarthy, L. A., et al. 2020. Polarized evanescent waves reveal trochoidal dichroism. *P. Nati. Acad. Sci.* 117(28): 16143–16148.

[42] Zhang, Y., et al. 2019. Metric-torsion duality of optically chiral structures. *Phys. Rev. Lett.* 122(20): 200201.

[43] Gorkunov, M. V., Antonov, A. A., and Kivshar, Y. S. 2020. Metasurfaces with maximum chirality empowered by bound states in the continuum. *Phys. Rev. Lett.* 125(9): 093903.

[44] Xie, F., et al. 2020. Phase-transition optical activity in chiral metamaterials. *Phys. Rev. Lett.* 125: 237401.

[45] Hutter, E., and Fendler, J. H. 2010. Exploitation of localized surface plasmon resonance. *Adv. Mater.* 16(19): 1685–1706.

[46] Gramotnev, D. K., and Bozhevolnyi, S. I. 2010. Plasmonics beyond the diffraction limit. *Nat. Photon.* 4(2): 83–91.

[47] Goerlitzer, E. S. A., et al. 2020. Chiral surface lattice resonances. *Adv. Mater.* 32(22): 2001330.

[48] Luk'Yanchuk, B., et al. 2010. The Fano resonance in plasmonic nanostructures and metamaterials. *Nat. Mater.* 9(9): 707.

[49] Guerrero-Martínez, A., et al. 2011. Intense optical activity from three-dimensional chiral ordering of plasmonic nanoantennas. *Angew. Chem. Int. Edit.* 50(24): 5499–5503.

[50] Oh, H. S., Liu, S., Jee, H. S., Baev, A., Swihart, M. T., and Prasad, P. N. 2010. Chiral poly(fluorene-alt-benzothiadiazole) (PFBT) and nanocomposites with gold nanoparticles: plasmonically and structurally enhanced chirality. *J. Am. Chem. Soc.* 132(49): 17346–17348.

[51] Song, C., et al. 2013. Tailorable plasmonic circular dichroism properties of helical nanoparticle superstructures. *Nano Lett.* 13(7): 3256.

[52] Merg, A.D., et al. 2015. Adjusting the metrics of 1-D helical gold nanoparticle super-structures using multivalent peptide conjugates. *Langmuir.* 31(34): 9492–9501.

[53] Qi, H., et al. 2011. Chiral nematic assemblies of silver nanoparticles in mesoporous silica thin films. *J. Am. Chem. Soc.* 133(11): 3728–3731.

[54] Che, S., et al. 2004. Synthesis and characterization of chiral mesoporous silica. *Nature.* 429(6989): 281–284.

[55] Xie, J., Duan, Y., and Che, S. 2012. Chiral nanoparticles: chirality of metal nanoparticles in chiral mesoporous silica. *Adv. Func. Mater.* 22(18): 3750–3750.

[56] Zhu, L., et al. 2013. Chirality control for in situ preparation of gold nanoparticle superstructures directed by a coordinatable organogelator. *J. Am. Chem. Soc.* 135(24): 9174–9180.

[57] Jung, S.H., et al. 2014. Chiral arrangement of achiral Au nanoparticles by supramolecular assembly of helical nanofiber templates. *J. Am. Chem. Soc.* 136(17): 6446–6452.

[58] Wu, X., et al. 2016. Propeller-like nanorod-upconversion nanoparticle assemblies with intense chiroptical activity and luminescence enhancement in aqueous phase. *Adv. Mater.* 28(28): 5907–5915.

[59] Lan, X., et al. 2018. DNA-guided plasmonic helix with switchable chirality. *J. Am. Chem. Soc.* 140(37): 11763–11770.

[60] Wang, M., et al. 2019. Reconfigurable plasmonic diastereomers assembled by DNA origami. *ACS Nano.* 13(12): 13702–13708.

[61] Yan, W., et al. 2012. Self-assembly of chiral nanoparticle pyramids with strong R/S optical activity. *J. Am. Chem. Soc.* 134(36): 15114–15121.

[62] Haynes, C. L., and Duyne, R. V. Dichroic optical properties of extended nanostructures fabricated using angle-resolved nanosphere lithography. *Nano Lett.* 3(7): 939–943.

[63] Zhang, G., Wang, D., and Hwald, H. M. 2007. Fabrication of multiplex quasi-three-dimensional grids of one-dimensional nanostructures via stepwise colloidal lithography. *Nano Lett.* 7(11): 3410–3413.

[64] Kosiorek, A., et al. 2010. Fabrication of nanoscale rings, dots, and rods by combining shadow nanosphere lithography and annealed polystyrene nanosphere masks. *Small.* 1(4): 439–444.

[65] Frank, B., et al. 2013. Large-area 3D chiral plasmonic structures. *Acs Nano.* 7(7): 6321–6329.

[66] Xie, S., et al. 2015. Scalable fabrication of quasi-three-dimensional chiral plasmonic oligomers based on stepwise colloid sphere lithography technology. *Nanoscale Res. Lett.* 10(1): 1–9.

[67] Nemiroski, A., et al. 2014. Engineering shadows to fabricate optical metasurfaces. *Acs Nano.* 8(11): 11061–11070.

[68] Zhang, G., and Wang, D. 2009. Colloidal lithography—the art of nanochemical patterning. *Chemistry–An Asian Journal.* 4(2): 236–245.

[69] Su, Y., et al. 2014. Design and fabrication of diverse three-dimensional shell-like nanostructures. *Microelectron. Eng.* 115(Mar): 6–12.

[70] Hou, Y., et al. 2013. Design and fabrication of three-dimensional chiral nanostructures based on stepwise glancing angle deposition technology. *Langmuir.* 29(3): 867–872.

[71] Larsen, G.K., et al. 2013. Hidden chirality in superficially racemic patchy silver films. *Nano Lett.* 13(12): 6228–6232.

[72] Tang, C., et al. 2020. Large-area cavity-enhanced 3D chiral metamaterials based on the angle-dependent deposition technique. *Nanoscale.* 12(16): 9162–9170.

[73] Wu, S., et al. 2013. Enhanced rotation of the polarization of a light beam transmitted through a silver film with an array of perforated S-shaped holes. *Phys. Rev. Lett.* 110(20): 207401.

[74] Plum, E., and Zheludev, N. I. 2015. Chiral mirrors. *Appl. Phys. Lett.* 106(22): 775–388.

[75] Yang, X., et al. 2019. Active perfect absorber based on planar anisotropic chiral metamaterials. *Opt. Express.* 27(5): 6801–6814.

[76] Yin, X., et al. 2015. Active chiral plasmonics. *Nano Lett.* 15(7): 4255–4260.

[77] Duan, X., et al. 2016. Hydrogen-regulated chiral nanoplasmonics. *Nano Lett.* 16(2): 1462–1466.

[78] Kan, T., et al. 2015. Enantiomeric switching of chiral metamaterial for terahertz polarization modulation employing vertically deformable MEMS spirals. *Nat. Commun.* 6(1): 1–7.

[79] Arbabi, E., et al. 2018. Full-Stokes imaging polarimetry using dielectric metasurfaces. *ACS Photon.* 5(8): 3132–3140.

[80] Rubin, N. A., et al. 2019. Matrix Fourier optics enables a compact full-Stokes polarization camera. *Science.* 365(6448): eaax1839.

[81] Dai, B., et al. 2016. Programmable artificial phototactic microswimmer. *Nat. Nanotechnol.* 11(12): 1087–1092.

[82] Liu, M., et al. 2010. Light-driven nanoscale plasmonic motors. *Nat. Nanotechnol.* 5(8): 570–573.

[83] Maggi, C., et al. 2015. Micromotors with asymmetric shape that efficiently convert light into work by thermocapillary effects. *Nat. Commun.* 6: 7855.

[84] Li, J., et al. 2015. Magneto–acoustic hybrid nanomotor. *Nano Lett.* 15(7): 4814–4821.

[85] Joh, H., and Fan, D. E. 2021. Materials and schemes of multimodal reconfigurable micro/nanomachines and robots: review and perspective. *Adv. Mater.* 33(39): 2101965.

[86] Hendry, E., et al. 2010. Ultrasensitive detection and characterization of biomolecules using superchiral fields. *Nat. Nanotechnol.* 5(11): 783–787.

[87] García-Guirado, J., et al. 2018. Enantiomer-selective molecular sensing using racemic nanoplasmonic arrays. *Nano Lett.* 18(10): 6279–6285.

[88] Zhao, Y., et al. 2017. Chirality detection of enantiomers using twisted optical metamaterials. *Nat. Commun.* 8: 14180.

[89] Ma, W., et al. 2013. Attomolar DNA detection with chiral nanorod assemblies. *Nat. Commun.* 4: 2689.

[90] Wu, X., et al. 2013. Unexpected chirality of nanoparticle dimers and ultrasensitive chiroplasmonic bioanalysis. *J. Am. Chem. Soc.* 135(49): 18629–18636.

[91] Song, M., et al. 2021. Nonlinear amplification of chirality in self-assembled plasmonic nanostructures. *ACS Nano.* 15(3): 5715–5724.

[92] Zhang, Q., et al. 2019. Unraveling the origin of chirality from plasmonic nanoparticle-protein complexes. *Science.* 365(6460): 1475–1478.

[93] Qu, Y., et al. 2020. Chiral near-fields induced by plasmonic chiral conic nanoshell metallic nanostructure for sensitive biomolecule detection. *J. Phys. Chem. C.* 124(25): 13912–13919.

[94] Hajji, M., et al. 2021. Chiral quantum metamaterial for hypersensitive biomolecule detection. *ACS Nano.* 15(12): 19905–19916.

[95] Kakkar, T., et al. 2020. Superchiral near fields detect virus structure. *Light Sci. Appl.* 9(1): 1–10.

[96] Liu, X., et al. 2021. Heralded entanglement distribution between two absorptive quantum memories. *Nature.* 594(7861): 41–45.

[97] Farshchi, R., et al. 2011. Optical communication of spin information between light emitting diodes. *Appl. Phys. Lett.* 98(16): 162508.

[98] Brandt, J. R., et al. 2016. Circularly polarized phosphorescent electroluminescence with a high dissymmetry factor from PHOLEDs Based on a platinahelicene. *J. Am. Chem. Soc.* 138(31): 9743–9746.

[99] Cohen, T. A. E. 2010. Optical chirality and its interaction with matter. *Phys. Rev. Lett.* 104(16): 163901.

[100] Yoo, S., and Park. Q. H. 2015. Chiral light-matter interaction in optical resonators. *Phys. Rev. Lett.* 114(20): 203003.

[101] Tang, Y., and Cohen, A. E. 2011. Enhanced enantioselectivity in excitation of chiral molecules by superchiral light. *Science.* 332(6027): 333–336.

[102] Sun, J., et al. 2020. Selectively depopulating valley-polarized excitons in monolayer MoS. *Nano Lett.* 20(7): 4953–4959.

[103] Zhang, W., et al. 2021. Plasmonic chiral metamaterials with sub-10 nm nanogaps. *ACS Nano.* 15(11): 17657–17667.

[104] Pendry, J. B. A chiral route to negative refraction. *Science.* 306(5700): 1353–1355.

[105] Zhang, S., et al. 2009. Negative refractive index in chiral metamaterials. *Phys. Rev. Lett.* 102(2): 023901.

[106] Zhao, R., et al. 2009. Repulsive Casimir force in chiral metamaterials. *Phys. Rev. Lett.* 103(10): 103602.

[107] Li, S., et al. 2016. Dual-mode ultrasensitive quantification of microRNA in living cells by chiroplasmonic nanopyramids self-assembled from gold and upconversion nanoparticles. *J. Am. Chem. Soc.* 138(1): 306–312.

[108] Fang, Y., et al. 2016. Hot electron generation and cathodoluminescence nanoscopy of chiral split ring resonators. *Nano Lett.* 16(8): 5183–5190.

[109] Hervés, P., et al. 2012. Catalysis by metallic nanoparticles in aqueous solution: model reactions. *Chem. Soc. Rev.* 41(17): 5577–5587.

[110] Hartland, G.V. 2011. Optical studies of dynamics in noble metal nanostructures. *Chem. Rev.* 111(6): 3858–3887.

[111] Gao, F., et al. 2017. A singlet oxygen generating agent by chirality-dependent plasmonic shell-satellite nanoassembly. *Adv. Mater.* 29(18): 1606864.

[112] Rafiei Miandashti, A., et al. 2020. Experimental and theoretical observation of photothermal chirality in gold nanoparticle helicoids. *ACS Nano.* 14(4): 4188–4195.

8 Emerging Two-Dimensional Materials and Their Applications in Detection of Polarized Light

Xiao Luo, Qing Liu, Huidong Yin, and Fucai Liu

CONTENTS

8.1 Introduction..208
8.2 Fundamental of 2D Materials 209
    8.2.1 Intriguing Properties of 2D Materials...209
    8.2.2 Synthetic Strategies of 2D Materials 210
        8.2.2.1 Top-Down Growth Method 211
        8.2.2.2 Bottom-Up Growth Method 213
8.3 Working Mechanism of Photodetector Based on 2D Material....................214
    8.3.1 Photoconductive Effect..214
    8.3.2 Photovoltaic Effect..215
    8.3.3 Photogating Effect 215
    8.3.4 Photothermoelectric Effect 215
8.4 Detection of Polarization of Light 215
    8.4.1 Linear Polarization Detection 215
        8.4.1.1 Traditional Way of Detecting Linear Polarization of Light with Anisotropic 2D Materials 215
        8.4.1.2 Black Phosphorus..216
        8.4.1.3 ReS_2...219
        8.4.1.4 Binary IV-VI Chalcogenides 222
        8.4.1.5 Layered Perovskite and Other 2D Systems 223
    8.4.2 Circular Polarization Detection 224
        8.4.2.1 Introduction of Circular Photogalvanic Effect (CPGE) 224
        8.4.2.2 CPGE Mechanisms 226
        8.4.2.3 Non-2D System for the Detection of Circular Polarization..226
        8.4.2.4 Topological Insulator..226
        8.4.2.5 Valley TMDC 227
        8.4.2.6 Weyl Semimetal..227

DOI: 10.1201/9781003202608-8

8.5   Summary and Perspective..228
Acknowledgement 229
Bibliography 229

8.1   INTRODUCTION

With the development of science and technology, modern communication enters the era of optical communication which can be briefly classified into two types: free-space optical communication and fiber-optic communication. For the latter, it is essential to enhance the speed of transmission and enlarge the bandwidth of the signal. One effective way is the wavelength division multiplexing, which can provide more than one hundred channels. Recently, Dr. Bill Corcoran *et al.* reported the world's fastest effective speed of 39.0 Tbit/s with 160 channels in practical optical communication systems [1].

New technologies which employ the spin or angular degrees of freedom of light could further expand the communication bandwidth [2]. The former is related with the approach of polarization-division multiplexing, which can effectively enlarge the bandwidth through transmitting the left- and right-circularly polarized light in the same optical fiber. However, conventional means that operate two orthogonal polarization states limit the further improvement of spectral efficiency, thus, the measure of manipulating four different linearly polarized states in a channel of wavelength division complexing was proposed [3]. For the latter, it's expected to break the restriction induced by the nonlinear effect in optical fiber. To date, the multiplex application of orbital angular momentum (OAM) has been demonstrated in areas ranging from free-space optics [4] to optical fibers [5,6], and even on a nanoring aperture array which sorts the modes of light at a chip [7,8]. These works focus on resolving the problem that the OAM information carried by light can produce mode cross-talk when it propagates over a long distance, where some external factors could influence the state of optical fiber, such as twist, strain and rotation. The methods adopted in these works include utilizing a particular vortex fiber and fabricating a non-interference multiplexing chip.

For realizing the ambition of increasing the capacity of optical fiber communication network by thousands upon thousands of times, it is pivotal to improve the performance of photodetector in favor of these different multiplexing methods. In contrast to the pure optical detection method, the electric detection method exhibits considerable merits, especially when considering the compatibility with integrated circuit (IC) technology. For instance, CMOS-compatible photodetectors that support single-chip photonic–electronic systems have gained considerable progresses as build blocks of optoelectronic IC for on-chip optical interconnection or communication systems [9,10]. Owing to inherent characteristics of stability, anisotropy, van der Waals (vdW) force at interlayers and so on, low dimensional materials show obvious superiorities for exploring information of light [11]. Currently, direct photocurrent detection has been proposed in two-dimensional (2D) materials, such as WTe_2, with special electronic band structures [12]. The applications for polarization imaging [13], infrared imaging sensors, and energy-efficient spin current devices [14] as well as exploitation of material's properties, have been elaborately established.

In this chapter, we mainly focus on the fundamental understanding and discussion on the emerging 2D materials including their basic properties, synthetic strategies, and applications in detection of polarized light.

8.2   FUNDAMENTAL OF 2D MATERIALS

The re-discovery of graphene in 2004 triggered a surge of interest of the novel 2D material family [15]. The past decades have witnessed an unprecedented explosion of scientific achievements in 2D vdW materials due to their unique physical characteristics and huge application potentials. 2D materials can be obtained through detaching their bulk counterparts, which are stacked by multi-platelets bonded via weak interlayer vdW forces. Representative 2D materials include graphene, h-boron nitride (BN), black phosphorus (BP) and various transition metal dichalcogenides (TMDCs) such as molybdenum disulphide or tungsten diselenide. These materials have been substantially applicated in the fabrication of the novel electronic, optoelectronic, and photonic devices as they demonstrate intriguing electronic structures, resulting in exotic optical, thermal, mechanical, vibrational, spin, and plasmonic properties [16,17].

8.2.1   INTRIGUING PROPERTIES OF 2D MATERIALS

Plenty of monolayer or few-layer materials exhibit a combination of unique and intriguing properties such as ferroelectricity, magnetism, and superconductivity, which are distinctive compared with their bulk counterparts. Taking the gap structure of graphene and TMDCs as examples, monolayer graphene demonstrates zero bandgap between the valence and conduction bands at Dirac points, allowing electrical conduction regardless of the fermi energy. The carrier mobility of graphene can reach an exceptionally large value over 10000 $cm^2 \cdot v^{-1} \cdot s^{-1}$ at room temperature [18]. On the other hand, bulk TMDCs usually exhibit indirect bandgaps while direct bandgaps can be observed in some monolayer TMDCs, which is an important feature for the fabrication of ultra-thin transistors or optoelectronic devices. Besides, with atom level thicknesses, 2D materials can be easily tuned by external local fields such as gate voltage induced electrical fields, magnetic fields, ferroelectric fields and so forth, facilitating the modulation of transport properties. Due to the quantum confinement, there exist sharp peaks in the density of states near the band edges of 2D materials, enhancing the light-matter interaction with photon energy close to the bandgap. Despite the thin thickness, the light absorption efficiencies of 2D materials can be substantially large so that the efficient generation of free electron-hole pairs is expected, indicating 2D materials could be competitive candidates in diverse photodetectors [19]. The naturally passivated surface of 2D material can be another beneficial characteristic. As the intralayers of 2D materials are bonded by strong covalent bonds while the interlayers are held together by weak vdW forces, the notorious crystal lattice mismatching obstacle occurring in the heterogeneous integration for conventional semiconductor materials doesn't exist in

2D materials [20,21]. Last but not least, atomically thin 2D layered materials demonstrate large specific surface areas, extraordinary elastic modulus, large mechanical strengths, and high flexibilities, making them suitable for bendable and wearable/portable devices with novel functionalities (more details can be found in Chapter 11) [22]. These unique features of 2D materials clearly explain why they are considered as extremely promising building blocks for the next generation electronics. However, there is lack of large-scale synthetic techniques for high quality 2D materials, which brings a significant barrier for the commercialization and practical application of 2D devices [23].

By stacking different monolayers vertically or laterally, vdW heterostructures can be built. vdW heterostructures can provide a versatile platform for exploring and further modulating the extraordinary properties of 2D materials. Various strategies such as external excitation, stacking sequence and crystallographic alignment could be utilized to form vdW heterostructures. The aligned angle between different layers was seldom considered in conventional vdW heterostructures. However, it was proved to be a new dimension to tune the energy band. The lattice mismatch and rotation angle between the twisted bilayer are the determining factors for the generated moiré patterns and commensurate structures. The moiré period is given by $b \approx a / \sqrt{\delta^2 + \theta^2}$, where δ is the lattice mismatch, θ is the relative twist angle, a is the monolayer lattice constant [24]. Usually, it ranges from several nanometers to ten of nanometers as shown in Figure 8.1 [25]. Early in 2011, Bistritzer and Macdonald theoretically predicted that when θ reaches close to 1.05°, a flat moiré band near Fermi energy could be achieved though the intrinsic flat band structure at low energies doesn't exist in graphene [26]. The twisted bilayer graphene with flat bands creates a novel platform for exploring exotic correlated systems and many-body quantum phases. Later in 2018, the unconventional superconducting and the magneto-transport phenomena were experimentally observed in twisted bilayer graphene with small twist angle ~1.1° by Cao *et al.* [27,28]. Recently, Cao *et al.* also demonstrated that the electric field tunable superconductivity could be obtained in alternating-twist magic-angle tri-layer graphene [29]. In addition to the twisted bilayer graphene with a magic angle, the twisted multilayer graphene with a large twisted angle of 30° also demonstrates many exotic phenomena associated with the interlayer coupling and the moiré potential [30]. Besides, graphene/BP and graphene/BN heterostructures with moiré superlattices have been constructed using the near-lattice matching of two crystals. It is worth noting that the emerging pseudo-Landau levels arising from moiré superlattices could also give rise to many intriguing characteristics including moiré excitons, moiré phonons, band-structure reconstructions, and other topological transitions, which were widely reported in TMDCs and TMDCs-based heterostructures. In summary, these novel phenomena are highly associated with the interlayer interactions as well as the direct modulations of momentum spaces of moiré superlattices.

8.2.2 Synthetic Strategies of 2D Materials

Researches have devoted continuous efforts to synthesize high quality 2D materials which are the prerequisites for scientific researches and industrial applications. Basically, the synthesis of 2D materials can be classified into two strategies, namely

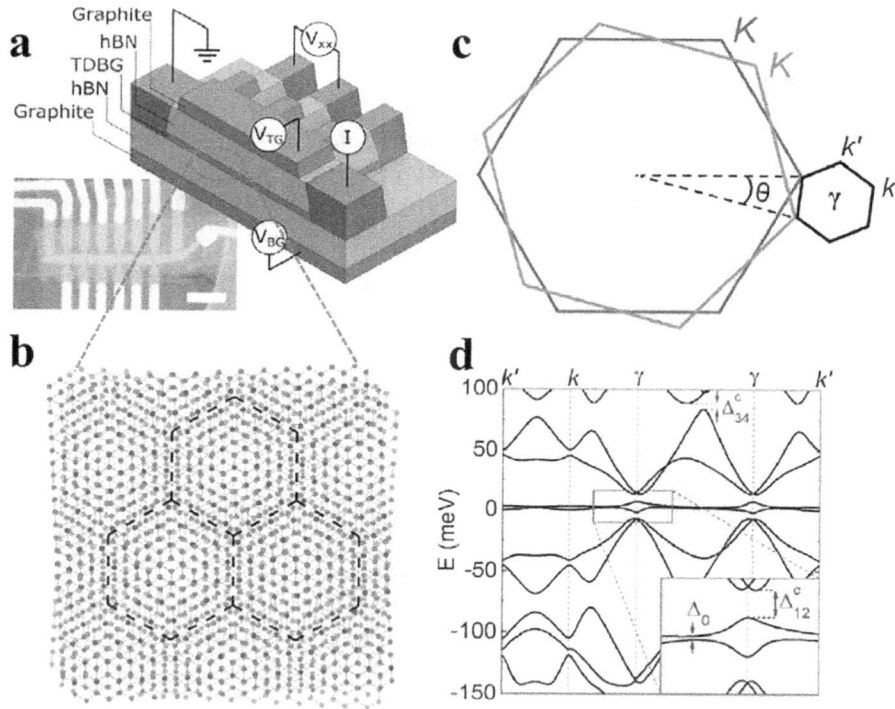

FIGURE 8.1 Electronic band structure of twisted double bilayer graphene (TDBG) as a function of twist angle and the moiré pattern as seen in TDBG.

Source: Reprinted with permission from Burg, G. W., et al. *Phys. Rev. Lett.* 123, 197702, 2019. Copyright 2019 American Physical Society.

top-down and bottom-up methods. In the top-down method, high-quality bulk crystals were first obtained by flux growth [31] or chemical vapor transport (CVT) method [32,33], and then atomically thin flakes were delaminated from the bulk crystals via various exfoliation technologies. In contrast, bottom-up method means to obtain vdW atomically thin flakes via the material growth methods that assemble atoms on substrates directly, including chemical vapor deposition (CVD), physical vapor deposition (PVD) and molecular beam epitaxy (MBE).

8.2.2.1 Top-Down Growth Method

In the top-down synthetic strategies, single- and few-layer vdW flakes are obtained by isolating the bulk crystals. The growth methods for bulk vdW single crystal can be mainly classified into two types: chemical vapor transport (CVT), and flux method. CVT reaction is a classic yet powerful technology in the growth of bulk vdW materials. A systematic research on CVT reaction was conducted by Schafer in the 1950s and 1960s. Subsequently, a large number of vdW bulk crystals have been synthesized through CVT reactions, including TMDCs, TMPX$_3$ (such as CuInP$_2$S$_6$,

FePS$_3$, CoPS$_3$, MnPS$_3$ and NiPS$_3$), halides (such as CrBr$_3$ and CrI$_3$), Fe$_3$GeTe$_2$, *etc.* As shown in Figure 8.2(a)–(b), a CVT reaction consists of three processes: sublimation, transport, and deposition. The raw materials are placed in the high temperature region and the crystal growth occurs at the other side with low temperature. One can get a comprehensive and systematic understanding of the CVT reactions by referring to the book entitled "Chemical Vapor Transport Reaction" in 2013 [34]. On the other hand, flux growth utilizes a high-temperature melt of inorganic compounds as the

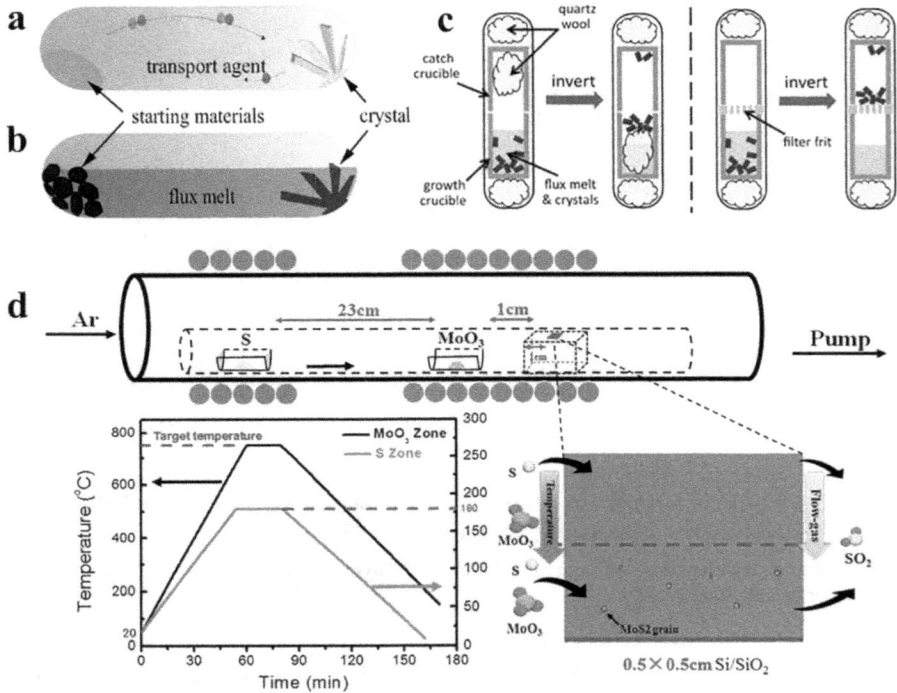

FIGURE 8.2   (a) Scheme diagrams of CVT.

Source: Reprinted with permission from Yan, J. -Q. et al. *Phys. Rev. Materials* 1, 023402, 2017. Copyright 2017 American Physical Society.

Flux growth methods with (b) horizontal and (c) vertical configuration for fabrication of 2D materials.

Source: Reprinted with permission from May, A. F. et al. *J. Appl. Phys.* 128, 051101, 2020. Copyright 2020 American Institute of Physics.

(d) Temperament curve of CVD reaction furnace and inset is a schematic view of SiO$_2$/Si substrate, illustrating the growth of MoS$_2$ under different temperatures and gas fluxes. The black and red curves indicate the temperature evolutions of central heating zone and S zone, respectively.

Source: Reprinted with permission from Zhou, D. et al. *Cryst. Growth Des.* 18, 1012–1019, 2018. Copyright 2018 American Chemical Society.

solvent for crystallization [32]. Common fluxes are the simple inorganic solids melting at conveniently low temperatures, such as Te, Sn, Bi, CsCl, KCl, NaCl, $AlCl_3$, *etc.* In order to form a suitably lower melting eutectic liquid, simultaneously using several inorganic compounds is necessary. A flux growth can be conducted both in horizontal and vertical configurations as shown in Figure 8.2(c) [35]. More detailed information about flux method can be found in the book entitled "Beginner's Guide to Flux Crystal Growth" in 2013 [36].

After obtaining the bulk materials, the next step to prepare atomic scale 2D flakes is the exfoliation, which can also be categorized into two methods: liquid phase exfoliation and mechanical exfoliation. Liquid phase exfoliation is suitable for the large-scale fabrication of mono- or few-layer flakes delaminated from bulk crystals in a specific solvent such as N-methyl-2-pyrrolidone (NMP) anddimethylformamide (DMF) [37]. Treatments such as thermal shock, ultrasonication and microwave heating are the necessary auxiliary means during the liquid exfoliation processing, where the solvents can play an important role in transferring shearing force to vdW materials. The liquid exfoliation processing can be further facilitated by ion intercalation/exchange, surface passivation, or oxidation by solvents [38]. Though suffering from a low yield, mechanical exfoliation still remains the core method to obtain high-quality mono- or few-layer flakes in scientific researches. In 2004, Geim and Novolosov conducted the mechanical exfoliation of graphite just by simply rubbing the graphite on a surface [15]. To be specific, scotch tape is utilized to pull apart the layers of bulk crystal and repeating the peeling off process multiple times can produce the mono- or few-layer flakes. The as-exfoliated layers from the crystal can be easily transferred onto various substrates, which are widely used in developing novel electronic devices. The successful mechanical exfoliation relies on a larger bonding strength between the outermost flakes and the substrate compared to the vdW forces of interlayer flakes. To uniformly improve the interactions between target vdW materials and substrates, advanced exfoliation methods such as crack-assisted exfoliation and metal-assisted mechanical exfoliation spring up. These mechanical exfoliation methods remain the sources of various 2D flakes of high qualities and cleanliness, which are suitable for fundamental characterizations and device fabrications.

8.2.2.2   Bottom-Up Growth Method

In the bottom-up growth method, 2D crystal networks are constructed using molecular precursors. As a common method, chemical vapor deposition (CVD) is introduced and developed to efficiently synthesize large-area 2D material films, though higher defect densities are expected compared with the exfoliated methods. Early in 2009, CVD on Cu foils has been used to synthesize the uniform large-area graphene film [39]. Since then, plenty of monolayer films, such as MoS_2 [40–43], WS_2 [44,45], ReS_2 [46] and corresponding vdW heterostructures [47,48] have been epitaxially grown from the reactions of corresponding vapors via this powerful tool. The growth process of conventional CVD can be divided into three stages: transportation, nucleation, and growth. At the stage of transportation, the gaseous precursors were formed from the sublimation of raw solid materials at high temperature and then transported to the nucleation positions by the carrier gas. At the stage of nucleation, the gaseous precursors diffuse to the substrate and nucleation initiates at random regions. At the stage of growth, the growth of individual domains accompanies with the continued reactions and aggregations of the gaseous precursors, and later these domains merge into a continuous thin layer, as shown in Figure 8.2(d) [49]. The synthesized product is usually polycrystalline, whose size can be up to wafer scale. With tuned growth conditions such as the symmetry of single crystal substrate, the gas flux, and the precursor ratio, the crystallinity, shape, and number of layers of the target flakes can be controlled. It should be noted that the thermal CVD is not suitable for all 2D materials due to the high sublimation temperatures of some source metal oxides (*i.e.*, WO_3). The salt-assistant CVD method is proposed to overcome this drawback as introducing suitable salts could in principle decrease the melting points of metal compounds or metal elements. Besides, volatile metal oxychlorides might be formed from the reactions of salts and metal-compounds, thereby accelerating the reaction rates with chalcogenides. By taking the advantage of halide-salt in the CVD system, Zhou *et al.* have successfully fabricated 47 compounds and heterostructures [50]. Moreover, the growth of metal phase 2D materials can also be accomplished via utilizing the salt-assisted CVD method.

8.3 WORKING MECHANISM OF PHOTODETECTOR BASED ON 2D MATERIAL

It is essential for detecting the degrees of freedom of light through the light-matter interaction. Photodetectors are the key components in direct electric detection, which convert incident photons with different energies and information into modulable electrical signals. Compared with conventional 3D bulk semiconductor materials, 2D crystals are not only significant for exploring new physical phenomena and performance under 2D limits but also have many novel applications in the optoelectronics due to their high carrier mobility, strong anisotropy/non-symmetry, suitable bandgap, good light absorption, and photoresponse properties. Thus, the 2D material-based photodetectors exhibit superior performance in broadband response, high-photoresponsivity and polarization-sensitive feature. The realization of photodetectors essentially relies on the effective and efficient generation of photocurrents. The physical mechanisms of photodetectors used in commercial applications can be classified into two kinds. One is related with photoemission, including photoconductive effect, photovoltaic effect and photogating effect [51,52]. The other is related with thermal effect which is not involving direct optical transition but due to the fact that the photo-generated carriers only reach midgap states and then relax back to lower bands, including photothermoelectric effect and photobolometric effect.

8.3.1 PHOTOCONDUCTIVE EFFECT

The photoconductive effect occurs because the photo-induced excess carriers could cause the increase of free carrier concentration, leading to a reduced resistance of semiconductor material. These excess carriers can be separated and extracted under the external applied bias to generate photocurrent.

8.3.2 PHOTOVOLTAIC EFFECT

The photovoltaic effect relies on a built-in electric field to separate the photo-generated electron-hole pairs, which are then collected at the opposite electrodes to form photocurrent. Photodetectors operating in the photovoltaic effect are called photodiodes, normally functioning at p-n or Schottky junctions. For 2D materials, the photovoltaic effect can be readily achieved via the heterostructures based on opposite-doped 2D semiconductors or by combining 2D semiconductors with metals.

8.3.3 PHOTOGATING EFFECT

The photogating arises from the photo-induced filling of localized states due to the existence of defects and impurities. If the localized states in the channel material trap one type of carriers such as electrons, subsequently, these states with negative charges can act as local gates, which will effectively modulate the conductivity of the channel via electrostatic interactions. The free holes in the channel can recirculate many times during the lifetime of the trapped electrons, leading to a higher gain and lower temporal response for photodetectors.

8.3.4 PHOTOTHERMOELECTRIC EFFECT

Photothermoelectric effect results from the Seebeck effect associated with the temperature gradient between different substances. In short, a photo-induced voltage can be produced through two different substances with different Seebeck coefficients. Generally, the photons are firstly absorbed and converted into heat by one side of the material, which results in a temperature difference between the two channel ends, and then lead to directional diffusion of carriers and lead to building a photothermoelectric voltage across the channel.

8.4 DETECTION OF POLARIZATION OF LIGHT

This section focuses on the aforementioned 2D materials and discusses their recent processes on photodetection for linear- and circular polarized light.

8.4.1 LINEAR POLARIZATION DETECTION

Linear polarization photodetectors can sense the polarization information in addition to the intensity and wavelength of incident light, which provide much richer optical data for precise target analysis and recognition than traditional light detection techniques [53]. Therefore, the detection of linear polarization is exceptionally useful in military and civilian fields, such as optical communication, remote sensing, and polarization imaging [54–56].

8.4.1.1 Traditional Way of Detecting Linear Polarization of Light with Anisotropic 2D Materials

In principle, linear polarization photodetection can be achieved by either combining a polarizer (or analogue with polarization selectivity) with a conventional photodetector, or utilizing anisotropic light absorbers with intrinsically asymmetric crystal structures or extrinsic geometric effects [57,58]. The strategy via integrating polarizer or metal nanostructures (such as plasmonic photonic cavities [59] and metal gratings [60]), would greatly increase the cost and complexity of photodetection system. On the other hand, some kinds of geometric anisotropic materials, such as one-dimensional (1D) nanowires [54,61], nanoribbons [62], and nanotubes [63], have been successfully demonstrated in polarized photodetection. However, these materials require complicated operation to pattern the devices and align photoactive channel due to their aspect ratios, thereby hindering practical applications. In contrast, the emerging 2D anisotropic materials are intrinsically sensitive to the linear polarization of incident light due to their asymmetric crystal structure [64–66], naturally rendering them promising for linear-polarization-sensitive photodetectors.

8.4.1.2   Black Phosphorus

Properties of BP. Black phosphorus (BP) is a single-element layered material with three crystalline structures: orthorhombic, simple cubic and rhombohedral [67]. Semiconducting orthorhombic BP, consisting of eight phosphorus atoms per cell unit, draws great attentions since its rediscovery as active layer with high carrier mobility and moderate bandgap for nanoelectronic and optoelectronic applications [68]. Analogous to the graphite structure, bulk BP is a layered allotrope with an interlayer spacing of 5.3 Å (Figure 8.3(a)). The resulting vdW interaction makes BP highly suitable for exfoliation. In a single-layer BP, each phosphorus atom covalently bonds to three neighboring atoms with two different bond lengths (2.244 Å and 2.224 Å) and angles (96.34° and 102.09°), leading to a puckered, honeycomb structure with two special directions: armchair and zigzag geometries along the x- and y-axis, respectively (see Figure 8.3(b)) [69,70]. These P-P covalent bonds take up all valence electrons of phosphorus, so BP is a semiconductor with a thickness-dependent bandgap. Unlike graphene, the puckered BP, formed by sp^3 hybridization, breaks the three-fold rotational symmetry of a flat honeycomb lattice, leading to a remarkable in-plane anisotropy of electrical, optical and phonon properties [64–66,71–74]. This is a new degree of freedom to manipulate BP-based devices and applications.

BP has an anisotropic energy band due to the low-symmetry in-plane structure [68,75]. Cui and coworkers utilized *ab initio* band calculations and angle-resolved photoemission spectroscopy (ARPES) to uncover the band dispersion anisotropy of BP [76]. According to the band structure calculations (Figure 8.3 (c)), one can find that the bandgap is located at the Γ point, and the energy band along the Γ-X direction (armchair) disperses more strongly than that of the Γ-Y direction (Zigzag), which indicates a smaller carrier effective mass along the armchair direction [69]. The electron (hole) effective masses along the armchair and zigzag direction were experimentally measured to be about 0.08 (0.08) and 1.00 (0.65) m_0 respectively, where m_0 denotes the free electron mass [72]. In addition, the constant-energy contours clearly showed the evolution of the anisotropic band dispersion at different energies with respect to the Fermi level. Thus, the electronic- and optical-anisotropy of BP are determined to be direct results of the anisotropic energy band structure, which supplies a viable approach for linear-polarization photodetection [64–66].

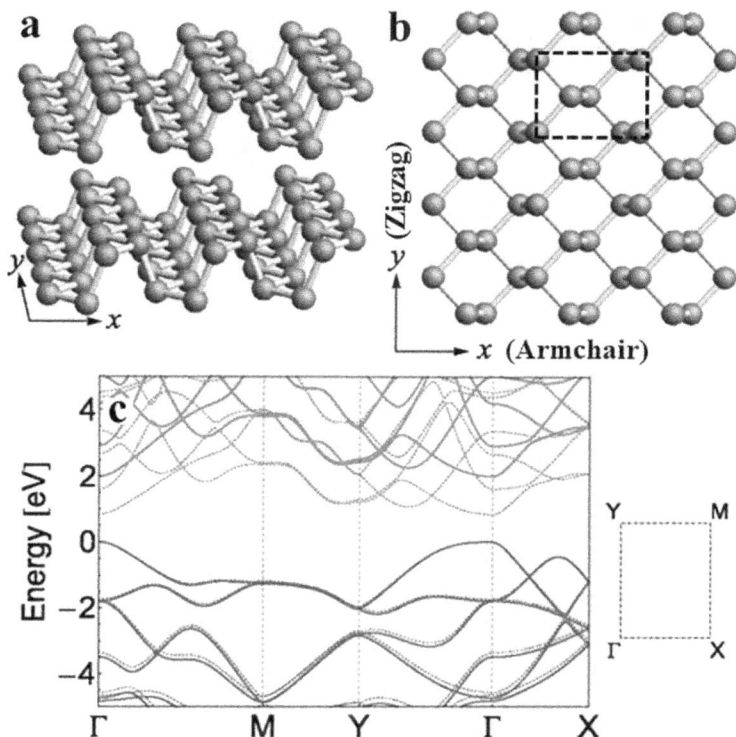

FIGURE 8.3 (a) The ball-stick model of few-layer BP. (b) Top view of the single-layer BP. The unit cell is marked by dashed line. (c) The density-functional-theory calculated (dashed lines) and GW-calculated (solid lines) band structures of monolayer BP. The top of the valence band is set to be zero.

Source: Reprinted with permission from Tran, V., et al. *Phys. Rev. B* 89, 235319, 2014. Copyright 2014 American Physical Society.

BP-based linear-polarization photodetection. BP is a potential candidate of linear-polarization photodetection due to its optoelectronic anisotropy with high carrier mobility (typically 100–1000 cm^{2}V^{-1}s^{-1}), tunable direct bandgap from 0.33 eV (bulk) to 2.0 eV (monolayer) spanning a wide range of the electromagnetic spectrum (from mid-IR to visible), and remarkable linear dichroism [72]. As shown in Figure 8.4(a), the calculated absorption of zigzag-polarized light (black line) starts from around 3.0 eV, while for armchair-polarized light (red line), the absorption starts at a much lower photon energy for monolayer, bilayer, trilayer, 10-layer, and bulk BP [77]. Moreover, the experimental results also identify the anisotropic absorbance for BP flakes (Figure 8.4(b)-(c)), promoting its use to build polarized photodetectors. Early in 2014, Xia *et al.* and Xu *et al.* firstly reported the polarized photocurrent responses of BP films, which triggered the study on linear-polarization photodetection [78,79]. In order to exclude the effect of two-fold polarization-dependent photocurrent from the

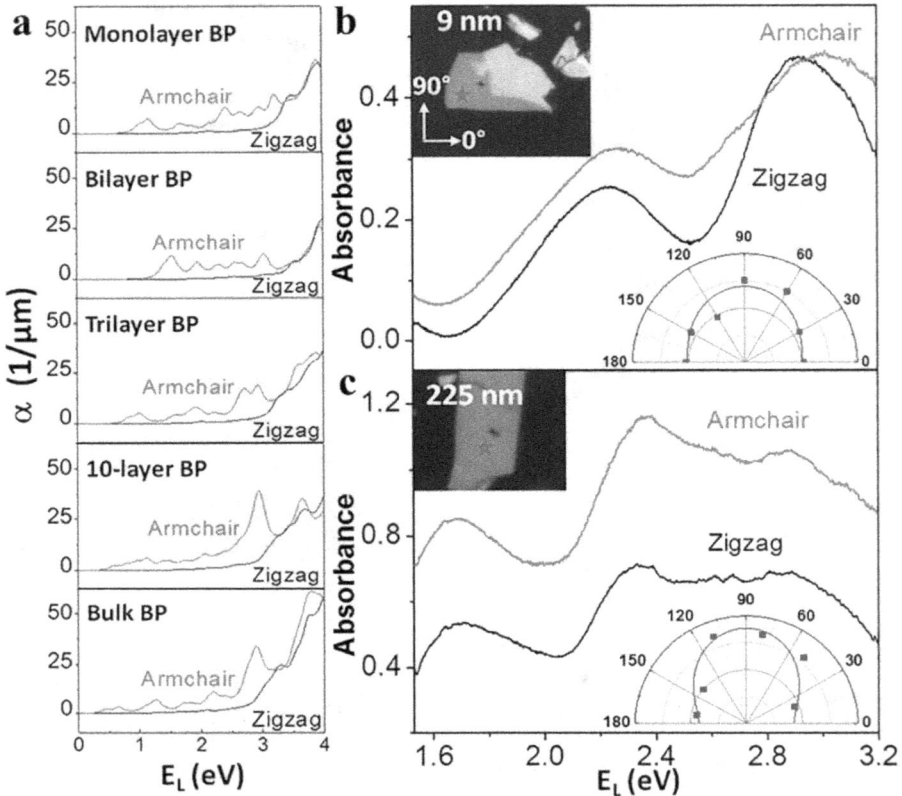

FIGURE 8.4 (a) Calculated absorption coefficient α as a function of photon energy for monolayer, bilayer, trilayer, 10-layer, and bulk BP. Typical absorbance spectra of a thin (b) and a thick (c) BP flake with incident light polarization along the armchair and zigzag directions.

Source: Reprinted with permission from Ling, X., et al. *Nano Lett.* 16(4), 2260–2267, 2016. Copyright 2016 American Chemical Society.

metal-BP geometry edge, Cui's group designed a ring-shaped metal electrode as the photocurrent collector, so that the photo-generated carriers can be collected isotropically [75]. By applying a small source-drain voltage (0.1 V), the device could work in the photo-thermoelectric regime, which helped to observe the intrinsic anisotropy of the BP flake. Under broadband illuminations (400 nm–1700 nm), spectrally resolved photoresponsivity was measured accurately, from which one can find that the values with 0° light polarization (defined as armchair direction) are much larger than that for 90° light polarization (zigzag direction), and the contrast ratio is as high as 3.5 at ~ 1200 nm. These observations directly verify the linear dichroism detection capability based on the anisotropic crystal orientation of BP. In addition, an ionic gel was utilized to tune the interfacial band bending of BP and realized a vertical p-n junction with greatly enhanced linear-polarization photodetection efficiency. Soon, Guo *et al.* achieved a BP mid-IR photodetector with a room temperature responsivity of 82 A/W at 3.39 μm, capable of achieving ultralow power detection (~ pW scale) and resolving incident light polarization [80]. This work opens a new avenue towards object detection that requires longwave polarization imaging, such as target contrast enhancement in hazy/foggy conditions [81].

However, the device performance is still limited by the large exciton binding energies and relatively high carrier recombination in the pristine BP polarized photodetectors [82]. To address these issues, Hu and coworkers constructed a vertical photogate heterostructure of BP-on-WSe$_2$, in which BP served as the photogate and WSe$_2$ acted as the conductive channel, achieving a highly polarization-sensitive IR photodetector [83]. Note that this device displayed a broadband photoresponse from visible to IR spectral range. The photoresponsivity can reach up to ~ 10^3 A/W under moderate illumination and bias conditions, which is two orders of magnitude higher than the previously reported BP photodetectors [76]. Particularly, because of the intrinsically linear dichroism of BP as photoactive layer, this device showed a better polarization photosensitivity with a contrast ratio of ~ 6. In parallel, Wang *et al.* have further demonstrated a vertically stacked BP-InSe p-n heterojunction for highly polarized and fast photoresponse [82]. From the normalized scanning photocurrent image (calculated as the anisotropy ratio $\gamma = (I_{0°}-I_{90°})/(I_{0°}+I_{90°})$), one can clearly see that the $\gamma \approx 0.8$ at the edge and remains only ≈ 0.3 in the junction inner region, which is mainly due to the different light absorptions and conversions originating from the anisotropic optical properties of BP and the heterostructured band alignment. Furthermore, the plasmonic resonance could also be introduced into BP-based photodetector, leading to a polarization-sensitive photoresponse to IR light with a polarization sensitivity of 8.7 [84]. The figure-of-merit was further improved to ~ 22 in the BP/MoS$_2$ heterostructured mid-IR photodiode [85].

8.4.1.3 ReS$_2$

Properties of ReS$_2$. Rhenium disulfide (ReS$_2$) is a 2D group VII transition metal dichalcogenide (TMDC), which has emerged as a potential star material for linear-polarization photodetection, owing to its unique anisotropic in-plane crystal structure [64–66,86,87]. ReS$_2$ possesses a unique 1T' crystal structure due to the asymmetrical Peierls distortion [86]. It is composed of two hexagonal planes of S atoms and an intercalated hexagonal plane of Re atoms, leading to two principle crystal axes, *i.e.*, the shortest b-axis and second-shortest a-axis in the basal plane [88]. The b-axis corresponds to the Re atomic chain. Figure 8.5(a) displayed the Raman modes of the ReS$_2$ monolayer, few layer and bulk film, respectively. Unlike the conventional TMDCs with highly hexagonal symmetries, ReS$_2$ possesses 15 vibrational modes which are associated with the fundamental Raman modes (A_{1g}, E_{2g}, and E_{1g}) coupled to each other and to acoustic phonons [89], manifesting the anisotropic crystal structure. The ARPES measurements further revealed the electronic band dispersion anisotropy of ReS$_2$ [89,90]. As shown in Figure 8.5(b)–(d), the equivalence of M1 and M2 points was broken in the Brillouin zone of ReS$_2$ hexagonal lattice, leading to obvious differences between the dispersions along Γ-M1 and Γ-M2 directions.

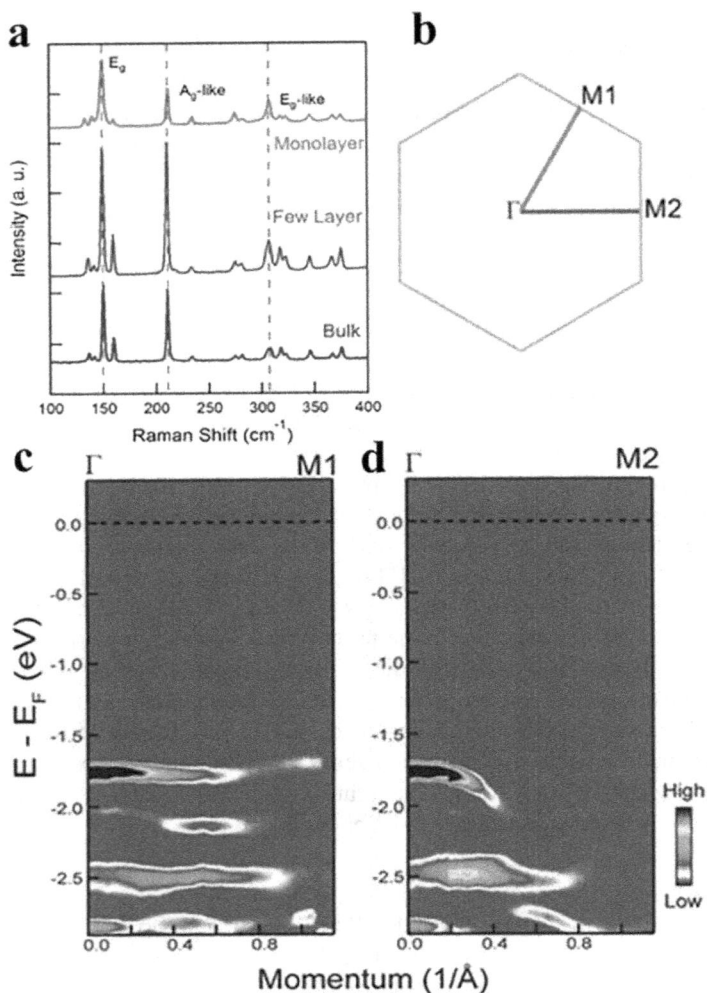

FIGURE 8.5 (a) Raman spectra of monolayer, few-layer and bulk ReS_2, (b) schematic of undistorted hexagonal Brillouin zone of ReS_2, and the corresponding band dispersion along (c) Γ-M1 and (d) Γ-M2 obtained by ARPES measurements.

Source: Reprinted with permission from Liu, F., et al. *Adv. Funct. Mater.* 26, 1169–1177, 2016. Copyright 2016 Wiley-VCH Verlag.

An *ab initio* band calculation indicated the direct bandgap semiconductor nature of ReS_2 regardless of whether it is monolayer, few-layer or bulk structure [88]. Thus, the anisotropic structure along with the direct bandgap property makes ReS_2 highly suitable for polarization-sensitive photodetection.

ReS_2-based linear-polarization photodetection. Although the electrical and optical anisotropies of ReS_2 have been widely demonstrated in the 2D material

community [91,92], the ReS_2 based polarized photodetector is still rarely exploited. Until 2015, Liu and coworkers synthesized high-quality ReS_2 single crystals through CVT, and then fabricated a simple but excellent phototransistor [89]. The ReS_2-based device showed an ultrahigh responsivity of 10^3 A/W with the electron mobility of 40 $cm^2V^{-1}s^{-1}$ and on/off ratio of 10^5. Furthermore, the anisotropic ReS_2 as the photoactive layer leads to linear dichroism. The dependence of photocurrent as a function of the incident polarization angle shows a clear polarization dependence. Note that the dependence of photocurrent on the incident light polarization is almost identical to that of ReS_2 absorption, as shown in Figure 8.6, which strongly demonstrates that ReS_2 can be used as an intrinsic linear dichroism media with highly polarized photo-absorption and photocurrent detection. Recently, 2D $MoTe_2/ReS_2$ vdW heterostructure was proposed to construct high-performance linear polarization-sensitive photodetector [93]. $ReSe_2$ is an analogue of ReS_2 with a highly similar 1T' phase and strong in-plane anisotropy [94]. Utilizing high-quality $ReSe_2$ nanosheets, Zhang et al. fabricated a photodetector with an excellent on/off current ratio exceeding 10^7 and ambipolar gate-tunable linear dichroism photodetection [95]. The photocurrent showed a noticeable evolution with the polarization angle changes from 0° to 90°, and reached its maximum value when the incident light is polarized along the b-axis (defined as 0°), and the anisotropy ratio γ was estimated to be ~ 0.5. In addition, a back-gate voltage can be used to tune the Fermi level of the $ReSe_2$ channel, and leads to an enhanced linear dichroism. Recently, Xu's group developed a CVD growth of large-scale 1T' alloy $ReS_{2x}Se_{2(1-x)}$ monolayer with a fully tunable bandgap from 1.31 eV to 1.62 eV [96]. The device showed anisotropic photoresponse depending on the polarization angle of the incident light. In another related study, Xu and coworker constructed a large-scale 1T' ReS_2-$ReSe_2$ lateral p-n junction by two-step epitaxial

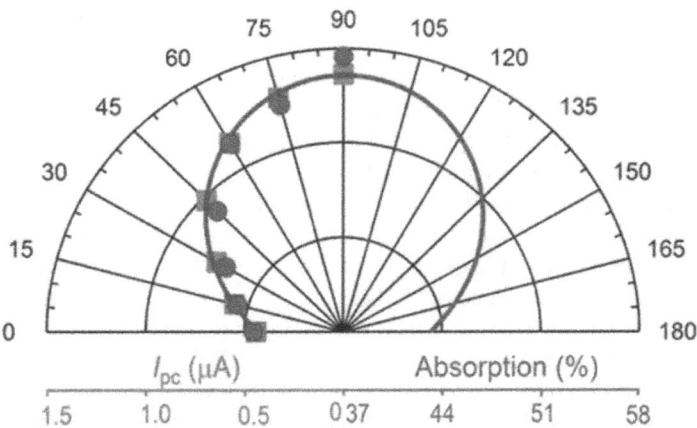

FIGURE 8.6 Polar plot of photocurrent and absorption measured under different polarization angle of green light.

Source: Reprinted with permission from Liu, F., et al. *Adv. Funct. Mater.* 26, 1169–1177, 2016. Copyright 2016 Wiley-VCH Verlag.

growth, and achieved linear dichroism photodetection in the 2D lateral heterostructure for the first time [97]. This device can work as a polarization-dependent photodiode, enabling the development of multi-functional optoelectronic devices.

8.4.1.4 Binary IV-VI Chalcogenides

The binary IV-VI monochalcogenides (MX, M = Ge, Sn; X = S, Se, *etc*) possess similar puckered orthorhombic crystal structure as BP, and thus lead to in-plane anisotropic physical properties [98]. The bandgaps of these materials range from 0.9 eV to 1.9 eV corresponding to the near-IR and part of visible regimes, which can be utilized for linear-polarization photodetection. As a representative semiconductor, GeSe-based polarization-sensitive devices have been widely exploited over the past years [58,99–101]. For instance, Wang *et al.* reported a shortwave near-infrared (SW-NIR) linear dichroism photodetectors by using few-layer GeSe as photoactive channel. The as-prepared device exhibited a moderate photoresponsivity (~ 4 A/W) with well-reproducible cycle along with the polarized angle and notable dichroic ratio of 2.16 [99]. To further improve the photoresponse, trap engineering has been applied to induce photoconductive gain with an ultrahigh photoresponsivity of 1.6×10^5 A/W, which surpasses that of many other 2D material photodetectors [53,100]. Meanwhile, the GeSe-based device also shows promising potential in linear-polarization photodetection with a dichroic ratio of 1.3. Given that the fundamentals of the optical anisotropy of GeSe are still lacking, Yang *et al.* systematically studied its anisotropic optical absorption, reflection, extinction, and refraction [101]. ARPES measurements experimentally demonstrated the strong band dispersion anisotropy of GeSe along the *x*- and *y*-axis, further revealing its intrinsic anisotropy. Recently, an innovative type II heterojunction of GeSe/MoS$_2$ was constructed to achieve a polarization-sensitive self-powered photodetector [58]. Such device exhibited a superior dichroic ratio and polarization sensitivity of 2.02 and 2.95, respectively. In addition, as important family members, GeS-, SnSe- and SnS-based linear-polarization photodetectors have led to considerable progress in the field requiring highly anisotropic properties, enabling further development of polarization-sensitive optoelectronic devices [102–107].

Other typical binary IV-VI chalcogenides, like GeS$_2$ and β-GeSe$_2$, also possess low-symmetric (monoclinic) crystal structures (Figure 8.7(a)) and in-plane anisotropic behaviors [64]. Given high ambient stabilities, large direct bandgaps, and absorption coefficients, GeS$_2$ and β-GeSe$_2$ are promising candidates for highly selective polarized photodetectors in the shortwave spectral range. In 2018, Yang and coworker systematically investigated the in-plane anisotropy of β-GeSe$_2$ thin flakes from structural, vibrational, electrical, and optical perspectives, and then fabricated a polarized photodetector [108]. As evidenced by the anisotropy of absorption, one can observe polarization angle-dependent linear dichroism of GeSe$_2$ flakes. The as-prepared device displayed excellent ambient stability and achieved a dichroic ratio up to 3.4 (Figure 8.7(b)). In order to improve spectral selectivity, monolayer GeSe$_2$ (bandgap, E_g = 2.96 eV) was prepared and applied for ultraviolet (UV) detection with a good cutoff wavelength of 405 nm [109]. Note that GeS$_2$ is more suitable for UV-polarized photodetection as it possesses the widest bandgap (> 3 eV) in the family of in-plane anisotropic 2D semiconductors so far. Adopting GeS$_2$ as the photoactive layer, the

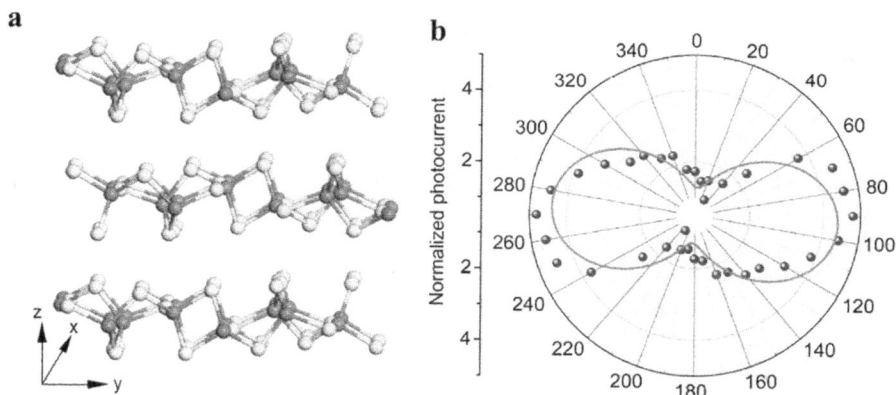

FIGURE 8.7 (a) Side view of crystal structure of monoclinic $GeSe_2$ or GeS_2. (b) Polar plot of the normalized polarized angle-resolved photocurrent of the $GeSe_2$ based photodetector.

Source: Reprinted with permission from Yang, Y., et al. *J. Am. Chem. Soc.* 140, 4150–4156, 2018. Copyright 2018 American Chemical Society.

first linear dichroic photodetector in UV region was achieved in Hu's Group with a dichroic ratio of 2.1 [110].

8.4.1.5 Layered Perovskite and Other 2D Systems

Recently, an emerging star material, 2D layered halide perovskite, was chosen as a novel class of anisotropic semiconductors to construct linear-polarization photodetectors [57,111–116]. In general, 2D perovskites, $(A')_m(A)_{n-1}B_nX_{3n+1}$ (where A' and A are cations; B is divalent metal and X is halogen), are obtained by chemically inserting organic cations (A') into the cubic prototype of 3D perovskite, ABX_3, sheets [117]. The resulting alternate alignment of inorganic perovskite multilayers and organic cation spacers leads to unique quantum-well structures, which enables intrinsic structural anisotropy along different crystallographic directions. In 2019, Li *et al.* reported a polarization-sensitive narrowband photodetector by using 2D (iso-$BA)_2PbI_4$ perovskite with no additional optical components [57]. The device showed competitive performance with a linear dichroic ratio of 1.56 at 552 nm and a full width at half maximum (FWHM) of 20 nm, which can be ascribed to the crystal anisotropy and the charge collection narrowing mechanism in 2D perovskites induced by self-trapped states. Following this work, the same group further revealed the anisotropy characteristics of free excitons and self-trapped excitons with different optical selection rules in 2D perovskites [118]. Various 2D hybrid perovskites, such as $(iBA)_2(MA)Pb_2I_7$ [111], $(FPEA)_2PbI_4$ [112], $(i\text{-}PA)_2CsAgBiBr_7$ [113], have been successfully exploited to demonstrate polarization-sensitive photodetection. Strikingly, these devices exhibited large dichroic ratios up to 2.1, rapid response times (hundreds of microsecond) and high detectivities to the linear polarized light. Such figure-of-merits are comparable to those of conventional 2D anisotropic materials, providing promising candidates for the polarized optoelectronic applications [65]. Interestingly, the organic cations (A') allow molecular freedom of dynamic motion, leading to spontaneous ferroelectric polarization in the 2D perovskites [115]. Thus, the combination of ferroelectricity and intrinsic anisotropy benefits highly efficient dissociation of photo-generated charges and improves device performances. A new 2D perovskite ferroelectric, $[CH_3(CH_2)_3NH_3]_2(CH_3NH_3)Pb_2Br_7$, has been reported by Sun and coworkers for polarization-sensitive shortwave photodetection [114]. One can see that an order-disorder phase transition leads to electric polarization along the crystallographic c-axis direction. The polarization-electric field (P-E) hysteresis loops further showed a direct evidence for switchable polarization detection, and the as-prepared device achieved a large dichroic ratio of 2. In addition, 2D perovskite ferroelectrics were also utilized to fabricate ultrasensitive polarized-light detectors with an extremely low detectable limit of 40 nW/cm² [115]. More recently, Ji et al. firstly reported a ferroelectricity-driven self-powered device towards polarized UV photodetection using wide bandgap 2D perovskite ferroelectrics. The device exhibited an impressive dichroic ratio up to 6.8, which is superior to those previously reported polarized photodetectors [116]. These results suggest that integrating ferroelectricity into 2D perovskite would pave a new pathway towards high-performance polarized-light detectors.

It is worth emphasizing that, in addition to the materials mentioned above, some other anisotropic 2D materials also play important roles in linear-polarization photodetection, which includes but not limited to some low-symmetry group IV-V compounds, group V_2-VI_3 compounds and transition metal chalcogenides [64]. For example, GeAs [119], $GeAs_2$ [120], SiP [121], GeP [122], Sb_2Se_3 [123], $PdSe_2$ [124,125], TiS_3 [126], ZrS_3 [127], and Ta_2NiS_5 [128] based polarized photodetectors have been successfully demonstrated during the past few years. The most important parameter, dichroic ratio, varies from 1.49 to 4.4 for different materials or spectral regions. The richness of low-symmetry 2D materials offers considerable possibilities for high-performance polarized photodetectors.

8.4.2 Circular Polarization Detection

Circularly polarized light (CPL) plays a key role in many photonic applications, such as optical communication, optical quantum computing and information processing of spin, which require high-performance circularly polarized photodetectors to fully detect and exploit the functions of CPL in these fields [129–131].

8.4.2.1 Introduction of Circular Photogalvanic Effect (CPGE)

The first prediction for circular photogalvanic effect (CPGE) in homogeneous crystals was made in 1978 by E. L. Ivchenko and G. E. Pikus [132]. They theoretically showed the CPGE in gyrotropic crystals (element Tellurium), and calculated the value of photocurrent for interband and intraband light absorption. Later in 1979, they firstly observed CPGE in bulk tellurium under an illumination of 10.6 μm circularly polarized light [133].

One theory that is well consistent with the hypothesis of V. M. Asnin et al. argued that CPGE appearednot only in gyrotropic crystals, but also in any arbitrary medium lacking symmetric center, such as ferroelectrics, piezoelectrics, and even in gases and liquids [134]. Therefore, researchers began to develop new structures and materials to achieve this effect. A large number of studies on CPGE have emerged in recent years [135–138].

In addition, with the development of spintronics (see Chapter 10), it is feasible to use the spin information of electrons. To implement this idea, it's very significant to detect spin polarization and relevant current in a kind of materials with low dimensions and strong spin-orbit interactions. Fortunately, the quantum well not only satisfies these conditions but also has the advantage of being independent of external bias field, motivating the development of CPGE in quantum Wells. During this period, many scholars studied the CPGE in quantum well systems. In particular, S. D. Ganichev and E. L. Ivchenko began to focus on the presence of CPGE in quantum wells that could be regarded as a system of inversion symmetry broken. They proposed two mechanisms including spin-galvanic effect and spin orientation-induced CPGE. The former is the result of the spin relaxation, while the latter is the asymmetric momentum distribution of the photocarriers under optical excitation of circularly polarized radiation.

Furthermore, S. D. Ganichev and W. Prettl suggested that the CPGE is a novel method for rapidly detecting the degree of circular polarization of light, especially in infrared wavelengths. In general, CPGE can be induced by three absorption mechanisms, including interband, intersubband, and intrasubband (free carrier absorption) transitions [139]. However, they only observed the last two current compositions in the middle- and far- infrared bands in n-type and p-type quantum wells. Thereafter, the investigation of interband absorption induced CPGE has been conducted in semiconductor heterojunctions, quantum wells and superlattices [140–142]. As is known to all, spin-orbit interaction arising from structural inversion asymmetry (SIA) which is beneficial to control the spin polarization of the electron. With these different heterojunction structures, we can manipulate the composition of the materials to implement a stronger SIA, which is also in flavor of obtaining the stronger spin-orbit interaction.

In recent years, the relationship between a reflection plane of a gyrotropic crystal and the CPGE was analyzed, and the influence of magnetic field on the photocurrents has also been verified [143]. Later in 2018, Zhang *et al.* shared a viewpoint that the photocurrent was caused by the three-band transition assisted by an intermediate band away from the Weyl cone [144]. The photocurrent irradiated by the circularly polarized light was referred to as CPGE. This theory was similar to the intraband transition induced by free-carrier absorption in the 1979's literature [130], which only took into consideration the splitting of subbands in the valence band due to spin-orbit interaction.

In general, the CPGE results from the interaction between the material asymmetry and the spin angular momentum of light. Therefore, it can be regarded as a critical detecting way to analyze the change of information of electrons or holes in the microscopic level. Furthermore, we can obtain more information of light by CPGE, such as amplitude, phase, and polarization of light. The section is aimed to reveal the basic principles of CPGE detection in different material systems such as quantum wells, TMDCs, topological insulators, Weyl semimetals, among others.

8.4.2.2   CPGE Mechanisms

The CPGE is well known as the result of spin-orbit interaction/coupling (SOC) due to the spatial inversion symmetry broken. There are two kinds of SOCs to help removing the spin degeneracy. The first one is the structural inversion asymmetry (SIA) exsisting in low dimensional structures, which is caused by structure asymmetries in heterostructures and 2D materials with spatial inversion asymmetries or under an electric field. Meanwhile, the SIA term has advantage of flexible tunability by uniaxial strain, interface defect, and external electric field, leading to a promoting development of spintronics in spin injection and spin control. The second one is the bulk inversion asymmetry (BIA), which mainly contributes to bulk crystals lacking inversion centers, *e.g.*, zinc blende structures [145].

8.4.2.3   Non-2D System for the Detection of Circular Polarization

V. M. Asnin *et al.* firstly observed the CPGE in a bulk tellurium, and proposed the photocurrent equation considering the average velocities of the photocarriers [133]. They argued that the CPGE was the result of the transformation of spin angular momentums and energies of photons into the drift velocities of photocarriers. However, a controversial issue is that they only showed the CPGE induced by free hole absorption.

The novel photogalvanic effect has attracted significant interests due to its simple experimental conditions and promising prospects in detection applications. An optical active crystal (broken inversion symmetry semiconductor) irradiated by a circular polarization light under an unbiased electric field will generate CPGE current with certain sign, depending on the degree of circular polarization. Later, Dhara *et al.* observed a complex circular photogalvanic effect in silicon nanowires, which made it possible to find the CPGE in other achiral materials through some analogous measures [146]. There is a difference here from previous studies, in which an electric field is utilized to break the inversion symmetry, and the magnitude and sign of CPGE are dependent on the level of the applied field. The results indicated that the CPGE was only produced at nanowire-metal contacts, rather than on nanowires. In fact, the laser energy (1.82 eV@680 nm) was above the indirect bandgap of Si nanowire, which made the optical excitation in the metal-nanowire Schottky junction an interband transition. An electric field along the nanowire growth direction broke the in-plane mirror symmetry, which can be regarded as a distinct chiral structure.

8.4.2.4   Topological Insulator

In 2010, the theoretical foundation and experimental phenomenon of topological insulators were reviewed by Hasan and Kane [147]. The protected metallic states emerge on the surfaces of topological insulators due to spin-orbit interactions and time reversal symmetries. The coupling between the angular momentum of circularly polarized light and the electron spin of topological insulator gives rise to CPGE due to the spin-momentum locking in surface states [148]. P. Hosur theoretically studied the CPGE in topological insulators, and he proposed a visual understanding about the CPGE dependence on the Berry curvatures of electron bands, which described the nonequilibrium absorption of electrons in the conducting surface states to photon momentums [149].

The relationship of CPGE on the properties of Dirac cone has been discussed by N. Gedik *et al.* in 2012 [150]. In general, it is believed that CPGE is originated from the asymmetric distribution of spin in the momentum space. Since the current is detected along y direction, the circular polarized dependent term should be maximum when light is incident in the x-z plane. In other words, the two spin polarizations induced by different circularly polarized lights should be presented to the planes which are parallel to the direction of electrodes, and the CPGE should be produced when the distributions of these opposite spin polarizations are asymmetric. Therefore, the polarization dependence of CPGE may be derived from the asymmetric absorption on both sides of the Dirac cone, which is distinct from previous perspectives that only account the non-degenerate valence bands in materials.

Additionally, C. Kastl *et al.* detected the ultrafast photocurrent response by the time-domain THz photocurrent spectroscopy with a picosecond time-resolution, and revealed the distinctness between photothermoelectric currents and CPGE. The photothermoelectric currents can be ascribed by a process that the hot carriers generated under laser excitation transmit along the gradient directions of the potentials (*e.g.*, thermopower and density gradient). The current is caused by the thermopower between Bi_2S_3 and Au contacts [151]. The peculiar properties of surface states have deepened physical understanding of current generation in the band transition due to spin polarization, which is derived from Dirac fermions' unique selection rule for circular polarized light. Moreover, CPGE can not only be used to detect the physical properties of materials, but also control the spin polarization currents.

8.4.2.5 Valley TMDC

In monolayer TMDCs, the giant energy level splitting emerges, which is induced by the spin-orbit interaction due to lack of spatial inversion symmetry in energy band structure [152]. The premise of CPGE is the circularly polarized dependence, endowed by spin coupled valley dependent optical selection rule due to inversion symmetry breaking in TMDCs, which has been theoretically investigated by Yao *et al.* in 2008 [153]. Based on their results, we may observe the CPGE in monolayer TMDCs without bias voltages, which can give rise to undesired photothermoelcetric current. In contrast, one can also apply an external electric field to break the inversion symmetries in bulk TMDCs. Yuan *et al.* [154] and Quereda *et al.* [155] both investigated the influence of external bias voltage, such as gate and drain-source voltages, on the CPGE. The former found that the CPGE current can be enhanced by two orders of magnitude through a perpendicular gate electric field in a multilayer WSe_2. As a result, they obtained a CPGE current of microamps level. While the latter analyzed the effect of a drain-source voltage, suggesting that the CPGE of a monolayer $MoSe_2$ with different drain-source voltages might arise from exciton interband transition, which is consistent with the previous work by Eginligil *et al.* [156].

8.4.2.6 Weyl Semimetal

Weyl semimetals hosting massless fermions are regarded as novel platforms to conduct electric charge due to its high mobility and massless nature. Note that, Ching-Kit Chan *et al.* in 2007 predicted that the large photocurrents in type-II Weyl semimetals are due to the combination of inversion symmetry breaking and the finite tilts of Weyl spectra [157]. In addition, many advantages promote Weyl semimetals for broadband and high-sensitivity detectors such as the linear dispersion of energy band, photo-carrier generation without bias electric field, and the adjustable magnitude with gate voltage.

Su-Yang Xu *et al.* and Zhurun Ji *et al.* has made great contributions in controlling the physical nature of Weyl semimetal through electrical or optical measurement [158,159]. In a dual-gated monolayer WTe_2 devices, the CPGE showed the dependence of both magnitude and sign on an out-plane electric field, while previous works in bulk crystal can only achieve magnitude modulation. The effect of out-plane electric field results in an in-plane polarity, which makes it possible to control the sign of CPGE. Meanwhile, the bulk $MoTe_2$, $Mo_{0.9}W_{0.1}Te_2$ and $Mo_{0.3}W_{0.7}Te_2$ devices for thicknesses of 100 - 300 nm also observed the CPGE under normal incidence at a 70 K temperature. A novel phenomenon emerged in these materials, which was referred to as spatially dispersive circular photogalvanic effect (s-CPGE) to distinguish it with conventional CPGE. Note that the presence of s-CPGE does not provide any new means to manipulate the CPGE, but describes a different role of optical field gradients in the radial direction.

Compared with experiments of CPGE on semiconductor heterostructures, quantum wells, topological insulators and TMDCs, researches on the microscopic mechanism in Weyl semimetals identify a semiclassical behavior of interband transition by considering the lowest conduction and highest valence bands in the vicinity of Weyl nodes, where the pure classical theory fails to describe the origin of CPGE. Thereafter, some scholars developed a semiclassical theory of the CPGE in noncentrosymmetric materials. Fernando de Juan *et al.* expanded this theory to Weyl semimetals and quantized the circular photogalvanic tensor [160]. Furthermore, they proposed to model the CPGE effect as the dipole moment of Berry curvature in 3D materials. In 2018, Su-Yang Xu *et al.* demonstrated the tunable Berry curvature dipole in a monolayer WTe_2 [158]. In particular, monolayer WTe_2, bulk $MoTe_2$, and any Weyl semimetal would allow CPGE at normal incidence, whereas bilayer graphene and monolayer MoS_2 can generate CPGE only with oblique incidence, which indicates that the Berry curvature dipole exists in monolayer WTe_2 rather than graphene or MoS_2 [161].

8.5 SUMMARY AND PERSPECTIVE

The multi-degree of freedom of photodetection empowers a plethora of applications in modern society including optical imaging, communication and sensing as well as quantum physics, *etc.* In addition to the conventional detecting methods of intensity and wavelength, observations of the polarization and phase information of light attract many interests in recent years. From synthetic-, structural- and electronic-properties of materials, we discuss various experiment methods and physical mechanisms in detecting linear-polarization, circular-polarization of incident light based on emerging 2D material devices and nanophononics. Meanwhile, we also have given a comprehensive survey on the recent development of 2D material photodetectors based diverse nanostructures. Up to date, the wafer-scale growth of 2D materials has make a great progress, further improvement of the crystal quality will make the commercialization of 2D material device available soon. The design of novel van der Waals heterostructure and twist samples would provide more functional optoelectronic device of 2D materials. Due to the unique optoelectronic properties associated with 2D materials, we have witnessed great successes in improving the detection performance of degrees of freedom of light, and look forward to future developments in commercial applications.

ACKNOWLEDGEMENT

This work is supported by the National Natural Science Foundation of China (62074025, 12161141015, 21903084) and the National Key Research & Development Program (2021YFE0194200), the Applied Basic Research Program of Sichuan Province (2021JDGD0026, 2021YJ0408), and Sichuan Province Key Laboratory of Display Science and Technology.

BIBLIOGRAPHY

[1] Corcoran, B., Tan, M. X., Xu, X. Y., Boes, A., Wu, J. Y., Nguyen, T. G., Chu, S. T., Little, B. E., Morandotti, R., Mitchell, A., Moss, D. J. 2020. Ultra-dense optical data transmission over standard fibre with a single chip source. *Nat. Commun.* 11(1): 2568.

[2] *Groundbreaking new technology could allow 100-times-faster internet by harnessing twisted light beams.* https://phys.org/news/2018-10-groundbreaking-technology-times-faster-internet-harnessing.html (accessed February 8, 2022).

[3] Chen, Z. Y., Yan, L. S., Pan, Y., Jiang, L., Yi, A. L., Pan, W., Luo, B. 2017. Use of polarization freedom beyond polarization-division multiplexing to support high-speed and spectral-efficient data transmission. *Light: Sci. Appl.* 6(2): e16207.

[4] Gibson, G., Courtial, J., Padgett, M. J., Vasnetsov, M., Pas'ko, V., Barnett, S. M., Franke-Arnold, S. 2004. Free-space information transfer using light beams carrying orbital angular momentum. *Opt. Express.* 12(22): 5448–5456.

[5] Bozinovic, N., Golowich, S., Kristensen, P., Ramachandran, S. 2012. Control of orbital angular momentum of light with optical fibers. *Opt. Lett.* 37(13): 2451–2453.

[6] Bozinovic, N., Yue, Y., Ren, Y., Tur, M., Kristensen, P., Huang, H., Willner, A. E., Ramachandran, S. 2013. Terabit-scale orbital angular momentum mode division multiplexing in fibers. *Science.* 340(6140): 1545–1548.

[7] Ren, H., Li, X., Zhang, Q., Gu, M. 2016. On-chip noninterference angular momentum multiplexing of broadband light. *Science.* 352(6287): 805–809.

[8] Yue, Z., Ren, H., Wei, S., Lin, J., Gu, M. 2018. Angular-momentum nanometrology in an ultrathin plasmonic topological insulator film. *Nat. Commun.* 9(1): 4413.

[9] Huang, B., Zhang, X., Wang, W., Dong, Z., Guan, N., Zhang, Z., Chen, H. 2011. CMOS monolithic optoelectronic integrated circuit for on-chip optical interconnection. *Opt. Commun.* 284(16): 3924–3927.

[10] Yang, W., Chen, J., Zhang, Y., Zhang, Y., He, J.-H., Fang, X. 2019. Silicon-compatible photodetectors: Trends to monolithically integrate photosensors with chip technology. *Adv. Funct. Mater.* 29(18): 1808182.

[11] Li, Z., Xu, B., Liang, D., Pan, A. 2020. Polarization-dependent optical properties and optoelectronic devices of 2D materials. *Research.* 2020: 5464258.

[12] Ji, Z. R., Liu, W. J., Krylyuk, S., Fan, X. P., Zhang, Z. F., Pan, A. L., Feng, L., Davydov, A., Agarwal, R. 2020. Photocurrent detection of the orbital angular momentum of light. *Science.* 368(6492): 763–767.

[13] Tong, L., Huang, X. Y., Wang, P., Ye, L., Peng, M., An, L. C., Sun, Q. D., Zhang, Y., Yang, G. M., Li, Z., Zhong, F., Wang, F., Wang, Y. X., Motlag, M., Wu, W. Z., Cheng, G. J., Hu, W. D. 2020. Stable mid-infrared polarization imaging based on quasi-2D tellurium at room temperature. *Nat. Commun.* 11(1): 2308.

[14] Schaibley, J. R., Yu, H., Clark, G., Rivera, P., Ross, J. S., Seyler, K. L., Yao, W., Xu, X. 2016. Valleytronics in 2D materials. *Nat. Rev. Mater.* 1: 16055.

[15] Novoselov, K. S., Geim, A. K., Morozov, S. V., Jiang, D.-e., Zhang, Y., Dubonos, S. V., Grigorieva, I. V., Firsov, A. A. 2004. Electric field effect in atomically thin carbon films. *Science.* 306(5696): 666–669.

[16] Sangwan, V. K., Hersam, M. C. 2018. Electronic transport in two-dimensional materials. *Annu. Rev. Phys. Chem.* 69: 299–325.

[17] Gupta, A., Sakthivel, T., Seal, S. 2015. Recent development in 2D materials beyond graphene. *Prog. Mater Sci.* 73: 44–126.

[18] Tsen, A. W., Brown, L., Levendorf, M. P., Ghahari, F., Huang, P. Y., Havener, R. W., Ruiz-Vargas, C. S., Muller, D. A., Kim, P., Park, J. 2012. Tailoring electrical transport across grain boundaries in polycrystalline graphene. *Science.* 336(6085): 1143–1146.

[19] Britnell, L., Ribeiro, R. M., Eckmann, A., Jalil, R., Belle, B. D., Mishchenko, A., Kim, Y.-J., Gorbachev, R. V., Georgiou, T., Morozov, S. V. 2013. Strong light-matter interactions in heterostructures of atomically thin films. *Science.* 340(6138): 1311–1314.

[20] Geim, A. K., Grigorieva, I. V. 2013. Van der Waals heterostructures. *Nature.* 499(7459): 419–425.

[21] Xia, F., Wang, H., Xiao, D., Dubey, M., Ramasubramaniam, A. 2014. Two-dimensional material nanophotonics. *Nat. Photon.* 8(12): 899–907.

[22] Lim, Y. R., Song, W., Han, J. K., Lee, Y. B., Kim, S. J., Myung, S., Lee, S. S., An, K. S., Choi, C. J., Lim, J. 2016. Wafer-scale, homogeneous MoS$_2$ layers on plastic substrates for flexible visible-light photodetectors. *Adv. Mater.* 28(25): 5025–5030.

[23] Novoselov, K. S., Mishchenko, A., Carvalho, A., Castro Neto, A. H. 2016. 2D materials and van der Waals heterostructures. *Science.* 353(6298): aac9439.

[24] Yu, H., Liu, G.-B., Tang, J., Xu, X., Yao, W. 2017. Moiré excitons: From programmable quantum emitter arrays to spin-orbit–coupled artificial lattices. *Sci. Adv.* 3(11): e1701696.

[25] Burg, G. W., Zhu, J., Taniguchi, T., Watanabe, K., MacDonald, A. H., Tutuc, E. 2019. Correlated insulating states in twisted double bilayer graphene. *Phys. Rev. Lett.* 123(19): 197702.

[26] Bistritzer, R., MacDonald, A. H. 2011. Moiré bands in twisted double-layer graphene. *Proc. Natl. Acad. Sci.* 108(30): 12233–12237.

[27] Cao, Y., Fatemi, V., Demir, A., Fang, S., Tomarken, S. L., Luo, J. Y., Sanchez-Yamagishi, J. D., Watanabe, K., Taniguchi, T., Kaxiras, E. 2018. Correlated insulator behaviour at half-filling in magic-angle graphene superlattices. *Nature.* 556(7699): 80–84.

[28] Cao, Y., Fatemi, V., Fang, S., Watanabe, K., Taniguchi, T., Kaxiras, E., Jarillo-Herrero, P. 2018. Unconventional superconductivity in magic-angle graphene superlattices. *Nature.* 556(7699): 43–50.

[29] Park, J. M., Cao, Y., Watanabe, K., Taniguchi, T., Jarillo-Herrero, P. 2021. Tunable strongly coupled superconductivity in magic-angle twisted trilayer graphene. *Nature.* 590(7845): 249–255.

[30] Deng, B., Wang, B., Li, N., Li, R., Wang, Y., Tang, J., Fu, Q., Tian, Z., Gao, P., Xue, J. 2020. Interlayer decoupling in 30° twisted bilayer graphene quasicrystal. *ACS Nano.* 14(2): 1656–1664.

[31] Wang, D., Luo, F., Lu, M., Xie, X., Huang, L., Huang, W. 2019. Chemical vapor transport reactions for synthesizing layered materials and their 2D counterparts. *Small.* 15(40): e1804404.

[32] Bugaris, D. E., zur Loye, H. C. 2012. Materials discovery by flux crystal growth: quaternary and higher order oxides. *Angew. Chem. Int. Ed.* 51(16): 3780–3811.

[33] Yan, J. Q., Sales, B. C., Susner, M. A., McGuire, M. A. 2017. Flux growth in a horizontal configuration: An analog to vapor transport growth. *Phys. Rev. Mater.* 1(2): 023402.

[34] Schmidt, P., Binnewies, M., Glaum, R., Schmidt, M. 2013. *Chemical Vapor Transport Reactions-Methods, Materials, Modeling.* Advanced Topics on Crystal Growth. Rijeka, Croatia: InTech.

[35] May, A. F., Yan, J., McGuire, M. A. 2020. A practical guide for crystal growth of van der Waals layered materials. *J. Appl. Phys.* 128(5): 051101.

[36] Tachibana, M. 2017. *Beginner's Guide to Flux Crystal Growth.* Tokyo, Japan: Springer.

[37] Coleman, J. N., Lotya, M., O'Neill, A., Bergin, S. D., King, P. J., Umar Khan, Young, K., Gaucher, A., De, S., Smith, R. J., Shvets, I. V., Arora, S. K., Stanton, G., Kim, H.-Y., Lee, K., Kim, G. T., Duesberg, G. S., Hallam, T., Boland, J. J., Wang, J. J., Donegan, J. F., Grunlan, J. C., Moriarty, G., Shmeliov, A., Nicholls, R. J., Perkins, J. M., Grieveson, E. M., Theuwissen, K., McComb, D. W., Nellist, P. D., Nicolosi, V. 2011. Two-dimensional nanosheets produced by liquid exfoliation of layered materials. *Science.* 331: 568–571.

[38] Nicolosi, V., Chhowalla, M., Kanatzidis, M. G., Strano, M. S., Coleman, J. N. 2013. Liquid exfoliation of layered materials. *Science.* 340(6139): 1226419.

[39] Li, X., Cai, W., An, J., Kim, S., Nah, J., Yang, D., Piner, R., Velamakanni, A., Jung, I., Tutuc, E. 2009. Large-area synthesis of high-quality and uniform graphene films on copper foils. *Science.* 324(5932): 1312–1314.

[40] Najmaei, S., Liu, Z., Zhou, W., Zou, X., Shi, G., Lei, S., Yakobson, B. I., Idrobo, J.-C., Ajayan, P. M., Lou, J. 2013. Vapour phase growth and grain boundary structure of molybdenum disulphide atomic layers. *Nat. Mater.* 12(8): 754–759.

[41] Shi, Y., Zhou, W., Lu, A.-Y., Fang, W., Lee, Y.-H., Hsu, A. L., Kim, S. M., Kim, K. K., Yang, H. Y., Li, L.-J. 2012. van der Waals epitaxy of MoS_2 layers using graphene as growth templates. *Nano Lett.* 12(6): 2784–2791.

[42] Lee, Y. H., Zhang, X. Q., Zhang, W., Chang, M. T., Lin, C. T., Chang, K. D., Yu, Y. C., Wang, J. T. W., Chang, C. S., Li, L. J. 2012. Synthesis of large-area MoS_2 atomic layers with chemical vapor deposition. *Adv. Mater.* 24(17): 2320–2325.

[43] Ji, Q., Zhang, Y., Gao, T., Zhang, Y., Ma, D., Liu, M., Chen, Y., Qiao, X., Tan, P.-H., Kan, M. 2013. Epitaxial monolayer MoS_2 on mica with novel photoluminescence. *Nano Lett.* 13(8): 3870–3877.

[44] Zhang, Y., Zhang, Y., Ji, Q., Ju, J., Yuan, H., Shi, J., Gao, T., Ma, D., Liu, M., Chen, Y. 2013. Controlled growth of high-quality monolayer WS_2 layers on sapphire and imaging its grain boundary. *ACS Nano.* 7(10): 8963–8971.

[45] Gao, Y., Liu, Z., Sun, D.-M., Huang, L., Ma, L.-P., Yin, L.-C., Ma, T., Zhang, Z., Ma, X.-L., Peng, L.-M. 2015. Large-area synthesis of high-quality and uniform monolayer WS_2 on reusable Au foils. *Nat. Commun.* 6(1): 1–10.

[46] Keyshar, K., Gong, Y., Ye, G., Brunetto, G., Zhou, W., Cole, D. P., Hackenberg, K., He, Y., Machado, L., Kabbani, M. 2015. Chemical vapor deposition of monolayer rhenium disulfide (ReS_2). *Adv. Mater.* 27(31): 4640–4648.

[47] Duan, X., Wang, C., Shaw, J. C., Cheng, R., Chen, Y., Li, H., Wu, X., Tang, Y., Zhang, Q., Pan, A. 2014. Lateral epitaxial growth of two-dimensional layered semiconductor heterojunctions. *Nat. Nanotechnol.* 9(12): 1024–1030.

[48] Gong, Y., Lei, S., Ye, G., Li, B., He, Y., Keyshar, K., Zhang, X., Wang, Q., Lou, J., Liu, Z. 2015. Two-step growth of two-dimensional $WSe_2/MoSe_2$ heterostructures. *Nano Lett.* 15(9): 6135–6141.

[49] Zhou, D., Shu, H., Hu, C., Jiang, L., Liang, P., Chen, X. 2018. Unveiling the growth mechanism of MoS_2 with chemical vapor deposition: From two-dimensional planar nucleation to self-seeding nucleation. *Crystal Growth & Design.* 18(2): 1012–1019.

[50] Zhou, J., Lin, J., Huang, X., Zhou, Y., Chen, Y., Xia, J., Wang, H., Xie, Y., Yu, H., Lei, J. 2018. A library of atomically thin metal chalcogenides. *Nature.* 556(7701): 355–359.

[51] Zheng, L., Zhongzhu, L., Guozhen, S. 2016. Photodetectors based on two dimensional materials. *Journal of Semiconductors.* 37(9): 091001.

[52] Wang, G., Zhang, Y., You, C., Liu, B., Yang, Y., Li, H., Cui, A., Liu, D., Yan, H. 2018. Two dimensional materials based photodetectors. *Infrared Physics & Technology.* 88: 149–173.

[53] Fang, J., Zhou, Z., Xiao, M., Lou, Z., Wei, Z., Shen, G. 2020. Recent advances in low-dimensional semiconductor nanomaterials and their applications in high-performance photodetectors. *InfoMat.* 2(2): 291–317.

[54] Wang, J., Gudiksen, M. S., Duan, X., Cui, Y., Lieber, C. M. 2001. Highly polarized photoluminescence and photodetection from single indium phosphide nanowires. *Science.* 293(5534): 1455–1457.

[55] Sun, Y., Xiong, J., Wu, X., Gao, W., Huo, N., Li, J. 2022. Highly sensitive infrared polarized photodetector enabled by out-of-plane PSN architecture composing of p-$MoTe_2$, semimetal-$MoTe_2$ and n-$SnSe_2$. *Nano Res.* 15(6): 5384–5391.

[56] Ahn, J., Ko, K., Kyhm, J.-h., Ra, H.-S., Bae, H., Hong, S., Kim, D.-Y., Jang, J., Kim, T. W., Choi, S., Kang, J.-H., Kwon, N., Park, S., Ju, B.-K., Poon, T.-C., Park, M.-C., Im, S., Hwang, D. K. 2021. Near-infrared self-powered linearly polarized photodetection and digital incoherent holography using $WSe_2/ReSe_2$ van der waals heterostructure. *ACS Nano.* 15(11): 17917–17925.

[57] Li, L., Jin, L., Zhou, Y., Li, J., Ma, J., Wang, S., Li, W., Li, D. 2019. Filterless polarization-sensitive 2d perovskite narrowband photodetectors. *Adv. Opt. Mater.* 7(23): 1900988.

[58] Xin, Y., Wang, X., Chen, Z., Weller, D., Wang, Y., Shi, L., Ma, X., Ding, C., Li, W., Guo, S. 2020. Polarization-sensitive self-powered type-II GeSe/MoS_2 Van der waals heterojunction photodetector. *ACS Appl. Mater. Interfaces.* 12(13): 15406–15413.

[59] Rosenberg, J., Shenoi, R. V., Krishna, S., Painter, O. 2010. Design of plasmonic photonic crystal resonant cavities for polarization sensitive infrared photodetectors. *Opt. Express.* 18(4): 3672–3686.

[60] Chen, E., Chou, S. Y. 1997. A novel device for detecting the polarization direction of linear polarized light using integrated subwavelength gratings and photodetectors. *IEEE Photonics Technology Letters.* 9(9): 1259–1261.

[61] Feng, J., Yan, X., Liu, Y., Gao, H., Wu, Y., Su, B., Jiang, L. 2017. Crystallographically aligned perovskite structures for high-performance polarization-sensitive photodetectors. *Adv. Mater.* 29(16): 1605993.

[62] Lim, S., Ha, M., Lee, Y., Ko, H. 2018. Large-area, solution-processed, hierarchical $MAPbI_3$ nanoribbon arrays for self-powered flexible photodetectors. *Adv. Opt. Mater.* 6(21): 1800615.

[63] He, X., Fujimura, N., Lloyd, J. M., Erickson, K. J., Talin, A. A., Zhang, Q., Gao, W., Jiang, Q., Kawano, Y., Hauge, R. H. 2014. Carbon nanotube terahertz detector. *Nano Lett.* 14(7): 3953–3958.

[64] Zhao, S., Dong, B., Wang, H., Wang, H., Zhang, Y., Han, Z. V., Zhang, H. 2020. In-plane anisotropic electronics based on low-symmetry 2D materials: Progress and prospects. *Nanoscale Advances.* 2(1): 109–139.

[65] Li, L., Han, W., Pi, L., Niu, P., Han, J., Wang, C., Su, B., Li, H., Xiong, J., Bando, Y., Zhai, T. 2019. Emerging in-plane anisotropic two-dimensional materials. *InfoMat.* 1(1): 54–73.

[66] Tian, H., Tice, J., Fei, R., Tran, V., Yan, X., Yang, L., Wang, H. 2016. Low-symmetry two-dimensional materials for electronic and photonic applications. *Nano Today.* 11(6): 763–777.

[67] Morita, A. 1986. Semiconducting black phosphorus. *Appl. Phys. A.* 39(4): 227–242.

[68] Li, L., Yu, Y., Ye, G. J., Ge, Q., Ou, X., Wu, H., Feng, D., Chen, X. H., Zhang, Y. 2014. Black phosphorus field-effect transistors. *Nat. Nanotechnol.* 9(5): 372–377.

[69] Tran, V., Soklaski, R., Liang, Y., Yang, L. 2014. Layer-controlled band gap and anisotropic excitons in few-layer black phosphorus. *Phys. Rev. B.* 89(23): 235319.

[70] Qu, G., Xia, T., Zhou, W., Zhang, X., Zhang, H., Hu, L., Shi, J., Yu, X.-F., Jiang, G. 2020. Property-activity relationship of black phosphorus at the nano-bio interface: From molecules to organisms. *Chem. Rev.* 120(4): 2288–2346.

[71] Sun, Z., Martinez, A., Wang, F. 2016. Optical modulators with 2D layered materials. *Nat. Photon.* 10(4): 227–238.

[72] Xia, F., Wang, H., Hwang, J. C. M., Neto, A. H. C., Yang, L. 2019. Black phosphorus and its isoelectronic materials. *Nat. Rev. Phy.* 1(5): 306–317.

[73] Wang, C., Zhang, G., Huang, S., Xie, Y., Yan, H. 2020. The optical properties and plasmonics of anisotropic 2D materials. *Adv. Opt. Mater.* 8(5): 1900996.

[74] Eswaraiah, V., Zeng, Q., Long, Y., Liu, Z. 2016. Black phosphorus nanosheets: synthesis, characterization and applications. *Small.* 12(26): 3480–3502.

[75] Wang, X., Jones, A. M., Seyler, K. L., Tran, V., Jia, Y., Zhao, H., Wang, H., Yang, L., Xu, X., Xia, F. 2015. Highly anisotropic and robust excitons in monolayer black phosphorus. *Nat. Nanotechnol.* 10(6): 517–521.

[76] Yuan, H., Liu, X., Afshinmanesh, F., Li, W., Xu, G., Sun, J., Lian, B., Curto, A. G., Ye, G., Hikita, Y., Shen, Z., Zhang, S.-C., Chen, X., Brongersma, M., Hwang, H. Y., Cui, Y. 2015. Polarization-sensitive broadband photodetector using a black phosphorus vertical p-n junction. *Nat. Nanotechnol.* 10(8): 707–713.

[77] Ling, X., Huang, S., Hasdeo, E. H., Liang, L., Parkin, W. M., Tatsumi, Y., Nugraha, A. R. T., Puretzky, A. A., Das, P. M., Sumpter, B. G., Geohegan, D. B., Kong, J., Saito, R., Drndic, M., Meunier, V., Dresselhaus, M. S. 2016. Anisotropic electron-photon and electron-phonon interactions in black phosphorus. *Nano Lett.* 16(4): 2260–2267.

[78] Xia, F., Wang, H., Jia, Y. 2014. Rediscovering black phosphorus as an anisotropic layered material for optoelectronics and electronics. *Nat. Commun.* 5(1): 4458.

[79] Hong, T., Chamlagain, B., Lin, W., Chuang, H.-J., Pan, M., Zhou, Z., Xu, Y.-Q. 2014. Polarized photocurrent response in black phosphorus field-effect transistors. *Nanoscale.* 6(15): 8978–8983.

[80] Guo, Q., Pospischil, A., Bhuiyan, M., Jiang, H., Tian, H., Farmer, D., Deng, B., Li, C., Han, S.-J., Wang, H., Xia, Q., Ma, T.-P., Mueller, T., Xia, F. 2016. Black phosphorus mid-infrared photodetectors with high gain. *Nano Lett.* 16(7): 4648–4655.

[81] Tyo, J. S., Rowe, M. P., Pugh, E. N., Engheta, N. 1996. Target detection in optically scattering media by polarization-difference imaging. *Appl. Opt.* 35(11): 1855–1870.

[82] Zhao, S., Wu, J., Jin, K., Ding, H., Li, T., Wu, C., Pan, N., Wang, X. 2018. Highly polarized and fast photoresponse of black phosphorus-inse vertical p-n heterojunctions. *Adv. Funct. Mater.* 28(34): 1802011.

[83] Ye, L., Wang, P., Luo, W., Gong, F., Liao, L., Liu, T., Tong, L., Zang, J., Xu, J., Hu, W. 2017. Highly polarization sensitive infrared photodetector based on black phosphorus-on-WSe₂ photogate vertical heterostructure. *Nano Energy.* 37: 53–60.

[84] Venuthurumilli, P. K., Ye, P. D., Xu, X. 2018. Plasmonic resonance enhanced polarization-sensitive photodetection by black phosphorus in near infrared. *ACS Nano.* 12(5): 4861–4867.

[85] Bullock, J., Amani, M., Cho, J., Chen, Y. Z., Ahn, G. H., Adinolfi, V., Shrestha, V. R., Gao, Y., Crozier, K. B., Chueh, Y. 2018. Polarization-resolved black phosphorus/molybdenum disulfide mid-wave infrared photodiodes with high detectivity at room temperature. *Nat. Photon.* 12(10): 601–607.

[86] Rahman, M. Z., Davey, K. R., Qiao, S. 2017. Advent of 2D Rhenium disulfide (ReS$_2$): Fundamentals to applications. *Adv. Funct. Mater.* 27(10): 1606129.

[87] Xiong, Y., Chen, H., Zhang, D. W., Zhou, P. 2019. Electronic and optoelectronic applications based on ReS$_2$. *Physica Status Solidi-Rapid Research Letters.* 13(6): 1800658.

[88] Liu, E., Fu, Y., Wang, Y., Feng, Y., Liu, H., Wan, X., Zhou, W., Wang, B., Shao, L., Ho, C.-H., Huang, Y.-S., Cao, Z., Wang, L., Li, A., Zeng, J., Song, F., Wang, X., Shi, Y., Yuan, H., Hwang, H. Y., Cui, Y., Miao, F., Xing, D. 2015. Integrated digital inverters based on two-dimensional anisotropic ReS$_2$ field-effect transistors. *Nat. Commun.* 6(1): 6991.

[89] Liu, F., Zheng, S., He, X., Chaturvedi, A., He, J., Chow, W. L., Mion, T. R., Wang, X., Zhou, J., Fu, Q., Fan, H. J., Tay, B. K., Song, L., He, R.-H., Kloc, C., Ajayan, P. M., Liu, Z. 2016. Highly sensitive detection of polarized light using anisotropic 2D ReS$_2$. *Adv. Funct. Mater.* 26(8): 1169–1177.

[90] Webb, J. L., Hart, L. S., Wolverson, D., Chen, C., Avila, J., Asensio, M. C. 2017. Electronic band structure of ReS$_2$ by high-resolution angle-resolved photoemission spectroscopy. *Phys. Rev. B.* 96(11): 115205.

[91] Lin, Y.-C., Komsa, H.-P., Yeh, C.-H., Björkman, T., Liang, Z.-Y., Ho, C.-H., Huang, Y.-S., Chiu, P.-W., Krasheninnikov, A. V., Suenaga, K. 2015. Single-layer ReS$_2$: two-dimensional semiconductor with tunable in-plane anisotropy. *ACS Nano.* 9(11): 11249–11257.

[92] Wang, Y. Y., Zhou, J. D., Jiang, J., Yin, T. T., Yin, Z. X., Liu, Z., Shen, Z. X. 2019. In-plane optical anisotropy in ReS$_2$ flakes determined by angle-resolved polarized optical contrast spectroscopy. *Nanoscale.* 11(42): 20199–20205.

[93] Ahn, J., Kyhm, J.-H., Kang, H. K., Kwon, N., Kim, H.-K., Park, S., Hwang, D. K. 2021. 2D MoTe$_2$/ReS$_2$ van der waals heterostructure for high-performance and linear polarization-sensitive photodetector. *ACS Photon.* 8(9): 2650–2658.

[94] Long, M. S., Wang, P., Fang, H. H., Hu, W. D. 2019. Progress, challenges, and opportunities for 2D material based photodetectors. *Adv. Funct. Mater.* 29(19): 1803807.

[95] Zhang, E., Wang, P., Li, Z., Wang, H., Song, C., Huang, C., Chen, Z.-G., Yang, L., Zhang, K., Lu, S., Wang, W., Liu, S., Fang, H., Zhou, X., Yan, H., Zou, J., Wan, X., Zhou, P., Hu, W., Xiu, F. 2016. Tunable ambipolar polarization-sensitive photodetectors based on high-anisotropy ReSe$_2$ nanosheets. *ACS Nano.* 10(8): 8067–8077.

[96] Cui, F., Feng, Q., Hong, J., Wang, R., Bai, Y., Li, X., Liu, D., Zhou, Y., Liang, X., He, X., Zhang, Z., Liu, S., Lei, Z., Liu, Z., Zhai, T., Xu, H. 2017. Synthesis of large-size 1T' ReS$_{2x}$Se$_{2(1-x)}$ alloy monolayer with tunable bandgap and carrier type. *Adv. Mater.* 29(46): 1705015.

[97] Liu, D., Hong, J., Wang, X., Li, X., Feng, Q., Tan, C., Zhai, T., Ding, F., Peng, H., Xu, H. 2018. Diverse atomically sharp interfaces and linear dichroism of 1T' ReS$_2$-ReSe$_2$ lateral p-n heterojunctions. *Adv. Funct. Mater.* 28(47): 1804696.

[98] Titova, L. V., Fregoso, B. M., Grimm, R. L. 2020. 5-*Group-IV monochalcogenides GeS, GeSe, SnS, SnSe.* Chalcogenide. Cambridge: Woodhead Publishing.

[99] Wang, X., Li, Y., Huang, L., Jiang, X., Jiang, L., Dong, H., Wei, Z., Li, J., Hu, W. 2017. Short-wave near-infrared linear dichroism of two-dimensional germanium selenide. *J. Am. Chem. Soc.* 139(42): 14976–14982.

[100] Zhou, X., Hu, X., Jin, B., Yu, J., Liu, K., Li, H., Zhai, T. 2018. Highly anisotropic GeSe nanosheets for phototransistors with ultrahigh photoresponsivity. *Adv. Sci.* 5(8): 1800478.

[101] Yang, Y., Liu, S., Wang, Y., Long, M., Dai, C., Chen, S., Zhang, B., Sun, Z., Sun, Z., Hu, C. 2019. In-plane optical anisotropy of low-symmetry 2D GeSe. *Adv. Opt. Mater.* 7(4): 1801311.

[102] Li, Z., Yang, Y., Wang, X., Shi, W., Xue, D., Hu, J. 2019. Three-dimensional optical anisotropy of low-symmetry layered GeS. *ACS Appl. Mater. Interfaces.* 11(27): 24247–24253.

[103] Hsueh, H., Li, J., Ho, C. 2018. Polarization photoelectric conversion in layered GeS. *Adv. Opt. Mater.* 6(4): 1701194.

[104] Tan, D., Zhang, W., Wang, X., Koirala, S., Miyauchi, Y., Matsuda, K. 2017. Polarization-sensitive and broadband germanium sulfide photodetectors with excellent high-temperature performance. *Nanoscale.* 9(34): 12425–12431.

[105] Zhang, C., Ouyang, H., Miao, R., Sui, Y., Hao, H., Tang, Y., You, J., Zheng, X., Xu, Z., Cheng, X. 2019. Anisotropic nonlinear optical properties of a SnSe flake and a novel perspective for the application of all-optical switching. *Adv. Opt. Mater.* 7(18): 1900631.

[106] Li, X.-Z., Xia, J., Wang, L., Gu, Y.-Y., Cheng, H.-Q., Meng, X.-M. 2017. Layered SnSe nano-plates with excellent in-plane anisotropic properties of Raman spectrum and photo-response. *Nanoscale.* 9(38): 14558–14564.

[107] Zhang, Z., Yang, J., Zhang, K., Chen, S., Mei, F., Shen, G. 2017. Anisotropic photoresponse of layered 2D SnS-based near infrared photodetectors. *J. Mater. Chem. C.* 5(43): 11288–11293.

[108] Yang, Y., Liu, S., Yang, W., Li, Z., Wang, Y., Wang, X., Zhang, S., Zhang, Y., Long, M., Zhang, G. 2018. Air-stable in-plane anisotropic GeSe$_2$ for highly polarization-sensitive photodetection in short wave region. *J. Am. Chem. Soc.* 140(11): 4150–4156.

[109] Yan, Y., Xiong, W., Li, S., Zhao, K., Wang, X., Su, J., Song, X., Li, X., Zhang, S., Yang, H. 2019. Direct wide bandgap 2D GeSe$_2$ monolayer toward anisotropic UV photodetection. *Adv. Opt. Mater.* 7(19): 1900622.

[110] Yang, Y., Liu, S., Wang, X., Li, Z., Zhang, Y., Zhang, G., Xue, D., Hu, J. 2019. Polarization-sensitive ultraviolet photodetection of anisotropic 2D GeS$_2$. *Adv. Funct. Mater.* 29(16): 1900411.

[111] Liu, Y., Wu, Z., Liu, X., Han, S., Li, Y., Yang, T., Ma, Y., Hong, M., Luo, J., Sun, Z. 2019. Intrinsic strong linear dichroism of multilayered 2D hybrid perovskite crystals toward highly polarized-sensitive photodetection. *Adv. Opt. Mater.* 7(23): 1901049.

[112] Li, M., Han, S., Teng, B., Li, Y., Liu, Y., Liu, X., Luo, J., Hong, M., Sun, Z. 2020. Minute-scale rapid crystallization of a highly dichroic 2D hybrid perovskite crystal toward efficient polarization-sensitive photodetector. *Adv. Opt. Mater.* 8(9): 2000149.

[113] Li, Y., Yang, T., Xu, Z., Liu, X., Huang, X., Han, S., Liu, Y., Li, M., Luo, J., Sun, Z. 2020. Dimensional reduction of Cs$_2$AgBiBr$_6$: A 2D hybrid double perovskite with strong polarization sensitivity. *Angew. Chem. Int. Ed.* 59(9): 3429–3433.

[114] Li, L., Liu, X., Li, Y., Xu, Z., Wu, Z., Han, S., Tao, K., Hong, M., Luo, J., Sun, Z. 2019. Two-dimensional hybrid perovskite-type ferroelectric for highly polarization-sensitive shortwave photodetection. *J. Am. Chem. Soc.* 141(6): 2623–2629.

[115] Wang, J., Liu, Y., Han, S., Ma, Y., Li, Y., Xu, Z., Luo, J., Hong, M., Sun, Z. 2021. Ultrasensitive polarized-light photodetectors based on 2D hybrid perovskite ferroelectric crystals with a low detection limit. *Sci. Bull.* 66(2): 158–163.

[116] Ji, C., Dey, D., Peng, Y., Liu, X., Li, L., Luo, J. Ferroelectricity-driven self-powered ultraviolet photodetection with strong polarization-sensitivity in a two-dimensional halide hybrid perovskite. *Angew. Chem. Int. Ed.* 132(43): 19095–19099.

[117] Mao, L., Stoumpos, C. C., Kanatzidis, M. G. 2019. Two-dimensional hybrid halide perovskites: Principles and promises. *J. Am. Chem. Soc.* 141(3): 1171–1190.

[118] Li, J., Ma, J., Cheng, X., Liu, Z., Chen, Y., Li, D. 2020. Anisotropy of excitons in two-dimensional perovskite crystals. *ACS Nano.* 14(2): 2156–2161.

[119] Zhou, Z., Long, M., Pan, L., Wang, X., Zhong, M., Blei, M., Wang, J., Fang, J., Tongay, S., Hu, W. 2018. Perpendicular optical reversal of the linear dichroism and polarized photodetection in 2D GeAs. *ACS Nano.* 12(12): 12416–12423.

[120] Li, L., Gong, P., Sheng, D., Wang, S., Wang, W., Zhu, X., Shi, X., Wang, F., Han, W., Yang, S. 2018. Highly in-plane anisotropic 2D GeAs$_2$ for polarization-sensitive photo-detection. *Adv. Mater.* 30(50): 1804541.

[121] Li, C., Wang, S., Li, C., Yu, T., Jia, N., Qiao, J., Zhu, M., Liu, D., Tao, X. 2018. Highly sensitive detection of polarized light using a new group IV-V 2D orthorhombic SiP. *J. Mater. Chem. C.* 6(27): 7219–7225.

[122] Li, L., Wang, W., Gong, P., Zhu, X., Deng, B., Shi, X., Gao, G., Li, H., Zhai, T. 2018. 2D GeP: An unexploited low-symmetry semiconductor with strong in-plane anisotropy. *Adv. Mater.* 30(14): 1706771.

[123] Zhao, M., Su, J., Zhao, Y., Luo, P., Wang, F., Han, W., Li, Y., Zu, X., Qiao, L., Zhai, T. 2020. Sodium-mediated epitaxial growth of 2D ultrathin Sb$_2$Se$_3$ flakes for broadband photodetection. *Adv. Funct. Mater.*: 1909849.

[124] Liang, Q., Wang, Q., Zhang, Q., Wei, J., Lim, S. X., Zhu, R., Hu, J., Wei, W., Lee, C., Sow, C. H. 2019. High-performance, room temperature, ultra-broadband photodetectors based on air-stable PdSe$_2$. *Adv. Mater.* 31(24): 1807609.

[125] Zhong, J., Yu, J., Cao, L., Zeng, C., Ding, J., Cong, C., Liu, Z., Liu, Y. 2020. High-performance polarization-sensitive photodetector based on a few-layered PdSe$_2$ nanosheet. *Nano Res.* 13(6): 1780–1786.

[126] Niu, Y., Frisenda, R., Flores, E., Ares, J. R., Jiao, W., De Lara, D. P., Sanchez, C., Wang, R., Ferrer, I. J., Castellanosgomez, A. 2018. Polarization-sensitive and broadband photodetection based on a mixed-dimensionality TiS$_3$/Si p-n junction. *Adv. Opt. Mater.* 6(19): 1800351.

[127] Wang, X., Wu, K., Blei, M., Wang, Y., Pan, L., Zhao, K., Shan, C., Lei, M., Cui, Y., Chen, B. 2019. Highly polarized photoelectrical response in vdW ZrS$_3$ nanoribbons. *Adv. Electron. Mater.* 5(7): 1900419.

[128] Li, L., Gong, P., Wang, W., Deng, B., Pi, L., Yu, J., Zhou, X., Shi, X., Li, H., Zhai, T. 2017. Strong in-plane anisotropies of optical and electrical response in layered dimetal chalcogenide. *ACS Nano.* 11(10): 10264–10272.

[129] Zhang, C., Wang, X., Qiu, L. 2021. Circularly polarized photodetectors based on chiral materials: A review. *Frontiers in Chemistry.* 9: https://doi.org/10.3389/fchem.2021.711488

[130] Duim, H., Loi, M. A. 2021. Chiral hybrid organic-inorganic metal halides: A route toward direct detection and emission of polarized light. *Matter.* 4(12): 3835–3851.

[131] Shang, X., Wan, L., Wang, L., Gao, F., Li, H. 2022. Emerging materials for circularly polarized light detection. *J. Mater. Chem. C.* 10(7): 2400–2410.

[132] Ivchenko, E. L., Pikus, G. E. 1978. New photogalvanic effect in gyrotropic crystals. *JETP Letters.* 27(11): 640–643.

[133] Asnin, V., Bakun, A., Danishevskii, A., Ivchenko, E., Pikus, G., Rogachev, A. 1979. "Circular" photogalvanic effect in optically active crystals. *Solid State Commun.* 30(9): 565–570.

[134] Belinicher, V., Sturman, B. I. 1980. The photogalvanic effect in media lacking a center of symmetry. *Soviet Physics Uspekhi.* 23(3): 199.

[135] Rohatgi-Mukherjee, K., Chaudhuri, R., Bhowmik, B. B. 1985. Molecular interaction of phenosafranin with surfactants and its photogalvanic effect. *J. Colloid Interface Sci.* 106(1): 45–50.

[136] Murthy, A., Reddy, K. 1983. Studies on photogalvanic effect in systems containing toluidine blue. *Solar Energy.* 30(1): 39–43.

[137] Ganichev, S., Ketterl, H., Prettl, W., Ivchenko, E., Vorobjev, L. 2000. Circular photo-galvanic effect induced by monopolar spin orientation in p-GaAs/AlGaAs multiple-quantum wells. *Appl. Phys. Lett.* 77(20): 3146–3148.

[138] Efanov, A., Entin, M. 1983. Photogalvanic effect in crystal with dislocations. *Physica Status Solidi (b)*. 119(2): 473–481.

[139] Ganichev, S. D., Ivchenko, E. L., Prettl, W. 2002. Photogalvanic effects in quantum wells. *Physica E-Low-Dimensional Systems & Nanostructures.* 14(1–2): 166–171.

[140] Cho, K. S., Chen, Y. F., Tang, Y. Q., Shen, B. 2007. Photogalvanic effects for interband absorption in AlGaN/GaN superlattices. *Appl. Phys. Lett.* 90(4): 041909.

[141] Yang, C. L., He, H. T., Ding, L., Cui, L. J., Zeng, Y. P., Wang, J. N., Ge, W. K. 2006. Spectral dependence of spin photocurrent and current-induced spin polarization in an InGaAs/InAlAs two-dimensional electron gas. *Phys. Rev. Lett.* 96(18): 186605.

[142] Bel'kov, V. V., Ganichev, S. D., Schneider, P., Back, C., Oestreich, M., Rudolph, J., Hagele, D., Golub, L. E., Wegscheider, W., Prettl, W. 2003. Circular photogalvanic effect at inter-band excitation in semiconductor quantum wells. *Solid State Commun.* 128(8): 283–286.

[143] Golub, L. E., Ivchenko, E. L., Spivak, B. Z. 2017. Photocurrent in gyrotropic Weyl semimetals. *JETP Letters.* 105(12): 782–785.

[144] Zhang, Y., Ishizuka, H., van den Brink, J., Felser, C., Yan, B., Nagaosa, N. 2018. Photogalvanic effect in Weyl semimetals from first principles. *Phys. Rev. B.* 97(24): 241118.

[145] Dresselhaus, G. 1955. Spin-orbit coupling effects in zinc blende structures. *Phys. Rev.* 100(2): 580.

[146] Dhara, S., Mele, E. J., Agarwal, R. 2015. Voltage-tunable circular photogalvanic effect in silicon nanowires. *Science.* 349(6249): 726–729.

[147] Hasan, M. Z., Kane, C. L. 2010. Colloquium: Topological insulators. *Rev. Mod. Phys.* 82(4): 3045–3067.

[148] Ando, Y. 2013. Topological insulator materials. *J. Phys. Soc. Jpn.* 82(10): 102001.

[149] Hosur, P. 2011. Circular photogalvanic effect on topological insulator surfaces: Berry-curvature-dependent response. *Phys. Rev. B.* 83(3): 035309.

[150] McIver, J. W., Hsieh, D., Steinberg, H., Jarillo-Herrero, P., Gedik, N. 2012. Control over topological insulator photocurrents with light polarization. *Nat. Nanotechnol.* 7(2): 96–100.

[151] Kastl, C., Karnetzky, C., Karl, H., Holleitner, A. W. 2015. Ultrafast helicity control of surface currents in topological insulators with near-unity fidelity. *Nat. Commun.* 6: 6617.

[152] Zhu, Z. Y., Cheng, Y. C., Schwingenschlogl, U. 2011. Giant spin-orbit-induced spin splitting in two-dimensional transition-metal dichalcogenide semiconductors. *Phys. Rev. B.* 84(15): 153402.

[153] Yao, W., Xiao, D., Niu, Q. 2008. Valley-dependent optoelectronics from inversion symmetry breaking. *Phys. Rev. B.* 77(23): 235406.

[154] Yuan, H., Wang, X., Lian, B., Zhang, H., Fang, X., Shen, B., Xu, G., Xu, Y., Zhang, S.-C., Hwang, H. Y. 2014. Generation and electric control of spin–valley-coupled circular photogalvanic current in WSe 2. *Nat. Nanotechnol.* 9(10): 851–857.

[155] Quereda, J., Ghiasi, T. S., You, J.-S., van den Brink, J., van Wees, B. J., van der Wal, C. H. 2018. Symmetry regimes for circular photocurrents in monolayer MoSe$_2$. *Nat. Commun.* 9(1): 3346.

[156] Eginligil, M., Cao, B., Wang, Z., Shen, X., Cong, C., Shang, J., Soci, C., Yu, T. 2015. Dichroic spin–valley photocurrent in monolayer molybdenum disulphide. *Nat. Commun.* 6(1): 7636.

[157] Chan, C.-K., Lindner, N. H., Refael, G., Lee, P. A. 2017. Photocurrents in Weyl semi-metals. *Phys. Rev. B*. 95(4): 041104.

[158] Xu, S. Y., Ma, Q., Shen, H. T., Fatemi, V., Wu, S. F., Chang, T. R., Chang, G. Q., Valdivia, A. M. M., Chan, C. K., Gibson, Q. D., Zhou, J. D., Liu, Z., Watanabe, K., Taniguchi, T., Lin, H., Cava, R. J., Fu, L., Gedik, N., Jarillo-Herrero, P. 2018. Electrically switchable Berry curvature dipole in the monolayer topological insulator WTe_2. *Nat. Phys*. 14(9): 900–906.

[159] Ji, Z. R., Liu, G. R., Addison, Z., Liu, W. J., Yu, P., Gao, H., Liu, Z., Rappe, A. M., Kane, C. L., Mele, E. J., Agarwal, R. 2019. Spatially dispersive circular photogalvanic effect in a Weyl semimetal. *Nat. Mater*. 18(9): 955–962.

[160] de Juan, F., Grushin, A. G., Morimoto, T., Moore, J. E. 2017. Quantized circular photo-galvanic effect in Weyl semimetals. *Nat. Commun*. 8(1): 15995.

[161] Ma, J., Gu, Q., Liu, Y., Lai, J., Yu, P., Zhuo, X., Liu, Z., Chen, J.-H., Feng, J., Sun, D. 2019. Nonlinear photoresponse of type-II Weyl semimetals. *Nat. Mater*. 18(5): 476–481.

9 Phase Change Materials

Qingyang Du

CONTENTS

9.1   Introduction 239
9.2   Fundamentals of Phase Change Materials 240
       9.2.1   Materials Properties 240
       9.2.2   Resonant Bonding..242
       9.2.3   Material Engineering 243
       9.2.4   Characterization Method 244
9.3   Phase Change Kinetics 245
       9.3.1   Kinetics Analysis 245
       9.3.2   Switching Method..247
9.4   Applications of Phase Change Materials 247
       9.4.1   Color Pixel, Display...247
       9.4.2   Reconfigurable Metamaterials and Metasurfaces 249
       9.4.3   TAP Coupler and Optical Switch...251
       9.4.4   Reconfigurable Photonic Network for On-Chip Computing 252
       9.4.5   Non-Volatile Storage Memory...254
       9.4.6   Volatile Phase Change Materials 255
9.5   Outlook 255
Acknowledgements...256
Bibliography 256

9.1 INTRODUCTION

Phase change materials (PCMs) are a family of materials that not only undergo phase change between crystalline and amorphous states easily, but also exhibit distinct optical property modifications. A PCM should meet the following criteria: (1) large optical contrast between the two states; (2) fast phase transition within sub-millisecond; (3) relatively low switching temperature; (4) non-volatility, that means both phases are stable over a long period of time without applying any external stimuli; and (5) endurance for reversible switching [1]. We want to emphasize here that even though some research papers also recognize VO_2 as PCMs for the reason that they show large refractive index change at elevated temperatures, yet such phase change is volatile and it returns to the original state when it cools down [2]. Therefore, in this chapter, we mainly refer chalcogenide glass based non-volatile material family as PCMs, while the volatile phase change materials such as VO_2 will be briefly discussed.

DOI: 10.1201/9781003202608-9

The industry has long been benefited from the large contrast in their optical constants (n and k, which denote the real and imaginary parts of material's refractive index) for PCMs to make various devices. From conventional phase change random access memory (PCRAM), such as CDs and DVDs in the 2000s [3], to the current hot spot of on-chip non-volatile memories, displays and active plasmonic and metasurfaces [4]. Here, in this chapter, we first focus on those materials' basics and their switching kinetics. Then we briefly introduce current applications and developments of PCMs.

9.2 FUNDAMENTALS OF PHASE CHANGE MATERIALS

9.2.1 MATERIALS PROPERTIES

PCMs have striking unique properties that in other material systems yet to be found. They exhibit significant changes in their electrical and optical properties upon crystallization. Surprisingly, this crystallization proceeds in an unconventional fast way. Detailed calculations have indicated that it is not possible for both metals and ionic insulators to provide such a large optical contrast between the amorphous and crystalline states [5], and thus semiconductors are the required form of PCMs. So far, the most widely studied PCMs are the Ge-Sb-Te (GST) ternary chalcogenide alloys. A typical phase diagram of this ternary phase is displayed in Figure 9.1(a). Depending on the bonding characteristics (discussed in 9.2.2), not all Ge-Sb-Te alloys are PCMs. Those portions exhibiting phase change properties are highlighted in the shaded region in Figure 9.1(a) [6].

Structurally, studies have found that PCMs in the crystalline states generally have a trigonal symmetry (space group R_3m) [7,8], in the form of hexagonal, cubic, or rhombohedral structure. Those compounds show an octahedral-like atomic

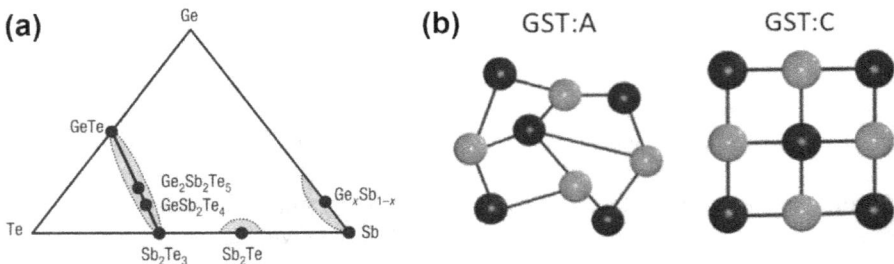

(a) Ge, GeTe, $Ge_2Sb_2Te_5$, $GeSb_2Te_4$, Ge_xSb_{1-x}, Te, Sb_2Te_3, Sb_2Te, Sb

(b) GST:A   GST:C

FIGURE 9.1 (a) Ternary phase diagram of Ge-Sb-Te. Phase change materials' compositions are indicated in the shaded region.

Source: Reprinted with permission from Lencer, D. et al. *Nat. Mater.*, 7, 972–977, 2008. Copyright 2008 Nature Publishing Group

(b) a schematic drawing of the bonding in amorphous and crystalline GST.

Source: Reprinted with permission from Rios, C. et al. *Nat. Photonics*, 9, 725–732, 2015. Copyright 2015 Nature Publishing Group.

arrangement, yet the vacancy concentration in a unit cell is significantly high [9]. For example, $GeSb_2Te_4$ has a 25% vacancy concentration for the cation sites while all the anion sites are occupied by Te. This unusually high vacancy concentration allows for resonant bonding that give rise to large property changes in PCMs. In the amorphous state, according to the covalent network theory [10], though the atoms lack long-range order, their local arrangement resembles that in the crystalline state, which differs mostly only in bond angles. Later, Kohara et al. conducted an X-ray diffraction study on the amorphous $Ge_2Sb_2Te_5$ and confirmed that its local structures were mostly inherited from the crystal state, such as the heteropolar bond and 90-degree bond angles, but the total pair correlation function significant differs with each other [11]. A 2D drawing of the structures of the crystalline and amorphous structure are illustrated in Figure 9.1(b) [12].

The optical properties of PCMs from 0 eV–2.5 eV are displayed in Figure 9.2 [4]. Large contrasts in both n and k are observed between the crystalline and amorphous states. The bandgaps of those materials coincide close to 0.6 eV, which is found to be a prerequisite for resonant bonding. At high energy regime, the photon energies are above the bandgap, electronic transitions are primarily dominated by inter-band

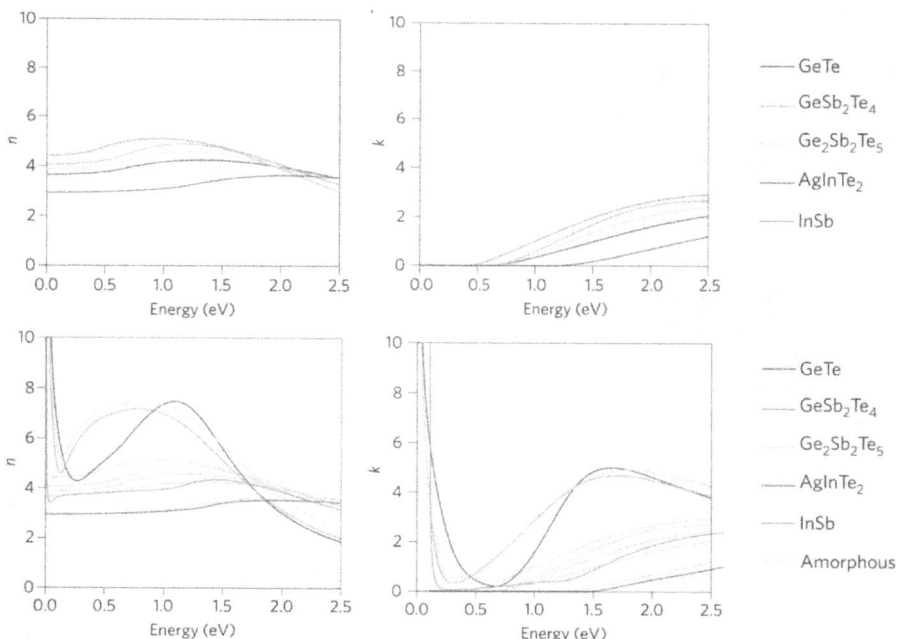

FIGURE 9.2   The comparison of the refractive index n and absorption coefficient k of various phase change materials in the crystalline (lower half) and the amorphous states (upper half).

Source: Adapted with permission from Wuttig, M. et al. *Nat. Photonics*, 11, 465–476, 2017. Copyright 2017 Nature Publishing Group.

transitions, the large n contrast are owing to the different bonding states between two phases. When photon energy decreases well below the bandgap, free carrier absorption and carrier-induced refractive index change shape the huge increases in both n and k in the crystalline state.

Similarly, electric conductance also exhibits significant enhancement upon crystallization. Under the amorphous state, lack of long-range order creates large scattering centers for electrons, prohibiting them from transporting. Whereas in the crystalline state, the relatively small bandgap allows certain distributions of free electrons in the conduction band. The free movements of those electrons are further enhanced by the resonant bonding, which together contribute to the huge increase in the electric conductance.

9.2.2 Resonant Bonding

Unlike oxide glasses, whose dielectric constants differ very little when undergo phase transitions from amorphous to crystalline state, the surprisingly huge contrasts of those in PCMs have caught increasing attentions to researchers. It is found that this unconventional phenomenon is primarily owing to the *resonant bonding* which is unique to PCMs [13]. This term, resonant bonding, describes the exotic condition that the covalent bonds formed in PCMs have an average of less than two valence electrons per bond. The overall interpretation is thus equivalent to several saturated valence bonds (two electrons per bond) "resonantly" co-exist between all possible bond sites [14]. A study by Luo et al. has pointed out that the number of valence electron in the s- and p-orbitals (N_{sp}) are above 4 for PCMs [15]. Semiconductors, like Si or GaAs, whose N_{sp} is strictly equal to 4, favors a sp^3 hybridization, forming a tetrahedrally bonded structure. However, for PCMs, whose N_{sp} is greater than 4, the anti-bonding state would be occupied, promoting a less stable bond for hybridization. Hence, in PCMs, the bonding is mainly contributed from the p-electrons, creating an octahedral bonding structure which is in good agreement of experimental observations. In addition, it is also found that most PCMs feature an average of 3 valence electrons per site, favoring p-orbital bonding rather than hybridization.

In terms of the chemical bonding language, quantitative measures, the "degree of ionicity" and the "degree of hybridization" are usually used to describe a chemical bond. The former one compares the difference in the electronegativity of the two bonding atoms. The latter, takes into account the s-electrons contribution to a certain bonding by comparing the valence radii of s- and p-orbitals. Compounds with large degrees of ionicity could have maintained the octahedral "rock-salt" like structures. However, the ionic nature makes the compounds large bandgaps which no longer fulfill the PCM requirement of being semiconductors. Likewise, increasing degree of hybridization leads to distortion of the chemical bonds and thus the resonant bonding diminishes. Therefore, it is not difficult to conclude that PCMs must comprise of atoms that forms bonds with both small degrees of ionicity and hybridization. Figure 9.3 compares the tendency towards ionicity and hybridization for common oxide and chalcogenide glasses [6]. It is quite indicative that all PCMs are located in the corner with small ionicity and hybridization.

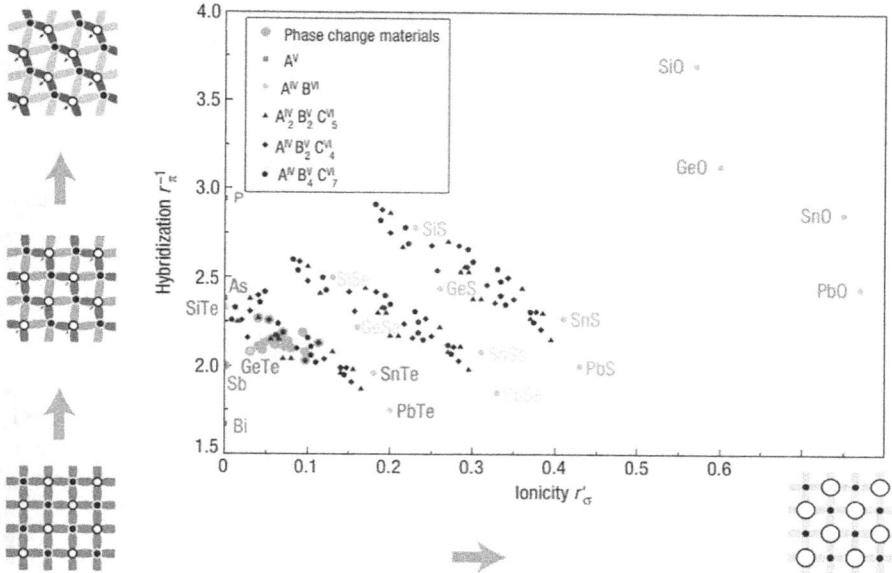

FIGURE 9.3 A map for degree of iconicity and hybridization of various materials. The location of PCMs family is indicated with green dots.

Source: Reprinted with permission from Lencer, D. et al. *Nat. Mater.*, 7, 972–977, 2008. Copyright 2008 Nature Publishing Group.

9.2.3 MATERIAL ENGINEERING

As discussed above, there are more than one groups of Ge-Sb-Te chacolgenide compounds that meet the resonant bonding criteria and thus have PCM properties. The largest one lies on the line between GeTe and Sb_2Te_3 phases in the ternary phase diagram. Replacing one type of atom with another introduces shifting of certain material properties. Based on applications, we could engineer the PCM to best match the requirement. For this family, replacing Ge with Sb atom shrinks the bandgap and increases electron polarizability hence yielding higher n, but in the meantime, it promotes bonding distortions that is not in favor of resonant bonding and thus the optical contrast between the two phases is reduced [16]. Another important parameter is optical loss. For most photonic applications, large optical loss is detrimental to devices. The loss in the PCM family mainly come from the free carrier absorption in the low photon energy regime and the interband transition for energies above the bandgap, which fortunately leaves behind a "transparency window" in between. This is the region where researchers are mostly interested in. Substituting Ge with Sb atom in PCMs allows tuning on this specific window. For example, Michel et al. discovered that $Ge_3Sb_2Te_6$ has minimal losses over the mid-IR and therefore suitable for applications in mid-IR active antenna [17].

Recently, it is discovered that replacing Te with Se could further enhance the amorphous state stability as Se is a better glass former than Te. Zhang et al. performed a systematic study of Se substitution of $Ge_2Sb_2Te_5$ PCM [18]. Their findings suggested that with the increasing concentration of Se, the bandgap widens, reducing the optical absorption in the near-IR range. The crystallization temperature also progressively increases, indicating a more stable amorphous phase. It is argued that $Ge_2Sb_2Se_4Te$ stands out as the best candidate for near-IR photonic applications as it not only inherited the resonant bonding character from GST, featuring a large Δn of 2.1, the large blue-shifted bandgap also guaranteed minimal optical losses in the near-IR, making it ideal for non-volatile photonic applications.

9.2.4 CHARACTERIZATION METHOD

X-ray Diffraction (XRD) XRD characterizes the structure of PCMs. The most obvious contrast of the two phases is that the crystalline phase shows distinct sharp peaks, which could be used to identify lattice spacing and symmetry group by comparing with standard data base. While the amorphous phase usually exhibits broad peaks. Detailed study of the XRD data on the amorphous phase could also yield bonding information in GeTe and $Ge_2Sb_2Te_5$ as suggested by Kohara [11].

Raman spectra Raman study of materials provides the chemical environment of certain bond. It characterizes the polarizability of its bonding configurations. As shown in the work of Zhang et al., a large shift of Raman peak is observed from 160 cm^{-1} to 120 cm^{-1} when $Ge_2Sb_2Se_4Te$ crystallizes [18]. It is worth noting that by combining Raman with the confocal apparatus, the examining area could go down to micron size. This approach offers a facile way to quickly determine the PCM phase at small scale.

Extended X-ray Absorption Fine Structure (EXAFS) Similar to XRD and neutron diffraction, EXAFS has been widely used in determining local structure of materials. The binding energy of a certain chemical bond is acquired when it matches the X-ray photon energy. Data analysis is usually combined with a Reverse Monte Carlo simulation to reconstruct the EXAFS spectrum. Though the tasks are formidable, studies with EXAFS have found that an octahedral bonding in the crystalline phase was transitioning to a tetrahedral bonding in the amorphous phase in GST [19–21].

High Resolution Transmission Electron Microscope (HRTEM) The best advantage of an HRTEM is that it gives direct observation of the crystal structure in real time, allowing for an in-situ characterization of samples. Customized TEM sample holder equipped with external bias source could conduct in-situ heating of the sample. The crystallization process is therefore observable under TEM imaging together with further confirmation from electron diffraction pattern. This method is extremely evidencing in characterizing a new unknown material as is presented in Ref. [18].

Ellipsometry The optical properties of PCM materials, *i.e.*, n and k, could be acquired from an ellipsometry scan of a thin film. This process also requires a numerical optical oscillator model to fit the raw data, however this task is much less challenging compared to EXAFS. A good model fit could accurately determine the refractive index n to the order of 0.01. Nevertheless, it is worthy pointing out that the ellipsometry is less sensitive to small k which should instead be measured with the waveguide loss or prism coupling for enhanced accuracy.

Hall effect Carrier type and concentration in a material largely determines its electrical properties. These two parameters could be readily extracted from the Hall effect measurement. Van der Pauw proposed a simplified method that leverages a 4-point probe placed on the perimeter on the sample and averages the hall voltage from each direction [22]. The voltage sign (+ or -) yields the carrier type while its value is used to derive the material's sheet resistance and carrier concentration.

9.3 PHASE CHANGE KINETICS

9.3.1 KINETICS ANALYSIS

The crystallization process of a material comprises of the nucleation and the growth steps. In the former process, the kinetic driving force is the Gibbs free energy difference between the crystalline and amorphous phases. In most practical cases, the nucleation of PCMs happens from a substrate, and therefore, it is the heterogeneous nucleation process that dominates. According to the classic thermodynamic theory, the total Gibbs energy for homogenous nucleation in the system is equal to the sum of the ΔG between the two phases and the energy needed to overcome the formation of a phase boundary surface:

$$\Delta G_{total} = -\Delta G_{ac} \times \frac{4}{3}\pi r^3 + \sigma_{ac} \times 4\pi r^2 \tag{9.1}$$

where ΔG_{ac} and σ_{ac} are the Gibbs free energy difference and the surface energy per unit respectively, r denotes the nuclei's radius. For the heterogeneous nucleation, a cap is formed on a substrate instead of a whole sphere. Consequently, the ΔG for the heterogeneous nucleation is reduced by a factor of $f(\theta)$ compared to the homogenous nucleation. The $f(\theta)$ is straightforward to acquire from the geometric relations. A schematic drawing of the heterogeneous nucleation is provided in Figure 9.4(a).

$$f(\theta) = \frac{(2+\cos\theta)(1-\cos\theta)^2}{4} \tag{9.2}$$

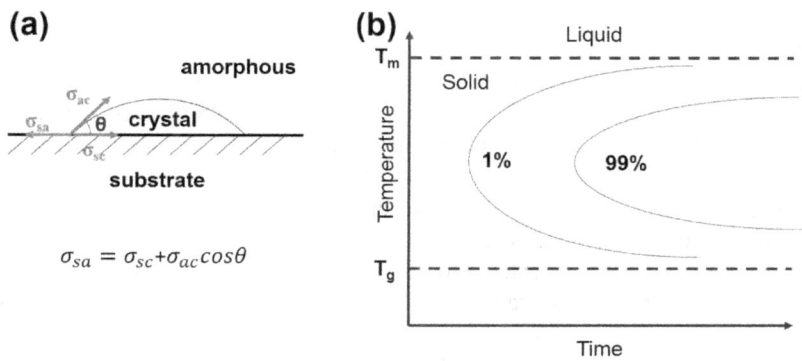

FIGURE 9.4 (a) An illustration of heterogeneous nucleation geometry; and (b) a typical TTT diagram of a material.

The second process is the nuclei growth. After the nuclei formed, the amorphous-crystal boundary gradually expands and the nuclei grows into bigger grains. In the classic model, the growth is described as a diffusion-reaction process. Target molecules diffuses into the vicinity of the grain boundary and then adsorb onto the surface, forming a chemical bond. The overall growth rate can be qualitatively expressed as:

$$\frac{dN}{dt} \propto \exp(\frac{-G_{diff}}{kT})\exp(\frac{-G_{nuc}}{kT}) \qquad (9.3)$$

where $\frac{dN}{dt}$ refers to the crystal growth rate, G_{diff} is the energy barrier for diffusion and G_{nuc} is the Gibbs energy for nucleation, which is negative when the temperature is below the melting temperature ($T < T_m$). This expression has provided the common kinetic characteristics of the growth process. When T is close to T_m, the diffusivities of atoms are high, but the marginal kinetic driving forces make most atoms perform a "touch and go landing" manner, limiting the growth speed. On the other hand, at low temperature, $T \ll T_m$, the large G_{nuc} drives the system towards crystallization, however, the low diffusivities make atoms frozen at their own locations, prohibiting crystals to grow. This growth character is usually presented by a Time-Temperature-Transformation, or TTT, diagram, as shown in Figure 9.4(b). In a TTT diagram, for $T > T_m$, the material is in a melt liquid state. Similarly, for temperature below the glass transition temperature ($T < T_g$), the atoms are not energetic enough to overcome the diffusion energy barrier, thus the material stays in the amorphous state. In between is the region where crystallization happens. The two parabolas indicate 1% crystallization and 99% crystallization, respectively. This TTT diagram provides crucial information about the switching thermal requirement of PCMs. To amorphize PCMs, the materials needs to go above its melting temperature, followed by a fast quenching step. On the contrary, to crystallize, the temperature should be kept at $T_g < T < T_m$ for long enough time to promote the crystal growth. The time and temperature together determine the degree of crystallization. The highest crystallization rate could be numerically found by balancing the two counter-acting driving forces in Eq. (9.3) or by drawing a tangential line from T_m to the desired degree of crystallization.

In order to leverage PCMs' phase change advantages into real applications, one key feature for PCM is that it should stay in the amorphous state at room temperature for more than 10 years long (data reliability), and should recrystallize below micro-second at elevated temperature (fast data writing). Studies has found that for most PCMs, the ratio between T_g and T_m ranges between 0.45 to 0.55 [23]. It clearly indicated that PCMs are made of poor glass former hence they recrystallize easily. In addition, strong evidence has suggested that the interfacial energy σ_{ac} between the amorphous and crystalline phases are rather low [24,25], reducing the required critical radius to form a nucleus. This enhanced kinetic driving force makes PCMs crystallize more easily and faster. Moreover, it is generally found that the recrystallization from a re-amorphized film is significantly faster than the as-deposited amorphous film [26,27]. Researchers have evidenced that subcritical nuclei exist even inside the melt-quenched amorphous thin film. This allows the recrystallization process avoid the slow nucleation step, and directly grow from the pre-existing nucleus that enable the unusually fast recrystallization behavior.

9.3.2 Switching Method

Switching of PCMs is purely a thermodynamic process which solely relies on the temperature profile. To realize reversible pixel-size switching on a large chip, two most commonly used methods are developed: opto-thermal switching and electro-thermal switching.

In opto-thermal switching, an external pulsed laser radiates on the PCM, which absorbs the light and generates intensive heat that rises its temperature. As discussed above, to recrystallize PCM, an elevated temperature between T_m and T_g is required and it is essential to hold at this temperature until crystallization completes. While for amorphization, a fast "melt-quench" method is leveraged to freeze the atoms at the amorphous state. Therefore, for opto-thermal switching, researchers utilize a pulse train at a mediated power level to prevent the temperature going above T_m and guarantees sufficient crystal growth. On the contrary, a high power short single pulse is deployed to transit the PCM back to its amorphous state [18].

Electro-thermal switching, on the other hand, utilizes an integrated heater to generate Joule heat when the electric current is flowed. Note that though the direct use of PCM itself as the electro-thermal heater is possible, researchers have found that a "filamentation" phenomena happens when the PCM size becomes bigger. A thin layer of PCMs crystallizes first and becomes significantly more conductive than its surrounding amorphous regions, preventing the continued generation of Joule heating [4]. With the integration of micro-heaters, the temperature profile of electro-thermal switching does not depend on the PCM state, which provides better reliability and temperature uniformity. Therefore, an integrated micro-heater is becoming a standard requirement for switching PCMs in photonic applications. Those micro-heaters could be made from metals [28], ITOs [29,30], doped silicons [31,32] and graphenes [33]. In general, micro-heaters should be placed close enough to the PCM to provide efficient heating but not too close to generate unwanted optical absorption losses. They also need to withstand the high-temperature and maintain their physical properties after numerous switching events. Details of an examplary heater design and its thermal profile could be found in Ref. [33] which will not be discussed in detail here.

9.4 APPLICATIONS OF PHASE CHANGE MATERIALS

9.4.1 Color Pixel, Display

The large optical contrast between the amorphous state and crystalline state in PCMs has attracted increasing attentions for researchers in the display field. The change in the refractive index induces a shift in the PCM film's reflectance spectrum thus changing its natural color. P. Hosseini et al. successfully demonstrated the first rewritable color pixel in 2014 [34]. They deployed an electrically induced phase change configuration by sandwiching a thin GST film between two transparent conductive ITO electrode layers, as shown in Figure 9.5(a). For pixel display application, a micro-manipulator scanned the top surface of the film and applied a voltage to those pixels to switch their phases. They have observed color changes in bulk thin film samples from 50 nm to 180 nm and demonstrated reflective display thin films both on a rigid and flexible substrate (Figure 9.5(b)). It is worth noting that

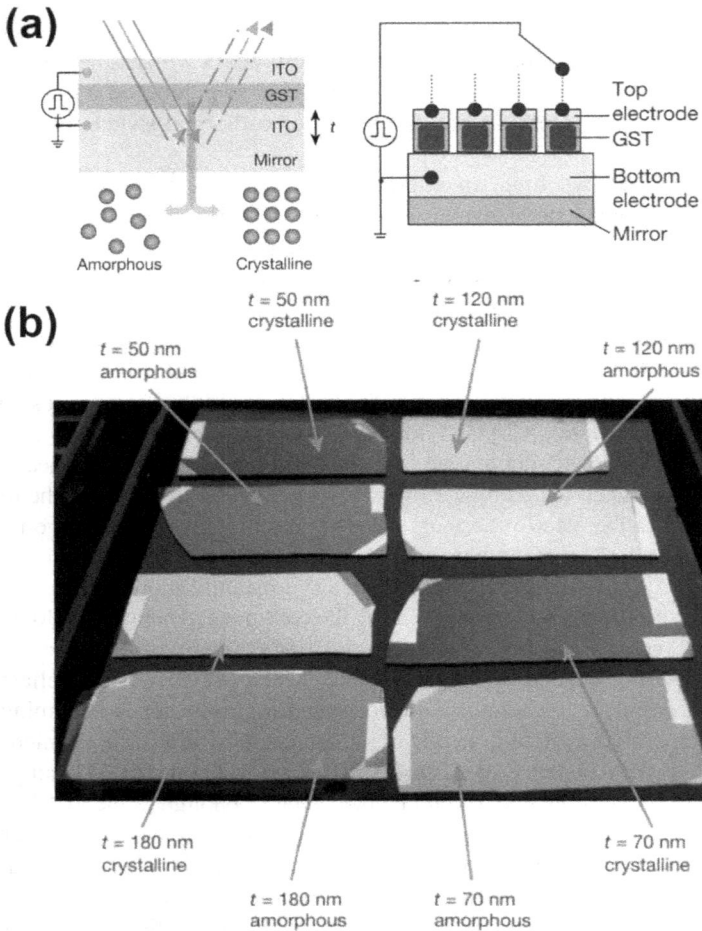

FIGURE 9.5 (a) Cross-sectional structure of electrically tunable display pixels; and (b) the illustration of different color realized by PCM in different state under various thickness.

Source: Adapted with permission from Hosseini, P. et al. *Nature*, 511, 206–211, 2014. Copyright 2014 Nature Publishing Group.

this work directly utilized the GST material itself as the heat generating element. As discussed in Section 9.3.5, "filamentation" phenomena tends to happen for thick and large films and subsequently promotes incomplete crystallizations. Luckily, for color pixel application, this phenomenon is effectively avoided as the pixel sizes are usually small. Another work instead uses an ultra-short high-power light pulse to induce heating in PCM [35]. Reversible switching was also demonstrated. In this work, GST PCM was sandwiched between layers of dielectric thin film stacks to enhance the angular robustness of the displayed color. The use of optical switching bypasses the "filamentation" effect and thus is suitable for large area applications.

9.4.2 Reconfigurable Metamaterials and Metasurfaces

As discussed in Chapter 5, metamaterials or metasurfaces refer to a specific group of artificial structures that produce unconventional optical properties. According to the generalized Snell's law, the reflection and transmission at a surface depends not only on the bulk material's refractive index of the incident and transmitting media, but also a new degree of freedom that leverages local surface topology which introduces a phase gradient of $[0, 2\pi]$ along the surface. This wide coverage of the entire phase map enables convenient and precise phase control of the electromagnetic wave, contributing to emerging light manipulating technologies such as LiDAR, zoom lens, phased array antennas etc.

Active control of metasurfaces has always been a hotspot in the current free space optics community. PCMs are widely recognized as a promising material to realized dynamic tuning of metasurfaces. B. Gholipour et al. first reported an all optical non-volatile switch based on PCM in 2013 [36]. A GST thin film was sandwiched between two ZnS/SiO$_2$ layers to prevent Au diffusion during high temperature and air degradation to the PCM. This entire film layer was places on a gold metamaterial plate. The switch ON/OFF was controlled by a high power single pulsed light at 660 nm to induce phase change. In this work, a maximum contrast ratio of 4 is achieved. However, degradation was observed after switching 50 cycles.

A few years later, Q. Wang et al. reported a multi-level reconfigurable metasurfaces based on PCMs [37]. The multi-level switching was achieved by carefully control the laser energy and the number of pulses to realize intermediate partial crystallization states. Using this new technology, several optical devices were successfully demonstrated. Interestingly, in this work, PCMs were the only material component for such devices. Their optical functionalities were brought by the large refractive index contrast between the amorphous and crystalline states of PCMs. The devices included but not limited to chromatically selective and chromatically corrected lens. Figures 9.6(a) to 9.6(c) provides the schematics of the switching apparatus, a GST film pattern to realize such lens as well as its experimental result, respectively. The lens achieved a focal spot separation of 17.8 μm as well as a

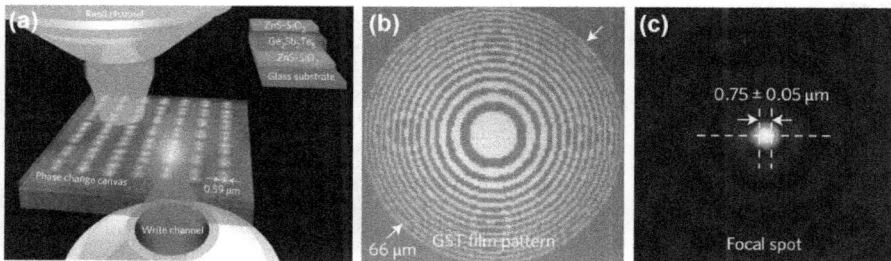

FIGURE 9.6   (a) Read-write apparatus of pattern generation; (b) Fresnel zone plate pattern of GST; and (c) experimental result of the focal spot.

Source: Reprinted with permission from Wang, Q. et al. *Nat. Photonics*, 10, 60–66, 2016. Copyright 2016 Nature Publishing Group.

coincidence for 730 nm and 900 nm light at a focal distance of 50 μm. In addition, reconfigurable Fresnel zone plates were also validated. The pattern was first written with 0.39 nJ pulse train, and then erased with 1.25 nJ single pulse to re-amorphize PCM. These regions were proven to be rewritable with 0.39 nJ pulse train again. This write-erase-write configuration enabled dynamic configuring of any surfaces with pixel by pixel accuracy.

More recently, Y. Zhang et al. reported the first experimental realization of a fully packaged electrical multi-level reconfigurable metasurface device [38]. A 3D drawing of such device as well as a photo of the bonded device is provided in Figures 9.7(a) and 9.7(b) respectively. In this work, 250 nm thick $Ge_2Sb_2Se_4Te$ (GSST) PCM meta-atoms were directly patterned on top of the doped silicon electro-thermal heater. The heater structure was carefully designed with a curved heater boundary to lower the required current density and improve temperature uniformity along the metasurface aperture region. Direct use of doped silicon as the heater also ensured high transmission (compared with metal) in the IR and fast quenching of PCMs. Gold metal contact was placed by the side of metasurface aperture and was wire-bonded to a printed circuit board to allow facile external control. Electric switching of PCMs were realized by applying a square wave voltage pulse. A 500 ms long, < 12 V low voltage pulse was applied to induce crystallization, while a 5 μs short, 20–23 V high voltage triggered amorphization. To achieve multi-level switching, the crystallization pulse voltage was set to vary between 10.5 V to 11.3 V with an increment step of 0.1 V that yielded partial intermediate crystallization. Leveraging this technology, ultra-broadband tuning from 1190 nm to 1680 nm of metasurface optical resonance was successfully demonstrated (Figure 9.7(c)). Apart from amplitude modulation, phase modulation through PCM metasurfaces were also validated. A phase-gradient Huygens' surface beam deflector was illustrated. In the OFF (amorphous) state, most light was directly reflected, while in the ON (crystalline) state, 24.8% incident light was deflected to 32°. This technology has provided a practical solution to facile control of both amplitudes and phases in active metasurfaces.

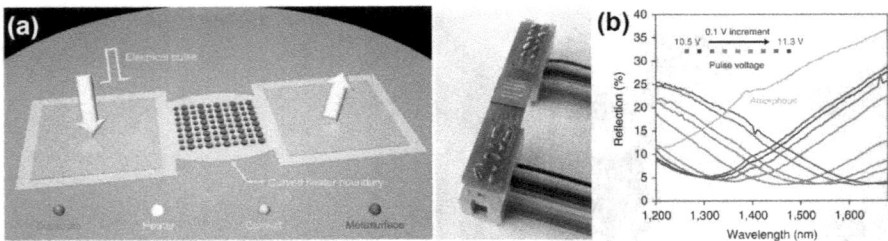

FIGURE 9.7 (a) Illustration of an electrically reconfigurable metasurface; (b) an image of the wire bonded device; and (c) experimental result of a spectrally continuously tunable reflector metasurface.

Source: Reprinted with permission from Zhang, Y. et al. *Nat. Nanotechnol.*, 16, 661–666, 2021. Copyright 2021 Nature Publishing Group.

9.4.3 TAP COUPLER AND OPTICAL SWITCH

The unique properties of PCM not only granted it with huge application potentials in free space optics, a large number of works with PCM in on-chip photonics also have been witnessed within the past decade. In this section, we will discuss the applications of TAP couplers and optical switches. A TAP coupler is a waveguide directional coupler device that couples small amount of light out from the bus waveguide for wafer-level testing, post-fabrication trimming and early stage error diagnosis. However, such device would permanently contribute to the total insertion loss of the entire system. Y. Zhang et al. have theoretically analyzed and experimentally demonstrated a transient TAP coupler that utilizes PCM [39]. In their design, PCM was carefully engineered to be placed right on top of the TAP waveguide, as drawn in Figure 9.8(a). The TAP waveguide geometry was also designed to have a similar waveguide effective index to the bus waveguide when the PCM was in the amorphous state. The operation principle of such TAP coupler was under a non-perturbative system. When PCM was in the amorphous state, the TAP waveguide met the phase matching condition and therefore coupled light out from the bus waveguide. On the contrary, when PCM crystallized, the huge refractive index difference yielded a large mismatch between the bus and TAP waveguide modes. Consequently, all light in the bus waveguide got directly transmitted, leaving behind minimal excess loss. In this

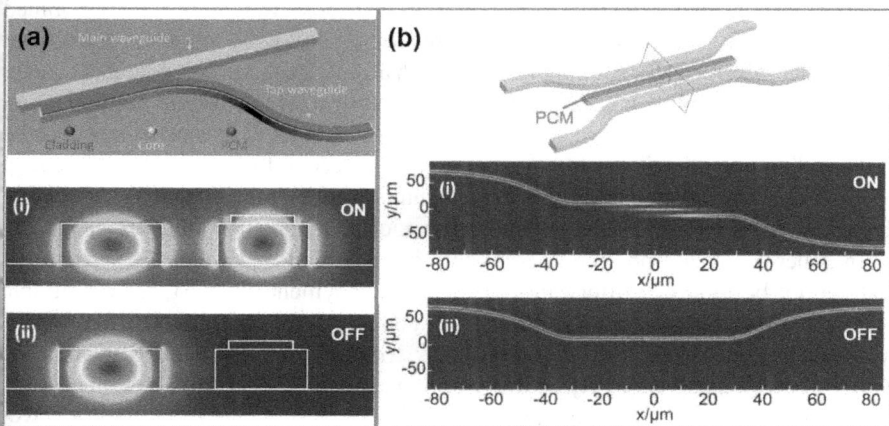

FIGURE 9.8 (a) The structure of TAP; subfigure (i) ON state, light couples to the TAP waveguide and (ii) OFF state, light is prohibited in the TAP waveguide due to large effective index contrast.

Source: Reprinted with permission from Zhang, Y. et al. *ACS Photonics*, 8, 1903–1908, 2021. Copyright 2021 American Chemical Society.

(b) the structure of an optical switch; subfigure (i) ON state, light couples to the other waveguide and (ii) OFF state, light stays in the through waveguide.

Source: Reprinted with permission from Zhang, Q. et al. *Opt. Lett.*, 43, 94–97, 2018. Copyright 2018 The Optical Society.

work, GSST PCM was chosen as it has a much lower absorption coefficient compared with GST in the near IR region for the amorphous state. Based on this design, they have successfully demonstrated a record low insertion loss of 0.01 dB on a SOI and SiN platform.

Similarly, an optical switch could be built with the same configuration. A schematic drawing is shown in Figure 9.8(b). A coupling waveguide was inserted between the two arms of a 3-dB coupler. In the "on" state, the PCM was amorphous and light coupled through this coupling waveguide to the other arm, while in the "off" state, the crystallized PCM hindered any coupling and thus the light stayed in the same arm. Numerical simulations with FDTD predicted such switch had a insertion loss of 0.32 dB and a crosstalk level of -32 dB, a significant improvement from the traditional Mach-Zender Interferometer (MZI) design [40].

9.4.4 RECONFIGURABLE PHOTONIC NETWORK FOR ON-CHIP COMPUTING

With the fast development of artificial intelligence and machine learning, the demand of fast computing and low energy consumption surges beyond electronics' limit. Most researchers are now focusing on a parallel and scalable photonic tensor core as the most promising candidate to address this challenge. A photonic tensor core takes photons, instead of electrons that generate excess heat during propagation, as the information carrier media. It mainly consists of photonic waveguide that delivers light and a weighting unit that modulates light intensity. All those weighting units combined together to form a weighting matrix, or more specifically in the language of artificial intelligence, a kernel. When light passes through a kernel, the incoming light is distributed, modulated, and then added back, which is equivalent to a matrix multiplication operation. With layers of kernels cascade together, a deep neural network could be built, realizing applications of computation, image recognition and natural language processing. Here we focus on the implementation of PCMs in such systems. The details of neural networks can be found in the relevant text books or literatures [41,42]

In general, when a photonic kernel chip is fabricated, the weight is determined and cannot be modified afterwards. With the deployment of PCM, reconfigurable photonic kernel emerges as the cutting-edge technology that enables deep neural networks on just one single chip. C. Wu et al. has demonstrated a photonic convolutional neural network using GST [41]. In this work, instead of using the absorption of GST, they took advantage of the huge refractive index contrast of GST and built a waveguide mode converter between TE_0 to TE_1, as shown in Figure 9.9(a). The weight, which is controlled by partial crystallization of GST, was added by converting certain percentage of light to the TE_1 mode. Using this configuration, the kernel matrix was successfully applied to edge detection and image recognition. Later this year, J. Feldmann demonstrated the first parallel on-chip convolutional processing [42]. They pioneered the use of comb teeth and the integration of comb into a photonic kernel. The kernel, as indicated in Figure 9.9(b), applied a GST on SiN waveguide configuration, which encoded the data by modulating the intensity of each comb teeth. The demonstration of such photonic tensor core enabled matrix vector multiplication that give rise to several applications such as image edge detection and digit recognition.

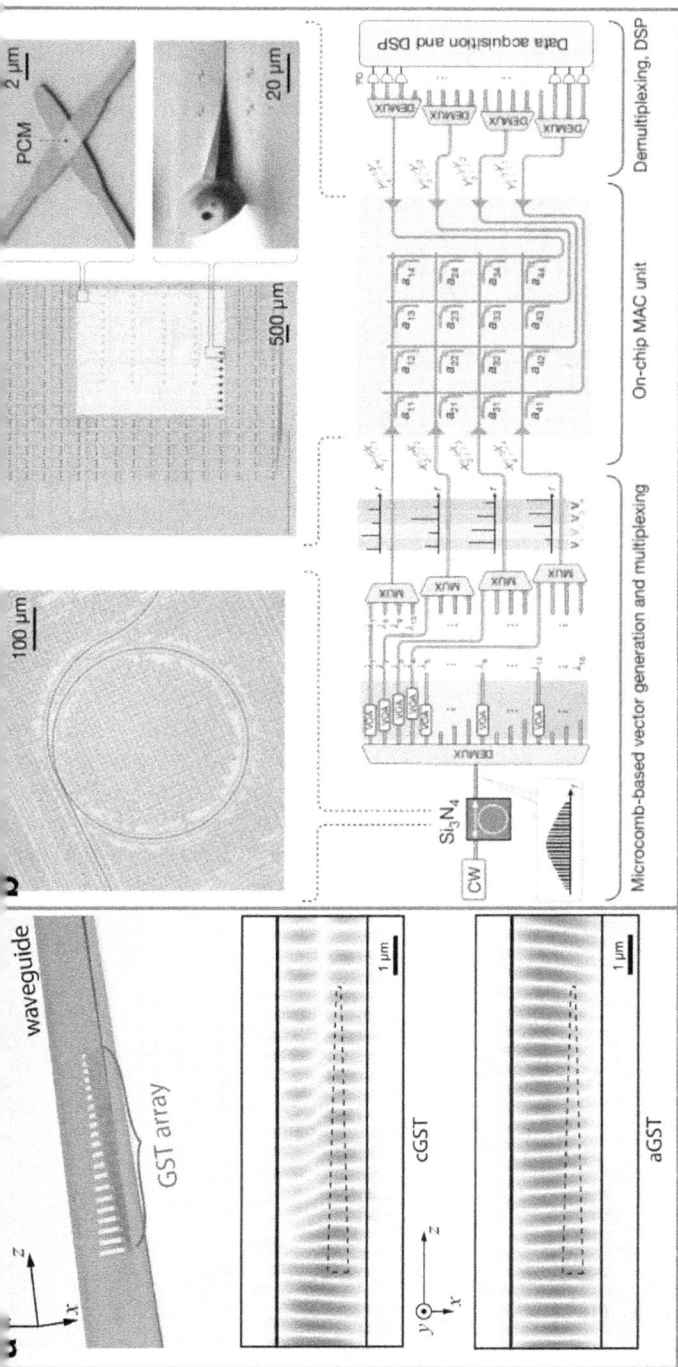

FIGURE 9.9 (a) Illustration of a GST based reconfigurable mode converter waveguide. When GST is in the crystalline state, it converts the mode from TE_0 to TE_1.

Source: Reprinted with from Wu, C. et al. *Nat. Commun.,* 12, 96, 2021. No permission is required. Copyright 2021 The authors.

(b) An integrated photonic tensor core. It comprises of a light source that passes through a high-Q ring resonator that generate frequency comb lines. These comb lines are sorted and combined by a multiplexing unit and directed to the matrix-vector-multiplication unit where the matrix elements are represented by the crystallinity of PCMs. Finally, the light is collected to yield the results.

Source: Reprinted with permission from Feldmann, J. et al. *Nature,* 589, 52–58, 2021. Copyright 2021 Nature Publishing Group.

9.4.5   Non-Volatile Storage Memory

Last but not least, PCMs are also extremely useful in photonic non-volatile storage memories. The crystalline and amorphous states in PCMs, which represent the bits 0 and 1 respectively, are extremely relevant for applications in on-chip memories. Though the idea seems quite straight forward, it requires a lot of efforts to success-fully combine PCMs to on-chip photonics. In 2014, C. Rios et al. has first demon-strated a GST on silicon nitride photonic memory platform [43]. In this work, they deposited a GST thin film and an ITO protective capping layer on top of a small section of a silicon nitride ring resonator. In the amorphous state, near-critical coupling condition was met and a resonance dip in the transmission spectrum was observed. While in the crystalline state, the large absorption changed the coupling condition, resulting in an all-pass transmission. However, it is worth noting that the phase change here was induced by heating on a hot-plate and this work is more a proof-of-concept which is still far from practical application. Later next year, the same team took a big step forward and demonstrated an all-photonic multi-level memory [29]. Instead of ring resonator, direct waveguide integration of PCM was implemented (Figure 9.10). This example directly took advantage of the absorption at different phase states of GST. The switching of GST, which writes and erases of the memory bit, was realized by transmitting a control light pulse that directly induce heat inside GST. To read this memory bit, a much weaker probe light pulse was used. The readout high transmission and low transmission state determined the bit to be either 1 or 0. This work also tested the reversible switching ability. It was validated that the system exhibited no signs of degradation after 50 write/erase cycles. On top of that, the partial crystallization property of GST was also taken advantage to realize a multi-level switching, where the intermediate transmission states were reliably reproduced during writing and erasing to realize an 8-level memory element. The successful demonstration of this prototype allows for com-pact non-volatile on-chip photonic storage which could significantly advance the optical data storage field.

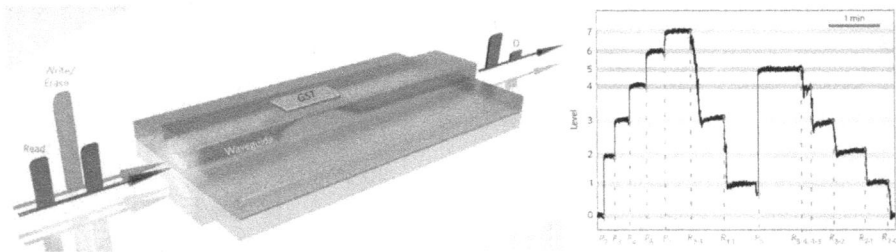

FIGURE 9.10   The structure of a waveguide integrated photonic memory unit (left) and the eight-level realization of erase and write in a single unit (right).

Source: Reprinted with permission from Rios, C. et al. *Nat. Photonics*, 9, 725–732, 2015. Copyright 2015 Nature Publishing Group.

9.4.6 VOLATILE PHASE CHANGE MATERIALS

In contrast to the phase change materials discussed above, there are another family of phase change materials, as represented by VO_2. Instead of going between crystalline-amorphous phase transitions, VO_2 exhibits metal-insulation transitions (MIT) where both phases are crystalline state. The room temperature state is monoclinic structured, optically transparent, and electrically resistive (VO_2:M), while at elevated temperature, it transits to a rutile structure, with high optical absorption and electric conductivity (VO_2:R) [46]. The phase transition is achieved by breaking the V-V dimer bond. During such phase change, the band structure of VO_2 shifts drastically, from a wide-bandgap material to a metal. The large optical absorption arises from the inter-band electronic transition for the metal state. The MIT does not involve crystallization kinetics and therefore it's much more easily accessible than chalcogenide glass-based phase change materials. Experiments has demonstrated that a transition temperature of around 68 °C for VO_2 [47], which is an order of magnitude lower than GST. This transition could be induced thermally [47], electrically [48] and optically [49]. However, it worth emphasizing that the MIT is volatile and the VO_2:R state will convert back to VO_2:M state regardless of the switching method.

The volatility of VO_2 excluded the possibility of applications in on-chip photonic memories. Yet, the more accessible temperature to induce phase transition endowed it with on-chip actively tunable switches and modulators. R.M. Briggs has integrated VO_2 with a silicon waveguide to build an on-chip photonic modulator [50]. In this work, a 2 µm long VO_2 patch was deposited on top of the silicon ring resonator, the on and off resonant state was realized by raising the temperature and cooling down. This work realized a 6.5 dB modulation depth and 2 dB insertion loss. Apart from thermally realized functionality, electrical actuation of such phase change is also demonstrated. Similar to GST, it also has been found switching larger area of VO_2 requires longer time, partial switching of VO_2 is also possible to realize applications for intermediate states [51]. P. Markov in 2015 demonstrated an electrically tunable VO_2 integrated silicon photonic switch. Their result indicated a record switching time less than 2 ns and inverse relaxation time of 3 ns. [51]

There is little doubt that VO_2 stand out as a significantly promising material for on-chip photonic modulators. The fast switching speed, low energy consumption compared to GST and small device footprint guaranteed VO_2 an indispensable role in PCM/Si hybrid on-chip modulator and switches. Though ultrafast modulator devices such as Gbps speed have yet to achieve, experimental results have evidenced such trend highly possible.

9.5 OUTLOOK

For most photonic applications of PCMs, a generic representation of the material's figure of merit (FOM) is expressed as:

$$FOM = \frac{\Delta n}{\Delta k} \qquad (9.4)$$

This equation decouples the contribution from the two optical constants and yields a universal rule of finding a "best" PCM, one that exhibits the largest refractive index change while maintains the minimal absorption loss. In addition to the material's FOM, the switching speed is another concern for PCM. It has been found that the critical cooling rate of GST is largely dependent on the thickness of the film. Though this challenge is addressable for on-chip photonics by making long waveguides, it still compromises the performances for most free-space applications where thicker films are required. Therefore, there is growing need to develop new PCMs or new switching methods for better performances [44]. Moreover, the endurances of those emerging low loss PCMs are still limited [45]. A failure proof PCM under numerous reversible switching cycles is an indispensable requirement for any practical application. Finally, the integration of PCM fabrication process into CMOS foundry lines represents the ability of scalable manufacturing that allows mass production and ensures high device quality. Though those challenges still lie ahead in the development of PCMs, there is little doubt that the PCM's unique properties could open the most potential solutions to non-volatile reconfigurable optical and photonic applications.

ACKNOWLEDGEMENTS

This work is supported by the Program for the Nanqiang Young Top Notch Talents of Xiamen University.

BIBLIOGRAPHY

[1] Wang, J.; Wang, L.; Liu, J., Overview of phase-change materials based photonic devices. *IEEE Access* 2020, 8, 121211–121245.

[2] Kats, M. A.; Sharma, D.; Lin, J.; Genevet, P.; Blanchard, R.; Yang, Z.; Qazilbash, M. M.; Basov, D.; Ramanathan, S.; Capasso, F., Ultra-thin perfect absorber employing a tunable phase change material. *Applied Physics Letters* 2012, 101, 221101.

[3] Wuttig, M.; Yamada, N., Phase-change materials for rewriteable data storage. *Nature Materials* 2007, 6, 824–832.

[4] Wuttig, M.; Bhaskaran, H.; Taubner, T., Phase-change materials for non-volatile photonic applications. *Nature Photonics* 2017, 11, 465–476.

[5] Raoux, S.; Wuttig, M., *Phase Change Materials: Science and Applications*. Springer, New York, NY, USA, 2010.

[6] Lencer, D.; Salinga, M.; Grabowski, B.; Hickel, T.; Neugebauer, J.; Wuttig, M., A map for phase-change materials. *Nature Materials* 2008, 7, 972–977.

[7] Karpinsky, O.; Shelimova, L.; Kretova, M.; Fleurial, J.-P., An X-ray study of the mixed-layered compounds of (GeTe) n (Sb2Te3) m homologous series. *Journal of Alloys and Compounds* 1998, 268, 112–117.

[8] Agaev, K.; Talybov, A., Electron-diffraction analysis of structure of $GeSb_2Te_4$. *Soviet Physics Crystallography* 1966, 11, 400.

[9] Nonaka, T.; Ohbayashi, G.; Toriumi, Y.; Mori, Y.; Hashimoto, H., Crystal structure of GeTe and $Ge_2Sb_2Te_5$ meta-stable phase. *Thin Solid Films* 2000, 370, 258–261.

[10] Zachariasen, W. H., The atomic arrangement in glass. *Journal of the American Chemical Society* 1932, 54, 3841–3851.

[11] Kohara, S.; Kato, K.; Kimura, S.; Tanaka, H.; Usuki, T.; Suzuya, K.; Tanaka, H.; Moritomo, Y.; Matsunaga, T.; Yamada, N., Structural basis for the fast phase change of $Ge_2Sb_2Te_5$: Ring statistics analogy between the crystal and amorphous states. *Applied Physics Letters* 2006, 89, 201910.

[12] Miller, K. J.; Haglund, R. F.; Weiss, S. M., Optical phase change materials in integrated silicon photonic devices. *Optical Materials Express* 2018, 8, 2415–2429.

[13] Shportko, K.; Kremers, S.; Woda, M.; Lencer, D.; Robertson, J.; Wuttig, M., Resonant bonding in crystalline phase-change materials. *Nature Materials* 2008, 7, 653–658.

[14] Lucovsky, G.; White, R., Effects of resonance bonding on the properties of crystalline and amorphous semiconductors. *Physical Review B* 1973, 8, 660.

[15] Luo, M.; Wuttig, M., The dependence of crystal structure of te-based phase-change materials on the number of valence electrons. *Advanced Materials* 2004, 16, 439–443.

[16] Wełnic, W.; Botti, S.; Reining, L.; Wuttig, M., Origin of the optical contrast in phase-change materials. *Physical Review Letters* 2007, 98, 236403.

[17] Michel, A.-K. U.; Chigrin, D. N.; Maß, T. W.; Schönauer, K.; Salinga, M.; Wuttig, M.; Taubner, T., Using low-loss phase-change materials for mid-infrared antenna resonance tuning. *Nano Letters* 2013, 13, 3470–3475.

[18] Zhang, Y.; Chou, J. B.; Li, J.; Li, H.; Du, Q.; Yadav, A.; Zhou, S.; Shalaginov, M. Y.; Fang, Z.; Zhong, H., Broadband transparent optical phase change materials for high-performance nonvolatile photonics. *Nature Communications* 2019, 10, 1–9.

[19] Kolobov, A. V.; Fons, P.; Frenkel, A. I.; Ankudinov, A. L.; Tominaga, J.; Uruga, T., Understanding the phase-change mechanism of rewritable optical media. *Nature Materials* 2004, 3, 703–708.

[20] Kolobov, A.; Fons, P.; Tominaga, J.; Ankudinov, A.; Yannopoulos, S.; Andrikopoulos, K., Crystallization-induced short-range order changes in amorphous GeTe. *Journal of Physics: Condensed Matter* 2004, 16, S5103.

[21] Jóvári, P.; Kaban, I.; Steiner, J.; Beuneu, B.; Schöps, A.; Webb, M., Local order in amorphous $Ge_2Sb_2Te_5$ and $GeSb_2Te_4$. *Physical Review B* 2008, 77, 035202.

[22] Philips'Gloeilampenfabrieken, O., A method of measuring specific resistivity and Hall effect of discs of arbitrary shape. *Philips Research Rep* 1958, 13, 1–9.

[23] Kalb, J.; Wuttig, M.; Spaepen, F., Calorimetric measurements of structural relaxation and glass transition temperatures in sputtered films of amorphous Te alloys used for phase change recording. *Journal of Materials Research* 2007, 22, 748–754.

[24] Kalb, J.; Spaepen, F.; Wuttig, M., Kinetics of crystal nucleation in undercooled droplets of Sb-and Te-based alloys used for phase change recording. *Journal of Applied Physics* 2005, 98, 054910.

[25] Friedrich, I.; Weidenhof, V.; Lenk, S.; Wuttig, M., Morphology and structure of laser-modified $Ge_2Sb_2Te_5$ films studied by transmission electron microscopy. *Thin solid Films* 2001, 389, 239–244.

[26] Coombs, J.; Jongenelis, A.; van Es-Spiekman, W.; Jacobs, B., Laser-induced crystallization phenomena in GeTe-based alloys. II. Composition dependence of nucleation and growth. *Journal of Applied Physics* 1995, 78, 4918–4928.

[27] Weidenhof, V.; Pirch, N.; Friedrich, I.; Ziegler, S.; Wuttig, M., Minimum time for laser induced amorphization of $Ge_2Sb_2Te_5$ films. *Journal of Applied Physics* 2000, 88, 657–664.

[28] Wang, Y.; Landreman, P.; Schoen, D.; Okabe, K.; Marshall, A.; Celano, U.; Wong, H.-S. P.; Park, J.; Brongersma, M. L., Electrical tuning of phase-change antennas and metasurfaces. *Nature Nanotechnology* 2021, 16, 667–672.

[29] Ríos, C.; Stegmaier, M.; Hosseini, P.; Wang, D.; Scherer, T.; Wright, C. D.; Bhaskaran, H.; Pernice, W. H., Integrated all-photonic non-volatile multi-level memory. *Nature Photonics* 2015, 9, 725–732.

[30] Kato, K.; Kuwahara, M.; Kawashima, H.; Tsuruoka, T.; Tsuda, H., Current-driven phase-change optical gate switch using indium–tin-oxide heater. *Applied Physics Express* 2017, 10, 072201.

[31] Zhang, H.; Zhou, L.; Lu, L.; Xu, J.; Wang, N.; Hu, H.; Rahman, B. A.; Zhou, Z.; Chen, J., Miniature multilevel optical memristive switch using phase change material. *ACS Photonics* 2019, 6, 2205–2212.

[32] Zheng, J.; Fang, Z.; Wu, C.; Zhu, S.; Xu, P.; Doylend, J. K.; Deshmukh, S.; Pop, E.; Dunham, S.; Li, M., Nonvolatile electrically reconfigurable integrated photonic switch enabled by a silicon PIN diode heater. *Advanced Materials* 2020, 32, 2001218.

[33] Ríos, C.; Zhang, Y.; Shalaginov, M. Y.; Deckoff-Jones, S.; Wang, H.; An, S.; Zhang, H.; Kang, M.; Richardson, K. A.; Roberts, C., Multi-level electro-thermal switching of optical phase-change materials using graphene. *Advanced Photonics Research* 2021, 2, 2000034.

[34] Hosseini, P.; Wright, C. D.; Bhaskaran, H., An optoelectronic framework enabled by low-dimensional phase-change films. *Nature* 2014, 511, 206–211.

[35] Schlich, F. F.; Zalden, P.; Lindenberg, A. M.; Spolenak, R., Color switching with enhanced optical contrast in ultrathin phase-change materials and semiconductors induced by femtosecond laser pulses. *ACS Photonics* 2015, 2, 178–182.

[36] Gholipour, B.; Zhang, J.; MacDonald, K. F.; Hewak, D. W.; Zheludev, N. I., An all-optical, non-volatile, bidirectional, phase-change meta-switch. *Advanced Materials* 2013, 25, 3050–3054.

[37] Wang, Q.; Rogers, E. T.; Gholipour, B.; Wang, C.-M.; Yuan, G.; Teng, J.; Zheludev, N. I., Optically reconfigurable metasurfaces and photonic devices based on phase change materials. *Nature Photonics* 2016, 10, 60–65.

[38] Zhang, Y.; Fowler, C.; Liang, J.; Azhar, B.; Shalaginov, M. Y.; Deckoff-Jones, S.; An, S.; Chou, J. B.; Roberts, C. M.; Liberman, V., Electrically reconfigurable non-volatile meta-surface using low-loss optical phase-change material. *Nature Nanotechnology* 2021, 16, 661–666.

[39] Zhang, Y.; Zhang, Q.; Ríos, C.; Shalaginov, M. Y.; Chou, J. B.; Roberts, C.; Miller, P.; Robinson, P.; Liberman, V.; Kang, M., Transient tap couplers for wafer-level photonic testing based on optical phase change materials. *ACS Photonics* 2021, 8, 1903–1908.

[40] Zhang, Q.; Zhang, Y.; Li, J.; Soref, R.; Gu, T.; Hu, J., Broadband nonvolatile photonic switching based on optical phase change materials: Beyond the classical figure-of-merit. *Optics Letters* 2018, 43, 94–97.

[41] Wu, C.; Yu, H.; Lee, S.; Peng, R.; Takeuchi, I.; Li, M., Programmable phase-change metasurfaces on waveguides for multimode photonic convolutional neural network. *Nature Communications* 2021, 12, 1–8.

[42] Feldmann, J.; Youngblood, N.; Karpov, M.; Gehring, H.; Li, X.; Stappers, M.; Le Gallo, M.; Fu, X.; Lukashchuk, A.; Raja, A. S., Parallel convolutional processing using an integrated photonic tensor core. *Nature* 2021, 589, 52–58.

[43] Rios, C.; Hosseini, P.; Wright, C. D.; Bhaskaran, H.; Pernice, W. H., On-chip photonic memory elements employing phase-change materials. *Advanced Materials* 2014, 26, 1372–1377.

[44] Zhang, Y.; Ríos, C.; Shalaginov, M. Y.; Li, M.; Majumdar, A.; Gu, T.; Hu, J., Myths and truths about optical phase change materials: A perspective. *Applied Physics Letters* 2021, 118, 210501.

[45] Delaney, M.; Zeimpekis, I.; Lawson, D.; Hewak, D. W.; Muskens, O. L., A new family of ultralow loss reversible phase-change materials for photonic integrated circuits: Sb_2S and Sb_2Se_3. *Advanced Functional Materials* 2020, 30, 2002447.

[46] Miller, K.; Haglund, R.; Weiss, S., Optical phase change materials in integrated silicon photonic devices: Review. *Optical Materials Express* 2018, 8, 2415–2429.

[47] Morin, F., Oxides which show a metal-to-insulatortransition at the Neel temperature. *Physical Review Letters* 1959, 3, 34–36.

[48] Kim, H.; Kim, B.; Choi, S.; Chae, B.; Lee, Y.; Driscoll, T.; Qazilbash, M.; Basov, D., Electrical oscillations induced by the metal-insulator transition in VO_2. *Journal of Applied Physics* 2010, 107, 023702.

[49] Wall, S.; Wegkamp, D.; Foglia, L.; Appavoo, K.; Nag, J.; Haglund, R.; Stähler, J.; Wolf, M., Ultrafast changes in lattice symmetry probed by coherent phonons. *Nature Communnications* 2012, 3, 721.

[50] Briggs, R.; Pryce, I.; Atwater, H., Compact silicon photonic waveguide modulator based on the vanadium dioxide metal-insulator phase transition. *Optics Express* 2010, 18, 11192–11201.

[51] Markov, P.; Marvel, R.; Conley, H.; Miller, K.; Haglund, R.; Weiss, S., Optically monitored electrical switching in VO_2. *ACS Photonics* 2015, 2, 1175–1182.

10 Magnetic and Spintronic Materials and Their Applications

Zheng Feng and Wei Tan

CONTENTS

10.1 Introduction..261
10.2 Fundamental Physics 262
    10.2.1 Magnetic and Spintronic Materials...262
    10.2.2 Magnetic and Spintronic Effects...263
10.3 Applications of Magnetic and Spintronic Materials.....................................266
    10.3.1 Generation of Optical, Infrared, and Terahertz Wave 266
    10.3.2 Manipulation of Optical, Infrared, and Terahertz Wave 274
    10.3.3 Detection of Optical, Infrared, and Terahertz Wave 276
10.4 Summary and Outlook...278
Acknowledgements...279
Bibliography 280

10.1 INTRODUCTION

The history of magnetism and magnetic materials is coeval with the history of sciences. In the past four decades, the integration of magnetic materials and other materials into micro-nano level, and the emergence of non-magnetic materials with rich spin properties, enable the use of the electron spin as well as its charge for high performance and new functional devices, which has prospered the development of spintronics [1]. The discovery of giant magnetoresistance (GMR) is regarded as the beginning of spintronics [2,3], and the exciting advancements in spintronic materials and physics in this field, such as magnetic tunnel junction (MTJ)/tunnel magnetoresistance (TMR) [4], exchange bias [5], spin-transfer torque (STT)/spin-orbit torque (SOT) [6,7], spin Hall effect (SHE)/inverse Hall effect effect(ISHE) [8], have been successfully applied in information industries, including high-density data storage, signal processor and magnetic sensor, etc. In addition, micro-nano magnetic and spintronic materials and related effects show great capability for optical, infrared, and terahertz (THz) applications. The coupling and interplay of photon, spins, and magnetism in magnetic and spintronic materials give rise to various interesting physical phenomena and effects, such as spin-polarized light-emitting diode effect, magneto-optical Faraday effect/Kerr effect, and light induced spin photocurrent, which extended a new research direction as opto-spintronics [9]. Recently,

DOI: 10.1201/9781003202608-10

THz spintronics [10], as the merge of the fields of spintronic and THz science and technology, has attracted increased interests based on the fact that the characteristic frequency or the characteristic time scale of many magnetic and spintronic effects (such as antiferromagnetic resonance, ferrimagetic resonance, ultrafast demagnetization, and ultrafast spin transport) lies in the THz range or in the picosecond time scale. In opto-spintronics and THz spintronics, optical, infrared, and THz waves act as effective and efficient tools for characterizing and manipulating magnetism and spin of materials and devices; on the other hand, magnetic and spintronic materials and devices pave new pathways for the generation, manipulation, and detection of optical, infrared, and THz waves.

10.2 FUNDAMENTAL PHYSICS

10.2.1 MAGNETIC AND SPINTRONIC MATERIALS

The traditional magnetic materials can be classified into the following three major groups: ferromagnet (FM), antiferromagnet (AFM) and ferrimagnet (FiM). As illustrated in Figure 10.1(a), the characteristic feature of ferromagnetic is the spontaneous magnetization, which is originated from the alignment of the magnetic moments in the same direction. The antiferromagnet has two equal but oppositely directed magnetic sub-lattices, while the ferrimagnet consists of two unequal oppositely directed magnetic sub-lattices, as illustrated in Figure 10.1(b)-(c). The typical examples of ferromagnetic materials are Fe, Co, Ni, Gd, Dy, FeNi, etc. The typical examples of antiferromagnetic materials are Cr, Mn, NiO, $MnFe_2$, FeMn, $IrMn_3$, and a-Fe_2O_3. The typical examples of ferrimagnetic materials are Fe_3O_4, YFe_5O_{12}(YIG), $BaFe_{12}O_{19}$, MnGa, and GdFeCo. In addition, various emergent spintronic materials and mico-nanosturctures have stepped onto the stage, such as spin valve [11], magnetic tunnel junction [4], and topological insulators [12], which are close related with the applications in the following sections. Figure 10.1(d) shows the the basic structure of a spin valve, which typically consists of two FM layer (a free FM layer and a pinned FM layer) separated by a nonmagnetic (NM) metal layer; the free FM layer easily changes its magnetization direction with a low magnetic field whereas the pinned FM layer has its magnetizaion pinned by the effictive exchange bias field from the neighboring AFM layer. Figure 10.1(e) shows the basic structure of a magnetic tunnel junction, which typically consists of two FM layer (a free FM layer and a pinned FM layer) separated by a tunnel insulator. The resistance of a spin valve and a magnetic tunnel junction both changes with the relative magnetization directions of the free and pinned FM layer due to spin dependent scattering, which will be discussion in the following section. Topological insulators are new states of quantum matter which have an insulating bulk and metallic surface due to a band inversion that is topologically protected against external perturbations; the rich spin related properties, such as locked spin and momentum, large spin-orbit coupling and light-spin interaction, result in new spintronic devices based on topological insulator. The typical examples of topological insulators are Bi_2Se_3, Bi_2Te_3, Sb_2Te_3, $Bi_{1-x}Sb_x$.

(a) Ferromagnet

(d) Spin valve

Free FM layer
NM metal layer
Pinned FM layer

(b) Antiferromagnet

(e) Magnetic tunnel junction

Free FM layer
Tunnel insulator
Pinned FM layer

(c) Ferrimagnet

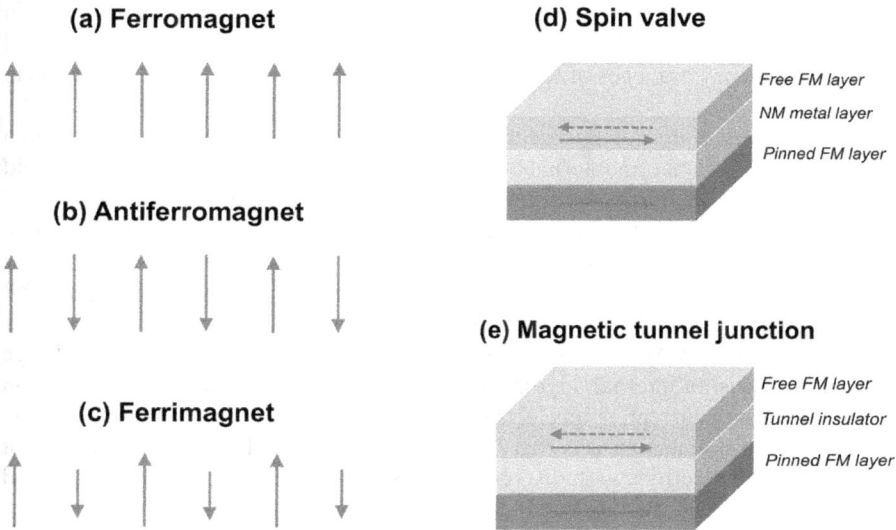

FIGURE 10.1   Schematic of (a) ferromagnet, (b) antiferroamgnet, (c) ferrimagnet, (d) spin valve, and (d) magnetic tunnel junction.

10.2.2   MAGNETIC AND SPINTRONIC EFFECTS

Magnetic dipole radiation & electric dipole radiation. According to classical electromagnetic theory, the emission of electromagnetic waves can be described by the wave equation of the electric field E derived from Maxwell's equations: $\nabla^2 E - \mu_0 \varepsilon_0 \dfrac{\partial^2 E}{\partial t^2} = \mu_0 \dfrac{\partial J}{\partial t}$, where J is the effective charge current density, $J = J_f + \nabla \times M + \dfrac{\partial P}{\partial t}$, J_f is the free charge density, M is the magnetization, P is the electric polarization, and ε_0 (μ_0) is the permittivity (permeability) in a vacuum. As can be seen from the wave equation above, there are three mechanisms that can contribute to electromagnetic wave radiation. Among them, the free current J_f and the electric polarization P radiate electromagnetic waves with the variation of time, which correspond to the electric dipole radiation. The magnetization M radiates electromagnetic waves with the variation of time, which corresponds to the magnetic dipole radiation. The radiation frequency corresponds to the time scale—for example, when the time scale is in picosecond, terahertz waves are radiated.

Ferro(Ferri-, Antiferro-)magnetic resonance. Under electromagnetic wave irradiation, magnetic vectors in magnetic materials will precess at a specific frequency determined by the external magnetic field and the magnetic properties of the materials, and meanwhile electromagnetic wave at the precession frequency is sharply absorbed. This is termed as magnetic resonance. Magnetic resonance exists in ferromagnetic, ferrimagnetic, and antiferromagnetic materials, which are named as ferromagnetic resonance, ferrimagnetic resonance, and antiferromagnetic resonance, respectively [13,14,15]. The ferromagnetic resonance frequencies are commonly in the microwave range, while the frequencies of the antiferromagnetic resonance and the high-frequency branches of ferrimagnetic resonance can be extended to the THz band.

Spin-orbit coupling. As illustrated in Figure 10.2(a), when an electron moves in a solid, it may experience an internal electric field from the lattice and/or external field from a bias voltage. In the rest frame of reference of the electron, the electric field appears in the form of an effective magnetic field. Therefore, the coupling between the momentum of the electron and the electron spin is equivalent to the coupling between an effective magnetic field and the electron spin. This is named spin-orbit coupling, which results in spin dependent scattering of electrons.

Charge current & Spin current. An electron has two degrees of freedom, charge and spin. Spin is an internal angular momentum of an electron. A flow of electron charge is a charge current, or an electric current. Similar to the charge current, a flow of spin angular momentum is called a spin current. A normal or pure charge current is the motion of electrons with disordered spins (random distribution of spin-up and spin-down electrons), as illustrated in Figure 10.2(b). Due to spin-dependent scattering, the charge current in ferromagnets will become spin polarized current that both charge and spin flow, as illustrated in Figure 10.2(c). Another type of spin current is pure spin current that flows without the accompaniment of charge current flow, as illustrated in Figure 10.2(d).

Spin Hall effect, Inverse spin Hall effect & Anormalous Hall effect. Spin Hall effect (SHE) is an effect in the bulk that pure charge currents are converted into transverse

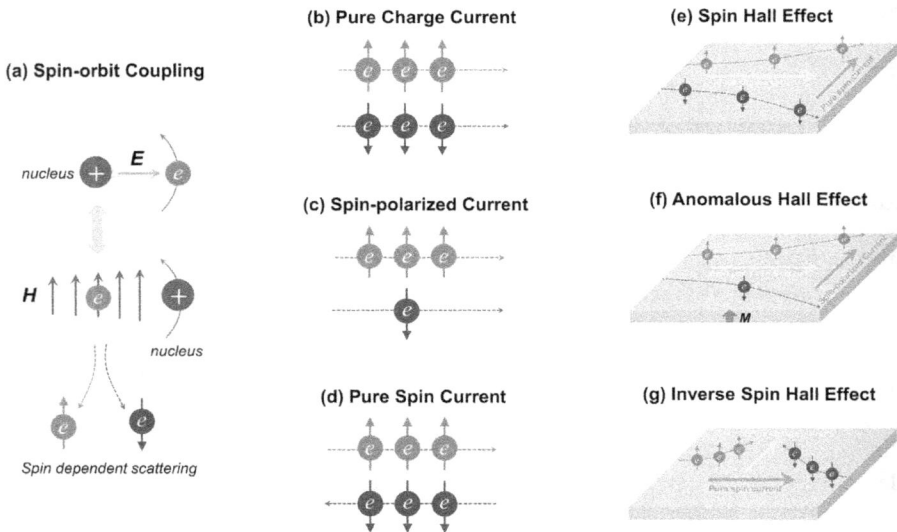

FIGURE 10.2 Schematic of (a) spin-orbit coupling, (b) pure charge current, (c) spin-polarized current, (d) pure spin current, (e) spin Hall effect, (f) anormalous Hall effect, and (g) inverse spin Hall effect.

pure spin currents due to the spin-orbit coupling, as shown in Figure 10.2(e). Inverse spin Hall effect (ISHE) is the reciprocal effect to the SHE, in which pure spin currents are converted into pure charge currents [8], as shown in Figure 10.2(f). The efficiency of the spin-charge interconversion can be quantified by a single material-specific parameter, i.e., the spin Hall angle. Anomalous Hall effect (AHE) is an effect that a charge current flowing in magnetic materials (which is a spin-polarized current) can be scattered and converted into a transverse spin polarized current which is perpendicular to the magnetization [16], as shown in Figure 10.2(g). AHE has the same mechanism with SHE and ISHE—spin-orbit coupling, and the parameter describing the conversion is named as anomalous Hall angle.

Rashba-Edelstein effect & Inverse Rashba-Edelstein effect. Rashba-Edelstein effect (REE) is an effect at the surface/interface that pure charge currents are converted into pure spin currents due to spin-orbit coupling, and Inverse Rashba-Edelstein effect (IREE) is the reciprocal effect to the REE [17]. REE/IREE usually occurs at the interfaces of metallic heterostructures, in topological insulators, two-dimensional electron gas system, and two-dimensional materials.

Spin transfer torque & spin-orbit torque. Spin transfer torque [6] is an effect that the interaction of spin polarized currents with the local magnetization lead to the angular momentum transfer between them, which results in a change of the magnetization orientation (oscillations or switching). Spin orbit torque [7] is a torque induced by spin currents which are generated from charge currents due to spin-orbit coupling, leading to the change of the magnetization orientation (oscillations or switching).

Spin pumping. Spin pumping [18] commonly refers to spin currents injection from a magnetic layer with a precessional magnetization at its resonance states into the adjacent nonmagnetic layer. The injected spin currents have a dc part and an ac part with the resonance frequency, and can be converted into dc and ac charge currents in the nonmagnetic layer via spin orbit coupling, respectively. Spin pumping effect was firstly observed in ferromagnetic/nonmagnetic heterostructures at microwave frequencies, and was also demonstrated in antiferromagnetic/nonmagnetic heterostructures at THz range [19,20].

Spin Seebeck effect. The spin Seebeck effect [21] is an effect that a spin voltage is generated by a temperature gradient in a magnetic layer, leading to the thermal injection of spin currents from the magnetic layer into an attached nonmagnetic layer. The injected spin currents can be converted into transverse charge currents via spin orbit coupling.

Magneto-optic effect. Magneto-optic effect refers to physical interaction between magneticmaterials (or materials in magnetic field) and light. Magneto-optic effect in magnetic materials mainly includes Faraday effect, magneto-optic Kerr effect (MOKE), Cotton-Mouton effect, and Voigt effect. Faraday effect depicts the light polarization rotation when a light through a material under a magnetic field which is applied in the direction of light propagation. MOKE describes the polarization changes when a linearly polarized light reflected from magnetic (or spintronic) materials. Cotton-Mouton effect and Voigt effect are both magneto-birefringence effects, in which the birefringence occurs when a magnetic field is applied perpendicular to the direction of light propagation. Due to magneto-optic effect, the magnetic materials can modulate and manipulate the properties of light, such as amplitude, phase, polarization, and spectrum.

Magnetic resistance. Magnetoresistance (MR) refers to the effect that the resistance changes when a magnetic field is applied. There are a variety of magnetoresistance effect, mainly including anisotropic magnetoresistance (AMR) [22], giant magnetoresistance (GMR) [2,3], tunnel magnetoresistance (TMR) [4], and colossal magnetoresistance (CMR) [23]. The AMR is an effect in single ferromagnetic or ferrimagnetic metals that the resistivity depends on the relative angle between the magnetization direction and the electric current direction. The GMR is a phenomenon that occurs in multilayered structures with ferromagnetic metallic layers separated by a non-magnetic metallic spacer, in which the resistance is low (or high) when the ferromagnetic metals are parallel- (or antiparallel-) magnetized. The TMR occurs in magnetic tunnel junction consists of two ferromagnetic layers separated by a thin insulator spacer; if a voltage is applied to the junction, electrons tunnel through the insulator space and the tunneling current (meanwhile the resistance) changes dramatically when the relative magnetization of the two ferromagnetic layers changes their alignment by magnetic field. Colossal magnetoresistance is an effect that occurs mostly in Mn-based perovskite oxides, which arises from the paramagnetic insulator-ferromagnetic metal phase transition under application of a magnetic field.

10.3 APPLICATIONS OF MAGNETIC AND SPINTRONIC MATERIALS

10.3.1 GENERATION OF OPTICAL, INFRARED, AND TERAHERTZ WAVE

Spin Light Emitting Diodes (Spin-LED). The core structure of Spin LED consists of two parts [24]: a spin-injector and a carrier recombination light-emitting region (active region), as shown in the Figure 10.3. The spin-injector injects spin-polarized

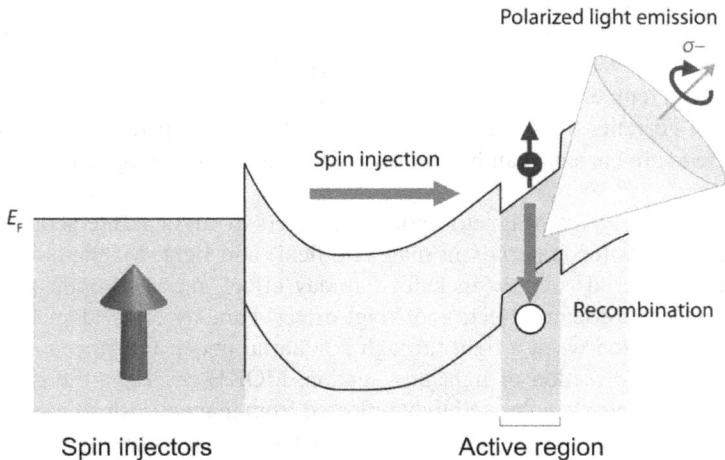

FIGURE 10.3   Schematic of spin light emitting diodes.

Source: Modified with permission from Taniyama, T. et al. NPG Asia Mater. 3, 65, 2011. Copyright 2011 Springer Nature.

electrons (or holes) into the active region, and recombined with the holes (electrons) according to the optical selection rule to generate left-handed or right-handed circularly polarized light. The polarizability of the emitted circularly polarized light is related to the spin polarizability of the injected electrons (holes). The active region of Spin LED mainly includes semiconductor bulk materials, semiconductor quantum well, and semiconductor quantum dots. The spin injection involves magnetic semiconductors, ferromagnetic materials, and magnetic semi-metals. In the earlier researches of the spin LED, magnetic semiconductors, such as GaMnAs, ZnMnSe, and $CdCr_2Se_4$, were mainly adopted. Magnetic semiconductors can avoid the conductance mismatch between the spin injection layer and the light-emitting region layer. However, due to the low Curie temperature of magnetic semiconductors, spin injection cannot be performed effectively at room temperature. For this reason, ferromagnetic metals with higher Curie temperature were studied for the spin injector. Nevertheless, the spin injection efficiency was still low due to the conductance mismatch between the metal and the semiconductor. Later, it was found that a barrier layer inserted between the ferromagnetic metal and the semiconductor to form "ferromagnet/barrier layer/semiconductor" structure can solve the problem of conductance mismatch, thereby greatly improving the spin injection efficiency. The ferromagnetic injectors in previous study mainly include Fe, Fe/AlO_x, Co/Al_2O_3, CoFe/Al_2O_3, and CoFeB/MgO. The magnetization of these materials is in the plane, and a large external magnetic field is required to make the outside magnetic field perpendicular to the film surface to meet the optical selection rule, so as to measure the emitted light and its circular polarization in the normal direction. Such external magnetic field is inconvenient in practical applications. Consequently, researchers sought the perpendicularly magnetized magnetic material to replace the in-plane magnetized magnetic material as the spin injector, in order to realize the spin luminescence without the external field. The magnetic materials of perpendicular magnetization mainly include: FeTb, MnGa, CoPt, Ta/CoFeB/MgO, and Mo/CoFeB/MgO. In the future, if the spin-orbit torque can be utilized to drive the reversal of the perpendicular magnetic anisotropy magnetization (magnetic moment) in the spin injector of the spin LED, it is expected to further improve the performance. In addition, new materials, such as low-dimensional materials with rich semiconductor properties and their heterostructures, can be applied to the active region of spin LEDs, and the emission frequency range can be modulated to further expand the applications.

Femtosecond (fs) laser induced THz pulse emission in single magnetic layer. Fs laser pulse could induce ultrafast demagnetization process in the time scale of picosecond in single ferromagnetic film, which was firstly discovered in Ni nano-film with the TR-MOKE (Time-resolved magneto-optical Kerr effect) technique by E. Beaurepaire and coworkers [25]. Later, they observed THz pulse emission from Ni film during the ultrafast demagnetization process excited by fs laser, as illustrated in Figure 10.4(a), and they claimed that the ultrafast demagnetization process acts as a time dependent magnetic dipole for THz pulse emission [26]. After that, numerous studies on fs laser induced THz pulse emission in various FM films have been investigated, including Fe, Co, FeNi, Co_2MnSn, CoFeB, FeMn, and FeMnPt. Most of them followed the ultrafast demagnetization picture, where the THz signals are symmetric under FM sample flipping (turned by 180°). However, Q. Zhang et al. observed an

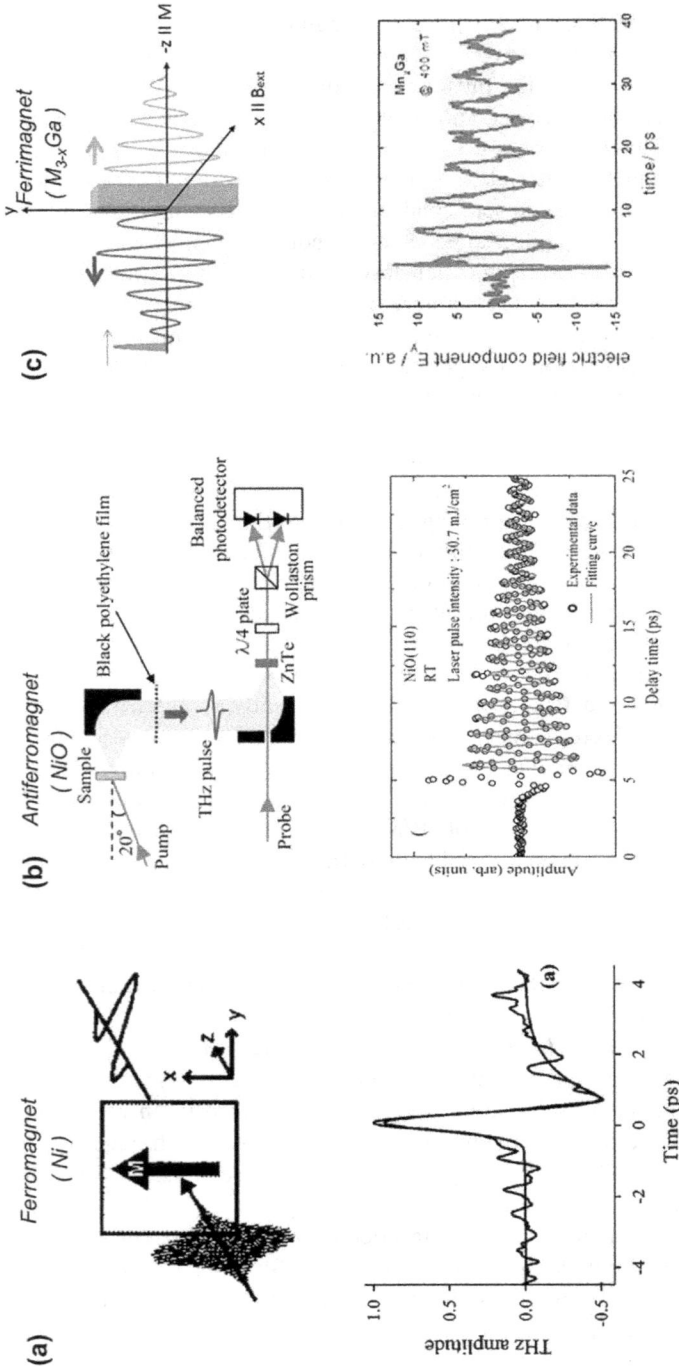

FIGURE 10.4 (a) Fs laser induced THz pulse emission from single ferromagnetic Ni film.

Source: Reprinted with permission from Beaurepaire, E. et al. *Appl. Phys. Lett.* 84, 3465, 2004. Copyright 2004 AIP Publishing

(b) Fs laser induced THz pulse emission with periodic oscillations from single antiferromagnetic NiO crystal.

Source: Reprinted with permission from Nishitani, J. et al. *Appl. Phys. Lett.* 96, 221906, 2010. Copyright 2010 AIP Publishing

(c) Fs laser induced THz pulse emission with periodic oscillations from single ferrimagnetic Mn$_{3-x}$Ga film.

asymmetric component in their experiments and they proposed an alternative mechanism for the observation: under fs laser illumination, a net backflow superdiffusive spin-polarized current is generated in the normal direction due to the different FM/dielectric interfaces, and subsequently converts to transverse transient charge current via the AHE in FM, thereby emitting THz pulse [27]. It suggests that there may exist various THz emission mechanisms in a single FM layer, and how they compete with each other remains unclear.

As mentioned before, the frequency of ferrimagnetic resonance and antiferromagnetic resonance is mostly in the terahertz range. As illustrated in Figure 10.4(b), J. Nishitani et al. reported that THz pulses with periodic oscillations are emitted from single NiO crystal induced by fs laser pulses, and the frequency of the oscillations coincide with the frequency of antiferromagnetic resonance [28]. They attributed such oscillations to the magnetic dipole radiation generated by fs laser induced antiferromagnetic resonance. Such behaviors have been observed in various antiferromagnetic materials, for example, MnO, $DyFeO_3$, $TmFeO_3$, $HoFeO_3$, $ErFeO_3$, $SmFeO_3$, and $FeBO_3$. In addition, THz pulses with periodic oscillations also have been observed from fs laser excited ferrimagnetic resonance in metallic ferrimagnetic $Mn_{3-x}Ga$ Heusler alloy nanofilms [29], as illustrated in Figure 10.4(c), and the oscillation frequency can be tuned precisely via the film composition.

Fs laser induced THz pulse emission in magnetic heterostructures. In 2013, T. Kampfrath et al. reported a novel type of THz emitter based on ultrafast spin-current to charge-current conversion process in ferromagnetic/nonmagnetic (FM/NM) nanofilms under an external magnetic field, which was then named as spintronic THz emitter [30]. The mechanism is schematically shown in Figure 10.5: i) a fs laser pulse impinges on the FM/NM heterostructure (e.g., Fe/Au, Fe/Ru), which excites electrons in the FM nanofilm to states above the Fermi energy; ii) the

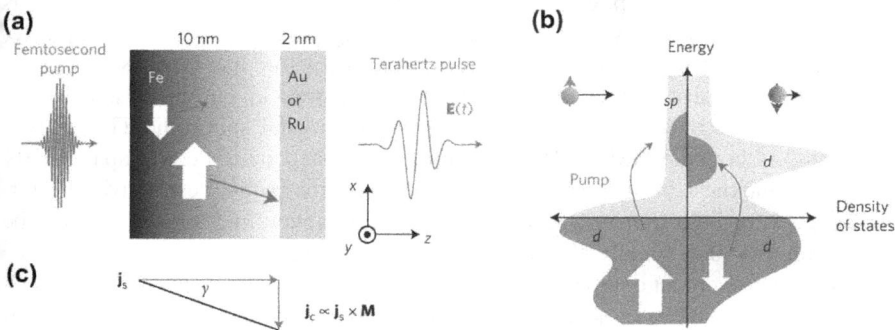

FIGURE 10.5 Schematic of fs laser induced THz pulse emission based on FM/NM heterostructure.

Source: Reprinted with permission from Kampfrath, T. et al. *Nat. Nanotechnol.* 8, 256, 2013. Copyright 2013 Springer Nature.

excited electrons diffuse to the NM nanofilm leading to a net current perpendicular to the interface, and the current is strongly polarized since the spin-up and spin-down electrons present distinct density, velocity, and lifetime; iii) such injected spin current J_s in the NM nanofilm converts into a charge current J_c due to the ISHE, which is perpendicular to J_s and the external magnetic field; iv) since such spin to charge conversion occurs at a picosecond timescale, the ultrafast current variation induced emission lies in the THz band. The spintronic THz emitter owns various advantages, such as easy fabrication and operation, scalability, and low cost, and the most important one is the ultra-large bandwidth (>10 THz), which presents significant superiority over the popular commercial THz emitter, such as photoconductive antenna and ZnTe crystal. However, the THz field amplitude as well as the efficiency of the original structure (Fe/Au and Fe/Ru) is only ~1% compared to that of ZnTe crystal. Whether the efficiency could be enhanced to a practical level becomes one of the key issues in the later studies.

In 2016, T. Seifert et al. presented a comprehensive study on improving the performance of spintronic THz emitter [31]. Several methods were proposed and proved to be efficient: i) selecting NM materials with larger spin Hall angle, such as Pt and W; ii) optimizing the composition of the FM/NM heterostructure—for example, CoFeB/Pt exhibits larger efficiency; iii) optimizing the thickness of the FM and NM layers, which commonly achieves the maximum value when the total thickness is around 6 nm; iv) utilizing both forward and backward spin current by constructing a NM/FM/NM sandwich structure, where the two NM layers are designed to possess opposite spin Hall angle, such as Pt and W. As a result, the optimized spintronic THz emitter provides highly enhanced efficiency that reaches the same level of commercial ZnTe crystal, and further provides much larger bandwidth (>15 THz vs ~3 THz). Besides, researchers found that the microstructural properties, including the crystallization, interfacial roughness, and interfacial intermixing also influence the efficiency, and then improvement method, such as annealing and roughness control, were developed.

Note that these methods were mostly inspired from the perspective of spintronics. Towards a further step, efforts from the perspective of light-matter interactions and THz emission process have been made. Z. Feng et al. elucidated that the laser absorption of the nanometer-thick spintronic heterostructure has its physical limitation ($\leq 50\%$). In order to break this restriction, they theoretically proposed and experimentally demonstrated a one-dimensional photonic-crystal spintronic THz emitter consisting of periodic NM/FM/NM nanofilms separated by dielectric spacers. By tuning the thickness of the dielectric spacer to a specific range, destructive interference occurs for both transmitted and reflected light waves, and thus enhancing the absorption to >80%. Finally, the strongest THz pulse emission amplitude presents a 1.7 times improvement compared to the bare structure [32]. Besides, Herapath et al. demonstrated that the laser absorption can be almost doubled via integrating the spintronic THz emitter with an optical cavity made from alternating dielectric overlayers of TiO_2 and SiO_2. In order to improve the THz emission process, hyper-hemispherical silicon lens and antenna were employed to couple with spintronic THz emitter, which showed excellent results especially for small laser spot size (~10 μm) [33].

Till now, spintronic THz emitter has come into practical application stage. The most direct example is to replace ZnTe crystal in the THz time-domain spectroscopy.

which has already been realized in several research labs. Another fascinating application is for super-resolution THz imaging, in which a near-field ghost imaging system is set up using a spintronic THz emitter excited by spatially encoded fs laser patterns. In such a configuration, the sensing distance (the distance between the emission plane and the object) plays a key role in achieving high resolution. In comparison with other THz emitters, spintronic THz emitter has only nanometer thickness, which makes the propagation and diffraction of THz waves in the emitter itself negligible. As a result, a resolution of 6.5 μm (1/100 wavelength) was achieved, and the configuration was termed as ghost spintronic THz emitter-array microscope system [34]. Besides, spintronic THz emitter offers the capability to be integrated with on-chip detector or metasurface, which can act as THz magneto-optic sensor/imager and label-free THz biosensor [35,36].

Most recently, novel spin related effects in magnetic heterostructures besides the earlier mentioned ultrafast diffusive hot electron process and the ISHE have emerged, which provide new manners for THz emission, as illustrated in Figure 10.6 [33].

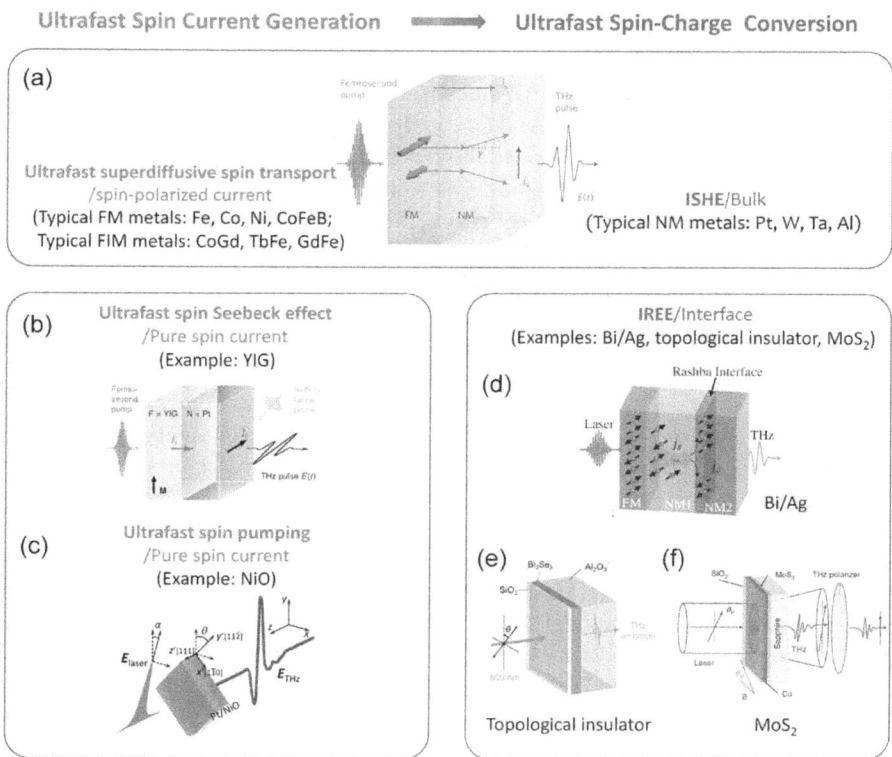

FIGURE 10.6   Summary of current mechanisms for ultrafast spin current generation and spin-charge conversion for THz pulse emission based on magnetic heterostructures.

Source: Reprinted with permission from Feng, Z. et al. *J. Appl. Phys.* 129, 010901, 2021. Copyright 2021 AIP Publishing.

On one hand, new magnetic material families, such as insulators and antiferromagnetic, were employed for ultrafast spin current generation. T. S. Seifert et al. observed THz emission from a magnetic insulator/NM heterostructure (YIG/Pt) according to the spin Seebeck effect, where the temperature gradient induces spin transport [37]. H. Qiu et al. reported ultrafast THz emission form an antiferromagnetic/NM heterostructure (NiO/Pt) at zero magnetic fields, where transient magnetization was induced through the magnetic difference frequency generation process in the NiO layer [38]. On the other hand, novel spin to charge conversion processes were investigated—for example, the inverse Rashba-Edelstein effect (IREE). Different from the ISHE that occurs in bulk material, the IREE emerges at the Rashba interface, including but not limit to the interfaces of metallic heterostructures (such as Ag/Bi and Cu/Bi), topological insulators (such as Bi_2Se_3), two -dimensional electron gas system (such as $LaAlO_3/SrTiO_3$), and two-dimensional materials (such as MoS_2). X. Wang et al. demonstrated ultrafast spin-injection and spin-to-charge conversion in Bi_2Se_3/Co heterostructure, and found that the conversion efficiency was temperature independent in Bi_2Se_3 as expected from the nature of surface states [39]. L. Cheng et al. observed an efficient spin current injection from Co to monolayer MoS_2. The MoS_2 monolayer with a bandgap filters only the high-energy carriers where the population is almost fully spin-polarized, and acts as a converter of the spin to charge current [40]. M. B. Jungfleisch et al. and C. Zhou et al. reported laser-induced broadband THz emission from CoFeB/Ag/Bi and Fe/Ag/Bi structures, respectively [41,42]. Both of them attributed it to the IREE, as the reversed Rashba interface between Ag/Bi and Bi/Ag structures can give rise to the opposite polarities of the THz waveforms. In contrary, J. Shen et al. designed a clear-cut experiment to identify the spin-to-charge mechanism, which exhibited that the ISHE rather than the IREE is the dominant mechanism [43]. They also pointed out that different atomic ratios of the Ag-Bi alloy at the interface may account for the results of spin-to-charge conversion signal sign change observed in various Ag/Bi bilayer samples before.

Fs laser induced THz pulse emission in other spintronic materials. The circular photogalvanic effect (CPGE) can generate spin-polarized photocurrents by absorbing circularly polarized photons according to the angular momentum selection rules. In principle, such effect can occur at a picosecond time scale when pumped by a circularly polarized fs laser pulse, and consequently emits THz pulses. Note that it can be controlled by the laser circular polarization without external electric fields or magnetic field. Different research groups have demonstrated CPGE induced helicity-dependent THz pulse emission in distinct families of spintronic materials, including topological insulators (Bi_2Se_3, Bi_2Te_3, Sb_2Te_3), Rashba-type polar semiconductor (BiTeBr), Weyl semimetal (WTe_2), and nodal-line semimetal (Mg_3Bi_2) [44–47]. Besides, helicity-dependent THz pulse emission also has been observed in Bi nanofilms, which is well known to be the basis of many topological insulators. However, the dominant origin of THz emission is attributed to the photoinduced ISHE: circularly polarized fs laser with a tilted incidence angle produces a gradient of in-plane spin polarization in the normal direction due to inverse Faraday effect, and an ultrafast spin current induced by the spin polarization gradient converts into a picosecond charge current via ISHE [48]. In addition, B. Guzelturk et al. observed THz pulse emission in multiferroic $BiFeO_3$ nano-films with spontaneously formed periodic stripe domains, which was expected to result from picosecond photocurrent due to the charge separation across the domain walls induced by fs laser pulse [49]. This work shows that control of the domain wall density could enable practical bias-free strong THz emitters based on the multiferroic nanofilms.

THz nano-oscillator based on magnetic heterostructures. R. Cheng et al. proposed a THz nano-oscillator based on an AFM/heavy metal heterostructure in a spin Hall geometry [50]. They showed that a dc spin current, which is created by a dc charge current flowing in the heavy metal layer due to the SHE, injects into the AFM layer and exerts a spin-orbit torque. It can sustain a stable oscillation of AFM magnetization vector, which results in an ac spin current back into the heavy metal layer by the virtue of antiferrmagnetic spin pumping, and converts into an ac charge current due to the ISHE, as illustrated in Figure 10.7. The frequency of the ac charge current is in the THz range, and can be tuned by the input dc current density. O. R. Sulymenko et al. proposed a THz spin Hall auto-oscillator based on a canted antiferromagnet/heavy metal heterostructure, where the THz emission is originated from the magnetic diploe radiation of the net magnetization vector oscillation induced by the spin orbit torque of heavy metal layer [51]. Theoretical calculations indicated that the radiation frequencies lie in the range of 0.05–2 THz, and the output power increases with the increment of frequency, which can exceed 1 μW at 0.5 THz. THz nano-oscillator based on spin-valve geometry and tunnel junction geometry were also proposed [51,52], where the magnetization vector oscillation is induced by spin transfer torque and the THz emission is originated from the retification of ac current and ac magnetoresistance. In addition, ferrimagnet and synthetic antiferromagnet were employed for the THz nano-oscillator [53,54]. Although these THz nano-oscillators based on magnetic heterostructures are only theoretically designed up to date, the demonstration of related mechanisms, such as microwave spintronic nano-oscillator and antiferromagnetic spin pumping, have already been experimentally realized, which boosts the confidence of researchers in realizing the THz nano-oscillator.

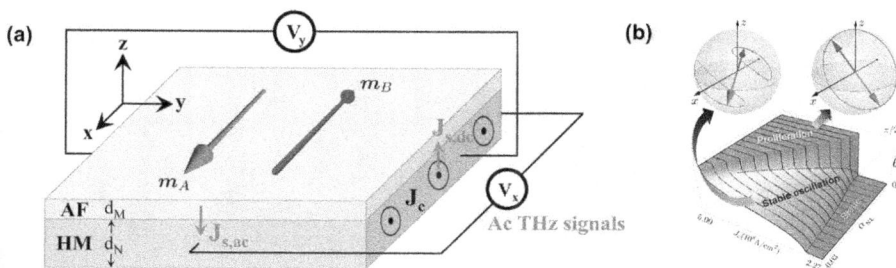

FIGURE 10.7 (a) Schematic of THz nano-oscillator based on an AFM/HM (heavy metal) heterostructure. (b) Phase diagram of spin current induced magnetization dynamics of AFM layer.

Source: Modified with permission from Cheng, R. et al. *Phys. Rev. lett.* 116, 207603, 2016. Copyright 2016 American Physical Society.

10.3.2 MANIPULATION OF OPTICAL, INFRARED, AND TERAHERTZ WAVE

Manipulation via magneto-optical materials. As mentioned in Section 10.2.2, the magnetic materials can modulate and manipulate the properties of light due to the magnetic-optic effects. Magneto-optical materials with giant magneto-optic effects have been found in many magnetic materials, such as rare-earth element doped garnets, rare-earth element-doped magneto optical glasses, rare-earth element-doped iron garnet films, and magnetic alloy films. Based on magneto-optical materials and related magneto-optical effects, fruitful magneto-optical devices have been created and developed for optical, infrared, and terahertz wave manipulation, including but not limit to magneto-optical modulator, switch, isolator, and circulator. Two of the most important devices are optical isolator and circulator, which are widely used in optical systems as nonreciprocal photonic components. The commercial optical isolator and circulator are fabricated based on liquid-phase epitaxy (LPE) grown magnetic garnet materials which possess the Faraday effect. However, the thickness of the magnetic garnet materials in these devices is several hundred micrometers which can hardly be integrated in a planar photonic waveguide, for the purpose of constructing an on-chip optical microsystem. In recent years, optical isolators and circulators have been integrated on silicon based on deposited YIG and Ce-doped YIG thin films (as illustrated in Figure 10.8) [55], and record high isolation ratio up to 40 dB and low insertion loss of 3 dB was reported at the wavelength of 1560.1 nm. In addition,

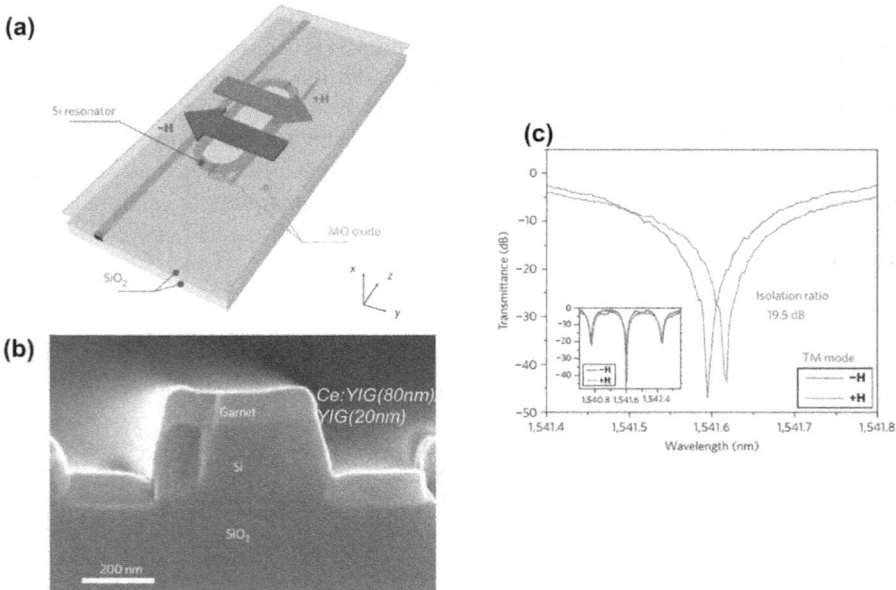

FIGURE 10.8 (a) Schematic of the on chip nonreciprocal optical isolator structure. (b) Transverse cross-sectional SEM image at the window section of the isolator. (c) Optical isolation performance of the isolator.

Source: Reprinted with permission from Bi, L. et al. *Nat. Photonics.* 5, 758, 2011. Copyright 2011 Springer Nature.

combining magneto-optical materials and photonic crystals can provide a completely different mechanism for light manipulation by the existence of photonic bandgaps, which result in many properties that are hardly obtained in bare magnetic materials, such as construction of topological nontrivial photonic states, giant magneto-optical effects, and huge magneto-optical modulation amplitude [56]. Combining magneto-optical materials and metamaterials also paves a way for engineering the permeability and permittivity tensors in metamaterials at arbitrary frequencies for light manipulation [57]. Meanwhile, some magneto-optical materials, like YIG film, InSb, and topological materials have also been investigated in the THz regime for realizing THz polarization conversion and non-reciprocal transmission [58,59]. Besides, THz manipulation was also realized with liquid-suspended magnetic ferrofluid Fe_3O_4 nanoparticles and ferrofluid-filled photonic crystals, which is based on magnetic induced birefringence and dichroism [60].

Manipulation via magneto-resistance materials. The THz transmission properties through materials are related to the materials' resistance—higher (lower) resistance state is typically accompanied by higher (lower) THz transmission. In principle, one can use magnetic field to manipulate THz wave propagation with magnetic and spintronic materials which have magnetic-resistance effect. Chau et al. have discovered that AMR can modulate the THz transmission in both ensemble subwavelength Co microparticles and dense micrometer-sized Co particles coated with Au microparticles. As illustrated in Figure 10.9(a), the AMR-based modulation

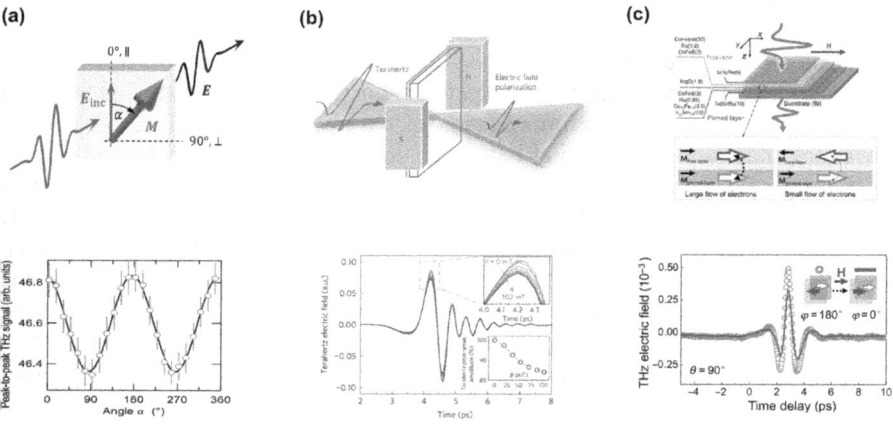

FIGURE 10.9 (a) THz modulation via AMR structure.

Source: Adopted from Nádvorník, L. et al. *Phys. Rev. X.* 11, 021030, 2021. No permission is required. Copyright 2021 the authors.

(b) THz modulation via GMR structure.

Source: Reprinted with permission from Jin, Z. et al. *Nat. Phys.* 11, 761, 2015. Copyright 2015 Springer Nature.

(c) THz modulation via TMR structure.

Source: Reprinted with permission from Jin, Z. et al. *Phys. Rev. Appl.* 14, 014032, 2020. Copyright 2020 American Physical Society.

of THz transmission has also been observed in single ferromagnetic layers [22], such as NiFe, Ni, and Co films. Z. Jin et al. have shown that the application of a magnetic field to a GMR structure induces a striking reduction of the THz transmission by about 20%, as illustrated in Figure 10.9(b) [61]. Later, TMR effect was also used for modulating THz transmission, and a maximum modulation depth of 60% was reached in MTJ films whose thickness is only 77.45 nm while the applied magnetic field was as low as 30 mT, as illustrated in Figure 10.9(c) [62]. The CMR-based modulation of THz transmission has also been achieved in epitaxial $La_{0.7}Sr_{0.3}MnO_3$ nanocomposites and single-phase thin films [63]. In addition, a novel route to achieve mid-infrared plasmon modulation by magnetic field with the GMR structures was demonstrated. And combining GMR structures with metasurfaces can archive magnetic modulation of far- and near-field infrared properties [64].

10.3.3 DETECTION OF OPTICAL, INFRARED, AND TERAHERTZ WAVE

Optical and infrared detection via spin photodiode. As discussed in Section 10.3.1, a spin-LED can generate circularly polarized light from the spin of electrons. Not surprisingly, the device can also work on the opposite principle—detecting the incident circularly polarized light, which makes the device act as "spin photodiode". The spin photodiode typically consists of a FM/(barrier layer)/semiconductor structure similar to spin-LED, and the mechanism is as follows: i) circularly polarized light irradiated on a spin photodiode excites spin-polarized carriers in the semiconductor layer; ii) the excited spin-up and spin-down electrons transport to the magnetic layer and are subjected to different resistances; iii) it results different electromotive forces according to the relations between their spin directions and the magnetization direction, which contains the information related to the light polarization state. A variety of materials and structures have been employed to build spin photodiode for room temperature operation—for example, GaAs-like direct band-gap semiconductors where spin-polarized electrons can be generated by optical orientation, MgO insulators which are ideal tunneling materials, and hybrid organic-inorganic materials which can replace the common FM contacts. However, the performance of the spin photodiode remains limited and the underlying physics has been far from being clearly understood during a long time. Just recently, a study by V. I. Safarov et al. revealed that the spin photo-signal is determined by a factor arising from the competition between tunneling into the ferromagnet and recombination with the holes, and suggested improving the spin photodiode performance by increasing the tunnel resistance [65].

Optical and infrared detection via magnetic heterostructures. In 2019, S. Kattel et al. demonstrated that the spin Seebeck effect (SSE) in Pt/YIG heterostructures can be used to detect light across a broad optical range—ultraviolet through visible to near-infrared (390 to 2200 nm) [66], as illustrated in Figure 10.10(a). Under light illumination, a temperature gradient will be built in Pt/YIG heterostructures which induces a pure spin current injecting into the Pt layer due to the spin Seebeck effect. And then the spin current converts into a charge current via the ISHE leading to a voltage which carries the information of the incident light and can be readily detected. The results showed that the spin-based device responsivity is remarkably flat across

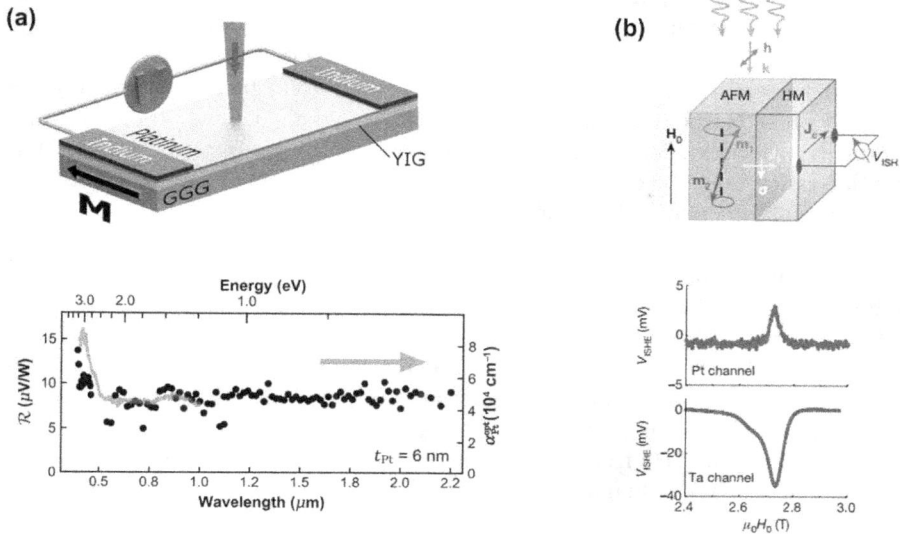

FIGURE 10.10 (a) Light detection based on spin Seebeck effect in Pt/YIG heterostructure.

Source: Reprinted with permission from Kattel, S. et al. *Phys. Rev. Appl.* 12, 034047, 2019. Copyright 2019 American Physical Society.

(b) THz detection based on antiferromagnetic spin pumping and ISHE in AFM/NM heterosturcture.

Source: Reprinted with permission from Li, J. et al. *Nature* 578, 70, 2020. Copyright 2020 Springer Nature.

a broad optical range which lies in the range of the charge-based Si-InGaAs light detector, and the performance may be competitive by enhancing the spin Seebeck effect in the future.

Terahertz detection via magnetic heterostructures. As mentioned in Section 10.2.2, the antiferromagnetic spin pumping is an effect in AFM/NM heterostructure that spin current is pumped out of the AFM layer into the adjacent NM metallic layer at the antiferromagnetic resonance, which was firstly proposed by R. Cheng et al [67]. When an ac continuous-wave THz radiation excites the antiferromagnetic resonance leading to antiferromagnetic spin pumping, the NM layer with large spin Hall angle will convert the dc part of the pumped spin current into a dc charge current. It indicates that AFM/NM heterostructure can act as a THz detector which converts the ac THz radiation into a detectable dc electric signal. In 2020, two groups realized this new electrical detection method independently. J. X. Li et al. reported that THz wave at 0.24 THz was electrically detected with heterostructures consisting of a uniaxial antiferromagnetic Cr_2O_3 crystal and a heavy metal (Pt or Ta in its β phase), as illustrated in Figure 10.10(b). The detected signals were reversed by switching the detector metal from Pt to Ta (meanwhile reversing the sign of the spin Hall angle), or by flipping the magnetic-field direction (meanwhile reversing the magnon chirality), which confirms the origination of antiferromagnetic spin pumping and ISHE [19]. P. Vaidya et al. observed electric signals due to THz induced antiferromagnetic spin pumping in MnF_2/Pt heterostructure, and confirmed that the signals depend on the chirality of the dynamical modes of the antiferromagnet. It supports the structure to act as a polarization sensitive detector, as it can be selectively excited by the handedness of the circularly polarized THz irradiation [20]. The theory developed by A. Safin et al. shows that the sensitivity of this THz detector can be enhanced by increasing the magnitude of the applied magnetic field, or by decreasing the thickness of the AFM layer. In addition, their estimations show that the sensitivity could be comparable to that of modern detectors based on the Schottky, Gunn, or graphene-based diodes [68]. Another type of THz detector based on an Pt/AFM/MgO/Pt antiferromagnetic tunnel junction was proposed by P. Y. Artemchuk et al., where the generation of the output dc voltage is originated from the mixing and rectification of the input THz frequency ac current and the THz frequency oscillations of junction's tunneling magnetoresistance, which ocurrs during antiferromagnetic resonance induced by under an external bias dc current [69]. Furthermore, P. Y. Artemchuk et al. proposed that spectrum analysis may be performed by using this antiferromagnetic tunnel junction [70].

Optical, infrared, and terahertz detection via other spintronic materials. As discussed in Section 10.3.1, THz pulse can be generated by fs laser pulse pump in spintronic materials based on circular photogalvanic effect (CPGE). Meanwhile, a dc photocurrent (accompanied by a dc voltage signal) can be generated by continuous-wave laser illumination due to CPGE, which makes these spintronic materials can also be used as light detectors. In addition, K. Ando et al. have demonstrated the direct conversion of light-polarization information into electric voltage by using the photoinduced ISHE in a Pt/GaAs hybrid structure [71]. The mechanism is as follows: i) a circularly polarized light generates spin-polarized carriers in the GaAs layer due to optical selection rules, which induce a spin current into the Pt layer through the interface; ii) the spin current is converted into an electric voltage based on ISHE, which is proportional to the degree of circular polarization of the illuminated light. Similar behaviors have also been realized in other metal/semiconductor structures, such as Pt/Ge, Pt/Si, and Au/InP. All of them work at room temperature without bias voltage and magnetic fields. Besides, J. Puebla et al. reported the light detection via the photoinduced Rashba spin-to-charge conversion at the Cu(111)/α-Bi_2O_3 interface [72].

10.4 SUMMARY AND OUTLOOK

With fruitful physical properties in the optical, infrared, and THz wave range, magnetic and spintronic materials pave new pathways for the generation, manipulation, and detection of electromagnetic wave in these spectra. In this chapter, the physical principles and typical effects of magnetic and spintronic materials have been introduced and the recent progresses of the related devices as well as their applications have been briefly reviewed. There are various mechanisms for electromagnetic wave generation based on spin related effects, such as Spin-LED for generating optical and infrared waves, THz pulses emission from a single magnetic layer, magnetic heterostructures, topological spintronic materials, and multiferroic nano-films while illuminated by fs laser pulses. Meanwhile, optical and infrared detection have been realized by spin photodiode, magnetic heterostructures, topological spintronic materials, and metal/semiconductor structures, and continuous-wave THz detection have been demonstrated by AFM/NM heterostructure. Besides, manipulation of optical, infrared, and terahertz wave have been realized via magneto-optical effects and magneto-resistance effect in various magnetic and spintronic materials. And more excitingly, great success has been achieved in the applications of the spintronic devices. One of the most remarkable examples is the spintronic THz emitter consisting of FM/NM heterostructure, which generates THz pulses with frequency up to 30 THz and the efficiency reaching the same level of the widely used commercial ZnTe crystals. It has already been applied to THz time-domain spectroscopy (THz-TDS), ghost microscope system, on-chip magneto-optic imager, and biosensing.

Looking to the future, it is clear that further improvement of the performance of the existent devices will have a strong impact on their applications for optical, infrared, and terahertz (THz) wave. Combining and introducing diverse physical effects and materials is a hopeful way to achieve this goal. For example, pure spin current with 100% spin-polarization from ISHE or IREE may be an more efficient spin injection source for spin-LED; integration of thermoelectric material and spintronic THz emitter may further improve the strength of THz emission, due to the superposition of the ultrafast photothermoelectric currents [73] and ultrafast spin-charge currents; introducing spin-orbit torque to the magneto-resistance based THz modulator will increase the modulation speed; by using alloys or insertion can enhance the spin Seebeck effect and further improve the light detection efficiency. At the same time, the previous studies show that several theoretically proposed devices for optical, infrared, and terahertz (THz) wave applications have excellent performances—for example, the THz nano-oscillator and THz nano-detector based on antiferromagnet. Although they haven't been realized up to date, the recent discoveries and experimental advancement give us more confidence to make them come true in the near future. We can expect that, the emergent success will continue to inspire the researchers to pursue better performance of the existent devices and realize/discover new devices/effects in the optical, infrared and THz wave range with magnetic and spintronic materials, and promote them into practical applications.

ACKNOWLEDGEMENTS

This work is supported by the National Key R&D Program of China (2021YFA1401400), the National Natural Science Foundation of China (NSFC) (62027807) and Science Challenge Project (TZ2018003).

BIBLIOGRAPHY

[1] Hirohata, A., Yamada, K., Nakatani, Y., Prejbeanu, I.-L., Diény, B., Pirro, P., Hillebrands, B. 2020. Review on spintronics: Principles and device applications. *J. Magn. Magn. Mater.* 509: 166711.

[2] Baibich, M. N., Broto, J. M., Fert, A., Van Dau, F. N., Petroff, F., Etienne, P., Creuzet, G., Friederich, A., Chazelas, J. 1988. Giant magnetoresistance of (001)Fe/(001)Cr magnetic superlattices. *Phys. Rev. Lett.* 61: 2472.

[3] Binasch, G., Grünberg, P., Saurenbach, F., Zinn, W. 1989. Enhanced magnetoresistance in layered magnetic structures with antiferromagnetic interlayer exchange. *Phys. Rev. B.* 39: 4828.

[4] Zhu, J.-G., Park, C. 2006. Magnetic tunnel junctions. *Mater. Today.* 9: 36.

[5] Nogués, J., Schuller, I. K. 1999. Exchange bias. *J. Magn. Magn. Mater.* 192: 203.

[6] Ralph, D. C., Stiles, M. D. 2008. Spin transfer torques. *J. Magn. Magn. Mater.* 320: 1190.

[7] Han, X., Wang, X., Wan, C., Yu, G., Lv, X. 2021. Spin-orbit torques: Materials, physics, and devices. *Appl. Phys. Lett.* 118: 120502.

[8] Sinova, J., Valenzuela, S. O., Wunderlich, J., Back, C. H., Jungwirth, T. 2015. Spin hall effects. *Rev. Mod. Phys.* 87: 1213.

[9] Dey, P., Roy, J. N., 2021. *Spintronics: Fundamentals and applications*, edited by P. Dey and J. N. Roy, Springer Singapore, Singapore, p. 163.

[10] Walowski, J., Münzenberg, M. 2016. Perspective: Ultrafast magnetism and THz spintronics. *J. Appl. Phys.* 120: 140901.

[11] Iusipova, I. A., Popov, A. I. 2021. Spin valves in microelectronics (a review). *Semiconductors.* 55: 1008.

[12] Hasan, M. Z., Kane, C. L. 2010. Colloquium: Topological insulators. *Rev. Mod. Phys.* 82: 3045.

[13] Feng, Z., Hu, J., Sun, L., et al. 2012. Spin Hall angle quantification from spin pumping and microwave photoresistance. *Phys. Rev. B.* 85: 214423.

[14] Cheng, R., Xiao, J., Niu, Q., Brataas, A. 2014. Spin pumping and spin-transfer torques in antiferromagnets. *Phys. Rev. Lett.* 113: 057601.

[15] Awari, N., Kovalev, S., Fowley, C., et al. 2016. Narrow-band tunable terahertz emission from ferrimagnetic Mn3-xGa thin films. *Appl. Phys. Lett.* 109: 032403.

[16] Nagaosa, N., Sinova, J., Onoda, S., MacDonald, A. H., Ong, N. P. 2010. Anomalous Hall effect. *Rev. Mod. Phys.* 82: 1539.

[17] Song, Q., Zhang, H., Su, T., Yuan, W., Chen, Y., Xing, W., Shi, J., Sun, J., Han, W. 2017. Observation of inverse Edelstein effect in Rashba-split 2DEG between SrTiO$_3$ and LaAlO$_3$ at room temperature. *Sci. Adv.* 3: e1602312.

[18] Tserkovnyak, Y., Brataas, A., Bauer, G. E. W. 2002. Spin pumping and magnetization dynamics in metallic multilayers. *Phys. Rev. B.* 66: 224403.

[19] Li, J., Wilson, C. B., Cheng, R., et al. 2020. Spin current from sub-terahertz-generated antiferromagnetic magnons. *Nature.* 578: 70.

[20] Vaidya, P., Morley Sophie, A., van Tol, J., Liu, Y., Cheng, R., Brataas, A., Lederman, D., del Barco, E. 2020. Subterahertz spin pumping from an insulating antiferromagnet. *Science.* 368: 160.

[21] Uchida, K., Takahashi, S., Harii, K., Ieda, J., Koshibae, W., Ando, K., Maekawa, S., Saitoh, E. 2008. Observation of the spin Seebeck effect. *Nature.* 455: 778.

[22] Nádvorník, L., Borchert, M., Brandt, L., et al. 2021. Broadband Terahertz probes of anisotropic magnetoresistance disentangle extrinsic and intrinsic contributions. *Phys. Rev. X.* 11: 021030.

[23] Ramirez, A. P. 1997. Colossal magnetoresistance. *J. Phys. Condens. Matter.* 9: 8171.

[24] Taniyama, T., Wada, E., Itoh, M., Yamaguchi, M. 2011. Electrical and optical spin injection in ferromagnet/semiconductor heterostructures. *NPG Asia Mater.* 3: 65.

[25] Beaurepaire, E., Merle, J. C., Daunois, A., Bigot, J. Y. 1996. Ultrafast spin dynamics in ferromagnetic nickel. *Phys. Rev. Lett.* 76: 4250.

[26] Beaurepaire, E., Turner, G. M., Harrel, S. M., Beard, M. C., Bigot, J.-Y., Schmuttenmaer, C. A. 2004. Coherent terahertz emission from ferromagnetic films excited by femtosecond laser pulses. *Appl. Phys. Lett.* 84: 3465.

[27] Zhang, Q., Luo, Z., Li, H., Yang, Y., Zhang, X., Wu, Y. 2019. Terahertz emission from anomalous hall effect in a single-layer ferromagnet. *Phys. Rev. Appl.* 12: 054027.

[28] Nishitani, J., Kozuki, K., Nagashima, T., Hangyo, M. 2010. Terahertz radiation from coherent antiferromagnetic magnons excited by femtosecond laser pulses. *Appl. Phys. Lett.* 96: 221906.

[29] Awari, N., Kovalev, S., Fowley, C., et al. 2016. Narrow-band tunable terahertz emission from ferrimagnetic Mn3-xGa thin films. *Appl. Phys. Lett.* 109: 032403.

[30] Kampfrath, T., Battiato, M., Maldonado, P., et al. 2013. Terahertz spin current pulses controlled by magnetic heterostructures. *Nat. Nanotechnol.* 8: 256.

[31] Seifert, T., Jaiswal, S., Martens, U., et al. 2016. Efficient metallic spintronic emitters of ultrabroadband terahertz radiation. *Nat. Photonics.* 10: 483.

[32] Feng, Z., Yu, R., Zhou, Y., et al. 2018. Highly efficient spintronic terahertz emitter enabled by metal–dielectric photonic crystal. *Adv. Opt. Mater.* 6: 1800965.

[33] Feng, Z., Qiu, H., Wang, D., Zhang, C., Sun, S., Jin, B., Tan, W. 2021. Spintronic terahertz emitter. *J. Appl. Phys.* 129: 010901.

[34] Chen, S.-C., Feng, Z., Li, J., et al. 2020. Ghost spintronic THz-emitter-array microscope. *Light Sci. Appl.* 9: 99.

[35] Bulgarevich, D. S., Akamine, Y., Talara, M., Mag-usara, V., Kitahara, H., Kato, H., Shiihara, M., Tani, M., Watanabe, M. 2020. Terahertz magneto-optic sensor/imager. *Sci. Rep.* 10: 1158.

[36] Bai, Z., Liu, Y., Kong, R., et al. 2020. Near-field Terahertz Sensing of HeLa cells and pseudomonas based on monolithic integrated metamaterials with a spintronic Terahertz emitter. *ACS Appl. Mater. Interfaces.* 12: 35895.

[37] Seifert, T. S., Jaiswal, S., Barker, J., et al. 2018. Femtosecond formation dynamics of the spin Seebeck effect revealed by terahertz spectroscopy. *Nat. Commun.* 9: 2899.

[38] Qiu, H., Zhou, L., Zhang, C., et al. 2021. Ultrafast spin current generated from an antiferromagnet. *Nat. Phys.* 17: 388.

[39] Wang, X., Cheng, L., Zhu, D., et al. 2018. Ultrafast spin-to-charge conversion at the surface of topological insulator thin films. *Adv. Mater.* 30: 1802356.

[40] Cheng, L., Wang, X., Yang, W., et al. 2019. Far out-of-equilibrium spin populations trigger giant spin injection into atomically thin MoS2. *Nat. Phys.* 15: 347.

[41] Jungfleisch, M. B., Zhang, Q., Zhang, W., Pearson, J. E., Schaller, R. D., Wen, H., Hoffmann, A. 2018. Control of Terahertz emission by ultrafast spin-charge current conversion at rashba interfaces. *Phys. Rev. Lett.* 120: 207207.

[42] Zhou, C., Liu, Y. P., Wang, Z., et al. 2018. Broadband Terahertz generation via the interface inverse rashba-edelstein effect. *Phys. Rev. Lett.* 121: 086801.

[43] Shen, J., Feng, Z., Xu, P., Hou, D., Gao, Y., Jin, X. 2021. Spin-to-charge conversion in ag/bi bilayer revisited. *Phys. Rev. Lett.* 126: 197201.

[44] Braun, L., Mussler, G., Hruban, A., Konczykowski, M., Schumann, T., Wolf, M., Münzenberg, M., Perfetti, L., Kampfrath, T. 2016. Ultrafast photocurrents at the surface of the three-dimensional topological insulator Bi2Se3. *Nat. Commun.* 7: 13259.

[45] Kinoshita, Y., Kida, N., Miyamoto, T., Kanou, M., Sasagawa, T., Okamoto, H. 2018. Terahertz radiation by subpicosecond spin-polarized photocurrent originating from Dirac electrons in a Rashba-type polar semiconductor. *Phys. Rev. B.* 97: 161104.

[46] Chen, M., Lee, K., Li, J., Cheng, L., Wang, Q., Cai, K., Chia, E., Chang, H., Yang, H. 2020. Anisotropic picosecond spin-photocurrent from weyl semimetal WTe 2. *ACS Nano.* 14: 3539.

[47] Tong, M., Hu, Y., Xie, X., Zhu, X., Wang, Z., Cheng, X. a., Jiang, T. 2020. Helicity-dependent THz emission induced by ultrafast spin photocurrent in nodal-line semimetal candidate Mg_3Bi_2. *Opto-Electron Adv.* 3: 200023.

[48] Hirai, Y., Yoshikawa, N., Hirose, H., Kawaguchi, M., Hayashi, M., Shimano, R. 2020. Terahertz emission from bismuth thin films induced by excitation with circularly polarized light. *Phys. Rev. Appl.* 14: 064015.

[49] Guzelturk, B., Mei, A. B., Zhang, L., Tan, L. Z., Donahue, P., Singh, A. G., Schlom, D. G., Martin, L. W., Lindenberg, A. M. 2020. Light-induced currents at domain walls in multiferroic $BiFeO_3$. *Nano. Lett.* 20: 145.

[50] Cheng, R., Xiao, D., Brataas, A. 2016. Terahertz antiferromagnetic spin hall nano-oscillator. *Phys. Rev. Lett.* 116: 207603.

[51] Sulymenko, O. R., Prokopenko, O. V., Tiberkevich, V. S., Slavin, A. N., Ivanov, B. A., Khymyn, R. S. 2017. Terahertz-frequency spin hall auto-oscillator based on a canted antiferromagnet. *Phys. Rev. Appl.* 8: 064007.

[52] Shukla, A., Rakheja, S. 2022. Spin-torque-driven Terahertz auto-oscillations in noncollinear coplanar antiferromagnets. *Phys. Rev. Appl.* 17: 034037.

[53] Lisenkov, I., Khymyn, R., Åkerman, J., Sun, N. X., Ivanov, B. A. 2019. Subterahertz ferrimagnetic spin-transfer torque oscillator. *Phys. Rev. B.* 100: 100409.

[54] Zhong, H., Qiao, S., Yan, S., Liang, L., Zhao, Y., Kang, S. 2020. Terahertz spin-transfer torque oscillator based on a synthetic antiferromagnet. *J. Magn. Magn. Mater.* 497: 166070.

[55] Bi, L., Hu, J., Jiang, P., Kim, D. H., Dionne, G. F., Kimerling, L. C., Ross, C. A. 2011. On-chip optical isolation in monolithically integrated non-reciprocal optical resonators. *Nat. Photonics.* 5: 758.

[56] Yang, Y., Liu, T., Bi, L., Deng, L. 2021. Recent advances in development of magnetic garnet thin films for applications in spintronics and photonics. *J. Alloys Compd.* 860: 158235.

[57] Yang, W., Liu, Q., Wang, H., et al. 2022. Observation of optical gyromagnetic properties in a magneto-plasmonic metamaterial. *Nat. Commun.* 13: 1719.

[58] Li, T.-F., Li, Y.-L., Zhang, Z.-Y., Yang, Q.-H., Fan, F., Wen, Q.-Y., Chang, S.-J. 2020. Terahertz faraday rotation of magneto-optical films enhanced by helical metasurface. *Appl. Phys. Lett.* 116: 251102.

[59] Wang, X. B., Cheng, L., Wu, Y., Zhu, D. P., Wang, L., Zhu, J.-X., Yang, H., Chia, E. E. M. 2017. Topological-insulator-based terahertz modulator. *Sci. Rep.* 7: 13486.

[60] Liu, X., Xiong, L., Yu, X., He, S., Zhang, B., Shen, J. 2018. Magnetically controlled terahertz modulator based on Fe_3O_4 nanoparticle ferrofluids. *J. Phys. D.* 51: 105003.

[61] Jin, Z., Tkach, A., Casper, F., et al. 2015. Accessing the fundamentals of magnetotransport in metals with terahertz probes. *Nat. Phys.* 11: 761.

[62] Jin, Z., Li, J., Zhang, W., et al. 2020. Magnetic modulation of Terahertz waves via spin-polarized electron tunneling based on magnetic tunnel junctions. *Phys. Rev. Appl.* 14: 014032.

[63] Lloyd-Hughes, J., Mosley, C. D., Jones, S. P., Lees, M. R., Chen, A., Jia, Q. X., Choi, E M., MacManus-Driscoll, J. L. 2017. Colossal Terahertz magnetoresistance at room temperature in epitaxial $La_{0.7}Sr_{0.3}MnO_3$ nanocomposites and single-phase thin films. *Nano. Lett.* 17: 2506.

[64] Armelles, G., Cebollada, A. 2020. Active photonic platforms for the mid-infrared to the THz regime using spintronic structures. *Nanophotonics.* 9: 2709.

[65] Safarov, V. I., Rozhansky, I. V., Zhou, Z., Xu, B., Wei, Z., Wang, Z.-G., Lu, Y., Jaffrès, H., Drouhin, H.-J. 2022. Recombination time mismatch and spin dependent photocurrent at a ferromagnetic-metal--semiconductor tunnel junction. *Phys. Rev. Lett.* 128: 057701.

[66] Kattel, S., Murphy, J. R., Ellsworth, D., Ding, J., Liu, T., Li, P., Wu, M., Rice, W. D. 2019. Broadband optical detection using the spin seebeck effect. *Phys. Rev. Appl.* 12: 034047.

[67] Cheng, R., Xiao, J., Niu, Q., Brataas, A. 2014. Spin pumping and spin-transfer torques in antiferromagnets. *Phys. Rev. Lett.* 113: 057601.

[68] Safin, A., Nikitov, S., Kirilyuk, A., Tyberkevych, V., Slavin, A. 2022. Theory of antiferromagnet-based detector of terahertz frequency signals. *Magnetochemistry.* 8: 26.

[69] Artemchuk, P. Y., Sulymenko, O. R., Prokopenko, O. V., 2019. A resonance-type terahertz-frequency signal detector based on an antiferromagnetic tunnel junction. *2019 IEEE 8th International Conference on Advanced Optoelectronics and Lasers*, p. 240.

[70] Artemchuk, P., Sulymenko, O., Louis, S., et al. 2020. Terahertz frequency spectrum analysis with a nanoscale antiferromagnetic tunnel junction. *J. Appl. Phys.* 127: 063905.

[71] Ando, K., Morikawa, M., Trypiniotis, T., Fujikawa, Y., Barnes, C. H. W., Saitoh, E. 2010. Photoinduced inverse spin-Hall effect: Conversion of light-polarization information into electric voltage. *Appl. Phys. Lett.* 96: 082502.

[72] Puebla, J., Auvray, F., Yamaguchi, N., Xu, M., Bisri, S. Z., Iwasa, Y., Ishii, F., Otani, Y. 2019. Photoinduced Rashba spin-to-charge conversion via an interfacial unoccupied state. *Phys. Rev. Lett.* 122: 256401.

[73] Lu, W., Fan, Z., Yang, Y., et al. 2022. Ultrafast photothermoelectric effect in Dirac semimetallic Cd3As2 revealed by terahertz emission. *Nat. Commun.* 13: 1623.

11 Soft and Flexible Materials and Their Applications

Yongbiao Wan, Zhiguang Qiu, and Chuan Fei Guo

CONTENTS

11.1 Introduction...286
11.2 Soft Substrate Materials..287
    11.2.1 PDMS 287
    11.2.2 PI..288
    11.2.3 Hydrogel 289
    11.2.4 Textile 290
    11.2.5 Paper 291
    11.2.6 Other Substrates..291
11.3 Flexible Active Materials..292
    11.3.1 Metal and Semiconductor NWs..292
        11.3.1.1 AgNWs...292
        11.3.1.2 CuNWs...293
        11.3.1.3 AuNWs...296
        11.3.1.4 Semiconductor NWs..296
    11.3.2 Carbon Materials 297
        11.3.2.1 Graphene 297
        11.3.2.2 CNT..298
    11.3.3 Conductive Polymer..299
        11.3.3.1 PEDOT:PSS..299
        11.3.3.2 PPy 300
        11.3.3.3 PANi 300
        11.3.3.4 CPC 302
11.4 Applications of Soft and Flexible Materials 302
    11.4.1 Flexible Transparent Electrode 302
    11.4.2 Flexible Photovoltaic Cell...306
    11.4.3 Flexible Photodetector 309
    11.4.4 Flexible Display...312
    11.4.5 Flexible Optical Security Label 316
11.5 Summary and Perspectives 321
Acknowledgments..322
Bibliography 322

DOI: 10.1201/9781003202608-11

11.1 INTRODUCTION

Since the inception of the world's first integrated circuit in 1959, the industrial revolutions on microelectronics technology have lasted for six decades. Although rigid electronics based on traditional silicon materials have been the mainstream for industrial and social applications, the increasing growth in intelligent production, daily wearables, and digital life demands flexibility in electronic systems. Flexible electronics, which enable the retention of functions of circuits and electric components when bent or stretched, are expected to introduce paradigm-shift changes to the electronic industry and extend the applications into a new era. Many advantages contribute to the attractiveness of flexible electronics. They are typically lighter, portable, inexpensive, and more robust compared with their rigid substrate counterparts. Figure 11.1 shows a remarkable and continued increase on the annual papers published related to flexible electronics from 430 (2005) to 4056 (2020), indicating a flourishing development of this field. On top of the academic values, flexible electronics are also quickly maturing into an extremely interesting discipline that could considerably benefit both industrial and social applications, for instance, flexible display [1], flexible sensing [2], flexible energy storage [3], and flexible detection [4].

A generic structure of flexible electronic devices contains a substrate and an active layer [2]. To make the structure flexible, all components must be capable of sustaining bending to some extent without losing their functions, which can hardly be fulfilled via conventional inorganic electronic materials with rigid structures. To empower high-performance flexible electronics, soft and flexible electronic materials with

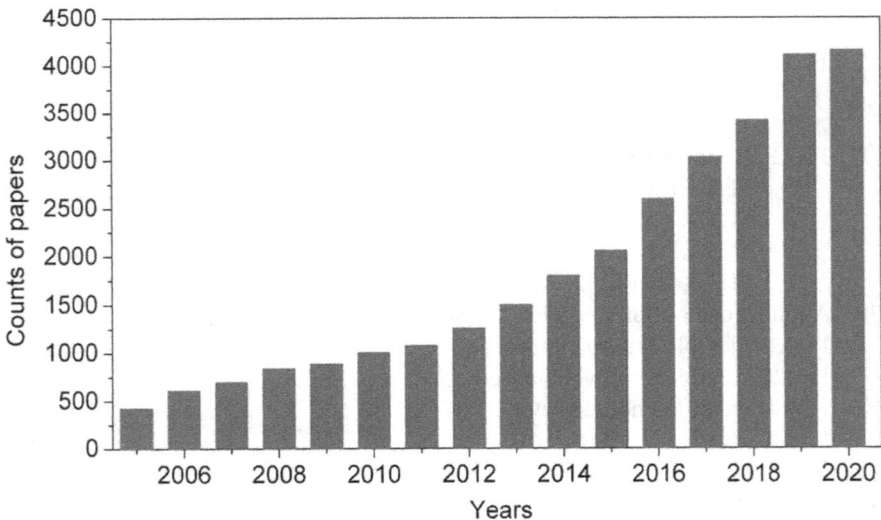

FIGURE 11.1 Number of published papers related to flexible electronics since 2005

Source: Data from Web of Science.

high mechanical compliance and favorable electrical properties that allow large-area processing are exigent.

This chapter focuses on the recent progress on soft and flexible materials, including both substrate and active layer components, and their applications in optoelectronics (*e.g.*, flexible transparent electrode, photovoltaic cell, photodetector, display, and optical security label). Moreover, some critical challenges and perspectives on developing soft and flexible materials are also presented in the latter part of this chapter.

11.2   SOFT SUBSTRATE MATERIALS

Flexible substrates, as both structural support and optical signal pathway, perform ever-increasing important functions in advanced optoelectronic devices. Appropriate materials with high mechanical flexibility and low roughness are essential in developing flexible electronics. This section introduces several soft substrate materials which are frequently applied to flexible electronics, *e.g.*, polydimethylsiloxane (PDMS), polyimide (PI), hydrogels, and textile (Figure 11.2) [5–12], *etc.*

11.2.1   PDMS

As a class of silicone, PDMS is the most widely used and well-studied substrate material for flexible electronics because it is simple to process, mechanically robust, highly transparent, chemically inert, biologically compatible, and has satisfactory thermal stability [2]. In particular, a PDMS membrane exhibits transmittance over 95% in optical bands and stretchability almost 100%, rendering it a promising candidate for flexible optoelectronic devices.

PDMS is made of a flexible Si–O backbone and a repeating $Si(CH_3)_2O$ unit [13]. The number of repeating units of $Si(CH_3)_2O$ typically defines molecular weight, thereby determining the viscoelastic properties of many PDMS-based materials. The Si–O–Si backbone endows PDMS elastomers with exceptional properties, such as high flexibility, thermal and electrical resistance, non-toxicity, non-flammability, and low bulk density. Typically, PDMS materials are formed by chemicals that are crosslinked via strong covalent bonds, and the crosslinking process advances by adding crosslinking curing agents. The strong covalent bonds among the molecular chains endow PDMS materials with superior thermal and mechanical stabilities, enabling the materials to exhibit recoverable features after tension, torsion, or compression. In addition, the mechanical properties of PDMS materials can be modulated by adjusting the amount of incorporated agents or the level of crosslinking temperature.

On top of that, mature technologies and manufacturing crafts have enabled the PDMS substrate with designable functionalities. Surface modification, such as charge physisorption or plasma irradiation, can form patterns with the adhesive or non-adhesive region on the PDMS surface to selectively bond the active layer materials [14]. Regulating the composition of prepolymers before polymerization can control their elasticity properties, transparency, and self-healing capabilities. Interestingly, PDMS prepolymers can mix with other active materials, allowing the fabrication of PDMS-based conductive composites. Furthermore, benefiting from the

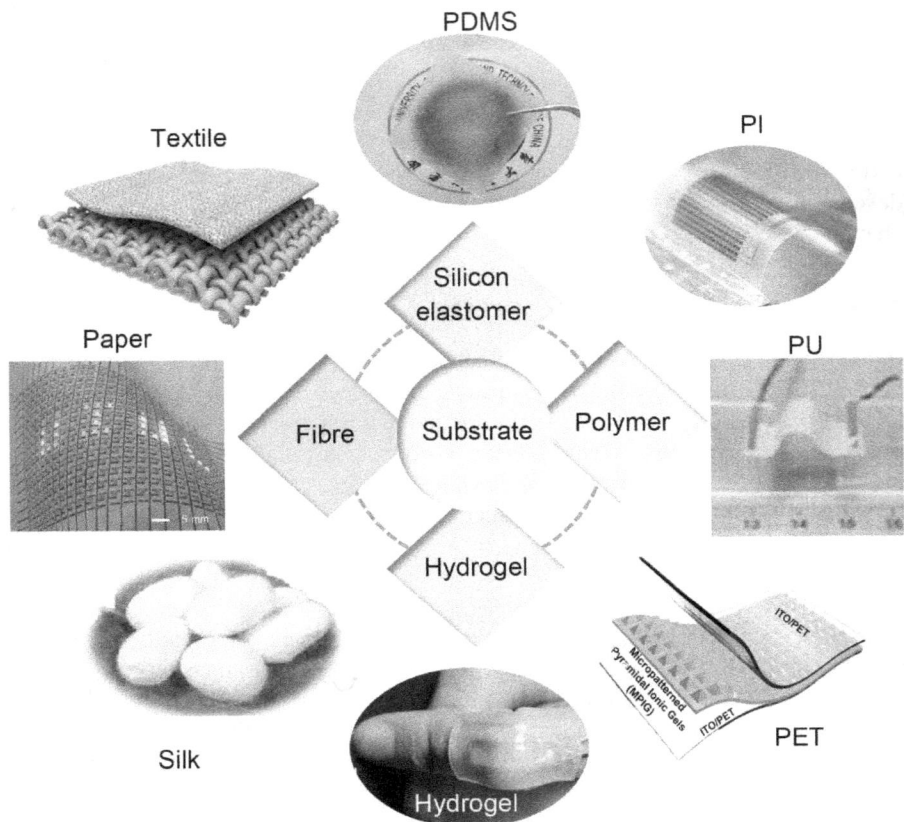

FIGURE 11.2   Schematic of material substrates for flexible devices. Clockwise from top: silicone elastomer ("PDMS" image: Reproduced with permission [5]. Copyright 2018, Wiley-VCH); PI ("PI" image: Reproduced with permission [8]. Copyright 2017, Springer Nature); Polymer ("PU" image: Reproduced under a Creative Commons license CC BY 4.0 [6]. Copyright 2021, the Authors); PET ("PET" image: Reproduced with permission [7]. Copyright 2017, ACS Publications); Hydrogel ("Hydrogel" image: Reproduced with permission [9]. Copyright 2014, Wiley-VCH); Fiber ("Silk" image: Reproduced with permission [10]. Copyright 2019, ACS Publications); Paper ("Paper" image: Reproduced with permission [11]. Copyright 2011, Wiley-VCH); Textile ("Textile" image: Reproduced with permission [12]. Copyright 2015, ACS Publications)

thermosetting process, a template-assisted strategy can easily endow PDMS with different surface structures, such as micropillar, micropyramid, and microdome [15].

11.2.2   PI

PI possesses excellent thermal stability, mechanical properties, and insulating feature It is regarded as one of the most promising flexible substrate candidates for industrial applications such as electronic packaging. Among various polymer substrates, PI exhibits great bendability, extremely low creep, and high mechanical strength, which are well-maintained even at a temperature up to 400 °C [16]. The high thermal stability of PI allows for the fabrication of refractory flexible electronic devices. Furthermore, PI is also resistant to weak acids and bases as well as commonly used organic solvents, such as ethanol and acetone. Nevertheless, it cannot recover under large strains, which prohibits it from being a highly stretchable substrate.

Different from transparent PDMS, conventional PI is a brown or brownish-yellow film because of the high crystallinity and stiffness of polymer backbones, strong intermolecular interaction through π–π interaction, and charge transfer complex formation,[8] which naturally leads to undesired absorption in the visible light range and severely degrades the transparency of PI-based flexible electronics [17]. For particular applications such as flexible transparent display and solar cell, PI materials with high transparency and low birefringent are demanded. In recent years, tremendous efforts have been devoted to reduce the yellowness and improve the transparency of PI film. This was achieved by incorporating fluorine, chlorine, sulfone groups, or unsymmetrical and bulky pendant units, as well as adopting the alicyclic moieties in the polymer structure [18]. Eventually, transparent and colorless PI films (CPI) and their analogues, including polyamideimide and polyetherimide, have been developed [18], which rapidly occupied the leading-edge position in the advanced polymer optical film industry. With such high technological contents, CPI films are suitable candidates for advanced flexible optoelectronic applications.

11.2.3 HYDROGEL

Hydrogel consists of three-dimensional (3D) polymer networks and considerable amounts of water molecules (> 50 wt%). It has been emerged as an ideal candidate for various flexible applications due to its considerable stretchability, biocompatibility, fatigue resistance, and reliability. The polymer network (size: ~10 nm) provides a solid elastic property, allowing the molecules to ingress into hydrogel to swell the networks. Owing to the substantial amount of water, hydrogel resembles the chemical and physical properties of liquid [19].

Hydrogel can be synthesized by polymerization techniques, such as bulk polymerization, solution polymerization/cross-linking, suspension or inverse-suspension polymerization, support grafting, and polymerization by irradiation. The classification of hydrogel products can be based on the various standards: (a) source: natural or synthetic hydrogel; (b) polymeric composition: homopolymeric hydrogels, copolymeric hydrogels, or multipolymer interpenetrating polymeric hydrogel; (c) configuration: amorphous (non-crystalline) or semicrystalline (a complex mixture of amorphous and crystalline phases, and crystalline); (d) cross-linking: chemical or physical cross-linking; (e) physical appearance: matrix, film, or microsphere; (f) network electrical charge: nonionic (neutral), ionic (*e.g.*, anionic or cationic), amphoteric electrolyte (ampholytic) containing both acidic and basic groups, or zwitterionic (polybetaines) containing both anionic and cationic groups. Moreover, the synthesis process can be categorized into either a one-step (polymerization and simultaneous reactions of multifunctional monomers) or multiple-step (polymerization of molecules followed by subsequent side reactions of their reactive groups) procedure. The large degree of freedom in processing allows engineers to flexibly tailored the properties of hydrogel polymer networks, including biodegradation, mechanical strength, conductivity, roughness, self-healing, anti-drying or freezing, and chemical and biological responses to stimulus [20].

Hydrogels have high deformability, excellent biocompatibility and robustness, and potential functionalization due to their ability to retain water or biological fluids, which are particularly beneficial for waste treatment, biomedicine, and biosensors. Among various kinds of hydrogels, double-network hydrogels have been demonstrated to exhibit extraordinary mechanical properties including high fracture resistance, high modulus, and excellent stretchability that are comparable to tendons, ligaments, and rubbers [21]. Since Wichterle and Lim introduced the first synthetic hydrogel in 1954, hydrogels have been applied to sealing, coal dewatering, biomolecule or cell separation, agriculture, soft actuators, soft robotics, displays, pharmaceuticals, regenerative medicines, wound dressing, drug delivery systems, and diagnostics. They are also used as electronic skins, hygienic products, biological adhesives, and food additives [20].

11.2.4 TEXTILE

Textiles have been in use for thousands of years and they have been regarded as the second human skin that ensures life quality. Humans are in direct contact with textiles in their daily life that may be in the form of clothing, pillows, bed sheets, bandages, woven bags, or even face masks. As human civilization advanced, textiles have been produced using a wide range of materials, from natural raw materials (*e.g.*, silk, wool, flax, and cotton) to artificially synthetic compounds (*e.g.*, polyester, nylon, acrylic, and aramid).

Textiles typically possess a number of unique characteristics, such as superior wearing comfort, excellent mechanical strength, softness (to accommodate complex deformation), lightweight, low cost, flexibility, and even foldability. Naturally, they are excellent substrate candidates for flexible electronics. Owing to their softness, flexibilities, and low Young's modulus, textiles can be deformed under a small external force. Furthermore, they have excellent fatigue resistance to maintain structural integrity when worn and washed. They are also capable of sustaining several millions of loading cycles under dry conditions. Mechanically, they can be bent, stretched, twisted, or sheared, thus exhibiting unique suitability to 3D drape. In terms of thermodynamics, the high porosity and large surface area of textiles keep human body warm in winter and cool in summer. Notably, many textile materials are biocompatible, biodegradable, and even bioabsorbable, which are properties highly desired for the interface between skin and electronics. An interesting work on biodegradable textile incorporating passive radiative cooling structures shows that woven metafabrics can achieve a superior refrigeration effect of 2–10 °C lower than the ambient temperature throughout the day [22].

With the rapid advancements in materials science and chemistry, textiles with versatile designabilities and superior properties are continually evolving for intelligent and flexible human-oriented electronic integration. As a soft substrate, textile architectures can be seamlessly integrated with diverse electrical functionality via the add-on or built-in technology without imposing additional burden to daily human activity. The combination of traditional textiles and electronic components is anticipated to spawn a revolutionary product—smart/electronic textiles [12].

11.2.5 PAPER

Paper, an inherently flexible material, is one of the oldest materials for storing and exchanging human information. Regular paper is produced by compressing cellulose fibers from wood into thin sheets [23]. The mesh network of cellulose in paper provides a unique set of properties that are attractive for flexible electronics as elaborated in the following [11]: (a) Paper is one of the most abundant materials on earth with extremely low cost; (b) Different from other flexible substrate materials, such as PDMS and PI, paper is an annually renewable material; (c) Simple fabrication methods and mass production are commercially available; (d) Paper can be easily folded or bent to form 3D structures without structural damage; (e) Paper possesses numerous porous internal architectures with a high surface-to-volume ratio, allowing convenient modification and absorption of functional materials; (f) Paper is lightweight which ensures the portability of flexible electronics; (g) Paper is biocompatible, biodegradable, and environmentally friendly; (h) Paper is capable of transporting liquids through its hydrophilic matrix by capillary action without the aid of external forces; (i) Wood cellulose is known to have piezoelectric properties, thereby endowing paper with a remarkable potential for piezoelectric electronics.

Although mechanical strength, flexibility, and customizability of paper are desirable traits for flexible substrate, several of its disadvantages, such as microscale surface roughness, opaqueness, and high porosity, have limited its usage in flexible electronics. For instance, the large pore sizes in paper result in poor thermal performance [24], restricting the exposure of the paper substrate to high temperatures during manufacturing. To overcome these obstacles, many researchers have invented new types of paper with reduced surface roughness and increased transparency. By combining different sizes and functional groups, this new paper substrate is also made of cellulose but with a high transparency (> 90%) and smooth (< 10 nm) surface [23], which might be beneficial in future flexible optoelectronics.

11.2.6 OTHER SUBSTRATES

Besides the aforementioned substrate materials, other commercial synthetic polymer materials, such as polyethylene terephthalate (PET) [7], polyethylene naphthalate (PEN) [25], and polyurethane (PU) [6], are suitable candidates for flexible electronics or optoelectronics.

One of the most diffused thermoplastic polymers available on the market is PET. It is bendable, lightweight, robust, and mechanically resistant to impact. However, because PET cannot be stretched under room temperature, it cannot be used in highly stretchable electronics. Regarding its optical property, semicrystalline PET is opaque (white), and amorphous PET is transparent; both can be used in flexible optical devices [26]. Similar to PET, PEN can also be utilized as a flexible substrate candidate because of its high strength and transparency, good thermal and chemical stability, superior gas barrier capability, and radiation resistance [25].

Another comparatively high-strength polymer that can be applied to the most durable and sustainable products is PU. It has high'elongation capacity, high energy absorption capacity, high resistance to aggressive environments, easy applicability, and versatility in products and applications. Further, it is thermally stable, chemically resistant, and cost-effective [27]. Beyond that, PU is capable of changing its microstructure to acquire different mechanical properties, including rigidity, elasticity, and flexibility. In particular, PU materials processed into soft elastomer or sponge are an excellent alternative substrate for flexible electronics.

11.3 FLEXIBLE ACTIVE MATERIALS

Active layer material is another important component in flexible electronic devices, which is capable of maintaining the device function when bent or stretched. To date, various kinds of materials including metal or semiconductor nanowires (NWs), carbon materials (including carbon nanotubes (CNTs) and graphene), conductive polymer, have been used as active materials for flexible electronics or optoelectronics.

11.3.1 METAL AND SEMICONDUCTOR NWS

The metals and metallic compound semiconductors used as active materials in traditional electronic devices suffer from rigid properties and poor mechanical compliance, prohibiting their integration with flexible and wearable systems. Though directly depositing metallic or semiconductor thin films onto a compliant substrate via printing, painting, sputtering, or other deposition strategies has been investigated for flexible electronic devices, they easily delaminate or crack when subjected to large stretchable loads and eventually lead to functional failures. To solve this problem, one-dimensional (1D) nanomaterials, also known as NWs, have been considered as the building blocks for active layers in flexible electronics due to their excellent mechanical flexibility and superior electrical properties. This section introduces several commonly used NW materials: silver NWs (AgNWs), copper NWs (CuNWs), gold NWs (AuNWs), Zinc oxide NWs (ZnONWs), and other semiconductor NWs.

11.3.1.1 AgNWs

Among various kinds of NWs, AgNWs are considered as one of the most promising active materials for stretchable and wearable electronics due to their high electrical and thermal conductivities, superior flexibility, and outstanding synthesis scalability and reproducibility. They have a high aspect ratio with diameters and lengths typically ranging from 10–200 nm and 5–100 μm, respectively, endowing them with unique mechanical flexibility. They can be easily dispersed in solvents and are thus compatible with low-cost, large-area, and solution-based fabrication methods, including polyol method, hydrothermal method, microwave-assisted process, electrochemical technology, ultraviolet irradiation technology, and template technology. Among these, polyol method is the most commonly used one because of its easy implementation in mass production and low cost.

Facile and large-area deposition methods of AgNWs have been developed for forming a conductive active network layer to satisfy the requirements of large-scale manufacturing of soft electronics. Wan et al. reported the production of a flexible microstructured AgNW film by depositing ultrathin AgNWs (diameter: 20 nm; length: 30 μm) onto a PDMS substrate with a bionic lotus structure [5]. The ultrathin AgNWs synthesized by the polyol method exhibited excellent mechanical compliance that can be uniformly deposited into the entire microstructure to form a conductive network (Figure 11.3(a)–11.3(c)) with a low sheet resistance (R_{sh}) below 10 $\Omega \cdot sq^{-1}$. In contrast, the commercially coarse AgNWs cannot uniformly cover the microcones due to their larger flexural rigidity, and they fall to the bottom of the microcones. Such a microstructured AgNW network enables the realization of robust capacitive flexible tactile sensors for human motion detection. Lee et al.[28] developed extremely long AgNWs for highly stretchable and highly conductive metal electrode. The high-aspect-ratio AgNWs were transferred onto a flexible substrate via vacuum filtration to realize a high-performance stretchable electrode with low R_{sh}, demonstrating a fully functioning highly stretchable LED circuit under a world record strain exceeding 460%.

11.3.1.2 CuNWs

Copper is 1000 times more abundant, 100 times less expensive and only 6% less conductive than Ag (i.e., second-most electrically conductive metal), which is also soft and ductile. Therefore, CuNWs have been extensively investigated as a potential active layer candidate for flexible electronics due to their preponderant optical, electrical, and mechanical properties. They have advantages over AgNWs in terms of low-cost and quick fabrication and can be synthesized by solution-phase chemical reduction, microwave-assisted process, self-catalytic growth, and electrospinning. Among these, low-temperature solution-phase synthesis is probably the most straightforward method and has been scaled up to produce gram-level amounts of CuNWs.

Nevertheless, CuNWs have some inherent drawbacks compared with AgNWs. In particular, they are limited by a relatively lower aspect ratio, higher tendency for oxidation, and poorer solvent dispersion. As demonstrated by Huang et al.[29], the mechanical strength of CuNWs decreases with the diameter where small-sized NWs tend to be elastic and large-sized NWs show a certain degree of ductility. To solve this problem, Ye et al.[30] proposed a rapid (30 min) ethylenediamine synthesis of CuNWs with aspect ratios as high as 5700 (diameter: 35 nm; length: 200 μm). Subsequently, transparent conducting films with a transmittance of > 95% and R_{sh} less than 100 $\Omega \cdot sq^{-1}$ have been produced. Moreover, post-processing steps, such as thermal welding, electrothermal welding, and plasmonic welding methods, have been exploited to remove oxide layers and enhance electrical conductivity. Park et al. [31] proposed a high-performance CuNW network (R_{sh}: ~17 $\Omega \cdot sq^{-1}$; transmittance: 88%) fabricated via plasmonic-tuned flash welding with ultrafast interlocking and

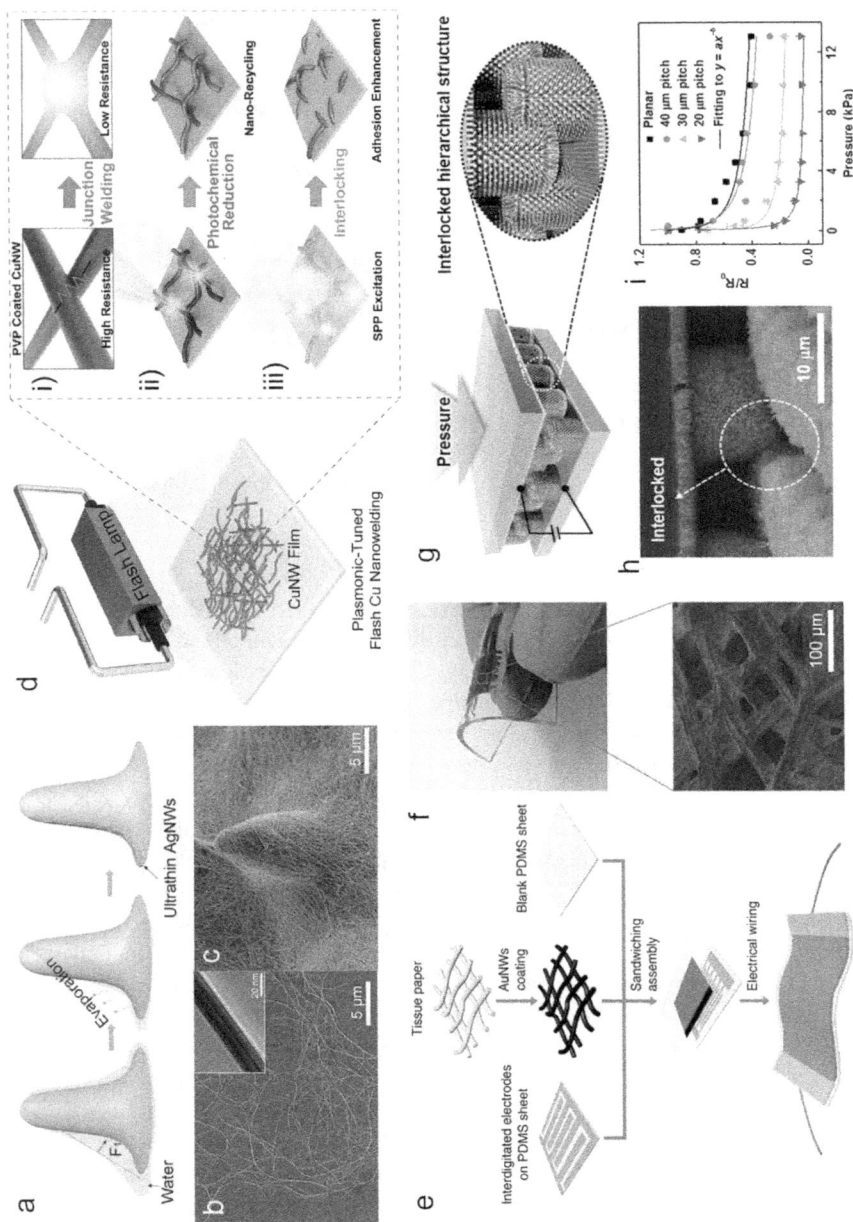

FIGURE 11.3 Metal or semiconductor NWs as active layers. (a) Schematic of ultrathin AgNW network process uniformly covering microstructures via deposition method. (b) SEM image of high-aspect-ratio AgNWs (inset, TEM image). (c) Morphology of microcone covered by AgNW network

(d) Schematic of CuNW plasmonic-tuned flash welding process.

Source: Reproduced with permission [31]. Copyright 2017. WILEY-VCH.

(e) Schematic of the fabrication process for AuNW-based sensor. (f) Photograph of the as-fabricated flexible device, and SEM image of AuNW-coated tissue fibers

Source: Reproduced under a Creative Commons license CC BY 4.0.[33] Copyright 2014, the Authors.

(g) Schematic structure of ZnONW-based piezoresistive device. (h) SEM image of interlocked ZnONWs on PDMS micropillar arrays. (i) Change in relative resistance as a function of normal pressure for flexible device

Source: Reproduced with permission [35]. Copyright 2015. WILEY-VCH.

photochemical reducing as shown in Figure 11.3(d) to significantly enhance the mechanical and chemical stabilities of CuNWs, largely benefitting the transparent resistive memory and touch screen panel applications.

11.3.1.3 AuNWs

AuNWs have attracted considerable research interest as active components of electronic devices due to their chemical stability, biocompatibility, and high electrical conductivity. Various synthesis methods, such as chemical reduction, lithography/lift-off processes, and electrodeposition, have been developed to promote the utilization of AuNWs. In 2008, Chen *et al.* produced ultrathin AuNWs (< 5 nm) using oleylamine or combined chemicals as both 1D growth templates and reducing agents [32]. Different from other NWs with straight lengths, ultrathin AuNWs are serpentine-like, similar to "polymer chains", due to their intrinsically small diameter (2 nm) and high aspect ratio (> 10000). Ultrathin AuNWs are inherently stretchable, so they are naturally suitable for stretchable electronic devices.

When used as active materials in wearable sensors, AuNWs exhibit excellent sensing performance. As demonstrated by Chen's group [33], a flexible and highly sensitive pressure sensor was fabricated by sandwiching ultrathin AuNW-impregnated tissue paper between two thin PDMS sheets (Figure 11.3(e) and 11.3(f)). By leveraging the contact resistance changes between the AuNW-tissue composite and the interdigitated electrodes underneath, the device can detect pressing forces as low as 13 Pa with a fast response time (< 17 ms), high sensitivity (> 1.14 kPa^{-1}), and high stability (more than 50000 loading-unloading cycles). Synergistically combining the superior sensing properties and excellent flexibility and robustness, AuNWs could enable real-time monitoring of blood pulses and even detection of light vibrations from music.

11.3.1.4 Semiconductor NWs

Semiconductor NWs have been extensively studied for over two decades. A wide range of semiconductor NWs (*e.g.*, groups V, III–V, and II–VI and their alloys) has been synthesized, which are more reliable and robust under bending conditions compared to their bulk counterparts due to the reduced amount of defects [34]. For example, the average Young's modulus of ZnONWs is similar to that reported for its bulk and thin-film values, while the ultimate strengths are higher 40 times than that of their bulk counterparts. Semiconductor NWs can be obtained by various methods, which are generally classified as either top-down or bottom-up strategies. Extraordinary progress has been attained for the exquisite synthetic control of a broad range of semiconductor NWs with tunable size, composition, heterostructure modulation, and electronic properties, creating a rich variety of versatile building blocks for flexible electronic devices.

Semiconductor NWs have also been demonstrated to act as highly sensitive materials for flexible electronics. For instance, the hierarchical micro-nanostructured ZnONW arrays on PDMS micropillars with an interlocked geometry have been used as an active layer for sensitive detection of both static and dynamic tactile stimuli (Figure 11.3(g)–11.3(i)) [35]. The interlocked hierarchical structures enable a stress-sensitive variation in the contact area in the interlocked ZnONWs and the efficient bending of ZnONWs, allowing flexible piezoresistive and piezoelectric pressure sensing for static and dynamic tactile strain monitoring.

11.3.2 CARBON MATERIALS

As the most abundant elements on earth, carbon materials have attracted increasing interest for flexible electronics because of their comprehensive superiority, such as intrinsic and structural flexibility, good electrical conductivity, high chemical and thermal stability, lightweight, mass production capability, and ease in chemical functionalization. This section briefly introduces the two most important carbon materials: graphene and CNTs.

11.3.2.1 Graphene

Since its discovery by Geim and Novoselov in 2004, graphene has quickly become a newly emergent candidate for flexible electronics owing to its high intrinsic carrier mobility, excellent mechanical properties, satisfactory optical transmittance, chemical inertness, and potential for large-scale production at low cost [36]. Graphene is a carbon allotrope composed of a monolayer of sp^2-hybridized atoms arranged in a two-dimensional (2D) honeycomb lattice. The C–C bond length of graphene is only 0.142 nm, and the connection among the carbon atoms of graphene is extremely strong. When an external force acts on graphene, the atomic surface inside the graphene could deform and bend to offset the external force, without inducing any atomic rearrangement or dislocation. Thus far, numerous methods including mechanical stripping, liquid phase stripping, chemical vapor deposition, epitaxial growth, and redox method, have been developed to achieve the low-cost and mass production of graphene sheets.

Graphene possesses a unique electronic structure with a slight overlap between valence and conductance bands that forms a zero-bandgap 2D semimetal, resulting in a number of extraordinary properties that are not observed in conventional materials. Previous studies have demonstrated that graphene has a strong ambipolar effect with large charge carrier concentrations up to 10^{13} cm^{-2} and high room-temperature mobilities of approximately 10000 cm$^2 \cdot$V$^{-1} \cdot$s^{-1} when the gate voltage is applied. Due to the existence of abundant delocalized electrons, suspended graphene exhibits extraordinary electronic properties. The carrier mobility can reach 2×10^5 cm$^2 \cdot$V$^{-1} \cdot$s^{-1} for carrier densities less than 5×10^9 cm^{-2} [37]. For thermal properties, graphene exhibits an exceptionally high in-plane thermal conductivity ≈ 5000 W$\cdot$m$^{-1} \cdot$K^{-1} and outstanding thermal stability at high temperatures (e.g., 250 °C during annealing and 1000 °C during vapor-based synthesis) [36], rendering it promising for efficient heat dissipation. Mechanically, graphene is elastic and incredibly strong with an in-plane tensile elastic strain of up to 25%, a Young's modulus of 1 TPa, a third-order elastic stiffness of approximately 2 TPa, and a shear modulus of up to 280 GPa. More importantly, the unique 2D honeycomb structure of graphene provides abundant active sites to react with functional groups on the base surface, empowering free tailoring of its electromechanical and electrochemical properties via various functional nanomaterials.

11.3.2.2 CNT

CNT is another carbon allotrope with a 1D cylindrical nanostructure as shown in Figure 11.4(a), which can be depicted as multiple nanotubes stacked inside each other (each nanotube layer is called a "wall"). Depending on the number of layers, they can be classified as single-walled CNTs (SWCNTs) or multiwalled CNTs (MWCNTs). The former is composed of a single-layer graphene sheet rolled into a hollow cylinder with a length of 1–50 μm and a diameter of 0.75–3 nm. The latter consists of multi-layer (2–50) graphene sheets similar to a coaxial cable, with a typical interlayer spacing of 0.34±0.01 nm, a diameter of 0.75–3 nm, and a length of 1–50 μm. CNTs have excellent mechanical properties (*e.g.*, Young's modulus and tensile strength of up to 1–2 TPa and 100 GPa), and they are inherently flexible, extremely stable, and cannot be easily damaged even under tensile and bending loads. Because of their hollowed structures, CNTs can absorb extra energy, resulting in excellent elastic properties with a high tensile strength of up to 40% strain without any plastic deformation [38]. In addition, CNTs have extraordinary electrical conductivities as high as 10^4–10^6 S·cm^{-1}, which are comparable to some metals. Their current-carrying capacities exceed 10^9 A·cm^{-2}, and their thermal conductivities are up to 3500 W·K^{-1}·m^{-1} [39]. More importantly, CNTs can stably maintain electrical properties under extreme and complex environments, which are beneficial for applications in aviation, aerospace, and navigation.

Although a single CNT can be used as a functional device for some applications, the adequate control of the CNT growth at desired locations is challenging, thereby hindering large-scale commercialization. Alternatively, a 2D macroscopic network, often referred to as a thin film, made of randomly distributed CNTs is considered

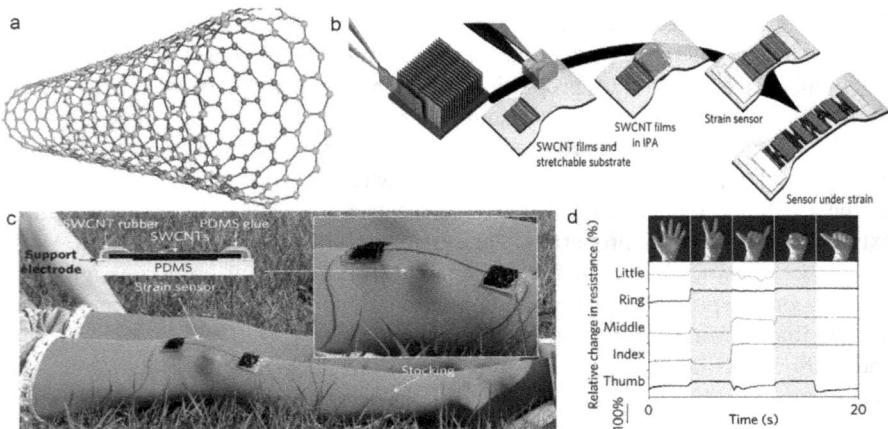

FIGURE 11.4   (a) Schematic of CNT structure.

(b) Steps in fabricating CNT strain sensor. (c) Photograph of bandage strain sensor fixed to stocking. (d) Application of sensors to data glove configurations.

Source: Reproduced with permission.[40] Copyright 2011. Springer Nature.

as a more practical active material. Compared with an individual CNT, thin CNT films can withstand a higher degree of strain because they can deform and recover by changing the orientation of each individual CNT in the film. This significantly reduces the tensile and compressive stresses applied directly onto a single CNT. Furthermore, thin CNT films can be easily fabricated via conventional lithography or printing processes, offering a viable approach for low-cost flexible devices. Hata *et al.*[40] reported a new type of stretchable CNT strain sensor based on a thin film of vertical single-well CNTs transferred onto a stretchable PDMS substrate (Figure 11.4(b)). Upon stretching, the device exhibited a remarkable stretchability (with a strain up to 280%) and satisfactory durability (1000 cycles at 150% strain), achieving effective human motion detection (Figure 11.4(c)) and data glove configuration (Figure 11.4(d)).

11.3.3 CONDUCTIVE POLYMER

Conductive polymers are special kinds of organic materials whose electrical and optical properties are similar to those of inorganic metals and semiconductors. They can be synthesized by simple, versatile, and cost-effective approaches, either chemically or electrochemically. Conductive polymers are recognized as excellent active materials owing to their extraordinary characteristics including satisfactory mechanical flexibility, lightweight, great compatibility with flexible solid supports, and charge transfer capability of conductive domains. The presence of π-electrons in the conjugated backbone is able to delocalize into a conduction band and then move freely within the unsaturated backbone to construct an electrical pathway for mobile charge carriers, resulting in a good conductivity of the polymer. Examples of conductive polymers that are commonly applied to flexible electronics include poly(3,4-ethylene dioxythiophene):polystyrene sulfonate (PEDOT:PSS), polypyrrole (PPy), polyaniline (PANi), and conductive polymeric composite (CPC).

11.3.3.1 PEDOT:PSS

PEDOT:PSS is a commercially available polyelectrolyte composed of positively charged conductive conjugated PEDOT and negatively charged insulating PSS. The PSS polymer anions can stabilize conjugated polymer cations in water and some polar organic solvents. The insoluble PEDOT short chain adheres to the water-soluble PSS long chain to form a grain by Coulomb force such that it can be steadily dispersed in water. Its conductivity depends on the processing method as well as other additives contained in the formula (commonly referred to as "secondary dopants"). For instance, electrical conductivity as high as 4380 S·cm^{-1} has been demonstrated for PEDOT:PSS through a post-treatment with sulfuric acid, much larger compared to < 1 S·cm^{-1} without any secondary dopant [41]. In addition, the superior transparency (80–95%) in the visible range and satisfactory flexibility of PEDOT:PSS render it an acceptable transparent electrode candidate in the roll-to-roll printing of optoelectronic devices, such as organic and perovskite solar cells, photoelectron spectroscopy, and organic LEDs. Moreover, PEDOT:PSS has been considered as a promising thermoelectric material owing to its excellent thermoelectric properties (ZT: ~0.42).

Remarkably, a PEDOT:PSS film shows notable mechanical properties with Young's modulus of 1–2.7 GPa, the tensile strength of 25–55 MPa, and strain at fracture of 3–5% [42]. It is expected that PEDOT:PSS might be used as an active layer material for next-generation flexible (or even stretchable) wearable electronic devices.

11.3.3.2 PPy

PPy is a conjugated polymer that displays high electrical conductivity, excellent environmental stability, simple fabrication, and ease in surface modification. It can be easily synthesized in large quantities by a variety of common organic solvents and water at room temperature. The conductivity of PPy films can reach up to 10^3 S·cm^{-1} depending on the types and amounts of dopants. Nevertheless, further processing for pure PPy, once synthesized, is complicated because highly cross-linked structures reduce fusibility and solubility. Besides, its crystalline is mechanically rigid, brittle, and insoluble, rendering PPy virtually impossible to be used alone. It must consistently be used together with other materials to form composites. Hence for applications in flexible active layers, most of the reported PPy films are formed by coating PPy on a support matrix, which determines the overall mechanical properties and machinability of composite films. Recently, Mao and co-workers developed a template-assisted interfacial polymerization method and successfully synthesized a soft and mechanically processable PPy membrane without modifying the monomers or using support materials [43]. The single-component PPy membrane presents superior softness and flexibility even under liquid nitrogen (−196 °C). Furthermore, this PPy film illustrates good mechanical processability which can be cut, tied, folded into a cubic box, laminated, and rolled into elastic tubes.

Highly stretchable polymer films have also been obtained by the nanoconfinement technique, which involves reducing the size of the polymer to nanometer scale [44]. Pan et al. reported an elastic PPy hydrogel with interconnected hollow sphere structures by redesigning a stiff and brittle PPy from its rigid conjugated-ring backbone into a microstructured conducting polymer [44]. Figure 11.5(a) schematically shows the multiphase reaction mechanism employed to achieve the hollow sphere morphology of PPy. After purification with deionized water, the PPy gel changed into a hydrogel (Figure 11.5(b)). The hollow sphere structure allows PPy to elastically deform and recover with an ultrasensitive response, as depicted in Figure 11.5(c).

11.3.3.3 PANi

The polymerization of PANi is based on the aniline monomer. It was first known as black aniline with different forms depending on its oxidation level. It has diverse structural forms, high environmental stability, low cost, and the ability to electrically switch between the conductive and resistive states by the doping/dedoping process [45].

The electrical activity of PANi originates from the p-electron conjugate structure in the molecular chain [45], where the p-bonded and p*-anti-bonded states form valence and conduction bands, respectively. This nonlocalized p-electron conjugate structure can form p-type and n-type conductive states by doping, leading to high conductivity. Different from the doping mechanism of cationic vacancies generated by other conductive polymers under the action of oxidants, the number of electrons

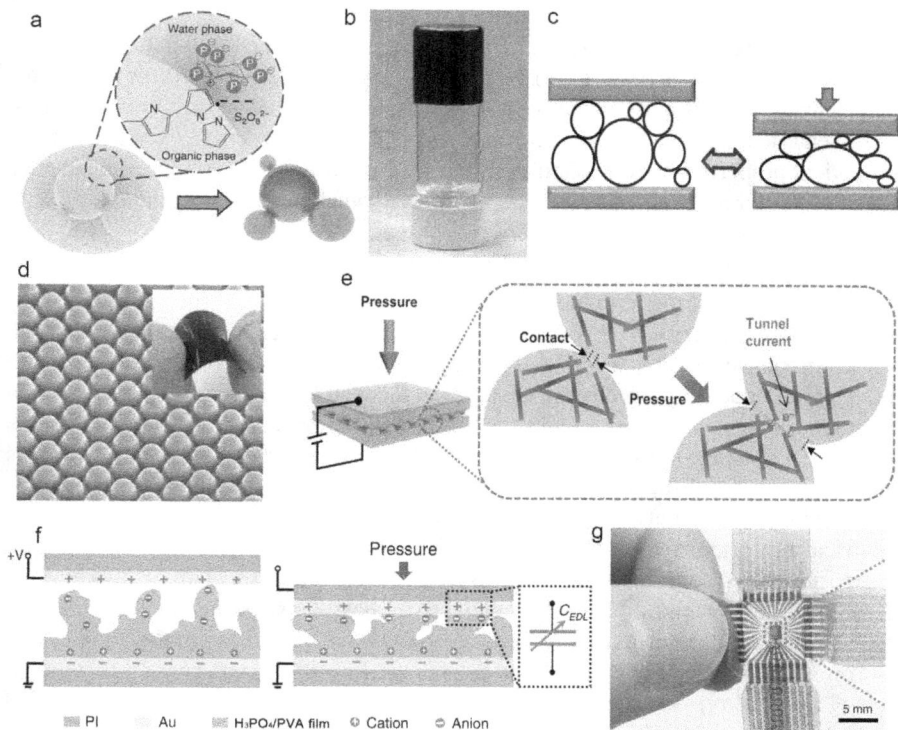

FIGURE 11.5 (a) Schematic of the multiphase reaction mechanism of PPy hydrogels with hollow sphere microstructures. (b) Photograph of PPy hydrogel. (c) Schematic of structural elasticity of hollow-sphere-structured PPy.

Source: Reproduced under a Creative Commons license CC BY 4.0 [44]. Copyright 2014, the Authors.

(d) SEM image of CNT/PDMS CPC film with microdome arrays (bottom diameter: ~5 μm; height: ~3.5 μm; pitch: 6 μm); inset is actual photograph of the composite film. (e) Schematic of the functional mechanism of CPC-based flexible electronic skin.

Source: Reproduced with permission [48]. Copyright 2014. ACS Publications.

(f) Schematic of graded intrafillable architecture of H_3PO_4/PVA film and mechanism of the sensor. (g) Actual photograph of sensor arrays.

Source: Reproduced under a Creative Commons license CC BY 4.0.[49] Copyright 2020, the Authors.

does not change during the doping process of PANi. However, H^+ and pair anions (such as Cl^-, sulfate, and phosphate) could enter the main chain through the decomposition of doped proton acid. It combines with N atoms in the amine and imine groups to form polar and bipolar delocalizations into the P bond of the entire molecular chain. This unique doping mechanism renders the doping and dedoping of PANi completely reversible. Its doping degree is affected by several factors, such as pH value and electric potential, which is reflected in a color change. Therefore, PANi also has electrochemical activity and electrochromic properties [46].

PANi also has unique optical properties. When irradiated by strong light, the electrons in the valence band of PANi are excited to the conduction band, resulting in additional electron-hole pairs, granting it intrinsic photoconductivity. At the same time, the electrons or holes at the impurity level are excited and change their conductivities, which leads to a significant photoelectric conversion effect. Moreover, excellent nonlinear optical properties are exhibited by PANi [47], making it potentially applicable for information storage, frequency modulator, optical switch, and optical computer.

11.3.3.4 CPC

By mixing an insulating polymer matrix with conductive fillers, such as CNTs, graphene, carbon black, metal NWs, or conductive filler particles, CPCs can be obtained. These composites exhibit several desirable traits for flexible electronics, such as high electrical conductivity, lightweight, superior mechanical behaviors, and corrosion resistance. The conductivity of CPCs arises from the formation of conductive paths of fillers within the polymer matrix, which is governed by many factors such as geometry, state of dispersion, abundance, and intrinsic properties of fillers. In addition, the interactions between fillers and matrix play an important role in the ultimate electrical performance of the composite. Therefore, it is critical to choose a proper composite preparation method that provides the desired level of filler distribution. Typically, the concentration of the conductive phase must be higher than the permeation threshold to form a continuous conductive network through the composite.

Tremendous advances in CPCs have been achieved for various flexible applications. Park *et al.* reported a flexible electronic skin using CPC with micro-interlocked microdome arrays [48]. The microstructured CPC was fabricated by printing a CNT/PDMS mixture from a silicon mold (Figure 11.5(d)). When pressure is applied, the microdomes in the top and bottom CPC electrodes come into contact to form the tunneling current, as shown in Figure 11.5(e). Consequently, satisfactory sensing results are achieved for monitoring the spatial pressure distribution of gas flow and detecting human breathing. Moreover, Guo *et al.* designed an ionic gel by filling phosphoric acid (H_3PO_4) into a polyvinyl alcohol (PVA) polymer [49]. The H_3PO_4/PVA film with graded intrafillable architecture (Figure 11.5(f)) was molded from sandpaper and used as an active layer for flexible pressure sensors (Figure 11.5(g)), which shows an ultra-broad-range high sensitivity (> 220 kPa^{-1}).

11.4 APPLICATIONS OF SOFT AND FLEXIBLE MATERIALS

11.4.1 FLEXIBLE TRANSPARENT ELECTRODE

Flexible transparent electrodes (FTEs) with high optical transmittance, low R_{sh}, and high flexibility are critical and indispensable components of emerging flexible optoelectronic devices. Conventional transparent electrodes are typically made of sputtered transparent conducting oxides, such as indium tin oxide (ITO) thin films, which possess excellent electrical conductivities and high optical transmittances (10–20 $\Omega \cdot sq^{-1}$ at 90% transmittance). However, transparent conducting oxides are brittle and often break or crack, hence they cannot be used in flexible photoelectronics where folding, stretching, twisting, or severe bending is required. In the past decade, the development of various emerging transparent conductive materials, such as metallic NWs or nanomeshes, graphene, CNT, and conductive polymers, brings new opportunities for high-performance FTEs.

Metallic NWs have been widely investigated to construct FTEs in the form of conductive networks with wire–wire junctions. Cui's group demonstrated an FTE composed of a free-standing gold nanotrough network via a process involving electrospinning and metal deposition (Figure 11.6(a) and 11.6(b)) [50]. The electrode exhibited superior optoelectronic performance (R_{sh}: ~ 2 $\Omega \cdot sq^{-1}$ at 90% transmission) and remarkable mechanical flexibility under stretching and bending stresses, showing its practical suitability for flexible touch screen devices and transparent conducting tapes. To achieve the low sheet resistance of metallic NW networks, the most straightforward method is to increase the density of metal NWs, but it suffers from a reduced light transmittance. Studies have been conducted to reduce the contact resistance among NWs via a welding technique to improve the conductivity of network electrodes without sacrificing optical transmittance. Park et al. welded NW junctions on a PET substrate via light-induced plasmonic interactions [51]. The welded AgNW electrodes exhibited 90% transmittance and R_{sh} of 5 $\Omega \cdot sq^{-1}$, realizing a transparent flexible energy harvester. Moreover, Liu et al.[52] reported a capillary force that could enable a cold welding strategy to fabricate AgNW networks for FTE. The nanoscale capillary force has a powerful driving capacity that can effectively realize self-limited cold welding of wire–wire junctions, as shown in Figure 11.6(c). It is noteworthy that the capillary-force-induced welding can be achieved by simply applying moisture to the AgNW sample without any technical support, such as "solder" or specific facilities. The moisture-treated AgNW films exhibit a significant decrease in R_{sh} with negligible changes in transparency (Figure 11.6(d)).

Metallic nanomeshes have also drawn considerable interest for FTEs. In 2014, Guo et al. proposed a highly stretchable Au nanomesh FTE on PDMS elastomers via grain boundary lithography [53]. Under one-time strain (~160%), or after 1000 cycles at 50% strain, the change in the R_{sh} of Au nanomeshes was modest, i.e., from 21 to 67 $\Omega \cdot sq^{-1}$. A high ratio of mesh size to wire width and unique structures with instability and out-of-plane deflection (Figure 11.6(e)) endow the FTE with high stretchability and excellent conductivity. After that, the same group combined Au nanomeshes with a pre-strain PDMS substrate to produce fatigue-free and super stretchable FTE (Figure 11.6(f)) [54]. The optimized topology and tuning adhesion of Au nanomeshes can significantly improve the stretchability and eliminate strain fatigue. As a result, the Au nanomesh FTE kept a low R_{sh} value and high transparency when the nanomesh was stretched to a strain level of 300% or exhibited no fatigue after 50000 stretches to a strain level of up to 150%. The Au nanomesh FTEs might be applied to implantable electronics because the nanomeshes could mechanically and biochemically match organs or tissues with minimal impact on human body.

Another fascinating approach to realize FTEs is to utilize carbon-based nanomaterials. Typically, CNT networks deposited by solution processes show random distributions of CNTs which can be stretched omnidirectionally. Lipomi et al. reported a skin-like FTE by spray-coating CNT random networks on a biaxial pre-strain PDMS substrate (Figure 11.6(g)) [55]. The CNT networks exhibited good stretchability up to 150%, a low R_{sh} value of 328 $\Omega \cdot sq^{-1}$ and a 79% optical transmittance, demonstrating

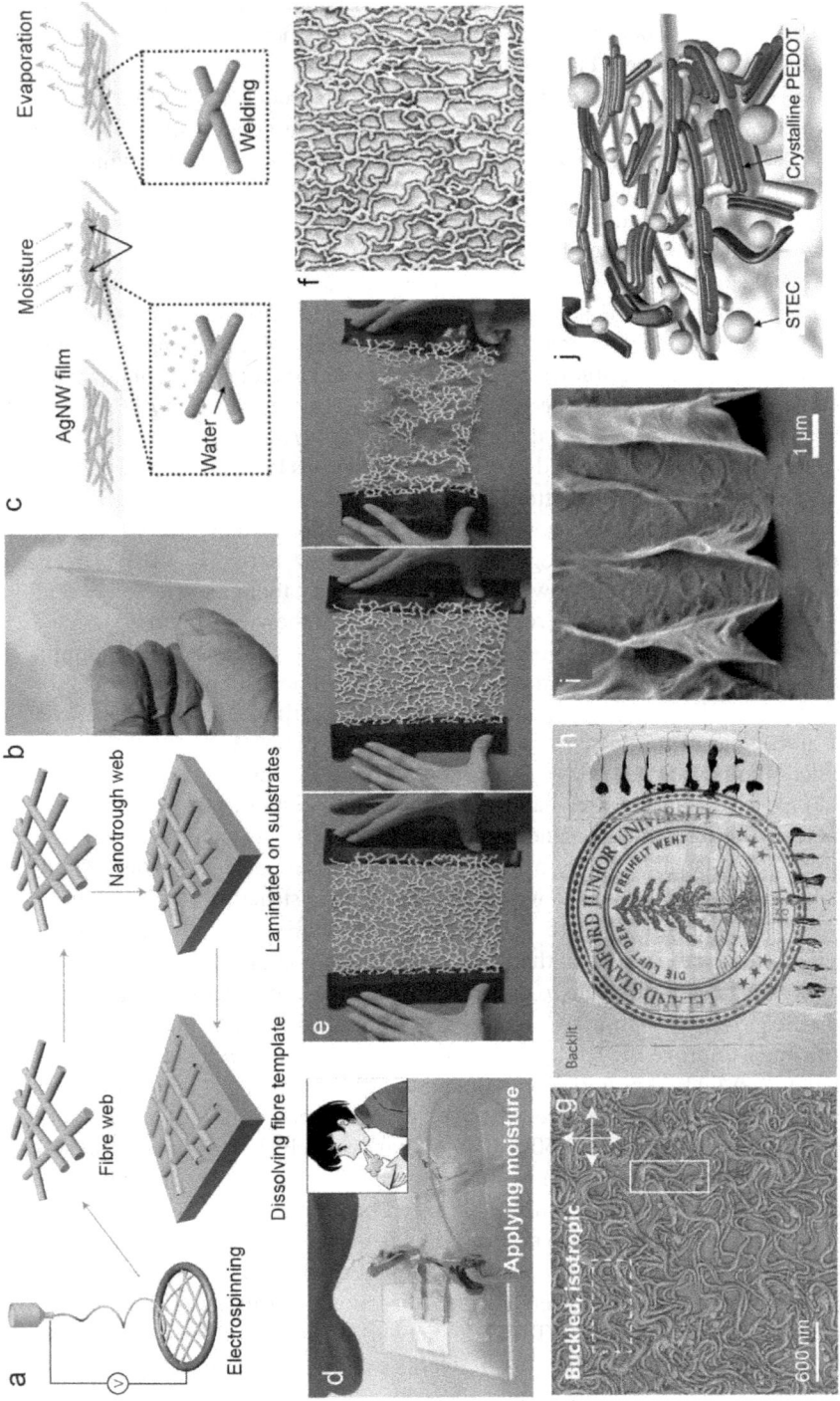

FIGURE 11.6 Applications of FTEs. (a) Schematic of process involving electrospinning and metal deposition for producing free-standing Au nanotrough networks. (b) Photograph of as-fabricated FTE with AuNWs networks

(c) Schematic of the cold-welding process for AgNW networks FTE. (d) Moisture-healing by breathing on damaged FTE with LED on.

Source: Reproduced with permission [52]. Copyright 2017, ACS Publications.

(e) Demonstration showing that pre-cut magnified Au mesh made of paper is considerably more stretchable.

Source: Reproduced under a Creative Commons license CC BY 4.0 [53]. Copyright 2014, the Authors.

(f) SEM image of Au nanomeshes on pre-strained PDMS.

Source: No permission required for PNAS authors.[54] Copyright 2015, National Academy of Sciences, USA.

(g) AFM phase image of CNT networks on biaxial pre-strain PDMS. (h) Image of pressure sensors made of CNT-based FTE

Source: Reproduced with permission [55]. Copyright 2011, Springer Nature.

(i) SEM image of wavy structures of graphene for FTE

Source: Reproduced with permission [57]. Copyright 2014, ACS Publications.

(j) Schematic of the structure of STEC-doped PEDOT:PSS.

Source: Reproduced under a Creative Commons license CC BY 4.0 [59]. Copyright 2017, the Authors.

excellent performance for flexible pressure and strain sensing (Figure 11.6(h)). Recently, Jiang *et al.* proposed a carbon-welded SWCNT network for FTEs [56]. Crossed SWCNTs were welded together using injection floating catalyst chemical vapor deposition method, yielding FTEs with a low sheet resistance of 41 $\Omega \cdot sq^{-1}$ at a 90% transmittance for flexible organic LEDs. Alternatively, graphene with wavy or serpentine structures has also been widely investigated for FTEs. Xu *et al.* reported the transfer of CVD-synthesized graphene onto a 50% pre-strained PDMS substrate [57]. When released, the resulting wavy structures (Figure 11.6(i)) exhibited stretchability up to their pre-strained states. Based on this FTE, a high-performance supercapacitor was produced with excellent stretchability (up to 40%), high optical transparency (72.9% transmittance), and outstanding electrochemical performance. Kim *et al.* demonstrated a stretchable (up to 106% at 18% pre-strain) and transparent serpentine graphene interconnected array, which was wrapped on a mechanically neutral plane using a polymer encapsulation layer. A low R_{sh} value of 480 $\Omega \cdot sq^{-1}$ at 90% transmittance was demonstrated with a four-layer graphene film.

Last but not least, FTEs based on conductive polymers have also been investigated. Zhang *et al.* reported a PEDOT:PSS/surfactant bilayer film by spin-coating a surfactant layer (glycerol monostearate) atop a PEDOT:PSS film [58]. This film exhibited a low R_{sh} value of 98 $\Omega \cdot sq^{-1}$ and ~80% transparency in the visible range, which was employed as the anode FTE for high-efficiency heterojunction polymer solar cell devices. Bao's group proposed another type of highly stretchable conducting polymer for FTEs, realized by doping dopants with the specific range of the stretchability and electrical conductivity (STEC) into PEDOT:PSS (Figure 11.6(j)) [59]. The polymer film FTE exhibited a high conductivity exceeding 4100 $S \cdot cm^{-1}$ at 100% strain and a low R_{sh} value of 59 $\Omega \cdot sq^{-1}$ at 96% transmittance, enabling an ultra-compact interconnection of flexible field-effect transistor arrays that is five times denser than the conventional lithographically patterned wavy interconnects.

11.4.2 FLEXIBLE PHOTOVOLTAIC CELL

The technique advances in smart devices and systems along with the increasing concerns on global warming highlight the necessity for portable, lightweight, green, and flexible energy sources. In this context, flexible photovoltaic cells with low cost, lightweight, foldability, and roll-to-roll fabrication are highly demanded to supply diverse electronic devices. Here, we introduce several applications of soft materials in flexible photovoltaic cells, such as flexible inorganic solar cells (ISCs), flexible dye-sensitized solar cells (DSSCs), flexible organic solar cells (OSCs), and flexible perovskite solar cells (PSCs).

The current main streams in the flexible solar cell market are made of inorganic materials, such as silicon, GaInP, and GaAs. Although inorganic materials are generally considered brittle, there are some strategies could enable the flexible ISCs, including the growth of amorphous silicon on flexible substrates or the transfer process of the active layers onto flexible substrates. Jo *et al.*[60] successfully presented highly flexible ISCs with amorphous silicon obtained via a transfer printing technique. The performance of the ISC device has been tested at a 50% strain without degradation as shown in Figure 11.7(a) and 11.7(b). Another interesting example was

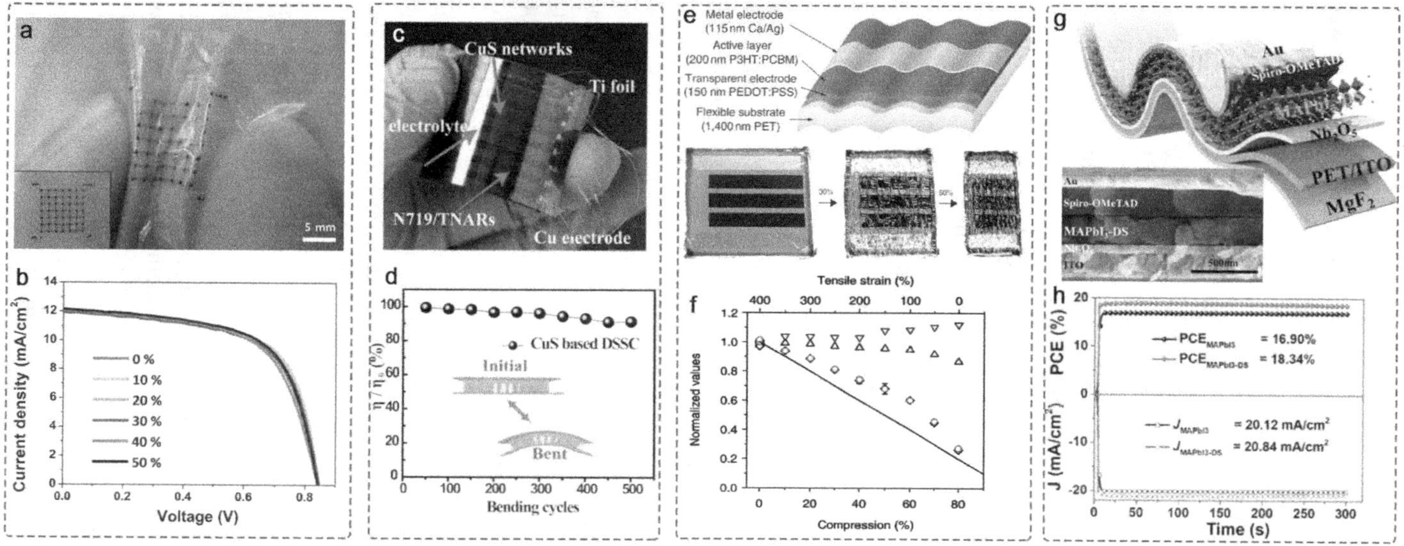

FIGURE 11.7 Four types of flexible photovoltaic cells. (a) Actual photograph of flexible ISC. (b) PCE performance under different strains.

Source: Reproduced with permission [60]. Copyright 2016, WILEY-VCH.

(c) Photograph of flexible CuS-based DSSC and (d) PCE property under cyclic bending loads

Source: Reproduced with permission [62]. Copyright 2018, Elsevier Publication.

(e) Schematic of the structure of flexible OSC and images of compression tests, and (f) behaviors under tensile strain and compression conditions.

Source: No permission required [64]. Copyright 2012, the Authors.

(g) Schematic structure of flexible PSC and sectional SEM image of the device, and (h) PCE of flexible PCS.

Source: Reproduced with permission [66]. Copyright 2018, WILEY-VCH.

reported by Rogers *et al.* who developed a stretchable ISC by bonding GaAs micro-cells onto a pre-strained structured substrate of PDMS [61]. This structure provides high stretchability without affecting the function of the device, with constant power conversion efficiency (PCE) of 12.5% even after 500 stretch cycles.

One of the newly emergent green energy sources is DSSC with a dye sensitizer as a light absorber. A flexible DSSC device typically consists of two transparent electrodes covering a cathode, an electrolyte, a highly sensitized photoanode. Recently, Xu *et al.* [62] developed a large-area and highly flexible transparent conductive CuS/PET film as the counter electrode for manufacturing flexible DSSC (Figure 11.7(c)) through colloidal crack pattern technology. The DSSC device shows superior resistance to temperature and pH variations with a PCE rate of approximately 4.54% and can maintain approximately 90% of its total PCE after 500 bending cycles, as shown in Figure 11.7(d). In addition, fiber-shaped flexible DSSCs have also been developed and integrated into textiles for stretchable optoelectronic applications. Hou *et al.*[63] demonstrated a fiber-like DSSC using the inexpensive commercial thread substrates with dip-coating conductive polymer PEDOT:PSS. The device showed a PCE of 4.8% and satisfactory electrochemical stability, rendering it potentially suitable for harsh environments.

Flexible OSCs mainly consist of a p-type electron donor and an n-type electron acceptor sandwiched between a cathode and an anode along with two charge transport layers. Many attempts have been explored to improve the flexibility, mechanical stability, and transparency of flexible OSCs. Kaltenbrunner and co-workers designed a wavy structure of Ca/Ag/P3HT:PBM/PEDOT:PSS/PET (Figure 11.7(e)) to realize an ultrathin, lightweight, and highly flexible organic solar cell with a maximum PCE of 3.9% and is capable of being uniaxially compressed by up to 80% [64]. After more than 20 full cycles of cyclic compression and stretching to 50%, the device showed only a marginal decrease in performance and no visible defect formation (Figure 11.7(f)). Someya's group reported a waterproof elastomer-coated OSC that simultaneously possessed high stretchability and stability in the water while maintaining high PCE and stretchability levels of 7.9% and 52%, respectively [65]. Moreover, the device PCE remained at 80% of its initial value even after 20 cycles of 52% mechanical compression under 100 minutes of water exposure.

The flexible PSC is another promising route for the development of a flexible power source due to its fully solid-state nature and low-temperature production procedure. Generally, flexible PSCs are constructed using flexible electrodes sandwiching a perovskite photoactive compound layer, where the perovskite compound chemical formula is ABX_3 (A, B, and X correspond to the monovalent cation, metal cation, and halide anion, respectively). Feng *et al.*[66] developed a highly flexible PSC by modifying the perovskite layer $MAPbI_3$ with dimethyl sulfide as new additives (Figure 11.7(g)). In principle, dimethyl sulfide could react with Pb^{2+} to form a chelating intermediate that can decelerate crystallization, leading to an increase in the grain size of the calcium titanium deposit and the reinforcement of its crystal structure. This additive can also diminish the trap density and promote the quality and function of the perovskite junction. As a result, the PCE increases to 18.40% with a satisfactory mechanical tolerance as shown in Figure 11.7(h). Huang *et al.* [67] demonstrated another flexible PSC by optimizing the device composition to overcome common defects of perovskite photovoltaic cells. By precisely controlling the thickness and morphology of the electron transport layer, the reflection of the ITO layer coated on the PEN substrate was significantly reduced. Meanwhile, the photon collection rate of the device was considerably enhanced. This strategy could also decrease the trap-state densities of perovskite films and charge transfer resistances, leading to an enhanced PCE of approximately 19.51% and robust bending resistance.

11.4.3 FLEXIBLE PHOTODETECTOR

As an indispensable component in electronics, flexible photodetectors (PDs) are highly desired to satisfy the demand of next-generation wearable electronics. Over the past years, tremendous efforts have been made to improve the key performance indicators of flexible PDs, such as on/off ratio, responsivity, response speed, detectivity, stability, and spectral response range [68]. A variety of functional materials have been explored to achieve accurate light detection covering the ultraviolet [6,69,70], visible [71,72], infrared [73–76], and terahertz regions [77,78]. Table 11.1 provides a few representative examples of flexible PDs, which are categorized in terms of the substrate/active material type, operation wavelength, as well as mechanical and electrical performance.

Flexible UV PDs are particularly useful in fire monitoring, biological and environmental sensing, and space exploration (see Chapter 1). Yan et al.[70] presented a stretchable UV PD consisting of a ZnONW percolation network channel and AgNW percolation network electrodes via a fast and effective lithographic filtration method. The device showed an on/off ratio of 188 and a response time of 30.3 s for 365 nm UV light. Moreover, it can function well at an applied strain level up to 100% with satisfactory stability and reproducibility (Figure 11.8(a) and 11.8(b)). In another work, Badhulika et al.[69] demonstrated a flexible deep-ultraviolet (DUV) PD made of 2D h-BN on a Cu foil substrate. The flexible DUV PD exhibits a responsivity value of 5.022 $A \cdot W^{-1}$, a response time of 0.2 s, and a specific detectivity of 6.1×10^{12} Jones, which are two to three orders higher than the other h-BN-based PDs. More interestingly, based on a well-defined ionic liquid-containing liquid crystal polymer and highly elastic PU composite fabric, a conversion mechanism of UV illuminance to mechanical stress, and then to electrical signal was proposed to realize a powerful device for UV monitoring and shielding [6]. The device can effectively maintain stable performance upon stretching, bending, and washing at 1000 testing cycles with 365 nm UV irradiation, enabling a robust UV security encoding application as shown in Figure 11.8(c) and 11.8(d).

Flexible visible PDs have wide applications in many fields, such as optical communication, spectrum analysis, and fluorescent biomedical imaging. Numerous functional materials such as ZnSe, CdS, V_2O_5, MoS_2, and perovskite, have been studied [68]. Kim et al. reported a flexible transparent large-area PD constructed of PEDOT:PSS/poly(4-vinylphenol)(PVP)/MoS_2/PEN, with a broadband detection range from 405 to 780 nm (Figure 11.8(e) and 11.8(f)) [71]. Such a flexible PD exhibited an on/off ratio of 10^3, a responsivity of 20 $mA \cdot W^{-1}$, and a detectivity of

TABLE 11.1

Summary of Various Flexible PDs in Terms of Materials, Target Wavelengths, and Device Performance Levels

Substrate	Active Layer	Wavelength	Electrical Data	Flexibility	Ref
Cu foil	h-BN	210 nm	R: ~5.022 A·W^{-1} Speed: ~200 ms D: ~6.1 × 10^{12} Jones	Bending cycles: ~500	[69]
PDMS	Ag/ZnO NWs	365 nm	On/off ratio: ~188 Speed: ~30.3 s	Tensile strain: ~100%	[70]
PU	Liquid crystalline polymer	365 nm	On/off ratio: ~270 Speed: ~5 s	Stretching: 30%, bending curvature: 200 m^{-1}, washing	[6]
PEN	MoS$_2$/PVP/ PEDOT:PSS	405–780 nm	On/off ratio: ~ 10^3 R: ~20 mA·W^{-1} D:~4.8 × 10^7 Jones	Bending cycles: ~10^3 Bending radius: ~5 mm	[71]
PEN	MoS$_2$/ perovskite/ rGO	660 nm	R: ~ 1.08 × 10^4 A·W^{-1}	Bending cycles: ~10^3 Bending radius: ~2 mm	[72]
PES	Ag$_2$Se	808 nm	D: ~7.14 × 10^9 Jones	Bending cycles: > 100 Bending radius: ~1.6 cm	[73]
Thin Si	MoS$_2$/Si	850 nm	R:~10.07 mA·W^{-1} D:~4.53 × 10^{10} Jones	Bending cycles: > 1000 Bending radius: ~5 mm	[74]
Paper	Bi$_2$Se$_3$/ Graphite	1064 nm	R: ~26.69 µA·W^{-1}	Bending cycles: ~ 1000 Bending angles: ~160°	[75]
PI	CNTs/C$_{60}$	1000–1400 nm	On/off ratio: ~ 23 R: ~19.4 A·W^{-1} Speed: ~2–4 ms	Bending cycles: ~10^3 Bending radius: ~2.5 mm	[76]
PET	Graphene	0.33–0.5 THz	R: ~2 V·W^{-1}	Strain: ~1.25%	[77]
PET	CNTs	0.14–39 THz	NEP ~242 pW·Hz$^{-1/2}$ for 1.4 THz	Bending, wearable	[78]

Abbreviations: R:~ responsivity; D:~ detectivity; NEP:~ noise equivalent power

4.8 × 10^7 Jones. The fully printed flexible PD was also able to maintain electrical characteristics under tensile strain and 1000 times bending loads. Wang and co-workers demonstrated a sensitive flexible solution-processed PD with a hybrid structure of the perovskite/MoS$_2$ bulk heterojunction as a photosensitizer and reduced graphene oxide (rGO) layer as a conducting channel (Figure 11.8(g) and 11.8(h)) [72], showing a high responsivity of 1.08 × 10^4 A·W^{-1}, high detectivity of 4.28 × 10^{13} Jones, a fast photoresponse time < 45 ms, and satisfactory flexibility when subjected to 1000 times cyclic bending tests with a bending radius of 2 mm.

For IR PD, Jang *et al.*[73] reported a high-detectivity device based on chalcogenide Ag$_2$Se nanoparticles deposited on a PES substrate. This device had a low power consumption and a high detectivity of 7.14 × 10^9 Jones at room temperature, which

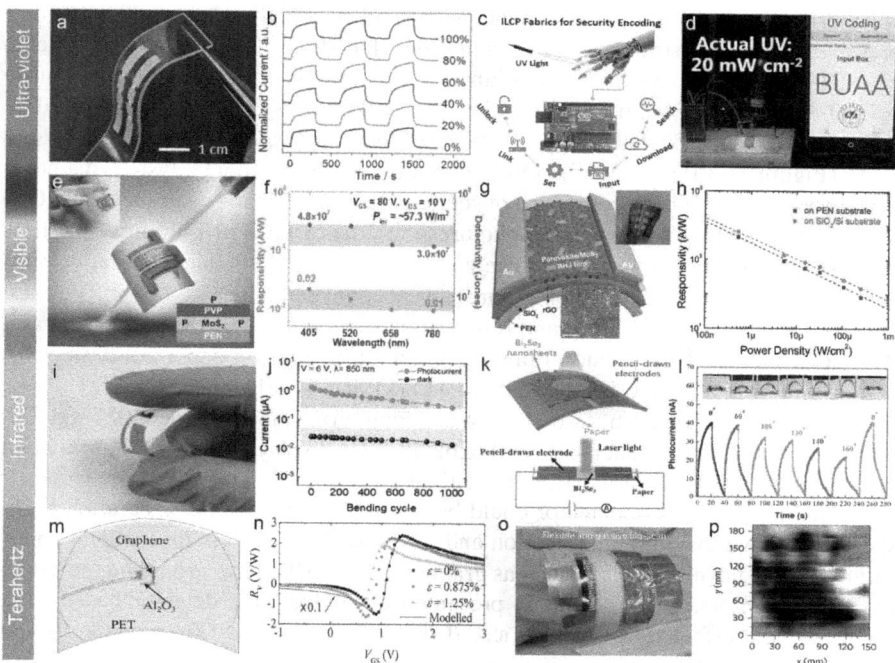

FIGURE 11.8   Flexible PDs covering from UV to THz. (a) Image of flexible UV PD and (b) UV sensing performance under stretching conditions at different strain levels.

Source: Reproduced with permission [70]. Copyright 2014, WILEY-VCH.

(c) Schematic of encoding application of flexible UV PD and (d) its coding application.

Source: Reproduced under a Creative Commons license CC BY 4.0 [6]. Copyright 2021, the Authors.

(e) Schematic of flexible visible PD and (f) its responsivity and detectivity under different ranges of visible light.

Source: Reproduced with permission [71]. Copyright 2017, ACS Publications.

(g) Schematic structure of flexible perovskite/MoS$_2$-based visible PD and (h) its responsivity to different light power densities.

Source: Reproduced with permission [72]. Copyright 2018, WILEY-VCH.

(i) Photograph of a flexible IR PD with 2D MoS$_2$/Si heterojunction and (j) its detection performance.

Source: Reproduced with permission [74]. Copyright 2021. RSC Publications.

(k) Schematic of flexible IR PD constructed of Bi$_2$Se$_3$ nanosheet, pencil-drawn graphite, and paper substrate and (l) its photoresponses under different degrees of bending tests.

Source: Reproduced with permission [75]. Copyright 2020, RSC Publications.

(m) Schematic of flexible graphene-based THz detector and (n) its responsivity to different strains.

Source: Reproduced with permission [77]. Copyright 2017, AIP Publications.

(o) Photograph of flexible CNT-based THz scanner and (p) its imaging result.

Source: Reproduced with permission [78]. Copyright 2016, Springer Nature.

is superior to conventional semiconductor PDs, albeit it had undergone 0.38% compressive and tensile strains. Choi *et al.*[74] developed an ultra-flexible and rollable near-IR PD with a 2D MoS_2/Si heterojunction via direct synthesis at low temperature (< 200 °C). Upon 850 nm radiation, the flexible PD shows a responsivity of 10.07 $mA \cdot W^{-1}$ and a specific detectivity of 4.53×10^{10} Jones, as well as good bending capability (Figure 11.8(i) and 11.8(j)). Other than that, Bao's group demonstrated a novel transistor-like flexible PD composed of semiconducting SWCNTs and fullerene (C_{60}) [76], with a remarkable broadband photoresponse from 1000 to 1400 nm. The flexible PD demonstrated a maximum responsivity of 200 $A \cdot W^{-1}$ at a low operating voltage of 1 V, a fast response time of 2–4 ms, an excellent detectivity of 1.17×10^9 Jones, and high mechanical flexibility and robustness. Moreover, 2D Bi_2Se_3 nanosheet as a light-sensitive layer with pencil-drawn graphite electrodes on paper substrates was demonstrated for flexible IR PD (Figure 11.8(k) and 11.8(l)) [75], which exhibited high photocurrent, excellent responsivity, and long-term stability under 1064 nm illuminance. Note that the pencil-drawn flexible PD also displayed superior stability and long-term durability under bending conditions.

Lastly, a flexible THz detector could be a key element for next generation high-speed indoor wireless communication and wearable THz sensors for medical applications. Some novel materials, such as graphene and CNTs, have promising potential for THz detectors. Yang *et al.*[77] presented a flexible THz detector based on a graphene field-effect transistor fabricated on a PET substrate (Figure 11.8(m) and 11.8(n)). At room temperature, it showed a voltage responsivity exceeding 2 $V \cdot W^{-1}$ and an estimated noise equivalent power (NEP) of less than 3 $nW \cdot Hz^{-1/2}$ at 487 GHz. During the bending tests, the responsivity only slightly decreased with increasing strain, indicating the robustness of the flexible device. Other than that, Kawano and co-workers introduced a wearable terahertz detector composed of macroscopic CNT films on a PET substrate [78], with a considerable NEP value of 242 $pW \cdot Hz^{-1/2}$ for 1.4 THz (Figure 11.8(o) and 11.8(p)). They demonstrated a high-performance THz imaging of bent materials and the passive imaging of a human hand with a wearable scanner.

11.4.4 FLEXIBLE DISPLAY

Flexible displays combine ultra-thinness and light weight, excellent mechanical flexibility, low power consumption, and widely adjustable saturation emission, offering new possibilities for a smart human-machine interface. The demand for flexible displays continues to grow explosively not only because of its vast market size but more importantly due to its infinite possibilities of wearable integrated platform as shown in Figure 11.9(a). The ideal flexible display should be skin-like, becoming part of the human body without creating discomfort [1]. Currently, diverse strategies have been implemented for flexible displays, including flexible LED, flexible electrochromic display (ECD), flexible electrophoretic display (EPD), and flexible photochromic display (PCD).

Flexible LED is one of the most extensively investigated devices for full-color display panels and eco-friendly lighting due to its superior color quality, wide

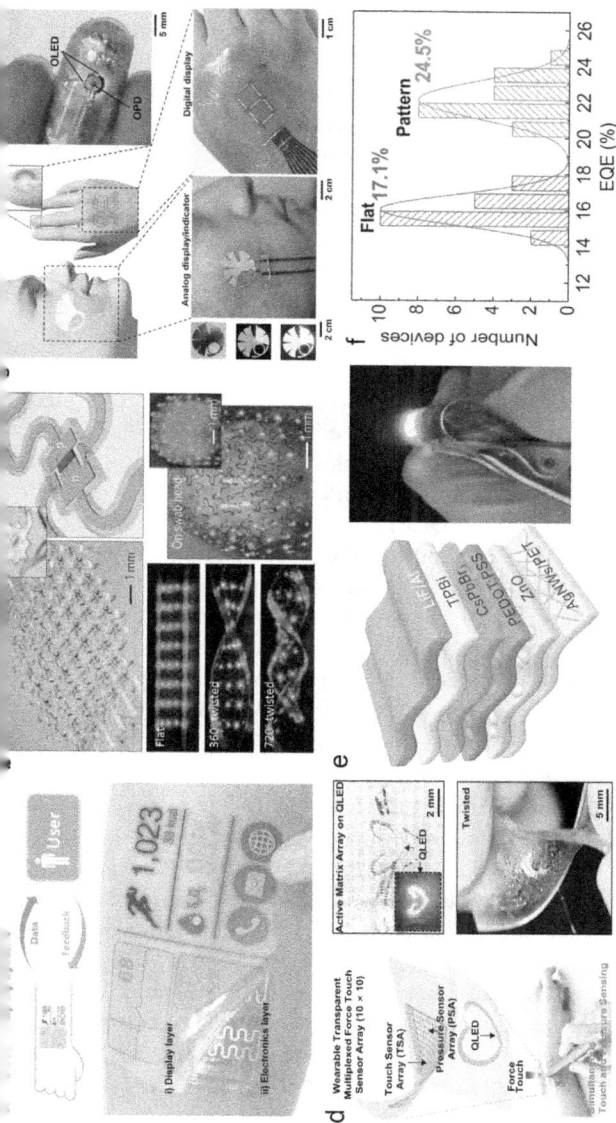

FIGURE 11.9 Flexible LED. (a) Schematic of the available smart flexible display system.

Source: Reproduced with permission [1]. Copyright 2018, WILEY-VCH.

(b) Images of stretchable inorganic LED structure and its illumination behaviors under different twisted loads as well as on swab head.

Source: Reproduced with permission [79]. Copyright 2010, Springer Nature.

(c) Smart electronic skin system comprising health-monitoring sensors, ultraflexible displays based on OLEDs.

Source: Reproduced under a Creative Commons license CC BY 4.0 [82]. Copyright 2016, the Authors.

(d) Schematic and image of transparent and wearable force touch sensor array integrated with skin-mounted QLED and image of twisted wearable devices on ultrathin QLEDs.

Source: Reproduced with permission [86]. Copyright 2017, WILEY-VCH.

(e) Schematic and image of flexible PeLED and (f) its EQE performance.

Source: Reproduced with permission [89]. Copyright 2020, ACS Publications.

viewing angle, mercury-free manufacture, and fascinating flexibility. In the past few years, various materials, processing techniques, and device architectures have been reported, including flexible inorganic LED, organic LED (OLED), flexible quantum dot LEDs (QLED), and flexible perovskite LEDs (PeLEDs).

Although many inorganic materials are intrinsically rigid, the stretchable inorganic display can be realized via first dividing the inorganic material into multiple micro-islands and then using serpentine-shaped interconnections to link all pixels. For instance, a stretchable display composed of 36 micro-inorganic LEDs with mechanically optimized layouts was fabricated by Rogers *et al.* (Figure 11.9(b)) [79]. Each pixel of LEDs on the PDMS substrate was connected with metal serpentine-shaped microribbons, and each microribbon formed electrical interconnections and/or structural bridges, which are capable of conformal integration to biomedicine and robotics.

The OLED is a newly emergent technology for flexible display. A typical flexible OLED consists of the flexible substrate, electron/hole transport layers, organic emitting layer, and electrodes. Considerable efforts have been devoted toward high-performance OLEDs in terms of luminance efficiency, color gamut, device stability, and fabrication techniques. Xu *et al.* developed transparent nanostructured metal–dielectric composite electrodes to realize flexible OLED with an external quantum efficiency (EQE) of 47.2% and power efficiency of 112.4 lm·W^{-1} [80]. Li *et al.* reported a multifunctional electrode architecture consisting of electrically conductive AgNWs, a nanopatterned ZnO outcoupling layer, and a hole injection polymer layer to attain a high-performance flexible OLED with a maximum EQE of 61.7% and power efficiency of 126.6 lm·W^{-1} [81]. Someya *et al.* demonstrated an ultraflexible and conformable three-color wearable OLED that introduces multiple functionalities, such as sensing and displays, on the human skin (Figure 11.9(c)) [82].

Quantum dots such as CdS, CdSe, ZnS, and InP, represents another acceptable candidate for flexible display emitting material due to their wide color gamut, high color purity, high brightness with low turn-on voltage, and ultrathin form factor [83]. Lim *et al.* engineered InP@ZnSeS heterostructured quantum dots to achieve environmentally benign flexible QLED with a maximum brightness of 3900 cd·m^{-2} [84]. Kim and co-workers reported a 4-inch full-color flexible QLED (320 × 240 pixels) using the transfer printing technique, which leverages the lifting layer to enable the reliable transfer of a quantum dot monolayer for layer-by-layer design [85]. Furthermore, flexible QLEDs were integrated with a multiplexed transparent touch sensor array as an input port of user intentions to realize a wearable multifunctional pressure-sensing and display (Figure 11.9(d)) [86].

Although PeLEDs are yet commercially matured, the multitude of compositional and structural variants enable the formation of charge-transport nanostructured perovskites materials and device processing with architectural innovations, endowing the flexible PeLED with considerable brightness and efficiency. Yu *et al.* developed a flexible printed organometal halide PeLED composed of ITO or CNTs (as the transparent anode), a printed perovskite composite film (as the emissive layer), and printed silver NWs (as the cathode) [87]. The device exhibited a low turn-on voltage of 2.6 V and a maximum luminance intensity of 21014 cd·m^{-2}. Lee and colleagues demonstrated a high-performance flexible PeLED using graphene/PET as an anode, which exhibited bright electroluminescence > 10000 cd·m^{-2}, high efficiency value of 18 cd·A^{-1}, and satisfactory bending stability [88]. Recently, Shen *et al.* presented an improved flexible CsPbBr3 PeLED structure (Figure 11.9(e)) based on interface engineering for energy-saving photon generation and enhanced light output coupling [89], achieving a record EQE of 24.5%, as shown in Figure 11.9(g).

Besides LED, ECDs based on the electrochromism are another competitive route for flexible displays, where certain materials can reversibly change their colors or optical properties through redox reactions under small external voltage or current [90]. Over the past few years, amazing advances in flexible ECDs have been made in the field of smart windows for architectural buildings, auto-dimming rear view mirrors, and flexible camouflage applications. Typically, ECD devices are sandwich structures that consist of transparent conductors, electrochromic layers, ion conducting electrolyte layers, and ion storage layers. For instance, a graphene-based flexible ECD constructed by multilayered graphene films and an electrolyte was developed by Polat *et al.*, showing high optical contrast and broadband optical modulation up to 55% in the visible and near-infrared ranges [91]. Lin and co-workers developed a non-heated roll-to-roll process to realize a continuous production of extra-large and transparent silver nanofiber (AgNF) network electrodes for flexible ECD (Figure 11.10(a))

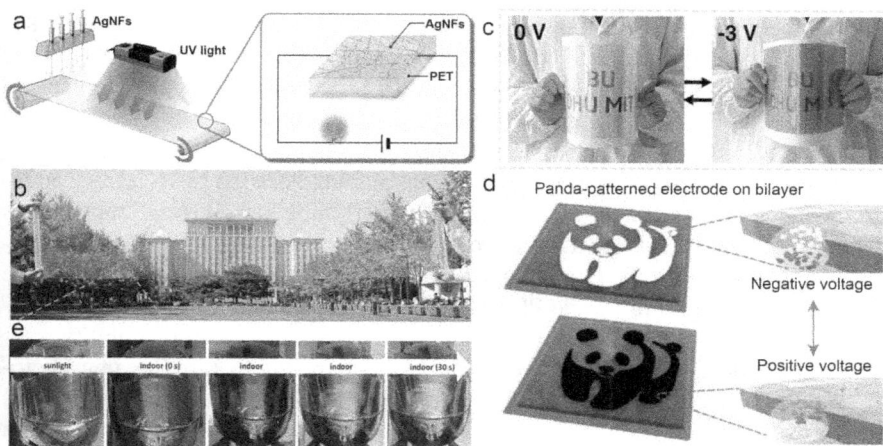

FIGURE 11.10 Flexible ECD, EPD, and PCD. (a) Schematic of the roll-to-roll process for the production of AgNF-based flexible ECDs and (b) actual product image.

Source: Reproduced with permission [92]. Copyright 2017, WILEY VCH.

(c) Images of electrochromic performance of sandwiched flexible ECD under bending state

Source: Reproduced under a Creative Commons license CC BY 4.0 [93]. Copyright 2021, the Authors.

(d) Schematic of the color change of a panda-patterned stretchable EPD.

Source: Reproduced with permission [94]. Copyright 2021, WILEY VCH.

(e) Images of flexible PCD attached to helmet face with discoloration kinetics.

Source: Reproduced with permission [97]. Copyright 2020, RSC Publications.

[92], which was successfully assembled into an A4-sized smart window with short switching time, satisfactory coloration efficiency and flexibility, as shown in Figure 11.10(b). In a recent work, Li *et al.*[93] reported a flexible and high-performance ECD with self-assembled 2D TiO_2/MXene heterostructures, exhibiting a fast coloration speed of ~ 1.71 s, excellent mechanical flexibility, and exceptional electrochromic efficiency under bending state (Figure 11.10(c)).

As one type of reflective displays, EPDs control the color change via moving charged pigment particles with the help of voltage driving. The different particles move toward the corresponding direction of the electric field changes, *e.g.*, the negatively charged white particles and positively charged black particles suspended in dielectric fluid migrate to the anode and cathode electrode to display the white state and the black state, respectively. More recently, Qiu *et al.* introduced a hydrogel–elastomer-based stretchable EPD via in situ adhesion decoration to form a strong bond between the electrode and display functional layer (Figure 11.10(d)) [94]. The hydrogel–elastomer hybrid structure possesses high optical transparency, biocompatibility, and robustness in electrophoretic displays. A prototype of a stretchable blackboard with re-writability has been demonstrated without using a complicated thin film transistor array, which is washable and can maintain its functionality for at least 43 days.

Lastly, photochromism depicts a reversible photo-induced transformation of a molecule between two isomers whose absorption spectra are distinguishably different, which has also been utilized for flexible PCD with reversible photo-switches capability. Wang *et al.* demonstrated a flexible fast-response PCD using photochromic gels formed by the mixtures of the PVA and TiO_2 nanoparticle [95]. The PCD exhibits long photoreversible switching cycles ($\geq$ 50 times), a short decoloration period of less than 8 s upon UV illumination, and recoloration in 16 min in ambient air and 140 s upon near-IR light illumination, implying potential applications in self-erasing rewritable media and colorimetric oxygen indicator. Chen and colleagues [96] integrated the inorganic lanthanide Er^{3+}-doped bismuth layer-structure ferroelectric $Na_{0.5}Bi_{2.5}Nb_2O_9$ and PI substrate to realize a flexible and rewritable nonvolatile PCD that can well sustain 10^5 bending cycles and maintain stability after many write-read-erase cycles. Other than that, commercial T-type organic photochromic dyes are embedded into polymer materials without affecting their optimal solution absorptions and isomerization kinetics, thus obtaining a flexible PCD film with a tunable and fast solution-like response [97]. The films are also highly transparent, recyclable, and scalable, and show enhanced fatigue resistance, which are highly suitable for different smart glass applications (Figure 11.10(e)).

11.4.5 FLEXIBLE OPTICAL SECURITY LABEL

Information security is extremely vital in modern society. The exponential growth of interconnected physical entities that generate massive data daily poses a severe challenge to information security, driving the development of novel security labels with high reliability and robustness. Flexible optical security labels (OSLs) have drawn considerable attention because they offer high security, satisfactory flexibility, and compatibility with wearable systems. Thus far, numerous approaches, such as fluorescence, Raman scattering, structural color, and laser speckle, have been exploited to realize flexible OSLs.

Fluorescence has been extensively applied to OSLs because of the readily detectable light emission [98]. Kim *et al.* proposed a flexible fluorescent OSL by introducing micropatternable triplet–triplet annihilation upconversion thin films composed of PU, PUA, PVA, and PET substrates [99]. By customizing the chromophore compositions and film patterns, such flexible OSL can store encoded data that are legible only under specific light sources, granting unlimited possibilities for anti-counterfeiting (Figure 11.11(a)). Wu *et al.* presented a rapid, convenient, and low-cost flexible fluorescent OSL using patterned perovskite film arrays on PET substrate [100], showing superior anti-counterfeiting performance with high modifiability, low reagent cost, fast authentication (overall time 12.17 s), and high encoding capacity (2.1×10^{623}). In addition, Guo *et al.*[101] exploited patterned AgNW film on a microstructured PDMS substrate (molded from sandpaper) for flexible OSL (Figure 11.11(b)), which shows high transmittance in the visible range (~95%) and high reflectance (60–70%) in the IR range. Such a device is invisible to the naked eye but visible through an IR camera, which also suits well for anti-counterfeiting applications (Figure 11.11(c)).

Raman scattering yields a unique vibrational fingerprint with a much narrower waveform than fluorescence emission, effectively minimizing the crosstalk between two Raman peaks close to each other. This property provides a non-invasive and highly sensitive route for flexible OSLs. Shan *et al.* reported a flexible and biocompatible PUF-based OSL using randomly embedded microdiamonds in silk fibroin films [102], where the silk fibroin films serve as the substrate and the Raman microdiamond signal serves as an excitation response. Remarkably, it exhibited a maximum encoding capability of 2^{10000} and satisfactory biosafety. Furthermore, surface-enhanced Raman scattering (SERS) with plasmonic structures presents an ideal platform for OSL. Ying and co-workers demonstrated a flexible patterned plasmonic metafilm OSL with a SERS barcode (Figure 11.11(d)) [103], which possesses excellent mechanical flexibility, robustness, and stability for anti-counterfeiting applications. Their group also reported another flexible plasmonic SERS metasurface for cryptographic applications [104]. The obtained butterfly and institution logo images represented a strong and reproducible SERS signal for information encryption (Figure 11.11(e)).

Structural colors induced by optical interference, such as Bragg diffraction caused by periodic structures and color polarization in birefringent materials, recently achieved a commercial success in OSL due to their inexpensive production and exceptional ease of perception. Zhang *et al.* embedded luminescent piezoelectric microparticles into a thermoplastic PU matrix to achieve waterproofness, flexibility/wearability, high stretchability, and multicolor emissions for anti-counterfeiting [105]. In a notable research by Song *et al.*, they presented a flexible, large-area covert polarization display based on ultrathin lossy nanocolumns with wide color selectivity [106]. As a demonstration, the hidden polarization display label was attached to daily objects with curved and wrinkled surfaces to reveal the hidden fast response code via polarization adjustment of indoor and outdoor environments, as shown in

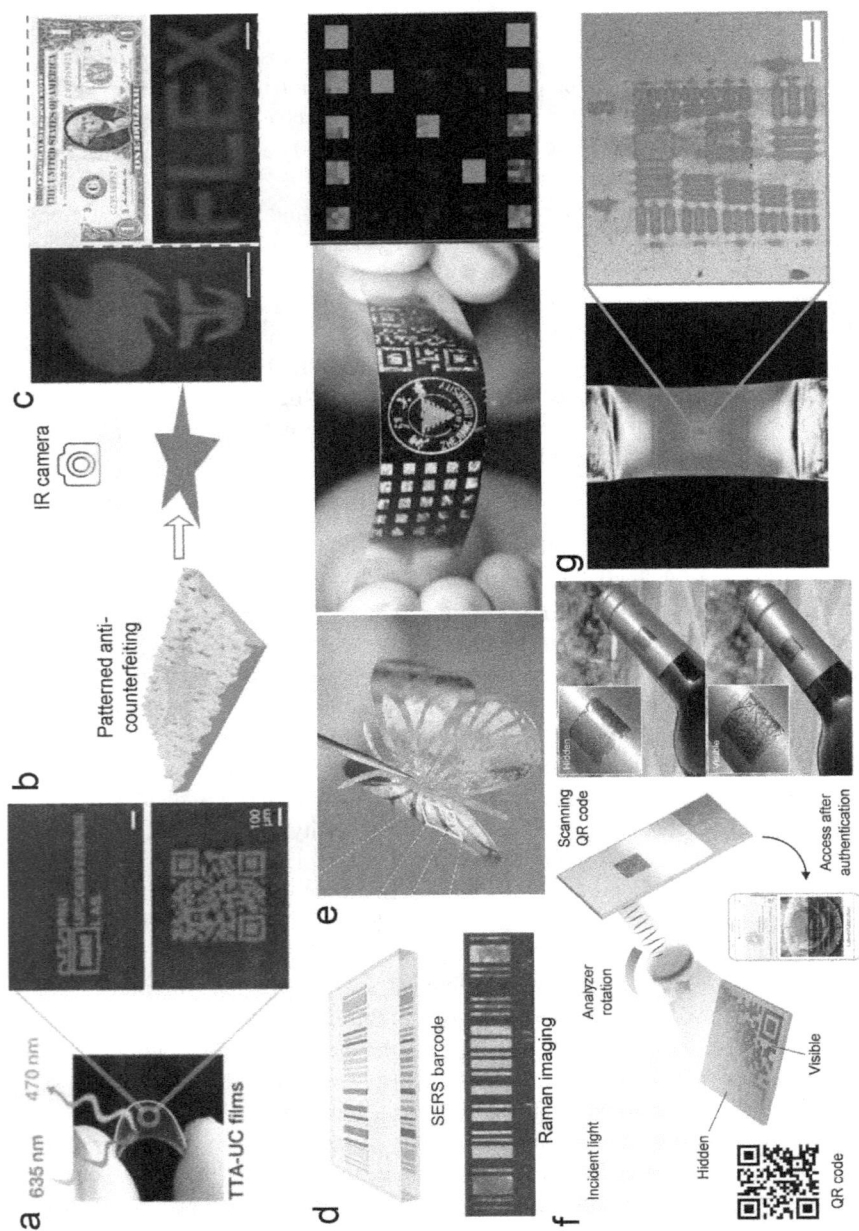

FIGURE 11.11 Applications of flexible OSLs. (a) Images of flexible fluorescent OSL and anti-counterfeiting demo.

(b) Schematic of flexible OSL based on microstructured AgNW film discriminated by IR camera. (c) Photographs of anti-counterfeiting with the logo "SUSTech," and the "FLEX" on paper money.

Source: Reproduced with permission [101]. Copyright 2020, Elsevier Publications.

(d) Schematic of SERS barcode and its Raman imaging.

Source: Reproduced with permission [103]. Copyright 2016, WILEY-VCH.

(e) Photographs of flexible SERS-based OSLs (with butterfly and logo shapes) and Raman imaging with a "Z" shape.

Source: Reproduced with permission [104]. Copyright 2016, WILEY-VCH.

(f) Schematic of authentication based on structural color OSL and its application in product anti-counterfeiting.

Source: Reproduced with permission [106]. Copyright 2019, WILEY-VCH.

(g) Images of stretchable OSL and its application in data storage.

Source: Reproduced with permission [107]. Copyright 2020, WILEY-VCH.

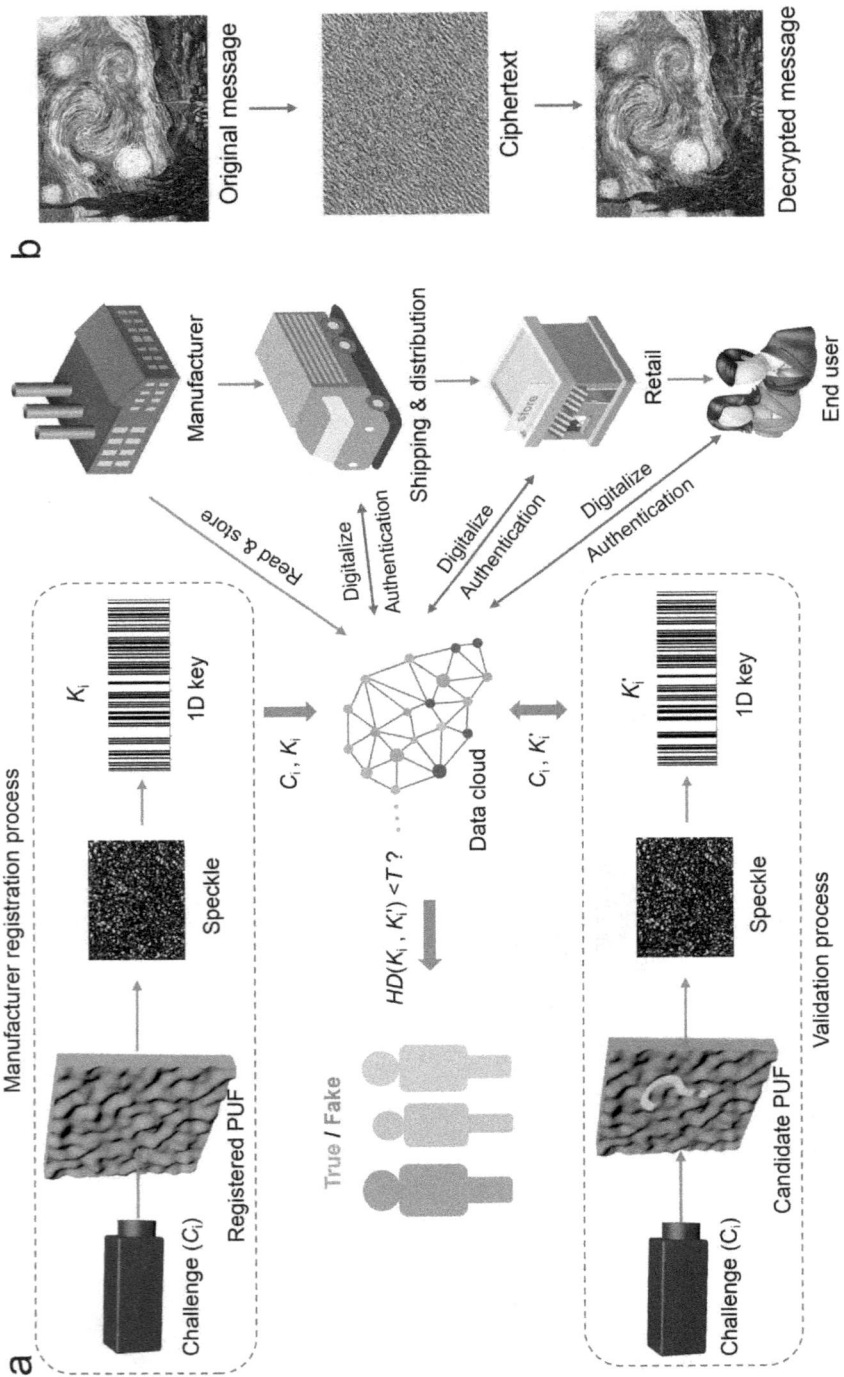

FIGURE 11.12 Flexible OSL based on bionic PUFs. (a) Schematic authentication protocol for product verification. (b) Demonstration for information encryption, communication, and decryption.

Figure 11.11(f). Xu and co-workers demonstrated a tunable structural color film based on the birefringence of stretched materials [107], which has considerable potential in display devices, anti-counterfeiting labels, and data storage (Figure 11.11(g)).

To identify a wide range of objects, OSLs based on the laser speckle of physical unclonable function (PUF) have been commonly used. In this regard, Wan and co-workers proposed the concept of bionic optical PUF using flexible microstructured PDMS film molded from natural plant tissue by a simple, low-cost, and environmentally friendly manufacturing process [108]. The laser speckle responses of the bionic PUFs were statistically proved to be intrinsically random, unique, unpredictable, and sufficiently robust for flexible OSL applications (Figure 11.12(a)). More interestingly, the proposed bionic PUFs can also be used for information encryption and communication, where an example of encryption and decryption of the painting "*The Starry Night*" is illustrated in Figure 11.12(b). In addition, the bionic PUF can also be used as a random number generator, where the laser speckle responses are fully extracted into random bit streams with a fast speed of up to 1.04 Gbit/s and a high extraction rate of 72%, which shows great potential in the application of flexible networked electronics [109].

11.5 SUMMARY AND PERSPECTIVES

So far, we briefly review the recent progress on soft and flexible materials covering the substrate materials and active layer materials, as well as their associated applications in optoelectronics, unambiguously exhibiting a bright future for next generation flexible and wearable electronics. Despite these marvelous achievements, there are several major challenges that remain to be solved, including:

1) Technological breakthroughs in material synthesis, preparation technology, and device integration are necessary for the commercialization of flexible electronics. The current lithography technology suffers from high processing cost, complex process, substantial material waste, and low throughput. In this regard, the printing manufacturing technologies (*e.g.*, 3D printing) are promising alternatives because of their superior advantages, such as universality, applicability, production sustainability, and convenient customization, promoting a low-cost and large-scale production of the functional flexible electronic device.

2) Continuous improvement on device performance is in continuous pursuit to satisfy long-term operation in extreme environments. Further investigations should focus on endowing soft materials with more functionalities and broaden the scope of applications. For instance, distinct features, such as biocompatibility, biodegradability, self-healing, robustness, and self-powering capability, are highly desired in applications of human-compatible microsystems, which is especially for implantable biomedical devices to achieve noninvasive healthcare treatments.

3) Integrated microsystems containing multifunctional soft materials and devices are highly demanded to fulfill the requirement of complex and

diverse applications. For example, sound, light, heat, humidity, magnetic field, force, and chemical information from a specific target can be simultaneously collected, processed, communicated, and actuated by a multifunctional integrated flexible system. Other than that, flexible solar cells or power generators provide options for energy supply, supported by a flexible super-capacitor as a large-capacity power storage device. All these efforts are anticipated to fully leverage the advantages of flexible materials and devices in the future.

ACKNOWLEDGMENTS

This work was financially supported by the "Guangdong Innovative and Entrepreneurial Research Team Program" under Contract No. 2016ZT06G587.

BIBLIOGRAPHY

[1] Koo, J. H., Kim, D. C., Shim, H. J., Kim, T. H., Kim, D. H. 2018. Flexible and stretchable smart display: materials, fabrication, device design, and system integration. *Adv. Funct. Mater.* 28:1801834.

[2] Wan, Y., Wang, Y., Guo, C. F. 2017. Recent progresses on flexible tactile sensors. *Mater. Today Phys.* 1:61–73.

[3] Wang, X., Lu, X., Liu, B., Chen, D., Tong, Y., Shen, G. 2014. Flexible energy-storage devices: design consideration and recent progress. *Adv. Mater.* 26:4763–4782.

[4] Cai, S., Xu, X., Yang, W., Chen, J., Fang, X. 2019. Materials and designs for wearable photodetectors. *Adv. Mater.* 31:1808138.

[5] Wan, Y., Qiu, Z., Hong, Y., Wang, Y., Zhang, J., Liu, Q., Wu, Z., Guo, C. F. 2018. A highly sensitive flexible capacitive tactile sensor with sparse and high-aspect-ratio microstructures. *Adv. Electron. Mater.* 4:1700586.

[6] Zheng, X., Jia, Y., Chen, A. 2021. Azobenzene-containing liquid crystalline composites for robust ultraviolet detectors based on conversion of illuminance-mechanical stress-electric signals. *Nat. Commun.* 12:4875.

[7] Cho, S. H., Lee, S. W., Yu, S., Kim, H., Chang, S., Kang, D., Hwang, I., Kang, H. S., Jeong, B., Kim, E. H., Cho, S. M., Kim, K. L., Lee, H., Shim, W., Park, C. 2017. Micropatterned pyramidal ionic gels for sensing broad-range pressures with high sensitivity. *ACS Appl. Mater. Interfaces.* 9:10128–10135.

[8] Fang, H., Yu, K. J., Gloschat, C., Yang, Z., Song, E., Chiang, C.-H., Zhao, J., Won, S. M., Xu, S., Trumpis, M. 2017. Capacitively coupled arrays of multiplexed flexible silicon transistors for long-term cardiac electrophysiology. *Nat. Biomed. Eng.* 1:0038.

[9] Sun, J. Y., Keplinger, C., Whitesides, G. M., Suo, Z. G. 2014. Ionic skin. *Adv. Mater.* 26:7608–7614.

[10] Wang, C., Xia, K., Zhang, Y., Kaplan, D. L. 2019. Silk-based advanced materials for soft electronics. *Acc. Chem. Res.* 52:2916–2927.

[11] Russo, A., Ahn, B. Y., Adams, J. J., Duoss, E. B., Bernhard, J. T., Lewis, J. A. 2011. Pen-on-paper flexible electronics. *Adv. Mater.* 23:3426–3430.

[12] Seung, W., Gupta, M. K., Lee, K. Y., Shin, K.-S., Lee, J.-H., Kim, T. Y., Kim, S., Lin, J., Kim, J. H., Kim, S.-W. 2015. Nanopatterned textile-based wearable triboelectric nanogenerator. *ACS Nano.* 9:3501–3509.

[13] Eduok, U., Faye, O., Szpunar, J. 2017. Recent developments and applications of protective silicone coatings: A review of PDMS functional materials. *Prog. Org. Coat.* 111:124–163.

[14] Wong, I., Ho, C.-M., 2009. Surface molecular property modifications for poly (dimethylsiloxane)(PDMS) based microfluidic devices. *Microfluid. Nanofluid.* 7:291.

[15] Park, J., Lee, Y., Ha, M., Cho, S., Ko, H. 2016. Micro/nanostructured surfaces for self-powered and multifunctional electronic skins. *J. Mater. Chem. B.* 4:2999–3018.

[16] Liaw, D.-J., Hsu, P.-N., Chen, W.-H., Lin, S.-L. 2002. High glass transitions of new polyamides, polyimides, and poly (amide-imide) s containing a triphenylamine group: Synthesis and characterization. *Macromolecules.* 35:4669–4676.

[17] Hasegawa, M., Horie, K. 2001. Photophysics, photochemistry, and optical properties of polyimides. *Prog. Polym. Sci.* 26:259–335.

[18] Tsai, C.-L., Yen, H.-J., Liou, G.-S. 2016. Highly transparent polyimide hybrids for optoelectronic applications. *React. Funct. Polym.* 108:2–30.

[19] Yang, C., Suo, Z. 2018. Hydrogel ionotronics. *Nat. Rev. Mater.* 3:125–142.

[20] Ahmed, E. M. 2015. Hydrogel: Preparation, characterization, and applications: A review. *J. Adv. Res.* 6:105–121.

[21] Turner, J. G., Og, J. H., Murphy, C. J. 2020. Gold nanorod impact on mechanical properties of stretchable hydrogels. *Soft Matter.* 16:6582–6590.

[22] Zeng, S., Pian, S., Su, M., Wang, Z., Wu, M., Liu, X., Chen, M., Xiang, Y., Wu, J., Zhang, M., Cen, Q., Tang, Y., Zhou, X., Huang, Z., Wang, R., Tunuhe, A., Sun, X., Xia, Z., Tian, M., Chen, M., Ma, X., Yang, L., Zhou, J., Zhou, H., Yang, Q., Li, X., Ma, Y., Tao, G. 2021. Hierarchical-morphology metafabric for scalable passive daytime radiative cooling. *Science.* 373:692–696.

[23] Zhu, H., Fang, Z., Preston, C., Li, Y., Hu, L. 2014. Transparent paper: Fabrications, properties, and device applications. *Energ. Environ. Sci.* 7:269–287.

[24] Khan, S. M., Nassar, J. M., Hussain, M. M. 2020. Paper as a substrate and an active material in paper electronics. *ACS Appl. Electron. Mater.* 3:30–52.

[25] Zardetto, V., Brown, T. M., Reale, A., Di Carlo, A. 2011. Substrates for flexible electronics: A practical investigation on the electrical, film flexibility, optical, temperature, and solvent resistance properties. *J. Polym. Sci. Pol. Phys.* 49:638–648.

[26] Nisticò, R. 2020. Polyethylene terephthalate (PET) in the packaging industry. *Polym. Test.* 90:106707.

[27] Somarathna, H., Raman, S., Mohotti, D., Mutalib, A., Badri, K. 2018. The use of polyurethane for structural and infrastructure engineering applications: A state-of-the-art review. *Constr. Build. Mater.* 190:995–1014.

[28] Lee, P., Lee, J., Lee, H., Yeo, J., Hong, S., Nam, K. H., Lee, D., Lee, S. S., Ko, S. H. 2012. Highly stretchable and highly conductive metal electrode by very long metal nanowire percolation network. *Adv. Mater.* 24:3326–3332.

[29] Xu, W.-H., Wang, L., Guo, Z., Chen, X., Liu, J., Huang, X.-J. 2015. Copper nanowires as nanoscale interconnects: their stability, electrical transport, and mechanical properties. *ACS Nano.* 9:241–250.

[30] Ye, S., Rathmell, A. R., Stewart, I. E., Ha, Y.-C., Wilson, A. R., Chen, Z., Wiley, B. J. 2014. A rapid synthesis of high aspect ratio copper nanowires for high-performance transparent conducting films. *Chem. Commun.* 50:2562–2564.

[31] Park, J. H., Han, S., Kim, D., You, B. K., Joe, D. J., Hong, S., Seo, J., Kwon, J., Jeong, C. K., Park, H. J. 2017. Plasmonic-tuned flash Cu nanowelding with ultrafast photochemical-reducing and interlocking on flexible plastics. *Adv. Funct. Mater.* 27:1701138.

[32] Chen, Y., Ouyang, Z., Gu, M., Cheng, W. 2013. Mechanically strong, optically transparent, giant metal superlattice nanomembranes from ultrathin gold nanowires. *Adv. Mater.* 25:80–85.

[33] Gong, S., Schwalb, W., Wang, Y. W., Chen, Y., Tang, Y., Si, J., Shirinzadeh, B., Cheng, W. L. 2014. A wearable and highly sensitive pressure sensor with ultrathin gold nanowires. *Nat. Commun.* 5:3132.

[34] Dasgupta, N. P., Sun, J., Liu, C., Brittman, S., Andrews, S. C., Lim, J., Gao, H., Yan, R., Yang, P. 2014. 25th anniversary article: Semiconductor nanowires-synthesis, characterization, and applications. *Adv. Mater.* 26:2137–2184.

[35] Ha, M., Lim, S., Park, J., Um, D.-S., Lee, Y., Ko, H. 2015. Bioinspired interlocked and hierarchical design of ZnO nanowire arrays for static and dynamic pressure-sensitive electronic skins. *Adv. Funct. Mater.* 25:2841–2849.

[36] Weiss, N. O., Zhou, H., Liao, L., Liu, Y., Jiang, S., Huang, Y., Duan, X. 2012. Graphene: An emerging electronic material. *Adv. Mater.* 24:5782–5825.

[37] Huang, X., Qi, X., Boey, F., Zhang, H. 2012. Graphene-based composites. *Chem. Soc. Rev.* 41:666–686.

[38] Obitayo, W., Liu, T. 2012. A review: Carbon nanotube-based piezoresistive strain sensors. *J. Sens.* 2012: 652438.

[39] Yu, L., Shearer, C., Shapter, J. 2016. Recent development of carbon nanotube transparent conductive films. *Chem. Rev.* 116:13413–13453.

[40] Yamada, T., Hayamizu, Y., Yamamoto, Y., Yomogida, Y., Izadi-Najafabadi, A., Futaba, D. N., Hata, K. 2011. A stretchable carbon nanotube strain sensor for human-motion detection. *Nat. Nanotechnol.* 6:296–301.

[41] Kim, N., Kee, S., Lee, S. H., Lee, B. H., Kahng, Y. H., Jo, Y. R., Kim, B. J., Lee, K. 2014. Highly conductive PEDOT: PSS nanofibrils induced by solution-processed crystallization. *Adv. Mater.* 26:2268–2272.

[42] Lang, U., Dual, J. 2007. *Mechanical properties of the intrinsically conductive polymer poly (3, 4-ethylenedioxythiophene) poly (styrenesulfonate)(PEDOT/PSS), Key Engineering Materials.* Trans Tech Publications Ltd, 345:1189–1192.

[43] Mao, J., Li, C., Park, H. J., Rouabhia, M., Zhang, Z. 2017. Conductive polymer waving in liquid nitrogen. *ACS Nano.* 11:10409–10416.

[44] Pan, L., Chortos, A., Yu, G., Wang, Y., Isaacson, S., Allen, R., Shi, Y., Dauskardt, R., Bao, Z. 2014. An ultra-sensitive resistive pressure sensor based on hollow-sphere microstructure induced elasticity in conducting polymer film. *Nat. Commun.* 5:3002.

[45] Kang, E., Neoh, K., Tan, K. 1998. Polyaniline: A polymer with many interesting intrinsic redox states. *Prog. Polym. Sci.* 23:277–324.

[46] DeLongchamp, D., Hammond, P. T. 2001 Layer-by-layer assembly of PEDOT/polyaniline electrochromic devices. *Adv. Mater.* 13:1455–1459.

[47] Mattes, B., Knobbe, E., Fuqua, P., Nishida, F., Chang, E.-W., Pierce, B., Dunn, B., Kaner, R. 1991. Polyaniline sol-gels and their third-order nonlinear optical effects. *Synthetic Met.* 343:183–3187.

[48] Park, J., Lee, Y., Hong, J., Ha, M., Jung, Y. D., Lim, H., Kim, S. Y., Ko, H. 2014. Giant tunneling piezoresistance of composite elastomers with interlocked microdome arrays for ultrasensitive and multimodal electronic skins. *ACS Nano.* 8:4689–4697.

[49] Bai, N., Wang, L., Wang, Q., Deng, J., Wang, Y., Lu, P., Huang, J., Li, G., Zhang, Y., Yang, J. 2020. Graded intrafillable architecture-based iontronic pressure sensor with ultra-broad-range high sensitivity. *Nat. Commun.* 11:209.

[50] Wu, H., Kong, D., Ruan, Z., Hsu, P.-C., Wang, S., Yu, Z., Carney, T. J., Hu, L., Fan, S., Cui, Y. 2013. A transparent electrode based on a metal nanotrough network. *Nat. Nanotechnol.* 8:421–425.

[51] Park, J. H., Hwang, G. T., Kim, S., Seo, J., Park, H. J., Yu, K., Kim, T. S., Lee, K. J. 2017. Flash-induced self-limited plasmonic welding of silver nanowire network for transparent flexible energy harvester. *Adv. Mater.* 29:1603473.

[52] Liu, Y., Zhang, J., Gao, H., Wang, Y., Liu, Q., Huang, S., Guo, C. F., Ren, Z. 2017. Capillary-force-induced cold welding in silver-nanowire-based flexible transparent electrodes. *Nano Lett.* 17:1090–1096.

[53] Guo, C. F., Sun, T., Liu, Q., Suo, Z., Ren, Z. 2014. Highly stretchable and transparent nanomesh electrodes made by grain boundary lithography. *Nat. Commun.* 5:3121.

[54] Guo, C. F., Liu, Q., Wang, G., Wang, Y., Shi, Z., Suo, Z., Chu, C.-W., Ren, Z. 2015. Fatigue-free, superstretchable, transparent, and biocompatible metal electrodes. *Proc. Natl. Acad. Sci. U. S. A.* 112:12332–12337.

[55] Lipomi, D. J., Vosgueritchian, M., Tee, B. C. K., Hellstrom, S. L., Lee, J. A., Fox, C. H., Bao, Z. 2011. Skin-like pressure and strain sensors based on transparent elastic films of carbon nanotubes. *Nat. Nanotechnol.* 6:788–792.

[56] Jiang, S., Hou, P.-X., Chen, M.-L., Wang, B.-W., Sun, D.-M., Tang, D.-M., Jin, Q., Guo, Q.-X., Zhang, D.-D., Du, J.-H. 2018. Ultrahigh-performance transparent conductive films of carbon-welded isolated single-wall carbon nanotubes. *Sci. Adv.* 4:eaap9264.

[57] Xu, P., Kang, J., Choi, J.-B., Suhr, J., Yu, J., Li, F., Byun, J.-H., Kim, B.-S., Chou, T.-W. 2014. Laminated ultrathin chemical vapor deposition graphene films based stretchable and transparent high-rate supercapacitor. *ACS Nano.* 8:9437–9445.

[58] Zhang, W., Zhao, B., He, Z., Zhao, X., Wang, H., Yang, S., Wu, H., Cao, Y. 2013. High-efficiency ITO-free polymer solar cells using highly conductive PEDOT: PSS/surfactant bilayer transparent anodes. *Energ. Environ. Sci.* 6:1956–1964.

[59] Wang, Y., Zhu, C., Pfattner, R., Yan, H., Jin, L., Chen, S., Molina-Lopez, F., Lissel, F., Liu, J., Rabiah, N. I. 2017. A highly stretchable, transparent, and conductive polymer. *Sci. Adv.* 3:e1602076.

[60] Nam, J., Lee, Y., Choi, W., Kim, C. S., Kim, H., Kim, J., Kim, D. H., Jo, S. 2016. Transfer printed flexible and stretchable thin film solar cells using a water-soluble sacrificial layer. *Adv. Energ. Mater.* 6:1601269.

[61] Lee, J., Wu, J., Shi, M., Yoon, J., Park, S. I., Li, M., Liu, Z., Huang, Y., Rogers, J. A. 2011. Stretchable GaAs photovoltaics with designs that enable high areal coverage. *Adv. Mater.* 23:986–991.

[62] Xu, Z., Li, T., Liu, Q., Zhang, F., Hong, X., Xie, S., Lin, C., Liu, X., Guo, W. 2018. Controllable and large-scale fabrication of rectangular CuS network films for indium tin oxide-and Pt-free flexible dye-sensitized solar cells. *Sol. Energ. Mater. Sol. C.* 179:297–304.

[63] Hou, S., Lv, Z., Wu, H., Cai, X., Chu, Z., Zou, D. 2012. Flexible conductive threads for wearable dye-sensitized solar cells. *J. Mater. Chem.* 22:6549–6552.

[64] Kaltenbrunner, M., White, M. S., Głowacki, E. D., Sekitani, T., Someya, T., Sariciftci, N. S., Bauer, S. 2012. Ultrathin and lightweight organic solar cells with high flexibility. *Nat. Commun.* 3:770.

[65] Jinno, H., Fukuda, K., Xu, X., Park, S., Suzuki, Y., Koizumi, M., Yokota, T., Osaka, I., Takimiya, K., Someya, T. 2017. Stretchable and waterproof elastomer-coated organic photovoltaics for washable electronic textile applications. *Nat. Energ.* 2:780–785.

[66] Feng, J., Zhu, X., Yang, Z., Zhang, X., Niu, J., Wang, Z., Zuo, S., Priya, S., Liu, S., Yang, D. 2018. Record efficiency stable flexible perovskite solar cell using effective additive assistant strategy. *Adv. Mater.* 30:1801418.

[67] Huang, K., Peng, Y., Gao, Y., Shi, J., Li, H., Mo, X., Huang, H., Gao, Y., Ding, L., Yang, J. 2019. High-performance flexible perovskite solar cells via precise control of electron transport layer. *Adv. Energ. Mater.* 9:1901419.

[68] Guan, X., Yu, X., Periyanagounder, D., Benzigar, M. R., Huang, J. K., Lin, C. H., Kim, J., Singh, S., Hu, L., Liu, G. 2021. Recent progress in short-to long-wave infrared photodetection using 2D materials and heterostructures. *Adv. Opt. Mater.* 9:2001708.

[69] Veeralingam, S., Durai, L., Yadav, P., Badhulika, S. 2021. Record-high responsivity and detectivity of a flexible deep-ultraviolet photodetector based on solid state-assisted synthesized hBN nanosheets. *ACS Appl. Electron. Mater.* 3:1162–1169.

[70] Yan, C., Wang, J., Wang, X., Kang, W., Cui, M., Foo, C. Y., Lee, P. S. 2014. An intrinsically stretchable nanowire photodetector with a fully embedded structure. *Adv. Mater.* 26:943–950.

[71] Kim, T.-Y., Ha, J., Cho, K., Pak, J., Seo, J., Park, J., Kim, J.-K., Chung, S., Hong, Y., Lee, T. 2017. Transparent large-area MoS_2 phototransistors with inkjet-printed components on flexible platforms. *ACS Nano.* 11:10273–10280.

[72] Peng, Z. Y., Xu, J. L., Zhang, J. Y., Gao, X., Wang, S. D. 2018. Solution-processed high-performance hybrid photodetectors enhanced by perovskite/MoS_2 bulk heterojunction. *Adv. Mater. Interfaces.* 5:1800505.

[73] Lee, W. Y., Ha, S., Lee, H., Bae, J. H., Jang, B., Kwon, H. J., Yun, Y., Lee, S., Jang, J. 2019. High-detectivity flexible near-infrared photodetector based on chalcogenide Ag_2Se nanoparticles. *Adv. Opt. Mater.* 7:1900812.

[74] Choi, J.-M., Jang, H. Y., Kim, A. R., Kwon, J.-D., Cho, B., Park, M. H., Kim, Y. 2021. Ultra-flexible and rollable 2D-MoS_2/Si heterojunction-based near-infrared photodetector via direct synthesis. *Nanoscale.* 13:672–680.

[75] Liu, S., Huang, Z., Qiao, H., Hu, R., Ma, Q., Huang, K., Li, H., Qi, X. 2020. Two-dimensional Bi_2Se_3 nanosheet based flexible infrared photodetector with pencil-drawn graphite electrodes on paper. *Nanoscale Adv.* 2:906–912.

[76] Park, S., Kim, S. J., Nam, J. H., Pitner, G., Lee, T. H., Ayzner, A. L., Wang, H., Fong, S. W., Vosgueritchian, M., Park, Y. J. 2015. Significant enhancement of infrared photodetector sensitivity using a semiconducting single-walled carbon nanotube/C_{60} phototransistor. *Adv. Mater.* 27:759–765.

[77] Yang, X., Vorobiev, A., Generalov, A., Andersson, M. A., Stake, J. 2017. A flexible graphene terahertz detector. *Appl. Phys. Lett.* 111:021102.

[78] Suzuki, D., Oda, S., Kawano, Y. 2016. A flexible and wearable terahertz scanner. *Nat. Photon.* 10:809–813.

[79] Kim, R.-H., Kim, D.-H., Xiao, J., Kim, B. H., Park, S.-I., Panilaitis, B., Ghaffari, R., Yao, J., Li, M., Liu, Z. 2010. Waterproof AlInGaP optoelectronics on stretchable substrates with applications in biomedicine and robotics. *Nat. Mater.* 9:929–937.

[80] Xu, L.-H., Ou, Q.-D., Li, Y.-Q., Zhang, Y.-B., Zhao, X.-D., Xiang, H.-Y., Chen, J.-D., Zhou, L., Lee, S.-T., Tang, J.-X. 2016. Microcavity-free broadband light outcoupling enhancement in flexible organic light-emitting diodes with nanostructured transparent metal–dielectric composite electrodes. *ACS Nano.* 10:1625–1632.

[81] Li, W., Li, Y. Q., Shen, Y., Zhang, Y. X., Jin, T. Y., Chen, J. D., Zhang, X. H.; Tang, J. X. 2019. Releasing the trapped light for efficient silver nanowires-based white flexible organic light-emitting diodes. *Adv. Opt. Mater.* 7:1900985.

[82] Yokota, T., Zalar, P., Kaltenbrunner, M., Jinno, H., Matsuhisa, N., Kitanosako, H., Tachibana, Y., Yukita, W., Koizumi, M., Someya, T. 2016. Ultraflexible organic photonic skin. *Sci. Adv.* 2:e1501856.

[83] Choi, M. K., Yang, J., Hyeon, T., Kim, D.-H. 2018. Flexible quantum dot light-emitting diodes for next-generation displays. *npj Flex. Electron.* 2:10.

[84] Lim, J., Park, M., Bae, W. K., Lee, D., Lee, S., Lee, C., Char, K. 2013. Highly efficient cadmium-free quantum dot light-emitting diodes enabled by the direct formation of excitons within InP@ZnSeS quantum dots. *ACS Nano.* 7:9019.

[85] Kim, T.-H., Cho, K.-S., Lee, E. K., Lee, S. J., Chae, J., Kim, J. W., Kim, D. H., Kwon, J.-Y., Amaratunga, G., Lee, S. Y. 2011. Full-colour quantum dot displays fabricated by transfer printing. *Nature Photon.* 5:176–182.

[86] Song, J. K., Son, D., Kim, J., Yoo, Y. J., Lee, G. J., Wang, L., Choi, M. K., Yang, J., Lee, M., Do, K. 2017. Wearable force touch sensor array using a flexible and transparent electrode. *Adv. Funct. Mater.* 27: 1605286.

[87] Bade, S. G. R., Li, J., Shan, X., Ling, Y., Tian, Y., Dilbeck, T., Besara, T., Geske, T., Gao, H., Ma, B. 2016. Fully printed halide perovskite light-emitting diodes with silver nanowire electrodes. *ACS Nano.* 10:1795–1801.

[88] Seo, H. K., Kim, H., Lee, J., Park, M. H., Jeong, S. H., Kim, Y. H., Kwon, S. J., Han, T. H., Yoo, S., Lee, T. W. 2017. Efficient flexible organic/inorganic hybrid perovskite light-emitting diodes based on graphene anode. *Adv. Mater.* 29:1605587.

[89] Shen, Y., Li, M.-N., Li, Y., Xie, F.-M., Wu, H.-Y., Zhang, G.-H., Chen, L., Lee, S.-T., Tang, J.-X. 2020. Rational interface engineering for efficient flexible perovskite light-emitting diodes. *ACS Nano.* 14:6107–6116.

[90] Mortimer, R. J. 1997. Electrochromic materials. *Chem. Soc. Rev.* 26:147–156.

[91] Polat, E. O., Balcı, O., Kocabas, C. 2014. Graphene based flexible electrochromic devices. *Sci. Rep.* 4:6484.

[92] Lin, S., Bai, X., Wang, H., Wang, H., Song, J., Huang, K., Wang, C., Wang, N., Li, B., Lei, M. 2017. Roll-to-roll production of transparent silver-nanofiber-network electrodes for flexible electrochromic smart Windows. *Adv. Mater.* 29:1703238.

[93] Li, R., Ma, X., Li, J., Cao, J., Gao, H., Li, T., Zhang, X., Wang, L., Zhang, Q., Wang, G. 2021. Flexible and high-performance electrochromic devices enabled by self-assembled 2D TiO$_2$/MXene heterostructures. *Nat. Commun.* 12:1587.

[94] Qiu, Z., Wu, Z., Zhong, M., Yang, M., Xu, J., Zhang, G., Qin, Z., Yang, B. R. 2021. Stretchable, washable, and rewritable electrophoretic displays with tough hydrogel–elastomer interface. *Adv. Mater. Technol.* 2100961.

[95] Gao, Z., Liu, L., Tian, Z., Feng, Z., Jiang, B., Wang, W. 2018. Fast-response flexible photochromic gels for self-erasing rewritable media and colorimetric oxygen indicator applications. *ACS Appl. Mater. Interfaces.* 10:33423–33433.

[96] Chen, H., Dong, Z., Chen, W., Sun, L., Du, X., Zhao, Y., Chen, P., Wu, Z., Liu, W., Zhang, Y. 2020. Flexible and rewritable non-volatile photomemory based on inorganic lanthanide: Oped photochromic thin films. *Adv. Opt. Mater.* 8:1902125.

[97] Torres-Pierna, H., Ruiz-Molina, D., Roscini, C. 2020. Highly transparent photochromic films with a tunable and fast solution-like response. *Mater. Horiz.* 7:2749–2759.

[98] Ma, T., Li, T., Zhou, L., Ma, X., Yin, J., Jiang, X. 2020. Dynamic wrinkling pattern exhibiting tunable fluorescence for anticounterfeiting applications. *Nat. Commun.* 11:1811.

[99] Hagstrom, A. L., Lee, H.-L., Lee, M.-S., Choe, H.-S., Jung, J., Park, B.-G., Han, W.-S., Ko, J.-S., Kim, J.-H., Kim, J.-H. 2018. Flexible and micropatternable triplet-triplet annihilation upconversion thin films for photonic device integration and anticounterfeiting applications. *ACS Appl. Mater. Interfaces.* 10:8985–8992.

[100] Lin, Y., Zhang, H., Feng, J., Shi, B., Zhang, M., Han, Y., Wen, W., Zhang, T., Qi, Y., Wu, J. 2021. Unclonable micro-texture with clonable micro-shape towards rapid, convenient, and low-cost fluorescent anti-counterfeiting labels. *Small.* 17:2100244.

[101] Wang, Y., Bai, N., Yang, J., Liu, Z., Li, G., Cai, M., Zhao, L., Zhang, Y., Zhang, J., Li, C. 2020. Silver nanowires for anti-counterfeiting. *J. Materiomics.* 6:152–157.

[102] Hu, Y. W., Zhang, T. P., Wang, C. F., Liu, K. K., Sun, Y., Li, L., Lv, C. F., Liang, Y. C., Jiao, F. H., Zhao, W. B. 2021. Flexible and biocompatible physical unclonable function anti-counterfeiting label. *Adv. Funct. Mater.* 31:2102108.

[103] Li, D., Tang, L., Wang, J., Liu, X., Ying, Y. 2016. Multidimensional SERS barcodes on flexible patterned plasmonic metafilm for anticounterfeiting applications. *Adv. Opt. Mater.* 4:1475–1480.

[104] Liu, X., Wang, J., Tang, L., Xie, L., Ying, Y. 2016. Flexible plasmonic metasurfaces with user-designed patterns for molecular sensing and cryptography. *Adv. Funct. Mater.* 26:5515–5523.

[105] Zhang, J. C., Pan, C., Zhu, Y. F., Zhao, L. Z., He, H. W., Liu, X., Qiu, J. 2018. Achieving thermo-mechano-opto-responsive bitemporal colorful luminescence via multiplexing of dual lanthanides in piezoelectric particles and its multidimensional anticounterfeiting. *Adv. Mater.* 30:1804644.

[106] Ko, J. H., Yoo, Y. J., Kim, Y. J., Lee, S. S., Song, Y. M. 2020. Flexible, large-area covert polarization display based on ultrathin lossy nanocolumns on a metal film. *Adv. Funct. Mater.* 30:1908592.

[107] Liu, C., Fan, Z., Tan, Y., Fan, F., Xu, H. 2020. Tunable structural color patterns based on the visible-light-responsive dynamic diselenide metathesis. *Adv. Mater.* 32:1907569.

[108] Wan, Y., Wang, P., Huang, F., Yuan, J., Li, D., Chen, K., Kang, J., Li, Q., Zhang, T., Sun, S., Qiu, Z., Yao, Y. 2021. Bionic optical physical unclonable functions for authentication and encryption. *J. Mater. Chem. C.* 9:13200–13208.

[109] Wan, Y., Chen, K., Huang, F., Wang, P., Leng, X., Li, D., Kang, J., Qiu, Z., Yao, Y. 2022. A flexible and stretchable bionic true random number generator. *Nano Res.* 15:4448–4456.

12 Piezoelectric Materials and Their Applications

Taiping Zhang and Qilin Hua

CONTENTS

12.1 Introduction...329
12.2 Basic Principles of Piezoelectricity 330
    12.2.1 Piezoelectric Effect...330
    12.2.2 Piezoelectric Materials..332
    12.2.3 Coupling Effects in Piezoelectric Materials 335
12.3 Applications of Piezoelectric Materials..337
    12.3.1 LEDs and PL Imaging 337
    12.3.2 Solar Cells...340
    12.3.3 Photodetectors...344
    12.3.4 Terahertz 345
    12.3.5 One-dimension Material Lasers..348
    12.3.6 Optomechanical Resonators 353
    12.3.7 Mechanoluminescence..356
12.4 Summary and Outlook..357
Acknowledgements...358
Bibliography 358

12.1 INTRODUCTION

Owing to the non-central-symmetric crystal structures, piezoelectric materials with intrinsic piezoelectricity are capable of responding to applied mechanical strains (or electric voltages) for sensing (or actuation) applications [1]. The piezoelectricity (*i.e.*, piezoelectric effect) is defined to describe the linear electromechanical interactions between the mechanical and electrical states under external mechanical or electric stimuli, including the direct and converse piezoelectric effects. Specifically, the direct piezoelectric effect enables piezoelectric materials to function as force-typed sensors, while the converse piezoelectric effect makes such materials promising for motor-typed actuators [1]. Piezoelectric materials can be found in many natural and synthesized materials, such as quartz, bone, and barium titanate ($BaTiO_3$), that can be intensively exploited in sensing and actuation applications [2]. More material types and the compositions have been continually developed to improve the piezoelectric charge/strain coefficients.

In recent years, the third-generation semiconducting materials (*e.g.*, ZnO, GaN, and AlN) are found to possess intrinsic piezoelectric properties, accompanied with

DOI: 10.1201/9781003202608-12

unique characteristics for optical, optoelectronic, and terahertz applications, arousing much attention in academia and industry communities. As firstly presented by Professor Zhong Lin Wang from Georgia Institute of Technology in 2010, combining the coupling piezoelectricity, photoexcitation, and semiconductor properties of such piezoelectric materials, novel physical effects can be observed as a result of strain-induced piezoelectric potential (piezopotential) modulated barrier heights for various charge carrier processes (including generation, separation, combination, and/or transport) [3]. In particular, piezophotonics is a two-way coupling effect of piezoelectricity and photoexcitation while piezophototronics is a three-way coupling effect of piezoelectricity, photoexcitation, and semiconductor properties, offering flexible strategies for enhancing device performance in optical and optoelectronic applications.

The piezoelectric materials used in optical and optoelectronic applications can be easily fabricated by conventional nanofabrication technologies in the forms of micro-nanostructures (*e.g.*, nanowire) or thin films. In this chapter, we focus on discussing the piezoelectric materials and the coupling physical effects (*e.g.*, piezophotonics, and piezophototronics) for optical/optoelectronic micro-nano devices. Some typical optical/optoelectronic micro-nano devices, such as light emitting diodes (LEDs) [4,5], solar cells [6], photodetectors [7], terahertz [8], one-dimension material lasers [9], optomechanical resonators [10], and mechanoluminescence [11], are discussed and further taken as examples for exploiting piezoelectric materials as sensors or actuators. Based on the fundamental principles of piezoelectricity, the optical/optoelectronic device performance can be effectively modulated via external-strain-induced piezoelectric polarization or electric-voltage-generated piezoelectric actuation, which promisingly open new windows for developing advanced intelligent systems. At last, a summary and future perspective are proposed to enlighten future research directions and practical applications of piezoelectric materials in the emerging photonics and optoelectronics.

12.2 BASIC PRINCIPLES OF PIEZOELECTRICITY

12.2.1 PIEZOELECTRIC EFFECT

Piezoelectric effect (or piezoelectricity) originates from the displacement of ionic charges within the inversion-symmetric crystal under external mechanical strain, which is firstly discovered by French physicists Jacques and Pierre Curie in 1880 [1]. Capable of describing the linear electromechanical interactions between the mechanical and electrical states, the piezoelectric effect is a physical process that exhibits direct or converse phenomena under mechanical or electric stimuli in a reversible manner, as schematically shown in Figure 12.1. The direct piezoelectric effect describes the generation of electrical polarization charges on the surfaces of piezoelectric materials in response to applied mechanical stresses (Figure 12.1(a)); Whereas the converse piezoelectric effect exhibits the generation of mechanical deformations in such materials under applied electric fields (Figure 12.1(b)). It is noteworthy that the piezoelectric effect is strongly dependent on the crystal orientation for applied mechanical strain or electric voltage. Additionally, the direct

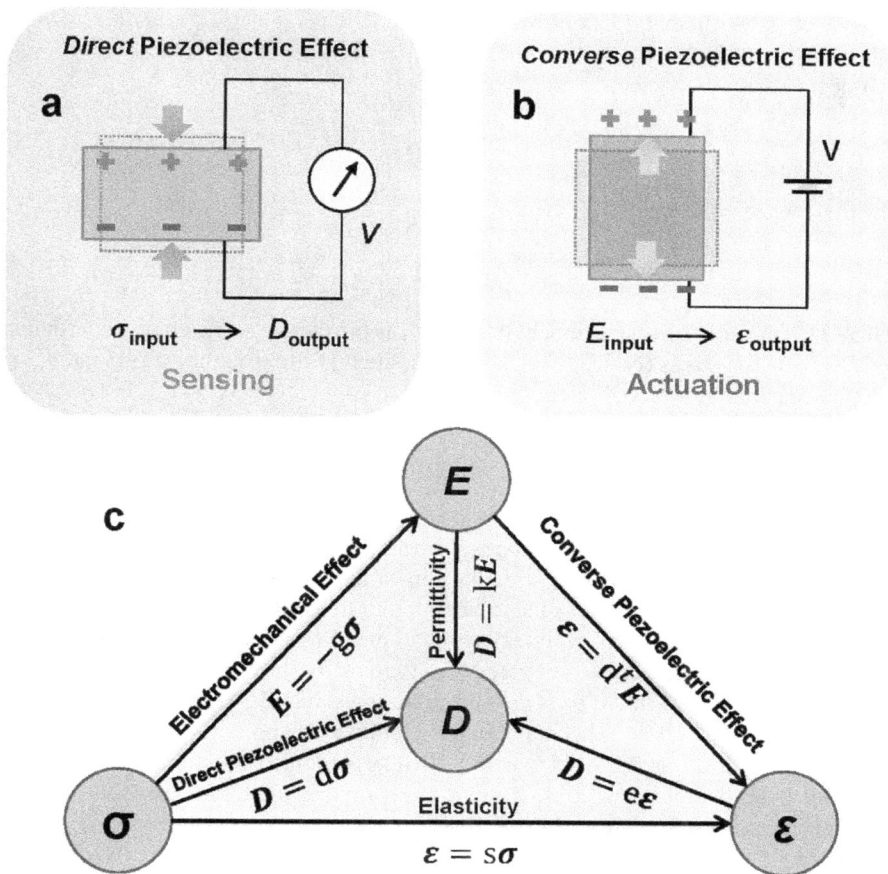

FIGURE 12.1 Definition and illustration of effects in piezoelectric materials. (a) Direct piezoelectric effect. (b) Converse piezoelectric effect. (c) Adapted Heckmann diagram that illustrating physical relations between electrical properties (electric field E, and electric displacement D) and mechanical properties (strain ε, and stress σ).

Source: Reprinted with permission from de Jong, M. *et al.* Scientific Data 2, 150053, 2015. Copyright 2015 Springer Nature.

piezoelectric effect enables piezoelectric materials for force-typed sensing (*e.g.*, force, pressure, vibration, or acceleration signals monitoring), while the converse piezoelectric effect makes such materials suitable for motor-typed actuators (*e.g.*, motor, sound, or ultrasonic outputs controlling).

Figure 12.1(c) illustrates the physical relations between electrical properties (electric field E, and electric displacement D) and mechanical properties (strain ε, and stress σ), which is adapted from the Heckmann diagram [12]. In general, the mathematical description of piezoelectricity shows a linearly relation, where the piezoelectric coefficient d is a third-rank tensor. The piezoelectric equations can be rigorously expressed as:

For the direct piezoelectric effect

$$D_k = d_{kij}\sigma_{ij} \tag{12.1}$$

For the converse piezoelectric effect

$$\varepsilon_{ij} = d_{kij}^T E_k \tag{12.2}$$

where D is the electric displacement (C/m^2), d_{kij} is the piezoelectric charge coefficient (pC/N), σ is the stress (N/m^2), E is the electric field (V/m), ε is the strain, and d_{kij}^T is the piezoelectric strain coefficient (pm/V).

12.2.2 PIEZOELECTRIC MATERIALS

Piezoelectric properties are observed in many natural materials, including Berlinite, cane sugar, quartz, Rochelle salt, topaz, tourmaline, and bone. Besides, some synthesized piezoelectric materials, e.g., barium titanate (BaTiO$_3$) and lead zirconate titanate (Pb(Zr,Ti)O$_3$, PZT), have also been found to exhibit improved piezoelectric charge/strain coefficients for practical applications of sensors or actuators [2].

Figure 12.2 schematically illustrates the atomic structures of some typical piezoelectric materials. The first type is the perovskite crystal. ABO$_3$ is the common chemical formula of perovskite, where A refers to a lanthanide or alkali earth-metal, and B refers to a transition-metal. Both O and A have larger ionic radii and are arranged in the close-packed cubic structures, while B has a smaller ionic radius to fill the octahedron central void. BaTiO$_3$ and Pb(Zr,Ti)O3 (PZT) are the typical perovskite-based piezoelectric materials with high electromechanical coupling characteristics. As shown in Figure 12.2(a), BaTiO$_3$ has a non-centrosymmetric tetragonal crystal structure. Under external stress, the BaTiO$_3$ crystal exhibits a strong electrical polarity due to the shift of Ti^{4+} within the tetragonal unit cell. PZT has a similar crystal structure as BaTiO$_3$ and possesses large piezoelectric coefficient. However, PZT are restricted for wide applications due to the toxicity of Pb. Concerning environmental-friendly processes, many lead-free piezoelectric materials in perovskite crystal structure have been developed, such as bismuth ferrite (BiFeO$_3$), lithium niobate (LiNbO$_3$), alkali niobate ((K,Na,Li)NbO$_3$), and alkaline bismuth titanate ((K,Na)$_{0.5}$Bi$_{0.5}$TiO$_3$) [2]. More information on perovskite materials can be found in Chapter 13.

The second type of piezoelectric material is the wurtzite crystal, which is a hexagonal crystal system with a tetrahedral coordination AB-type composition. In such a non-central-symmetric structure, the A atoms are in hexagonal close-packed arrangement with B atoms occupying the tetrahedral void. Many piezoelectric materials in II-VI and III-V groups have wurtzite crystal structures, e.g., ZnO, GaN, AlN, InN, and CdS [3]. Figure 12.2(b) shows the typical wurtzite crystal structure of GaN. A tetrahedral unit cell is composed of one Ga^{3+} (or N^{3-}) ion surrounded by four N^{3-} (or Ga^{3+}) ions. Initially, the centers of anion and cation coincide in the tetrahedron. When applying a strain on the cell, the centers of anion and cation would show a

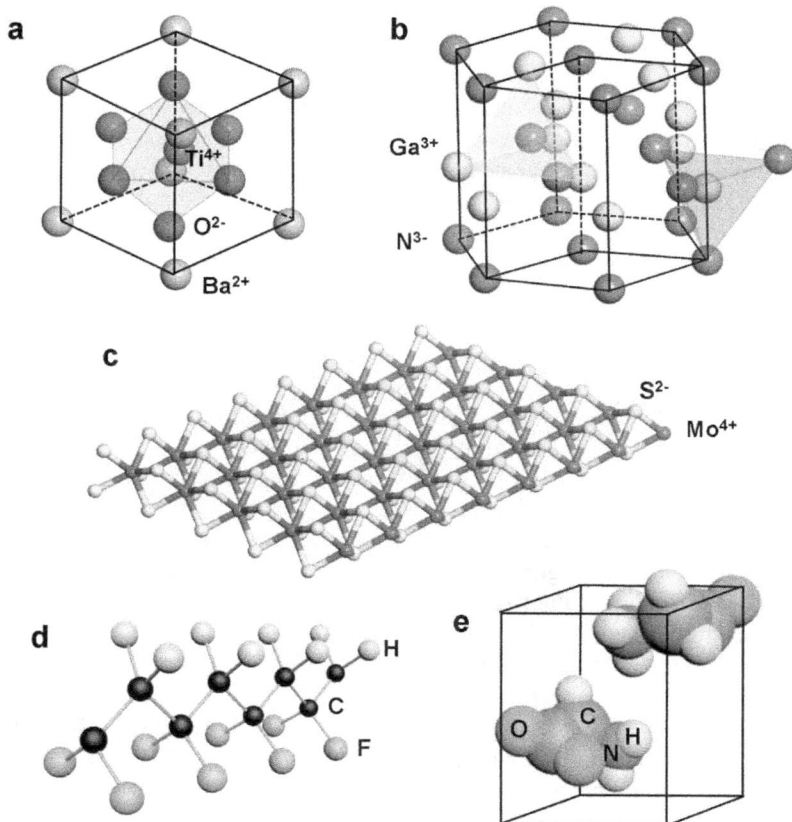

FIGURE 12.2 Schematic illustration of atomic structures of typical piezoelectric materials. (a) perovskite $BaTiO_3$, (b) wurzite GaN, (c) 2D MoS_2, (d) β-PDVF, and (e) β-glycine.

relative shift and cause the piezoelectric polarization along the c-axis [13]. Using the same principle, wurtzite-based ZnO has been intensively studied in piezoelectric properties for pressure sensing, energy harvesting, and biological applications. Besides, wide bandgap AlN with strong electromechanical coupling is commonly used as sensors or resonators. In particular, AlN doped with a certain amount of Sc will significantly enhance the piezoelectric charge coefficients for high-performance piezoelectric devices [14].

Both perovskite and wurtzite crystal structures belong to piezoelectric ceramics that suffer from inherent disadvantages of high-temperature processes, high rigidity, and brittleness. To improve the mechanical tolerance, perovskite or wurtzite piezoelectric materials are proposed to be fabricated in micro/nano-structures in various morphologies (*e.g.*, zero-dimensional (0D): nanoparticle, and quantum dot; one-dimensional (1D): nanowire, nanobelt, and nanotube) to obtain outstanding piezoelectric properties and excellent flexibilities. Furthermore, reducing the thicknesses of some two-dimensional (2D) piezoelectric materials, which possess non-centrosymmetric characteristics in one direction, into nanometer-scale could also enhance the piezoelectric properties. Representative 2D piezoelectric materials include black phosphorus, boron nitride, carbon nitride, and monolayer transition metal dichalcogenide. For instance, a monolayer MoS_2 is constructed from a Mo plane that is sandwiched between two S planes, thereby forming a triangular prism structure with Mo atoms in the center (Figure 12.2(c)). When subjected to external stress, the Mo^{4+} and S^{2-} are displaced to produce an electric dipole polarization on the surface of the material, thereby providing the material with piezoelectricity [2].

Other than these, because of the intrinsically asymmetrical molecular structure and orientation, some piezoelectric polymers are observed with an electrical polarization via molecular dipole reorientation under mechanical stimuli. Polyvinylidene difluoride (PVDF) is one of the most common piezoelectric polymers composed of ($-CH_2CF_2-$) monomers. The molecular dipoles are generated from the difference in electronegativity between hydrogen and fluorine atoms. In general, PVDF exhibits five forms of crystalline phases (α, β, γ, δ, ϵ), among which the α- and β- phases are the most common ones. The α-phase of PVDF consists of dipoles in reverse parallel order, thereby showing little piezoelectricity. In contrast, the dipoles in the β-phase are arranged in parallel, providing high dipole moments per unit cell and superior piezoelectric properties (Figure 12.2(d)). Moreover, several PVDF copolymers (e.g., polyvinylidene fluoride trifluoro ethylene [P(VDF-TrFE)]) have been developed with superior crystallinities, good flexibilities, and large electromechanical coupling effects. In addition to PVDF and its copolymers, the piezoelectric effects have also been found in other polymers featured with excellent mechanical flexibilities, good fatigue resistances, and piezoelectric performances without the need of high electric fields, e.g., poly-l-lactic acid, polyacrylonitrile, poly-β-hydroxybutyrate, polyvinyl chloride, and odd-numbered nylon. It should be noted that most piezoelectric polymers in general exhibit relatively low piezoelectric charge coefficients compared to inorganic materials, resulting in lower levels of charge generation.

The piezoelectric effect is also found in many biomolecules (e.g., amino acids, peptides, and proteins) and biological tissues (e.g., bones, ligaments, tendons, skins, and hairs), showing high biocompatibilities, stable piezoelectric coefficients, and dielectric properties. When subjected to mechanical stimuli, biomolecular piezoelectric materials generate surface charge polarizations or electric fields, which could be beneficial for physiological functions, such as tissue growth, wound healing, and regeneration. As a basic unit of biomolecules, amino acids are composed of a carboxyl group (-COOH), amino group ($-NH_2$) and variable side chains, all of which are attached to a central carbon atom. The difference between various amino acids depends on the structures of their side chains. Taking glycine as an example, under different crystallization conditions, glycine forms three kinds of crystal structures, namely α, β, and γ. The α-glycine crystal exhibits crystallographic symmetry and therefore lacks piezoelectricity, while both the β-glycine (Figure 12.2(e)) and γ-glycine have non-centrosymmetric crystal structures and exhibit ferroelectric properties. The piezoelectric charge coefficient d_{16} of β-glycine, exceeding the magnitudes of γ-glycine, can reach approximately 178 pm/V, which is comparable to conventional organic piezoelectric materials. With amino acids as the basic units, peptides and proteins are constructed to provide structure-dependent piezoelectric

properties. Featured with a dipeptide composed of two phenylalanine, diphenylala-
nine (FF) can be self-assembled into nanowires or nanotubes, exhibiting pronounced
piezoelectricity due to the non-centrosymmetric hexagonal structure. Similarly, certain proteins (*e.g.*, collagen) exhibit a piezoelectric polarization response, as a result
of the asymmetric spatial structures. In addition, some plant tissues (*e.g.*, lignocel-
lulosic molecules) also show the piezoelectric effects [2].

12.2.3 COUPLING EFFECTS IN PIEZOELECTRIC MATERIALS

Figure 12.3 illustrates the coupling effects of piezoelectricity, photoexcitation, and
semiconductor properties in piezoelectric materials, especially for the third-generation
semiconducting materials. The novel physical effects caused by strain-induced piezoelectric potential (piezopotential) are observed to exhibit three kinds of coupling
effects, including piezotronics (piezoelectricity-semiconductor coupling), piezophotonics (piezoelectric-photoexcitation coupling) and piezophototronics (piezoelectricity-
semiconductor-photoexcitation). The core of these coupling effects relies on the
piezopotential created by the piezoelectric materials [3].

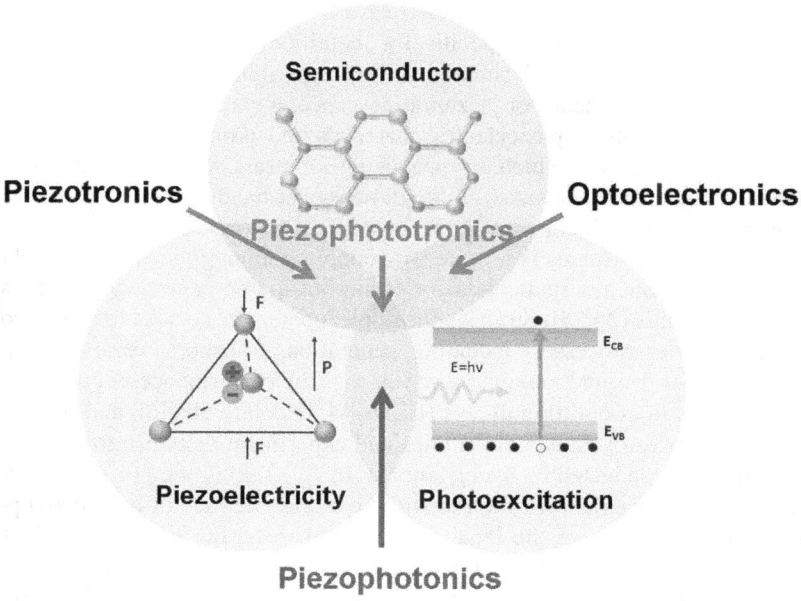

FIGURE 12.3 Coupling effects of piezoelectricity, photoexcitation, and semiconductor
properties in piezoelectric materials. Piezotronics is a coupling effect by piezoelectricity
and semiconductivity; Piezphototronics is a coupling effect by piezoelectricity, photoexcitation and semiconductivity; Piezophotonics is a coupling effect by piezoelectricity and
photoexcitation.

Source: Reprinted with permission from Dai, X. *et al.* Journal of Applied Physics 131, 010903, 2022.
Copyright 2022 AIP Publishing.

First of all, piezotronics is defined as that strain-induced piezopotential in piezoelectric materials, which could act as a virtual gate bias to tune/control the transport process of charge carriers for novel intelligent devices [15]. The fundamental principle of piezotronics was introduced by Wang in 2007 [16]. Specifically, ZnO nanowires were transferred on a flexible substrate with two Ag electrodes. Upon the bending of the substrate, the nanowire shows stretched or compressed states that lead to the generation of piezopotential along the c-axis. Subsequently, the strain-induced piezopotential modified the Schottky barrier height (SBH) at the interface between ZnO and Ag. The positive piezopotential at one end will reduce the SBH, while the negative piezopotential at the other end will conversely increase it. As a result, the output performance of devices based on piezoelectric materials could be significantly modulated by such piezopotential. To date, a series of electronic devices have been demonstrated based on the piezotronic effect, such as strain-gated field-effect transistors [17], strain sensors [18], electromechanical memristors [19,20], sensory synapses [21], and strain-controlled power devices [22], showing broad and promising applications in flexible electronics, smart sensor networks, human-machine interfaces, and advanced robotics [3].

Secondly, piezophotonics is a two-way coupling effect between the piezoelectric polarization and photoexcitation properties, where the strain-induced piezopotential is capable of modulating/controlling the relevant optical process for the generation and recombination of photons. Specifically, metal ions would act as activators that respond to photoexcitation and subsequent light emission. This phenomenon is also called mechanoluminescence or piezoluminescence. It contributes to understand the physical fundamentals of piezoelectric materials and promoting a broad range of applications in photonics, which are promising for stress sensing, structural health diagnosis, non-destructive analysis, three-dimensional handwriting, magnetic-optical sensing, energy harvesting, biomedicine, novel light source, and display [23].

Lastly, piezophototronics is a three-way coupling among piezoelectric polarization, photoexcitation, and semiconductor properties in non-central-symmetric semiconductor materials [24,25]. Strain-induced piezopotential is capable of controlling the charge carrier processes, including generation, transport, separation, and/or recombination, at the junction regions to improve the performance of optoelectronic devices, such as light-emitting diodes (LEDs) [4,5], solar cells [6], and photodetectors [7,26]. A series of theoretical studies and experiment demonstrations of piezophototronics are conducted by Wang since 2010 [7]. Taking the p-n junction for an example, minority carriers tend to redistribute at the interface of n-type and p-type semiconductors to balance the local electric field, resulting in a charge depletion region. A modification in the local band by the strain-induced piezopotential could modulate the diffusion and recombination processes of electron-hole pairs. Positive piezoelectric charges at the junction could lower the energy band whereas the negative piezoelectric charges could raise the energy band near the junction region [3]. Owing to the strong capability in modifying SBH or depletion region by piezoelectric charges, the piezophototronic effect can provide an insightful guidance for developing novel optoelectronic devices in smart display, interactive light source, and renewable energy.

12.3 APPLICATIONS OF PIEZOELECTRIC MATERIALS

12.3.1 LEDs and PL Imaging

Many types of LEDs and photoluminescence (PL) devices based on piezoelectric materials have been intensively reported [3,5,27], such as n-ZnO/p-GaN, n-ZnO/p-polymer, n-ZnO/p-Si, n-ZnO/p-SiGe, and InGaN/GaN quantum well (QW). In such devices, light emissions can be significantly enhanced by introducing external strains due to the piezophototronic effect [4,24,28]. Yang *et al.* illustrated that the emission intensity of the n-ZnO nanowire/p-GaN film-based LED can be enhanced by applying a load of compressive strain, with an enhancement factor of 17 at a 0.093% strain and a bias of 9 V as shown in Figure 12.4(a) and 12.4(b) [4]. The local band diagram at the interface of the p-n junction is obviously modified due to the strain-induced piezopotential (Figure 12.4(c)) along the *c*-axis of the ZnO nanowire. The enhanced illumination performance of the nanowire LED with increased compressive strains is obviously demonstrated in Figure 12.4(d). In addition, ZnO nanowires can be vertically grown on the GaN film substrate, and form well-controlled nanowire array LEDs. Based on the piezophototronic effect, changes in light intensity of piezoelectric micro-nano LEDs are capable of recording the loading force condition, showing a promising potential in force imaging application [5].

Moreover, highly ordered micro-nano LED arrays can be obtained by UV or electron beam lithography, such as ZnO/Si heterojunction [29,30], ZnO/SiGe heterojunction [31,32], ZnO/p-polymer LED [33], ZnO/organic light emitting diode (OLED) hybrid device [34] and InGaN/GaN multiple quantum wells (MQWs) nanopillars [27,28]. Pan et al. first reported a high-resolution pressure distribution imaging based on n-ZnO nanowire/p-GaN LED array in 2013 [5]. The well-ordered ZnO nanowire array was patterned by electron beam photolithography and followed by a low-temperature hydrothermal method on the GaN film/sapphire substrate. Each nanowire worked as a single-pixel LED, and its blue light corresponds to the near band edge (NBE) emission of ZnO. The distance between neighbouring ZnO nanowire was 4 μm, corresponding to an ultrahigh pixel resolution of 6350 dpi. The light emission of the single-pixel ZnO nanowire LED is significantly enhanced with the increase of applied compressive strain, arriving at 300% at a 0.15% strain. Furthermore, the pressure mapping test was performed by adding pressure on the top of the LED array through a hard convex pattern of letter "PIEZO" on a sapphire substrate while the LED array device was lighted. Initially, there is no obvious difference between the light intensities of the nanowire LEDs without pressures. As the loading pressure gradually increases, the electroluminescence of the nanowire LEDs under the letter pattern was enhanced and became clearly stronger than the one without pressure. As a result, the pronounced shape of letter "PIEZO" became clear when the compressive strain reached 0.15% as shown in Figure 12.4(e). The mechanism of the light intensity enhancement is consistent with the earlier described single ZnO nanowire/GaN LED. The piezopotential in the compressed ZnO nanowire would result in a deformation of band structure at the junction region and form a dip at the interface of ZnO/GaN, and such distorted band is helpful to trap holes at the junction area. Therefore, the carrier injection rate and recombination efficiency at the

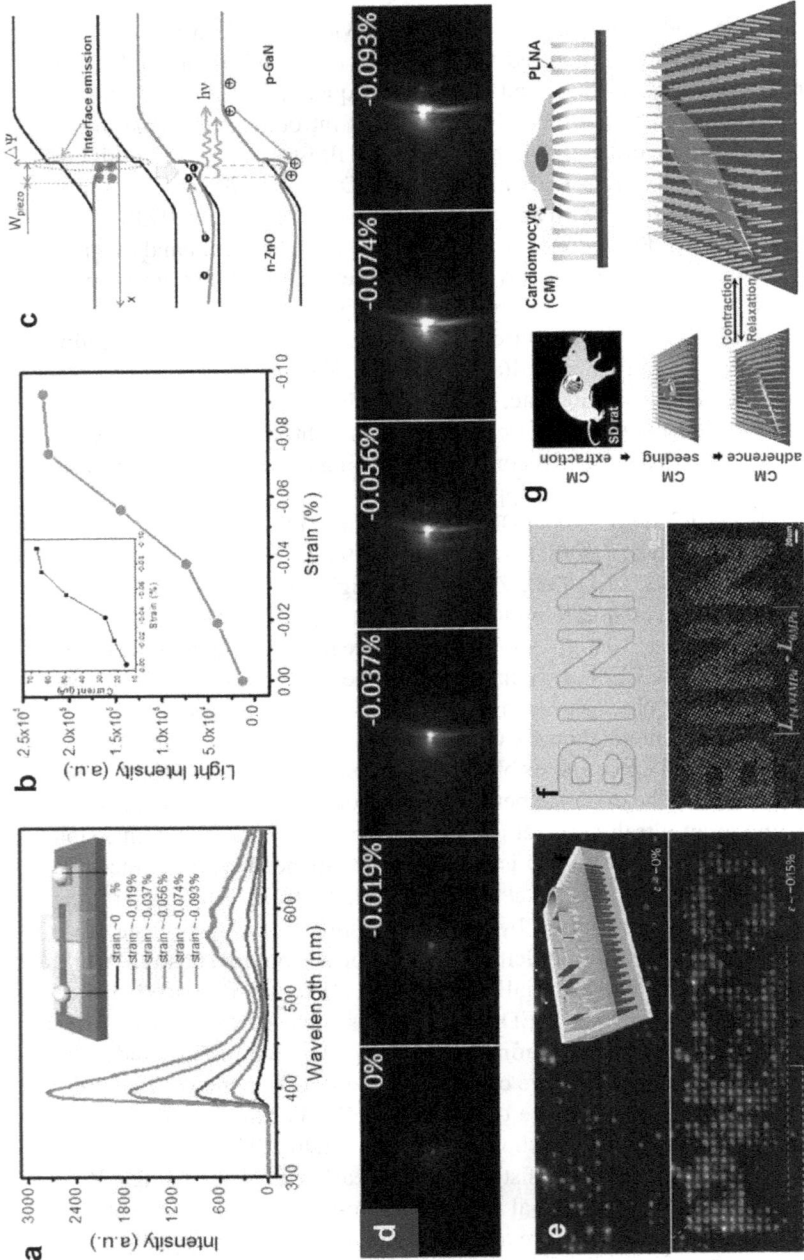

FIGURE 12.4 Piezophototronics in micro-nanowire LEDs and PL imaging. (a) Emitted light spectra of a n-ZnO nanowire/p-GaN film LED at different strain (at 9 V bias). (b) Integrated emission light intensities derived from (a) under compressive strain. (c) Schematic illustration of band diagram of the n-ZnO nanowire/p-GaN film LED without/with compressive strain conditions. (d) Optical images illustrating the enhanced light emission of the n-ZnO nanowire/p-GaN film LED with the increase of compressive strain.

(e) Optical images illustrating pressure distribution imaging of "piezo" mark with electroluminescence enhancement under strains of 0 and –0.15%. The inset is a schematic illustration of the operation principle of pressure distribution imaging of the ZnO nanowire LED array.

Source: Reprinted with permission from Pan, C. *et al.* Nature Photonics 7, 752–758, 2013. Copyright 2013 Springer Nature.

(f) High-resolution PL imaging of pressure/strain distribution in a vertical-aligned InGaN/GaN nanopillar array. BINN stamp (Top), and derived strain-tuned PL imaging at a strain of 14.94 MPa (bottom).

Source: Reprinted with permission from Peng. M. ACS Nano 9, 3143–3150, 2015. Copyright 2015 American Chemical Society.

(g) Schematic illustration of dynamic living cell traction force real-time imaging in the InGaN/GaN nanopillar array.

Source: Reprinted with permission from Zheng, Q. *et al.* Science Advances 7, eabe7738, 2021. Copyright 2021 American Association for the Advancement of Science.

junction are increased, leading to an enhanced light emission intensity of LED. In the follow-up studies, researchers have applied the piezophototronic effect induced band structure deformation to more kinds of LED arrays and realized effective and efficient pressure mapping, including Si microwires/ZnO nanofilm LED array [29], ZnO nanowires/SiGe film LED array [32], ZnO nanowires/p-polymer LED array [33], and CdS nanorods/organic hybrid LED array [35]. Meanwhile, researchers have fabricated the nanowire LED arrays to flexible devices and achieved flexible high-resolution pressure mapping [30,33,35]. See Chapter 11 for more details on flexible devices.

In addition to the LED devices, the piezoelectric materials can also achieve high-resolution pressure/strain mapping by a method of photoluminescence (PL), which is also called PL imaging [27]. Peng *et al.* has developed a dynamic force imaging with a vertical InGaN/GaN MQWs nanopillar array [28]. The PL intensity of the nanopillars decreases along with the increase of applied pressure. Figure 12.4(f) shows the derived strain-tuned PL imaging at a strain of 14.94 MPa. The mechanism of the decreased PL intensity can be attributed to the external stress-induced piezopotential, which works as a reverse bias voltage on the MQW nanopillar and contributes to the increase of the built-in electric field and depletion region width. Ultimately, the increased built-in electric field quickly separated the photogenerated electron-hole pairs in the PL process, leading to the decrease of radiative recombination probability and the PL intensity.

On top of that, the nanopillar array (also named as piezophototronic light nano-antenna, PLNA) can be used to realize dynamic real-time imaging of living cell traction force (CTF) [36], as shown in Figure 12.4(g). Via advanced top-down nano-fabrication technology, the pixel resolution of the PLNA achieved 31750 dpi. A spatial resolution of 800 nm and a temporal resolution of 333 ms of the living cell (*i.e.*, beating cardiomyocyte) have been demonstrated for force mapping, measuring the CTF in a range of 0.17 μN to 10 μN with a sensitivity of 15 nN/nm. Such piezoelectric nanowire array will contribute to high-resolution pressure/strain mapping with high sensitivity and reliability, showing promising applications in human-machine interface, visualized pressure-mapping systems, and smart sensor networks.

12.3.2 SOLAR CELLS

Solar cells are one of the most promising renewable energies for the era of carbon neutralization, which directly converted light into electricity. To date, a large number of different material systems, such as silicon, gallium arsenide, gallium nitride, copper indium gallium diselenide, and perovskite, have been proposed to fabricate highly efficient photovoltaic devices. Among them, the perovskite solar cell has been reported to achieve a world record efficiency of 25.5% [37], which is approaching the Shockley-Queisser limit ($\approx$ 30%). In general, the output performance of solar cells can be optimized by various common techniques such as material selection, structure design, and interfacial engineering. Beyond that, since the strain-induced piezopotential can modulate the separation and transport processes of photogenerated carriers based on the principle of piezophototronics, it can be naturally utilized to regulate the efficiency of solar cells [6,38,39].

Figure 12.5(a) shows a schematic illustration of a flexible ZnO-based perovskite solar cell, which is designed in a structure of Au/Spiro-OMeTAD/CH$_3$NH$_3$PbI$_3$/ZnO nanowires/ITO. The wurtzite-structured ZnO nanowire array acts as the electron-transfer layer, and is capable of modulating the interfacial processes of photogenerated carriers [6]. I-V characteristics of the device in Figure 12.5(b) illustrated the enhanced output performance by increasing the tensile strain. Under a tensile strain of 1.88%, the power conversion efficiency (PCE) shows a pronounced increase from 9.3% to 12.8%, and the short-circuit current density (J_{sc}) also has an obvious increase from 17.8 mA/cm^2 to 23.5 mA/cm^2. Determined by the quasi-Fermi levels, the open-circuit voltage (V_{oc}) also exhibits a slight increase as J_{sc} increase. As a result, the relative efficiency of the perovskite solar cell has been increased by about 40% at a tensile strain of 1.88%, which clearly verified the advantage of piezophototronic effect in boosting the performance of solar cells. The underlying mechanism is due to the strain-induced piezopotential of ZnO nanowires after applying an external strain on the device, which is capable of modulating the interfacial energy band structure. As shown in the inset of Figure 12.5(c), under the tensile strain (device upward bending), it will generate positive piezopotential along the polarity direction of the ZnO nanowires, leading to the reduction of interfacial barriers between ZnO nanowires and perovskite and thus causing an enhanced performance of solar cells. Conversely, the device performance will show a decreased trend upon the compressive strain, because of the increase of interfacial barriers. As the device bended downward, the relative efficiency of the perovskite solar cell has been decreased by about 16.3% at a compressive strain of 1.52% [6].

Boxberg et al. talked about the wurtzite InAs/InP core-shell nanowire solar cells with piezoelectric coupling effect [40]. The numerical simulation results of the effective piezoelectric charge density showed discontinuous characteristics at the core-shell interface (Figure 12.5(d)), because of the discontinuous permittivity of materials. The large-area InAs/InP core-shell nanowire solar cell (Figure 12.5(e)) was proposed to efficiently separate photogenerated carriers due to the axial piezopotential originating from the inner strain of core-shell nanowires, leading to an effective modulation on the photocurrent. The associated band diagram of strained piezoelectric core-shell nanowire is illustrated in Figure 12.5(f). Such strained core-shell nanowires with inherent piezoelectricity will contribute to the enhanced solar energy conversion.

In addition to the strained core-shell nanowires, strained InGaN/GaN MQWs epitaxial films are attracting much attention for high-performance solar cells, especially for the concentrated ones. Under the piezophototronic effect, the PCE of the InGaN/GaN MQWs solar cell can be enhanced by applying external strains. Figure 12.5(g) illustrates the calculated energy band diagram of the GaN/InGaN/GaN heterostructure solar cell at different strain conditions. With the increase of external strain (from 0 to 0.134%), the conduction band E_c of InGaN/GaN heterointerface shows an upward trend, while the valance band E_v of GaN/InGaN heterointerface conversely exhibits a downward trend. The external strain induced piezoelectric charges (i.e., piezo-charges) produced at the InGaN/GaN interfaces can compensate the piezo-charges induced by the lattice mismatch stress of InGaN well. As a result, both electron and hole wave functions move toward the quantum well, which would improve

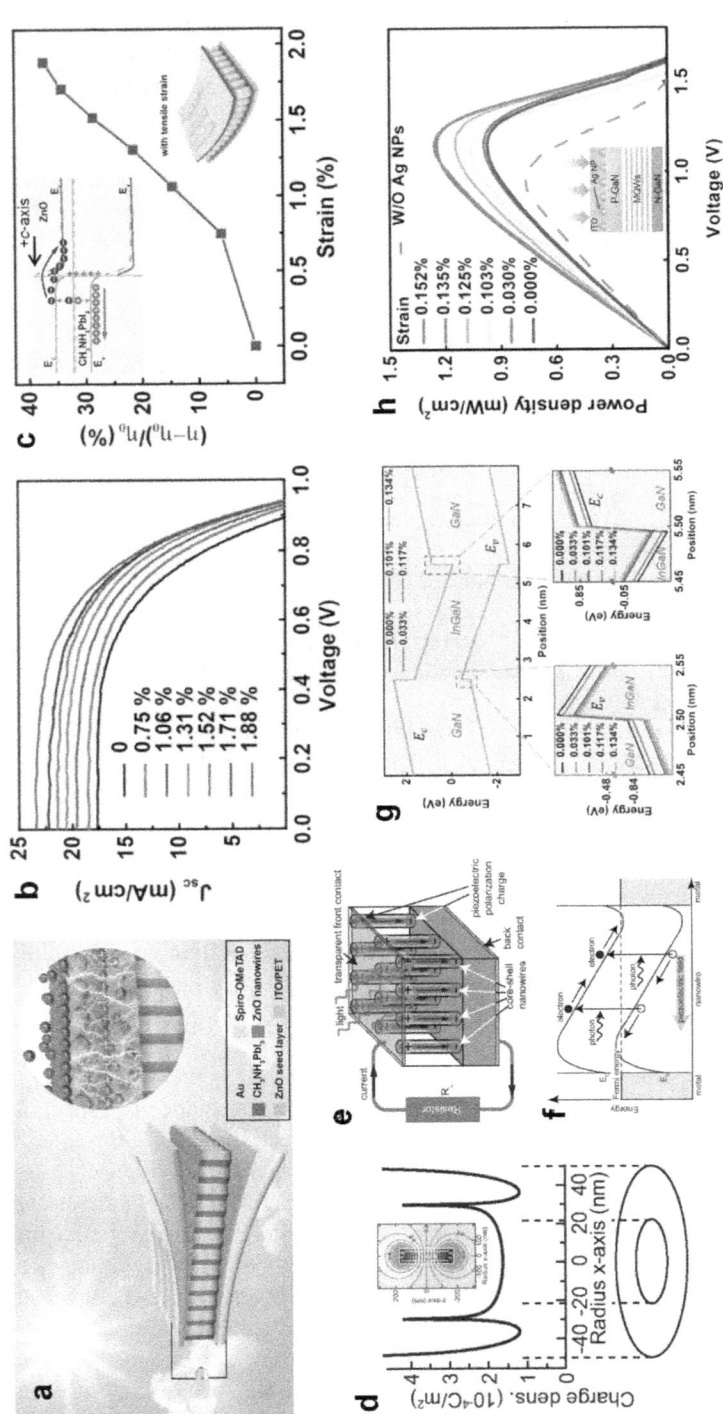

FIGURE 12.5 Piezophototronics in solar cells. (a) Schematic illustration of flexible ZnO-based perovskite solar cell. (b) Enhanced output performance of the perovskite solar cell at different tensile strains. (c) Enhanced efficiency of the perovskite solar cell at different tensile strain. The upper inset shows the corresponding energy band diagrams at tensile strain. The bottom inset shows the schematic illustration of the perovskite solar cell at tensile strain.

(d) Calculated effective surface charge density of wurtzite InAs/InP core-shell nanowire in a vacuum. The inset shows the corresponding piezoelectric potential. (e) Schematic illustration of InAs/InP core-shell nanowire solar cell. (f) Schematic illustration of band diagram of strained piezoelectric core-shell nanowire.

Source: Reprinted with permission from Boxberg, F. *et al.* Nano Letters 10, 1108–1112, 2010. Copyright 2010 American Chemical Society.

(g) Calculated energy band diagram of the GaN/InGaN/GaN heterostructure solar cell at different strain conditions.

Source: Reprinted with permission from Jiang, C. *et al.* ACS Nano 11, 9405–9412, 2017. Copyright 2017 American Chemical Society.

(h) The relation between power density and applied voltage in the InGaN/GaN MQW solar cell with Ag nanoparticles under external strain. The inset shows the schematic structure of the solar cell with plasmonic Ag nanoparticles.

Source: Reprinted with permission from Jiang, C. *et al.* Nano Energy 57, 300–306, 2019. Copyright 2019 Elsevier.

the optical absorption and produce more photogenerated carriers [38]. Moreover, by integrating the InGaN/GaN MQWs film with Ag plasmonic nanoparticles, the J_{sc} and PCE are increased by 40% and 66%, respectively, at a compressive strain of 0.152%, proving the possibility of coupling piezophototronics and plasmonics [39]. In summary, the piezophototronic effect offers an effective strategy to enhance the optoelectronic properties of piezoelectric materials-based solar cells.

12.3.3 PHOTODETECTORS

Photodetectors, as an important part of modern optoelectronic devices, have been widely applied in various emerging fields such as wearable electronics, healthcare monitoring, human-machine interface, and artificial intelligence applications [26]. The device structures and semiconductor material properties are the determining factors of photodetector performance. In recent years, photodetectors based on various piezoelectric materials, including wurtzite-structured third-generation semiconductor materials (e.g., ZnO, GaN and CdS), organic-inorganic halide perovskite materials (e.g., $CH_3NH_3PbI_3$ and $CsPbBr_3$), and Si-based p-n junctions (e.g., p-Si/n-ZnO), have been demonstrated and undergone rapid development [3]. Additionally, one-dimensional nanowire arrays and two-dimensional ultra-thin structures are also widely adopted to improve the spatial resolution, surface/volume ratio and mechanical properties of the photodetector devices [26,41]. Again, the piezophototronic modulation method through mechanical deformation can be implemented to modify the photodetection performance since it could effectively alter the band structures of contact region and/or junction interface [42].

Figure 12.6(a) shows the I-V curves of ZnO nanowire based ultraviolet (UV) photodetector in response to different strains under an illumination intensity of 2.2×10^{-5} W/cm^2 and the insert schematically illustrates the measurement setup [7]. The device was designed by a metal-semiconductor-metal structure of Ag/ZnO micro-nanowire/Ag with two Schottky contacts. Under positive bias, the photocurrent of ZnO micro-nanowires increase gradually as the external strains increase from -0.36% (compressive strain) to 0.36% (tensile strain). Furthermore, the relationship between the light intensity and the absolute current at different strains is depicted in Figure 12.6(b), which indicates that the absolute photocurrent could be effectively enhanced by the piezophototronic effect under low-power illumination. Other than that, the relative photoresponsivity changes of single crystals $CH_3NH_3PbI_3$ photodetector at different pressures or power densities are shown in Figure 12.6(c). When the illumination power density is below 3.641 mW/cm^2, the photoresponsivity of the device will increase monotonously as the external compressive stress increases. However, when the compression pressure reaches 43.48 kPa and the power density exceeds 3.641 mW/cm^2, the relative photoresponsivity changes will become saturated. For even larger pressure, the photoresponsivity will decrease, which is caused by giant-piezopotential-induced a potential well for trapping the photogenerated electrons at the interfacial barrier [43]. To further expand the absorption range of nanowires, many composite nanowire structures were developed, such as carbon-fiber/ZnO/CdS double-shell nanowires used for detecting visible/UV light (372–548 nm) [44]

(Figure 12.6(d)). It is found that the photoresponsivity of the device increases mono-tonically with the increase of compressive stress, and continuously decreases with the increase of tensile stress. The responsivity increases by 40% - 60% at a -0.38% compressive strain and decreases by 8% - 20% at a 0.31% tensile strain, respectively. In 2014, Wang *et al.* demonstrated a Si-based p-n junction photodetector by utilizing the external strain induced piezoelectric polarization in a n-ZnO layer to modulate the optoelectronic process in the p-Si layer [45]. Figure 12.6(e) shows the resultant changes of photoresponsivity of p-Si/n-ZnO based photodetectors at different strains and illumination intensities. When the compressive strain reaches -0.01%, the rela-tive changes of photoresponsivity will reach its maximum value.

On top of that, large-scale photodetectors based on integrated nanowire arrays with high photoresponsivity are considered for practical applications. Figure 12.6(d) shows the structure diagram of the nanowire array UV photodetectors. Each device consists of 32 × 40 ZnO nanocolumns, which constitute the pixel points of the pho-todetector. The ZnO nanowires were synthesized on the ITO layers by a hydrother-mal method and the spaces between nanowires are filled by SU-8 photoresist. The device has excellent sensitivity and stability, and shows good optical imaging perfor-mance [26]. The piezophototronic enhancement of the illuminated nanowire array (at 1.38 mW cm^{-2}) in response to applied pressures is displayed in Figure 12.6(f). With the increase of applied pressure from 0 to 40.83 MPa, the relative total current variation of the illuminated device increases linearly from 0% to 67%. The result indicates that the device can be well regulated by the piezophototronic effect and has a better imaging capability under high pressure condition.

12.3.4 TERAHERTZ

As introduced in Chapter 1, Terahertz (THz) in general refers to a special electro-magnetic wave band with frequencies from 0.3 THz to 3 THz, which possesses great promise for applications in electronic information, life sciences, communications radar, and national security [46–48]. In recent years, many significant progresses have been made in the research of THz devices based on GaAs/AlGaAs materials. However, due to the relatively low conduction band offsets (CBO) of GaAs/AlGaAs materials, new material systems, especially piezoelectric materials, such as InAs/AlSb, ZnO/ZnMgO, InGaN/GaN, GaN/AlN, and GaN/AlGaN, have been considered for next generation room-temperature THz devices [49]. Among these materials, the GaN-based semiconductor compounds are most attractive for broadband THz appli-cations, because of their unique properties of large bandgaps, high CBO energies, high longitudinal optical phonon (LO) energies, and high resistances to large break-down electric fields [8]. Up to present, various kinds of GaN-based THz sources, such as quantum cascade lasers (QCLs), quantum resonant-tunneling diodes (RTDs), and Gunn diodes, have been demonstrated [50].

Under the piezophototronic effect, the strong electric field generated by exter-nal strain in the wurtzite nitride semiconductor (*e.g.*, GaN) can serve as "gate volt-age" to effectively modulate the operation wavelength of THz devices, due to their large spontaneous and piezoelectric polarizations [51]. Especially in QW or MQW

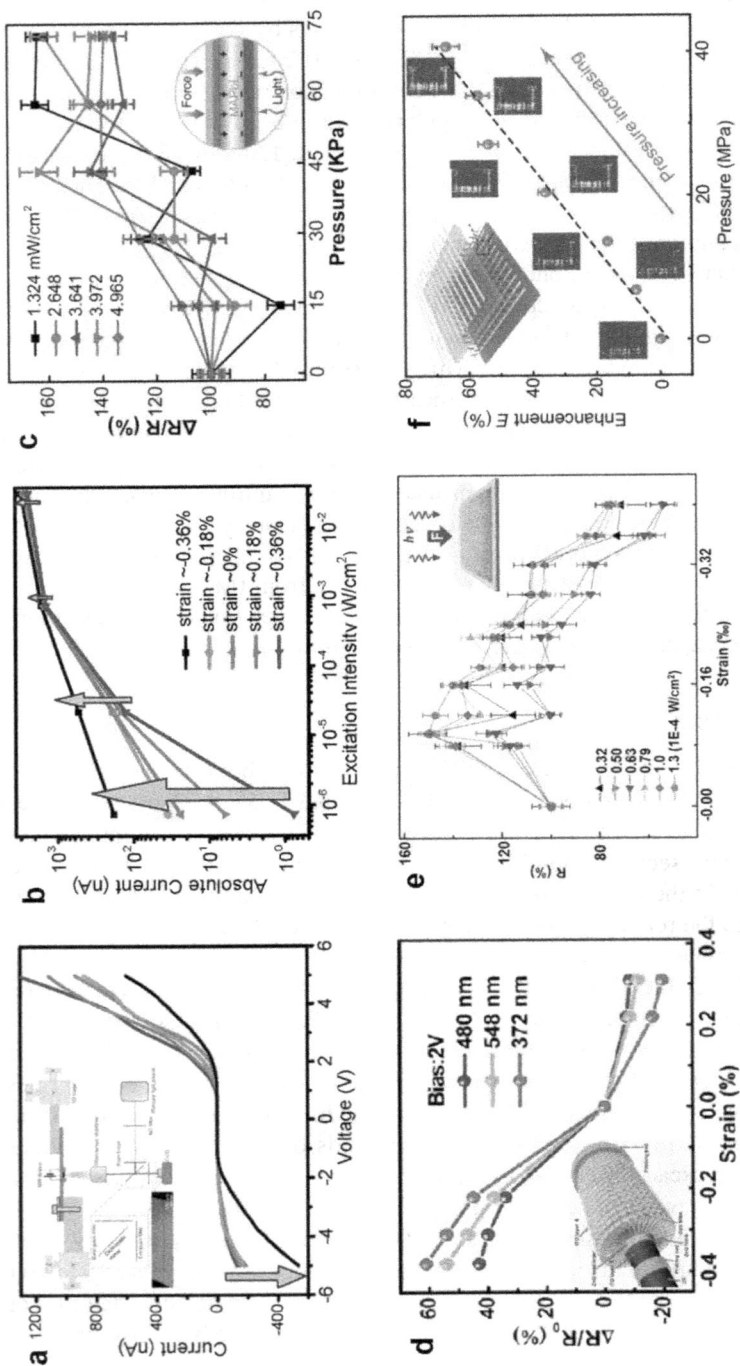

FIGURE 12.6 Piezophototronics in photodetectors. (a) I–V curves of ZnO nanowire UV photodetector in response to different strain under an excited light intensity of 2.2×10^{-5} W cm^{-2}. Inset shows a customized measurement setup for piezopotential-tuned photodetector. (b) The relation between excitation intensity and absolute current at different strain.

Source: Reprinted with permission from Yang, Q. *et al.* ACS Nano 4, 6285–6291, 2010. Copyright 2010 American Chemical Society.

(c) Relative changes of photoresponsivity of single crystals $CH_3NH_3PbI_3$ photodetector at different pressures or power densities.

Source: Reprinted with permission from Lai, Q. *et al.* ACS Nano 12, 10501–10508, 2018. Copyright 2018 American Chemical Society.

(d) Relative changes of photoresponsivity of carbon-fiber/ZnO-CdS microwire-based visible/UV photodetector at different strain.

Source: Reprinted with permission from Zhang, F. *et al.* ACS Nano 7, 4537–4544, 2013. Copyright 2013 American Chemical Society.

(e) Relative changes of photoresponsivity of p-Si/n-ZnO based photodetectors at different strain and illumination intensity.

Source: Reprinted with permission from Wang, Z. *et al.* ACS Nano 8, 12866–12873, 2014. Copyright 2014 American Chemical Society.

(f) Piezophototronic enhancement of ZnO nanowire photodetector array in response to applied pressures (at 1.38 mW cm^{-2}).

Source: Reprinted with permission from Han, X. *et al.* Advanced Materials 27, 7963–7969, 2015. Copyright 2015 John Wiley & Sons.

structures, the significantly large polarization charges at the interfaces of nitride heterostructures are capable of generating internal electric fields greater than 100 kV/cm across the active regions, which open the possibility of vertical THz emission based on the internal electric field induced transport [8,51].

Woodward *et al.* demonstrated the enhanced THz generation in an InGaN/GaN heterostructure composed of a coherently strained 200 nm thick InGaN layer on an n-type GaN template [8]. The THz emission signal from the InGaN/GaN heterostructure is 1.5 times than that from the bulk InAs sample (Figure 12.7(a)). The dominant THz generation mechanism is attributed to the improved electron acceleration in the internal electric fields generated by the piezoelectric polarization charges at the heterointerface (Figure 12.7(b)). In addition, Liu *et al.* reported that the wavelength of THz detectors can also be modulated by the piezophototronic effect with the wurtzite AlGaN/GaN QW structures [51]. Figure 12.7(c) depicts the conduction-band profiles of $Al_{0.1}Ga_{0.9}N/GaN/Al_{0.05}Ga_{0.95}N$ QW under applied strain from 0 to 5.6%. The potential profile becomes flattered while the barrier height becomes higher under applied external strains. The flatter potential profile is due to the piezoelectric field caused by the external strain, which could partially compensate the internal polarized electric field. The higher barrier height can localize more bound states in well and step barrier layer and induce more bound-bound transitions. As shown in Figure 12.7(d), that absorption energy first decreases when the strain is below 1.4%, and then increases for a larger strain. Figure 12.7(e) shows the variation of absorption energy in the THz range to external strain, which reaches the maximum value at a strain of 4.2%. When the strain is beyond 4.2%, the absorption energy goes into the far-infrared region. The optical absorption capability of QW devices can be evaluated by quantum efficiency. Figure 12.8(f) shows the quantum efficiency versus external strain. The quantum efficiency firstly increases and then decreases as the external strain increases, which indicates that the performance of the AlGaN/GaN QW can be flexibly controlled by an external strain, providing a feasible way to design highly sensitive THz detectors by the piezo-phototronic effect. In addition, AlGaN/GaN QWs are also proposed for absorption of THz radiation due to their much higher LO energies ($E_{LO} > 90$ meV) than those of conventional semiconductors (~ 36 meV) [49,52]. The piezophototronic effect, as an effective means of regulating the energy band of quantum wells, is expected to provide a new way to enhance the performance of THz devices.

12.3.5 ONE-DIMENSION MATERIAL LASERS

Semiconductor one-dimension (1D) material laser, such as micro-nanowire and nanobelt lasers, have been extensively studied for nanophotonic and optoelectronic systems due to their scaled sizes, large local coherent outputs, and efficient waveguiding. Micro-nanowire lasers were first realized with II-VI and III-V gain materials, such as ZnO [53] and GaN [54]. Each micro-nanowire is capable of functioning as both a waveguide along the axial direction and a Fabry-Pérot (F-P) cavity between the two end facets for optical amplification. The micro-nanowires can be synthesized through mature processing technology of epitaxial crystal growth. Having a crystal structure with regular geometrical configuration is critical for the micro-nanowire working as a microcavity in the lasing process. With proper material compositions

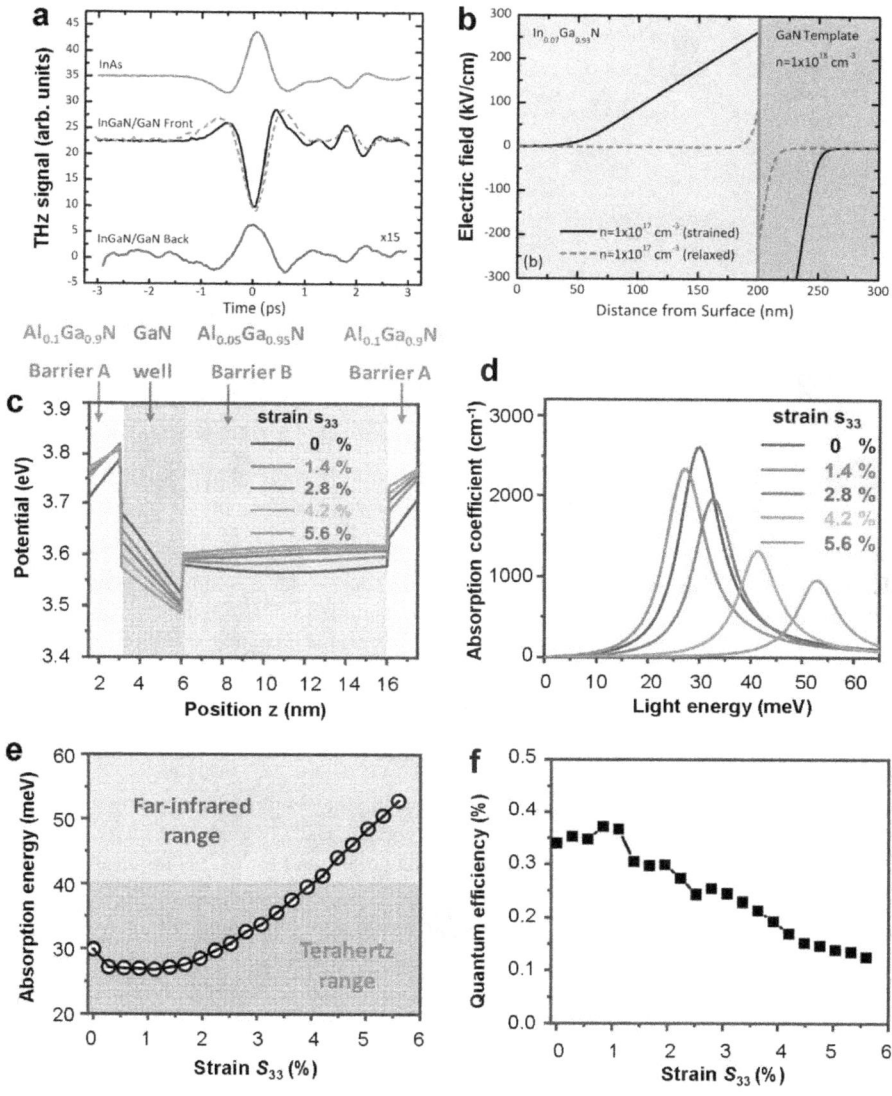

FIGURE 12.7 Terahertz generation and detection. (a) THz emission from InAs versus In$_{0.07}$Ga$_{0.93}$N/GaN heterostructure pumped at 400 nm. (b) Simulated electric field strength of In$_{0.07}$Ga$_{0.93}$N/GaN at the coherently strained and fully relaxed conditions.

Source: Reprinted with permission from Woodward, N. *et al.* Applied Physics Letters 100, 191110, 2012. Copyright 2012 AIP Publishing.

(c) Conduction-band potential profiles under various strains. (d) Absorption spectrum under various strains. (e) Absorption energy against the strain. (f) Quantum efficiency versus the external strain.

Source: Reprinted with permission from Liu, N. *et al.* Nano Energy, 65, 104091, 2019. Copyright 2019 Elsevier.

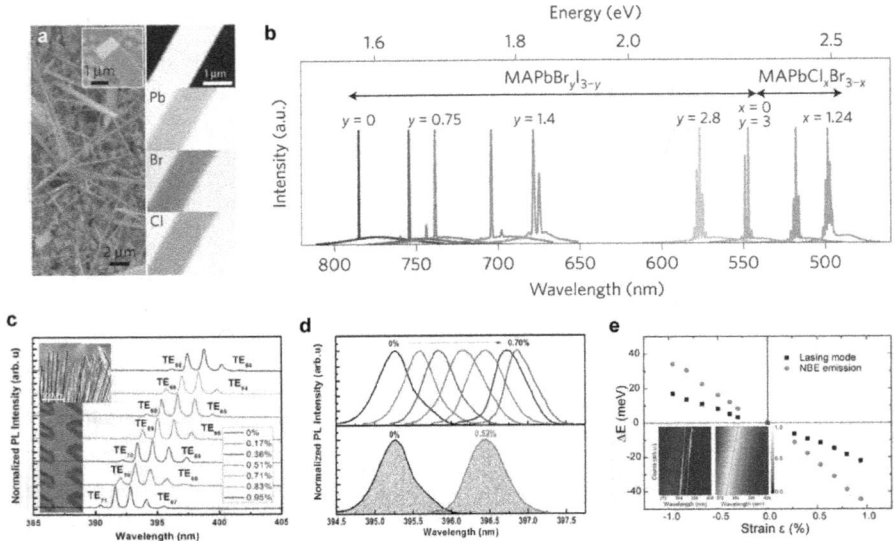

FIGURE 12.8 Piezophototronics in micro-nanowire lasers. (a) Scanning electron microscopy (SEM) image of mixed lead halide perovskite nanowires ($CH_3NH_3PbCl_{1.24}Br_{1.76}$) (left), and energy dispersive X-Ray spectroscopy (EDS) elemental mapping of Pb, Cl and Br in a typical single-crystal nanowire (right). (b) Tunable lasing characteristics from the mixed perovskite nanowire lasers by changing the compositions of Br and I.

Source: Reprinted with permission from Zhu, H. *et al.* Nature Materials 14, 636–642, 2015. Copyright 2015 Springer Nature.

(c) Tunable lasing spectra of ZnO microcavity by changing bending degrees. The insets are SEM morphology of ZnO micro-nanowires, and optical image of ZnO microwire laser. (d) Strain-tunable single-mode shift of ZnO whispering-gallery mode (WGM) lasing in 4.02 μm microcavity.

Source: Reprinted with permission from Lu, J. *et al.* ACS Nano, 12, 11899–11906, 2018. Copyright 2018 American Chemical Society.

(e) Photon energy changes in blue-/red-shift of near-band edge (NBE) emission and lasing mode under compressive and tensile strains. The inset shows the dynamic strain regulation of stimulated and spontaneous emission mapping in the ZnO microresonator.

Source: Reprinted with permission from Lu, J. *et al.* Materials Today 24, 33–40, 2019. Copyright 2019 Elsevier.

and physical geometries, the micro-nanowires can pump lasing wavelength ranging from UV to near-infrared regions.

In recent years, lead halide perovskite shows an unprecedented capability of broadband tunable wavelength for lasing applications by controlling its chemical stoichiometry [9]. Figure 12.8(a) and 12.8(b) show a mixed lead halide perovskite nanowire ($CH_3NH_3PbCl_{1.24}Br_{1.76}$) with tunable lasing characteristics. The resultant scanning electron microscopy (SEM) image and the corresponding energy dispersive

X-Ray spectroscopy (EDS) elemental mappings in Figure 12.8(a) clearly showed a single-crystal nanowire with smooth end facets, offering an excellent F-P laser cavity. The multi-color tunable laser wavelength ranging from near-infrared to blue can be achieved in such single-crystal perovskite nanowire at room temperature, by altering the compositions of bromide (Br) and chloride (Cl) or iodide (I) in the precursor solution (Figure 12.8(b)). Nevertheless, some key problems, including chemical stabilization, heat management, and electrical injection, should be further solved for a practical high-performance lead halide perovskite nanowire laser.

In addition to the controlled material composition, a novel method by using piezo-photonics to modulate optical constant or laser mode in the piezoelectric materials is developed [55,56]. Several years ago, researchers found that the optical bandgap of micro-nanowire with wurtzite crystal structure (*e.g.*, hexagonal ZnO and GaN) can be turned linearly to the applied stress, which was observed through the wavelength shift of the near-band-edge (NBE) emission spectrum [57,58]. In the conception of piezophotonics, a two-way coupling of photoexcitation and piezoelectricity, the refractive index of wurtzite micro-nanowire can be modulated by the piezoelectric polarization when a uniaxial strain applies on it. As a result, the variation of the refractive index will in turn modify the resonance of the F-P cavity as well as the ultimate laser mode [55,56,59,60].

Piezoelectric effect modulated microwire lasing was first reported in 2018 [56]. As shown in Figure 12.8(c), the lasing wavelength of ZnO mircowire can be dynamically modulated by the piezoelectric effect in the wurtzite crystal structure. The hexagonal ZnO microwires were synthesized on a Si substrate by a high-temperature vapor-transport method. The diameter of the microwires was several microns and the length of the microwires were a few millimeters to one centimeter. Such microwire supports a whispering gallery mode (WGM) lasing at room temperature, which was confirmed by the interference fringes of the optical image [61]. The relationship between the lasing performance and the uniaxial strain was measured by a light-force coupled measurement system, where an individual microwire was fixed on a flexible substrate by the epoxy resin glue to prevent slippage in the bending experiment. The lasing of the microwire was optically pumped by a focused pulsed laser through an objective of a microscope, and the emission of the microwire was collected by the same objective and analyzed by a spectral measurement system. Figure 12.8(d) shows the shift of laser mode of ZnO microwire with a diameter of 4.02 μm at various strains [56], where the uniaxial strain was applied on the microwire by bending the flexible substrate via controlling the two 3D manual displacement stages. Note that the same light-force coupled measurement system was also employed in the studies of piezoelectric effect modulated lasing of GaN and perovskite microwires.

The results demonstrated that the lasing mode wavelength of the microwire experienced a redshift along with the increase of tensile strain, and the emitted photon energy displayed a linear relationship with respect to the tensile strain. This finding is somewhat contradicted to the intuition that a blueshift of WGM lasing is expected since the diameter of the microwire should decrease under the tensile strain. Therefore, the redshift of lasing was attributed to a strain polarization induced refractive index variation. To prove this point, Lu *et al.* found that the wavelength of the WGM lasing of ZnO microwire displays a linear shift corresponding to the uniaxial strain, whereby blue-shift for compressive strain and red-shift for tensile strain [55], as shown in Figure 12.8(e). They compared the wavelength variations of the lasing modes and the NBE emissions of the ZnO microwires, and found that under the same strain, the photon energy shift of the former was greater than that of the latter. This result suggested that unlike the wavelength shift of the NBE emission, which was caused by the strain-induced energy bandgap variation, the wavelength shift of WGM lasing was essentially caused by piezoelectric polarization induced refractive index variation.

Based on the piezoelectric polarization effect tuned WGM lasing, a dynamic regulating of microwire lasing can be realized. Lu *et al.* reported a dynamic regulating of single-mode lasing up to 9.3 nm ranging from 386.06 nm to 395.36 nm with the uniaxial strain ranging from -0.96% to 0.94% for a 1.7 µm diameter ZnO microwire [55]. The piezoelectric polarization effect induced microwire lasing variation was also applied in the strain sensor area. The high-quality factor (Q factor) lasing modes can display an obvious shift with a tiny strain and significantly improve the resolution of the strain sensor. Pan's group reported a strain sensor based on the WGM lasing shift of a 3 µm diameter GaN microwire. The full-width-at-half-maximum (FWHM) of such WGM lasing was only 0.36 nm, corresponding to a high Q factor of ≈ 1000. With this narrow FWHM, the lasing peak was able to be completely separated under a small tensile strain from 0% to 0.5%, which is much better than the intrinsic PL peak of the microwire. The minimum resolution in strain reached up to 0.16% for the microwire lasing strain sensor.

In recent years, laser based on perovskite micro-nano materials was reported and has drawn more and more attention [60]. The perovskite materials have intrinsic piezoelectric and ferroelectric properties. Therefore, the piezoelectric polarization effect is able to induce variation for perovskite micro-nanomaterial-based lasing as well. In 2019, Yang *et al.* studied the piezoelectric induced dynamic modulation of perovskite WGM microwire lasing [60]. The material of the microwire laser was $CsPbBr_3$, which is an inorganic perovskites material. Different from ZnO and GaN microwire laser, which have a hexagonal wurtzite crystal structure, the perovskite microwire has an orthorhombic crystal structure. Under uniaxial strains, the lasing mode displays the same wavelength shift trends as the ZnO and GaN microwire lasing, whereby blueshift under compressive strain and redshift under tensile strain. With the uniaxial strain ranging from -0.96% to 0.94%, the lasing mode wavelength was able to be dynamic modulated up to 8.5 nm. This study indicates that the piezoelectric polarization effect induced refractive index variation is also valid in other piezo material systems.

The piezoelectric polarization effect can also modulate lasing of other 1D nanomaterials and other type modes. In 2019, Ma *et al.* reported piezoelectric polarization effect modulated F-P mode of CdS nanobelt lasing [62]. The CdS nanobelt had a wurtzite crystal structure, whose F-P cavity was formed by to opposite side face. There were two types of CdS nanobelt which had been studied. One had a parallelogram shape and the [001] crystal orientation was parallel to the long axis of the nanobelt. The other had a ladder shape and the [001] crystal orientation was perpendicular to the long axis of the nanobelt. Under uniaxial strain along the long axis of the nanobelt, researchers found that the lasing mode of the parallelogram shape nanobelt displayed redshift under tensile strain. In contrast, for the ladder shape belt, the lasing mode displays weak blueshift under tensile strain. Consider the [001] crystal orientations of the two types of nanobelts, for the parallelogram shape nanobelt, uniaxial strain can obviously induce piezoelectric polarization, which could cause variation on the refractive index. On the other hand, for the ladder shape nanobelt, uniaxial strain can hardly induce piezoelectric polarization. Instead, the microcavity only had shape change caused by the Poisson effect. This research deepened the understanding of the piezoelectric polarization induced refractive index variation, and provided solid guidance in modulating the lasing mode via piezo materials.

12.3.6 OPTOMECHANICAL RESONATORS

Optomechanical resonators exploit light-matter interactions using optical cavities under the mediation of radiation pressures, which is very useful for building integrated photonics [63]. Electromagnetic waves (photons) and high-frequency mechanical vibrations (phonons) are the two primary types of optomechanical resonances. Capable of modulating the mechanical motion at both mesoscopic and quantum levels, micro/nano-optomechanical resonators have been demonstrated for important applications in sensing, information processing, time/frequency metrology, and quantum physics [64]. In recent years, diverse optomechanical structures, including freestanding nanostructures (e.g., sphere, disk, cantilever, and membrane) and periodic arrays [65], have been developed in different material systems, such as SiO_2, Si_3N_4, Si, GaAs, AlN, GaP, and diamond, due to their large thermal conductivities and low optical absorptions. Generally, optomechanic microcavity primarily depends on the device material that demand not only high optical transparency and large refractive index for high-quality and strong-confined optical mode, but also large acoustic velocity and low material damping for high-frequency and high-quality mechanical resonance [64].

Piezoelectric materials, e.g., AlN, can provide suitable platforms in optomechanical coupling and enabling quantum transduction of photons between microwave and optical frequency domains. The interaction of piezoelectric transduction with optical cavities is used to explore the quantum state transfer in optomechanical crystals and enhance interactions with microwave signals. Figure 12.9(a) illustrates the piezoelectric AlN ring optomechanical resonator, consisting of centrally anchored acoustic and photonic resonators, that enables the production of optical modulation in WGM photonic resonance [66]. The two AlN rings are functionalized as an acoustic contour mode resonator and a WGM photonic resonator, respectively. When a radio frequency (RF) excitation is applied to the electrodes of the acoustic resonator ring, it can produce vibrations in the acoustic resonator ring and be laterally transferred to the photonic resonator ring by a coupling spring. Meanwhile, the telecom C-band light is coupled into the photonic ring, and the cavity is biased to a wavelength corresponding to photonic resonances. Upon the excitation of the RF signal, piezoelectrically induced mechanical deformations to the cavity path length show the frequency modulation of the photonic resonator. Piezoelectric excitation is the dominant source of actuation force in the photonic resonator, which is opposed to the circulating intracavity energy generated optical force.

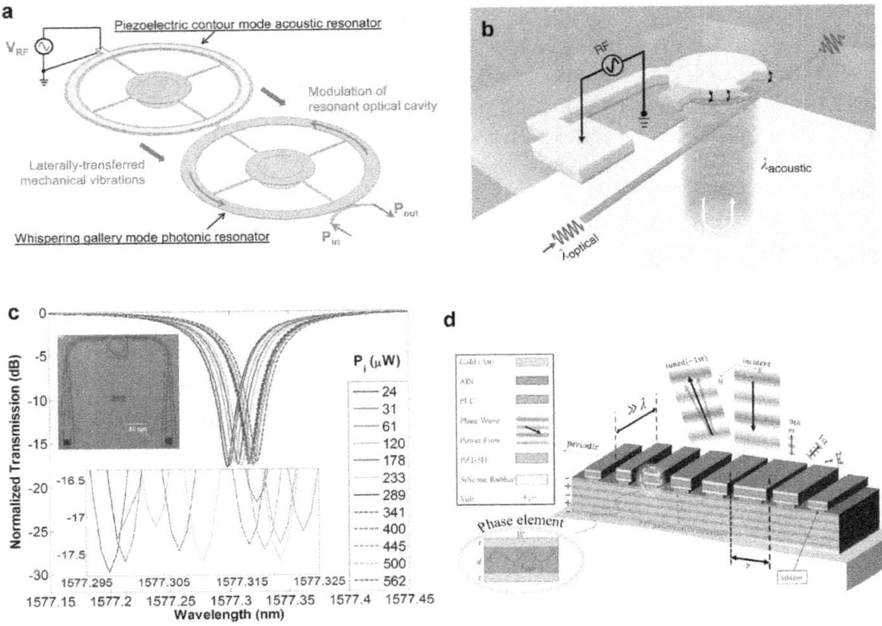

FIGURE 12.9 Optomechanical resonators. (a) Schematic illustration of piezoelectric AlN contour mode optomechanical resonators, which consists of centrally anchored acoustic and photonic resonators.

Source: Reprinted with permission from Ghosh, S., and Piazza, G. Optics Express 23, 15477–15490, 2015. Copyright 2015 Optical Society of America.

(b) Schematic illustration of high-overtone bulk acoustic resonator with the modulation of piezoelectric AlN actuator.

Source: Reprinted with permission from Tian, H. *et al.* Nature Communications 11, 3073, 2020. Copyright 2020 Springer Nature.

(c) Normalized transmission wavelength shift of piezoelectric AlN ring resonator with the increase of input powers. The insets show optical image of the disk/waveguide structure with gratings (top), and the enlarged transmission spectra (bottom).

Source: Reprinted with permission from Ghosh, S. and Piazza, G. Journal of Applied Physics 113, 016101, 2013. Copyright 2013 AIP Publishing.

(d) Schematic illustration of electrically tunable metasurface by piezoelectric actuation.

Source: Reprinted with permission from Mekawy, A. *et al.* IET Optoelectronics 13, 134–138, 2019 Copyright 2019 John Wiley and Sons.

In addition, integrated photonic devices combining piezoelectric actuators and low loss Si_3N_4 waveguides can provide effective acousto-optic modulations of Si_3N_4 microring resonators using high-overtone bulk acoustic wave resonances. As schematically shown in Figure 12.9(b), a disk-shaped AlN piezoelectric actuator placed directly on top of the Si_3N_4 microring resonator is capable to excite sub-micrometer acoustic waves that could transmit vertically into the substrate [10]. Trapped inside a F-P-like acoustic cavity that is formed by the top and bottom surfaces of the entire substrate, a rich family of acoustic resonant modes is efficiently excited at the microwave frequencies up to 6 GHz. The bulk acoustic standing waves contribute to enhance the stress fields around the Si_3N_4 waveguides and modulating the refractive index via the stress-optical effect. The coupling between the vertical acoustic waves and the in-plane optical circuits makes it possible for independent optimization of the actuator and optical components. Specifically, the Si_3N_4 waveguides buried deeply inside the oxide cladding can preserve low optical losses. The high lateral acoustic mode confinement further enables low cross-talk and compact integration. Moreover, many electro-optical materials, including piezo-materials (e.g., PZT, and $LiNbO_3$), and 2D materials (e.g., graphene, and monolayer WS_2), have been used to fabricate hybrid integrated photonic devices. However, there are still some limitations need to be addressed for developing a practical integrated photonic platform, including process complexity and poor compatibility, optical loss, and dispersion.

Piezoelectric AlN resonator possessing high optical quality factor enables a promising method for engineering optomechanical devices. As shown in Figure 12.9(c), an integrated photonic microdisk resonator based on c-axis AlN illustrates a red-shift of transmission wavelength with the increase of optical input power (P_i), due to the change of disk heating and refractive index [67]. The relatively constant extinction ratio across coupled powers indicates that there is no nonlinear absorption changing the quality factor of the disk. The high extinction ratio observed in the critically coupled-resonator also remains stable and the device behaves linearly even at high power levels. Combined with piezoelectric and piezo-optic characteristics, such piezoelectric materials enable improved quality factors of the devices and integrating features such as oscillators and modulators to construct RF-photonic mixed systems.

Metasurfaces can exploit strong subwavelength light-matter interactions, which empower the generation of anomalous wave propagation and show important applications in real-time imaging, wireless power transfer, and optical communication. See Chapter 5 for detailed review on metasurfaces. Previous reports on regulating metasurface functions by using common non-electrical tuning techniques (e.g., thermal actuation, and mechanical actuation), which bring difficulties in the integration of metasurfaces into optoelectronic devices [68]. To resolve this problem, electrical tuning techniques have been introduced to achieve tunable metasurface behavior. Piezoelectric actuation is one of the promising electrical tuning techniques, where optimizing piezoelectric materials with large piezoelectric coefficients and maximized tolerant strains are critical for improving electrical reconfigurable characteristics of photonics and optoelectronics. Figure 12.9(d) shows the piezoelectric tuning of a metasurface in a planar configuration. The grating device operating in the reflection mode consists of an array of phase elements in [0, 2π] range and a piezoelectric PZT substrate to modulate the metasurface periodicity (i.e., distances between the gratings), leading to a tuned reflected light angle. Such design takes advantage of piezoelectric materials to conduct voltage-tuned metasurfaces and shows promising applications in free-space optical communication and optical modulation [68].

12.3.7 MECHANOLUMINESCENCE

Mechanoluminescence (ML) is a kind of light emission phenomenon induced by external mechanical excitation (*e.g.*, grinding, stretching, compressing, or shaking) on a solid, and is also called piezoluminescence for piezoelectric materials. Metal ions as activators in ML are responsible for photoexcitation and illustrate the piezo-photonic effect. Specifically, lanthanide ions and transition-metal ions, whose luminescence covering a broad optical spectrum from UV to infrared regions, are the two primary metal ions used as activators for piezophotonics [23]. To date, metal ion-doped ZnS, CaZnOS, $SrAl_2O_4$, or $LiNbO_3$ have drawn much attention for developing high-performance mechanoluminescent devices in stress sensing, energy harvesting, displaying, and other flexible/stretchable optoelectronics [11].

Strain-induced piezopotential would play a crucial role in tuning the local band structure, and thus lead to the modulation of light emission. In Figure 12.10(a), the physical mechanism of piezoelectric semiconductor Er^{3+} doped CaZnOS (CaZnOS: Er^{3+}) is described for the ML behavior. CaZnOS as the host material, whose inner crystal piezopotential enables the tilt of conduction and valence bands in CaZnOS: Er^{3+} with a load of stress, exhibits green light emission from the mechanolumines-cent process [69]. Trapped electrons from upper electron defect states are easy to be detrapped and transited to the conduction band. Then nonradiative recombination occurs between detrapped electrons and holes by transferring energy to Er^{3+} ground state of $^4I_{15/2}$ to populate the higher $^4F_{7/2}$ level. The nonradiative relaxation of Er^{3+} ions from the excited $^4F_{7/2}$ level to lower atomic levels and the following radiative decays to the ground state together results in visible light emissions. According to the density functional theory, two primary luminous mechanisms can be identified: i) the formation of a deep or shallow energy trap induced by doped ions in the bandgap; and ii) the energy transitions of doping ions from excited state to the ground state [70].

Figure 12.10(b) illustrates the energy conversion between the piezoelectric poten-tials and the intrinsic defect levels for photon generations in the piezophotonic mate-rials. External mechanical stimuli can be converted into an electric polarization field as the M-load mode. Subsequently, the induced energy input gain shows an offset in the spatial charge separation and the charging of defects at a very fast speed (charg-ing mode). Consequently, it would present a simultaneous process of relaxation and recombination at a relatively high energy, leading to the photon emission (release mode). This phenomenon could be attributed to the charge alternation pair (CAP) formed by the intrinsic defect complex, illustrating the electronic transition process in a different form of energy conversion via correlated pairs [71].

A large number of emerging applications in mechanoluminescent or electrolumi-nescent systems have been envisioned. Especially, the typical piezoelectric materials Cu- or Mn-doped ZnS have aroused wide interests in developing the novel lumi-nescent systems, including the wind-driven light source (Figure 12.10(c)) [72], 3D handwriting (Figure 12.10(d)) [73], and multicolor display textile (Figure 12.10 (e)) [74]. However, there remain some key issues that should be addressed for an in-depth understanding of the strain-induced piezopontential on the photonic coupling effect and the generalized energy-conversion mechanism.

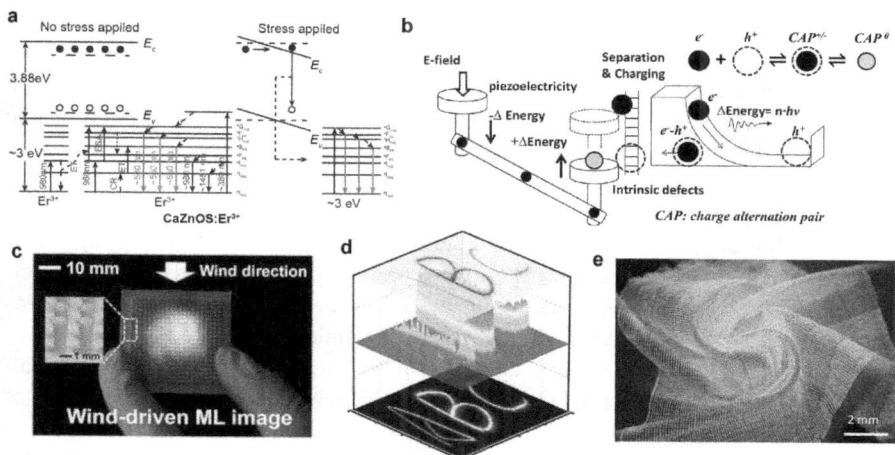

FIGURE 12.10 Mechanoluminescence. (a) Luminescent principle of Er-doped CaZnOS piezoelectric materials.

Source: Reprinted with permission from Zhang, H. *et al.* The Journal of Physical Chemistry C 119, 28136–28142, 2015. Copyright 2015 American Chemical Society.

(b) Schematic energy conversion between piezoelectric potential and intrinsic defect levels for charge separation and charging.

Source: Reprinted with permission from Huang, B. *et al.* Nano Energy 47, 150–171, 2018. Copyright 2018 Elsevier.

(c) Optical images of wind-driven mechanoluminescent display with the use of gas flow.

Source: Reprinted with permission from Jeong, S. M. *et al.* Energy & Environmental Science 7, 3338–3346, 2014. Copyright 2014 Royal Society of Chemistry.

(d) Signature pressure mapping in ZnS:Mn.

Source: Reprinted with permission from Wang, X. *et al.* Nano Research 11, 1967–1976, 2018. Copyright 2018 Tsinghua University Press and Springer-Verlag GmbH Germany.

(e) Optical image of electroluminescence of multicolour display textile with Cu or Mn doped ZnS.

Source: Reprinted with permission from Shi, X. et al. Nature 591, 240–245, 2021. Copyright 2021 Springer Nature.

12.4   SUMMARY AND OUTLOOK

This chapter briefly reviewed the recent progress of piezoelectric materials in typical optical/optoelectronic applications, including the basic principles of piezoelectricity and performance modulation of various devices. With the unique multi-coupling physical effects in the piezoelectric semiconductors, both piezophotonics and piezophototronics are attracting enormous research interests for device optimizations with tunable functionalities. The core of the piezophotonics and piezophototronics lies in exploiting the strain-induced piezopotential to tune/control the carrier processes, such as generation, transport, separation, and combination, leading to the enhancement of device performance. Moreover, by using piezoelectric materials as sensors or actuators, external-strain-induced piezoelectric polarization or electric-voltage-generated piezoelectric actuation can provide viable approaches to design advanced intelligent systems. This chapter discussed various optical/optoelectronic micro-nano devices, including LEDs, solar cells, photodetectors, terahertz, lasers, optomechanical resonators, and mechanoluminescence.

In the future, new materials with larger piezoelectric charge/strain coefficients should be developed through proper material selection, composition design, and interfacial engineering for highly sensitive force-typed sensors and highly controllable motor-typed actuators. In addition, the future development of piezophotonics and piezophototronics would mainly focus on continually improving optical/optoelectronic performance of devices and developing multifunctional flexible optical/optoelectronic devices. For instance, the InGaN/GaN MQWs film is capable of enhancing its photoluminescence intensity under the piezophototronic effect by exploiting a stress solidification strategy. In addition, the piezoelectric materials with the intrisinc piezoelectrity can contribute to provide additional functionalities for modulating optical and optoelectronic devices and making these devices more intelligent in the post-Moore era. Nevertheless, there are a large amount of works needed to be conducted to reveal the in-depth mechanical, electronic, and photonic coupling effects of piezoelectric materials from molecular, atomic, and/or electron-hole level, so as to promote their practical applications in photonics and optoelectronics.

ACKNOWLEDGEMENTS

This work is supported by the National Natural Science Foundation of China (NSFC, Grant No. 61904012).

BIBLIOGRAPHY

[1] Manjón-Sanz, A. M., and Dolgos, M. R. 2018. Applications of Piezoelectrics: Old and New. *Chemistry of Materials*. 30: 8718–8726.

[2] Xu, Q., Gao, X., Zhao, S., et al. 2021. Construction of Bio-Piezoelectric Platforms: From Structures and Synthesis to Applications. *Advanced Materials*. 33: 2008452.

[3] Pan, C., Zhai, J., and Wang, Z. L. 2019. Piezotronics and Piezo-Phototronics of Third Generation Semiconductor Nanowires. *Chemical Reviews*. 119: 9303–9359.

[4] Yang, Q., Wang, W., Xu, S., and Wang, Z. L. 2011. Enhancing Light Emission of ZnO Microwire-Based Diodes by Piezo-Phototronic Effect. *Nano Letters*. 11: 4012–4017.

[5] Pan, C., Dong, L., Zhu, G., et al. 2013. High-Resolution Electroluminescent Imaging of Pressure Distribution Using a Piezoelectric Nanowire LED Array. *Nature Photonics*. 7: 752–758.

[6] Sun, J., Hua, Q., Zhou, R., et al. 2019. Piezo-Phototronic Effect Enhanced Efficient Flexible Perovskite Solar Cells. *ACS Nano*. 13: 4507–4513.

[7] Yang, Q., Guo, X., Wang, W., et al. 2010. Enhancing Sensitivity of a Single ZnO Micro-/Nanowire Photodetector by Piezo-Phototronic Effect. *ACS Nano*. 4: 6285–6291.

[8] Woodward, N., Gallinat, C., Rodak, L. E., et al. 2012. Enhanced THz Emission from c-Plane $In_xGa_{1-x}N$ Due to Piezoelectric Field-Induced Electron Transport. *Applied Physics Letters*. 100: 191110.

[9] Zhu, H., Fu, Y., Meng, F., et al. 2015. Lead Halide Perovskite Nanowire Lasers with Low Lasing Thresholds and High Quality Factors. *Nature Materials*. 14: 636–642.

[10] Tian, H., Liu, J., Dong, B., et al. 2020. Hybrid Integrated Photonics Using Bulk Acoustic Resonators. *Nature Communications*. 11: 3073.

[11] Wang, X., Peng, D., Huang, B., Pan, C., and Wang, Z. L. 2019. Piezophotonic Effect Based on Mechanoluminescent Materials for Advanced Flexible Optoelectronic Applications. *Nano Energy*. 55: 389–400.

[12] de Jong, M., Chen, W., Geerlings, H., Asta, M., and Persson, K. A. 2015. A Database to enable Discovery and Design of Piezoelectric Materials. *Scientific Data*. 2: 150053.

[13] Hua, Q., Cui, X., Ji, K., Wang, B., and Hu, W. 2021. Piezotronics Enabled Artificial Intelligence Systems. *Journal of Physics: Materials*. 4: 022003.

[14] Jena, D., Page, R., Casamento, J., et al. 2019. The New Nitrides: Layered, Ferroelectric, Magnetic, Metallic and Superconducting Nitrides to Boost the GaN Photonics and Electronics Eco-System. *Japanese Journal of Applied Physics*. 58: SC0801.

[15] Zhang, Y., Liu, Y., and Wang, Z. L. 2011. Fundamental Theory of Piezotronics. *Advanced Materials*. 23: 3004–3013.

[16] Wang, Z. L. 2007. Nanopiezotronics. *Advanced Materials*. 19: 889–892.

[17] Wu, W., Wen, X., and Wang Zhong, L. 2013. Taxel-Addressable Matrix of Vertical-Nanowire Piezotronic Transistors for Active and Adaptive Tactile Imaging. *Science*. 340: 952–957.

[18] Zhou, J., Gu, Y., Fei, P., et al. 2008. Flexible Piezotronic Strain Sensor. *Nano Letters*. 8: 3035–3040.

[19] Liu, H., Hua, Q., Yu, R., et al. 2016. A Bamboo-Like GaN Microwire-Based Piezotronic Memristor. *Advanced Functional Materials*. 26: 5307–5314.

[20] Hua, Q., Sun, J., Liu, H., et al. 2020. Flexible GaN Microwire-Based Piezotronic Sensory Memory Device. *Nano Energy*. 78: 105312.

[21] Hua, Q., Cui, X., Liu, H., et al. 2020. Piezotronic Synapse Based on a Single GaN Microwire for Artificial Sensory Systems. *Nano Letters*. 20: 3761–3768.

[22] Zhang, S., Ma, B., Zhou, X., et al. 2020. Strain-Controlled Power Devices as Inspired by Human Reflex. *Nature Communications*. 11: 326.

[23] Hao, J., and Xu, C.-N. 2018. Piezophotonics: From Fundamentals and Materials to Applications. *MRS Bulletin*. 43: 965–969.

[24] Liu, Y., Niu, S., Yang, Q., et al. 2014. Theoretical Study of Piezo-Phototronic Nano-LEDs. *Advanced Materials*. 26: 7209–7216.

[25] Huang, X., Du, C., Zhou, Y., et al. 2016. Piezo-Phototronic Effect in a Quantum Well Structure. *ACS Nano*. 10: 5145–5152.

[26] Han, X., Du, W., Yu, R., Pan, C., and Wang, Z. L. 2015. Piezo-Phototronic Enhanced UV Sensing Based on a Nanowire Photodetector Array. *Advanced Materials*. 27: 7963–7969.

[27] Dai, X., Hua, Q., Sha, W., Wang, J., and Hu, W. 2022. Piezo-Phototronics in Quantum Well Structures. *Journal of Applied Physics*. 131: 010903.

[28] Peng, M., Li, Z., Liu, C., et al. 2015. High-Resolution Dynamic Pressure Sensor Array Based on Piezo-Phototronic Effect Tuned Photoluminescence Imaging. *ACS Nano*. 9: 3143–3150.

[29] Li, X., Chen, M., Yu, R., et al. 2015. Enhancing Light Emission of ZnO-Nanofilm/Si-Micropillar Heterostructure Arrays by Piezo-Phototronic Effect. *Advanced Materials*. 27: 4447–4453.

[30] Li, X., Liang, R., Tao, J., et al. 2017. Flexible Light Emission Diode Arrays Made of Transferred Si Microwires-ZnO Nanofilm with Piezo-Phototronic Effect Enhanced Lighting. *ACS Nano*. 11: 3883–3889.

[31] Zhang, T., Liang, R., Dong, L., et al. 2015. Wavelength-Tunable Infrared Light Emitting Diode Based on Ordered ZnO Nanowire/Si1–xGexalloy Heterojunction. *Nano Research*. 8: 2676–2685.

[32] Liang, R., Zhang, T., Wang, J., and Xu, J. 2016. High-Resolution Light-Emitting Diode Array Based on an Ordered ZnO Nanowire/SiGe Heterojunction. *IEEE Transactions on Nanotechnology*. 15: 539–548.

[33] Bao, R., Wang, C., Dong, L., et al. 2015. Flexible and Controllable Piezo-Phototronic Pressure Mapping Sensor Matrix by ZnO NW/p-Polymer LED Array. *Advanced Functional Materials*. 25: 2884–2891.

[34] Bao, R., Wang, C., Peng, Z., et al. 2017. Light-Emission Enhancement in a Flexible and Size-Controllable ZnO Nanowire/Organic Light-Emitting Diode Array by the Piezotronic Effect. *ACS Photonics*. 4: 1344–1349.

[35] Bao, R., Wang, C., Dong, L., et al. 2016. CdS Nanorods/Organic Hybrid LED Array and the Piezo-Phototronic Effect of the Device for Pressure Mapping. *Nanoscale*. 8: 8078–8082.

[36] Zheng, Q., Peng, M., Liu, Z., et al. 2021. Dynamic Real-Time Imaging of Living Cell Traction Force by Piezo-Phototronic Light Nano-Antenna Array. *Science Advances*. 7: eabe7738.

[37] Min, H., Lee, D. Y., Kim, J., et al. 2021. Perovskite Solar Cells with Atomically Coherent Interlayers on SnO2 Electrodes. *Nature*. 598: 444–450.

[38] Jiang, C., Jing, L., Huang, X., et al. 2017. Enhanced Solar Cell Conversion Efficiency of InGaN/GaN Multiple Quantum Wells by Piezo-Phototronic Effect. *ACS Nano*. 11: 9405–9412.

[39] Jiang, C., Chen, Y., Sun, J., et al. 2019. Enhanced Photocurrent in InGaN/GaN MQWs Solar Cells by Coupling Plasmonic with Piezo-Phototronic Effect. *Nano Energy*. 57: 300–306.

[40] Boxberg, F., Søndergaard, N., and Xu, H. Q. 2010. Photovoltaics with Piezoelectric Core–Shell Nanowires. *Nano Letters*. 10: 1108–1112.

[41] Wu, W., Han, X., Li, J., et al. 2021. Ultrathin and Conformable Lead Halide Perovskite Photodetector Arrays for Potential Application in Retina-Like Vision Sensing. *Advanced Materials*. 33: 2006006.

[42] Liu, Y., Yang, Q., Zhang, Y., Yang, Z., and Wang, Z. L. 2012. Nanowire Piezo-phototronic Photodetector: Theory and Experimental Design. *Advanced Materials*. 24: 1410–1417.

[43] Lai, Q., Zhu, L., Pang, Y., et al. 2018. Piezo-Phototronic Effect Enhanced Photodetector Based on CH3NH3PbI3 Single Crystals. *ACS Nano*. 12: 10501–10508.

[44] Zhang, F., Niu, S., Guo, W., et al. 2013. Piezo-phototronic Effect Enhanced Visible/UV Photodetector of a Carbon-Fiber/ZnO-CdS Double-Shell Microwire. *ACS Nano*. 7: 4537–4544.

[45] Wang, Z., Yu, R., Wen, X., et al. 2014. Optimizing Performance of Silicon-Based p-n Junction Photodetectors by the Piezo-Phototronic Effect. *ACS Nano*. 8: 12866–12873.

[46] Liu, H. C., Song, C. Y., SpringThorpe, A. J., and Cao, J. C. 2004. Terahertz Quantum-Well Photodetector. *Applied Physics Letters*. 84: 4068–4070.

[47] Yang, Y., Yamagami, Y., Yu, X., et al. 2020. Terahertz topological Photonics for On-Chip Communication. *Nature Photonics*. 14: 446–451.

[48] Manjappa, M., and Singh, R. 2020. Materials for Terahertz Optical Science and Technology. *Advanced Optical Materials*. 8: 1901984.

[49] Kiarash, A. 2017. Review of GaN-Based devices for Terahertz Operation. *Optical Engineering*. 56: 1–14.

[50] Acharyya, A. 2021. Gallium Nitride-Based Solid-State Devices for Terahertz Applications. *Advanced Materials for Future Terahertz Devices, Circuits and Systems.* 727: 9.

[51] Liu, N., Hu, G., Dan, M., et al. 2019. Piezo-Phototronic Effect on Quantum Well Terahertz Photodetector for Continuously Modulating Wavelength. *Nano Energy.* 65: 104091.

[52] Delga, A. 2020. *Mid-infrared Optoelectronics* (Eds: E. Tournié, L. Cerutti), Woodhead Publishing, pp. 337–377.

[53] Huang Michael, H., Mao, S., Feick, H., et al. 2001. Room-Temperature Ultraviolet Nanowire Nanolasers. *Science.* 292: 1897–1899.

[54] Choi, H.-J., Johnson, J. C., He, R., et al. 2003. Self-Organized GaN Quantum Wire UV Lasers. *The Journal of Physical Chemistry B.* 107: 8721–8725.

[55] Lu, J., Yang, Z., Li, F., et al. 2019. Dynamic Regulating of Single-Mode Lasing in ZnO Microcavity by Piezoelectric Effect. *Materials Today.* 24: 33–40.

[56] Lu, J., Xu, C., Li, F., et al. 2018. Piezoelectric Effect Tuning on ZnO Microwire Whispering-Gallery Mode Lasing. *ACS Nano.* 12: 11899–11906.

[57] Wei, B., Zheng, K., Ji, Y., et al. 2012. Size-Dependent Bandgap Modulation of ZnO Nanowires by Tensile Strain. *Nano Letters.* 12: 4595–4599.

[58] Fu, X., Liao, Z.-M., Liu, R., et al. 2015. Strain Loading Mode Dependent Bandgap Deformation Potential in ZnO Micro/Nanowires. *ACS Nano.* 9: 11960–11967.

[59] Peng, Y., Lu, J., Peng, D., et al. 2019. Dynamically Modulated GaN Whispering Gallery Lasing Mode for Strain Sensor. *Advanced Functional Materials.* 29: 1905051.

[60] Yang, Z., Lu, J., ZhuGe, M., et al. 2019. Controllable Growth of Aligned Monocrystalline CsPbBr3 Microwire Arrays for Piezoelectric-Induced Dynamic Modulation of Single-Mode Lasing. *Advanced Materials.* 31: 1900647.

[61] Czekalla, C., Sturm, C., Schmidt-Grund, R., et al. 2008. Whispering Gallery Mode Lasing in Zinc Oxide Microwires. *Applied Physics Letters.* 92: 241102.

[62] Ma, W., Lu, J., Yang, Z., et al. 2019. Crystal-Orientation-Related Dynamic Tuning of the Lasing Spectra of CdS Nanobelts by Piezoelectric Polarization. *ACS Nano.* 13: 5049–5057.

[63] Ghorbel, I., Swiadek, F., Zhu, R., et al. 2019. Optomechanical Gigahertz Oscillator Made of a Two Photon Absorption Free Piezoelectric III/V Semiconductor. *APL Photonics.* 4: 116103.

[64] Lu, X., Lee, J. Y., and Lin, Q. 2015. High-Frequency and High-Quality Silicon Carbide Optomechanical Microresonators. *Scientific Reports.* 5: 17005.

[65] Czerniuk, T., Brüggemann, C., Tepper, J., et al. 2014. Lasing from Active Optomechanical Resonators. *Nature Communications.* 5: 4038.

[66] Ghosh, S., and Piazza, G. 2015. Piezoelectric Actuation of Aluminum Nitride Contour Mode Optomechanical Resonators. *Optics Express.* 23: 15477–15490.

[67] Ghosh, S., and Piazza, G. 2013. Photonic Microdisk Resonators in Aluminum Nitride. *Journal of Applied Physics.* 113: 016101.

[68] Mekawy, A., Khalifa, M., Ali, T. A., and Badawi, A. H. 2019. *IET Optoelectronics*, Vol. 13, Institution of Engineering and Technology, pp. 134–138.

[69] Zhang, H., Peng, D., Wang, W., Dong, L., and Pan, C. 2015. Mechanically Induced Light Emission and Infrared-Laser-Induced Upconversion in the Er-Doped CaZnOS Multifunctional Piezoelectric Semiconductor for Optical Pressure and Temperature Sensing. *The Journal of Physical Chemistry C.* 119: 28136–28142.

[70] Huang, B., Peng, D., and Pan, C. 2017. "Energy Relay Center" for Doped Mechanoluminescence Materials: A Case Study on Cu-Doped and Mn-Doped CaZnOS. *Physical Chemistry Chemical Physics.* 19: 1190–1208.

[71] Huang, B., Sun, M., and Peng, D. 2018. Intrinsic energy Conversions for Photon-Generation in Piezo-Phototronic Materials: A Case Study on Alkaline Niobates. *Nano Energy*. 47: 150–171.

[72] Jeong, S. M., Song, S., Joo, K.-I., et al. 2014. Bright, Wind-Driven White Mechanoluminescence from Zinc Sulphide Microparticles Embedded in a Polydimethylsiloxane Elastomer. *Energy & Environmental Science*. 7: 3338–3346.

[73] Wang, X., Ling, R., Zhang, Y., et al. 2018. Oxygen-Assisted Preparation of Mechanoluminescent ZnS:Mn for Dynamic Pressure Mapping. *Nano Research*. 11: 1967–1976.

[74] Shi, X., Zuo, Y., Zhai, P., et al. 2021. Large-Area Display Textiles Integrated with Functional Systems. *Nature*. 591: 240–245.

13 Hybrid Perovskite Materials and Their Applications

Ru Li and Xue Liu

CONTENTS

13.1 Atomic Insights into Crystal Structures of Hybrid Perovskite Materials 363
13.2 Electrical and Optical Properties 365
13.3 Polarons in Hybrid Perovskites and Their Impact on Carrier Dynamics 368
13.4 High-Efficiency Perovskite Solar Cells 370
    13.4.1 Crystal Structure Consideration 370
    13.4.2 Material Chemistry 372
    13.4.3 Process Engineering 374
    13.4.4 Device Physics 376
    13.4.5 More beyond Shockley-Queisser Limit 379
13.5 Hybrid Perovskite Light Emitting Diode 379
    13.5.1 Luminescence Mechanism and Underlying Photophysics 380
    13.5.2 Light Management 381
13.6 Stable Issue of Hybrid Perovskite 382
13.7 Dimension Reduction in Hybrid Perovskite 384
13.8 Industry Application and Future of Hybrid Perovskite 385
Acknowledgements 388
Bibliography 388

13.1 ATOMIC INSIGHTS INTO CRYSTAL STRUCTURES OF HYBRID PEROVSKITE MATERIALS

Traditional perovskite is a calcium titanium oxide mineral with chemical formula of $CaTiO_3$, whose name is also applied to the class of compounds which have the same type of crystal structure as $CaTiO_3$, known as the perovskite structure. In recent ten years, the emerging organic/inorganic hybrid perovskite materials have gain intensive research interests, they share similar cubic crystal structures with conventional minerals $CaTiO_3$, both of which can be chemically formulated as ABX_3, where A and B are positively charged cations while X is negatively charged anion. More specifically, one B-site atom and six X-coordinated atoms form an octahedron sub-lattice, and the A-site atom sits within the cavity formed by adjacent corner-shared octahedrons. Figure 13.1 shows the typical hybrid perovskite $MAPbI_3$, where

DOI: 10.1201/9781003202608-13

363

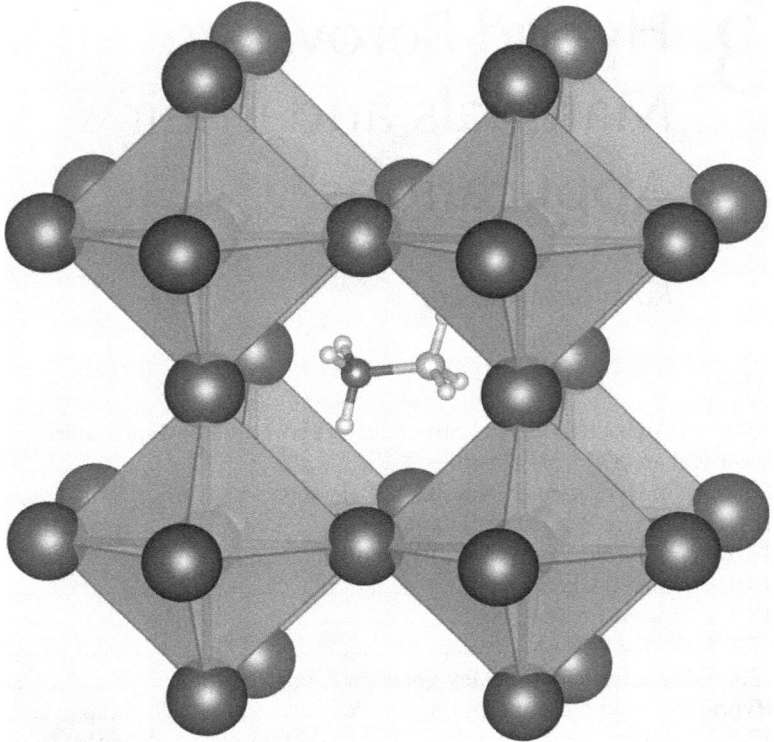

FIGURE 13.1 Schematic of the crystal structure of organic/inorganic hybrid perovskite MAPbI$_3$, MA$^+$ sits within the cavity formed by PbI$_6$ octahedrons.

Source: Adopted from C. Eames et. al., Nature Communications 6, 7497 (2015). No permission is required. Copyright 2015 the Authors.

MA$^+$ standards for the CH$_3$NH$_3$$^+$ organic cation and sits within the cavity formed by PbI$_6$ octahedrons, atom Pb is located at the center of the octahedron. In general, the formation of hybrid perovskite is determined by the geometric tolerance factor t, $t = (r_A + r_X) / \left[\sqrt{2} (r_B + r_X) \right]$, where, r_A, r_B and r_X are the effective ionic radii for A, B and X ions, respectively [1]. An ideal cubic crystal structure is obtained when $t = 1$, otherwise the octahedral structure would be distorted. For the most studied hybrid perovskites, their geometric tolerance factors lie in the range of 0.813–1.107 [2], for example, the hybrid perovskite MAPbI$_3$ has a geometric tolerance factor of 0.83 at room temperature. Experimental results show that at relatively high temperature, the highest symmetry cubic crystal structure (α phase, $Pm\bar{3}m$) is favourable. As temperature goes down, the tetragonal crystal structure (β phase, I_4/mcm) forms through the rotation of PbI$_6$ octahedra around the c-axis. Further cool down the perovskite will lead to the orthorhombic crystal structure (γ phase, $Pnma$) transition, where the PbI$_6$ octahedra is tilted out of the ab plane. The transition temperature between different phases is strongly dependent on their chemical compositions.

The extraordinary optoelectronic properties of MAPbI$_3$, such as broad spectral absorption, long charge-carrier lifetime, high defect tolerance, and small exciton binding energy (≈ 50 meV) [3–5], are deeply rooted in PbI$_6$ formed octahedron structure. For the A-site cation MA$^+$, it is surrounded by twelve nearest-neighbor iodide ions with weak interactions, which requires only a small energy to rotate randomly. As a result, the randomly rotated dipole induces random potential on an inorganic sub-lattice PbI$_6$ octahedron, further leading to a so-called dynamic disorder [6]. In addition, the corner-shared PbI$_6$ octahedras are easy to deform in the presence of excess electrons or holes that are induced by light absorption, and large-size electron/hole polarons are formed due to the strong electron-phonon interactions [7,8]. It shows that the formation of polarons efficiently screens the electron-electron and electron-defect scatterings, which leads to an extremely low recombination rate and long carrier lifetime [9].

13.2   ELECTRICAL AND OPTICAL PROPERTIES

The band structures of β and γ phases of MAPbI$_3$ are similar to that of α phase, which means that the Pb-I-Pb bond angle distortions do not significantly change the electronic structures. For the sake of illustration, the results of α phase is shown in Figure 13.2. First-principle calculation shows that MAPbI$_3$ is a direct bandgap semiconductor with a valence band maximum (VBM) and a conduction band minimum (CBM) at R point (Figure 13.2(a)). Different from most cations whose outer s orbitals

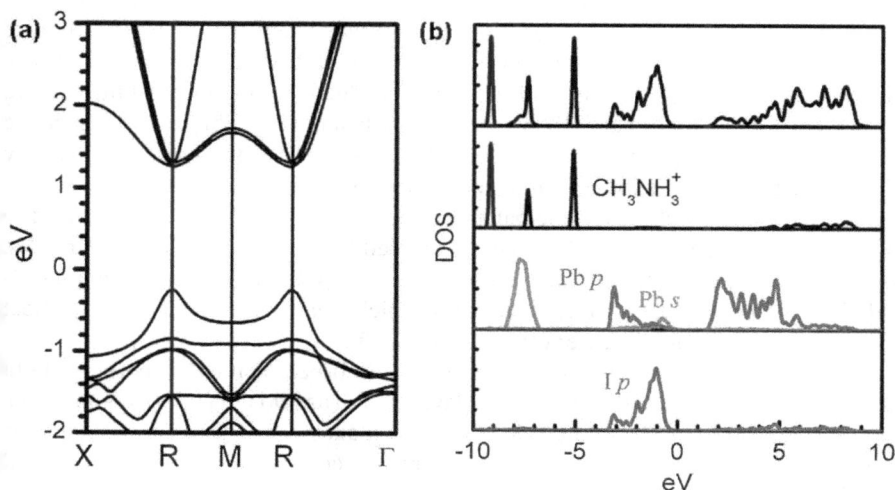

FIGURE 13.2   Electronic band structure of MAPbI$_3$ calculated from density functional theory at the PBE level. (a) The band structure along with the high symmetry point in the first Brillouin zone. (b) The projected density of states onto the chemical elements.

Source: Reprinted with permission from W.-J. Yin et. al., Applied Physics Letters 104, 063903 (2014). Copyright 2014 AIP Publishing.

are empty, Pb has an occupied $6s$ orbital, which is below the top of valence bands of perovskites. This so-called lone pair of s electrons in Pb often gives rise to unusual properties [10–12]. The (partial) density of states (DOS, Figure 13.2(b) shows that the VBM has a strong Pb s-state and I p-state antibonding character, whereas the CBM is almost contributed from the Pb p-state, which reflects the unique dual nature (ionic and covalent) of halide perovskite's electronic structure. The states contributed by the organic molecule are far from the band edges, which means that the organic molecule does not play a direct role in determining the fundamental electronic structure. A-site cations are intitially thought to only help stabilizing the lattice, however, it is later found that the π-conjugated organic cations could affect the perovskite frontier orbitals, and reconfigure the resultant electronic band edges [13].

The effective mass of the electron (hole) is approximately fitted by $\dfrac{1}{m^*} = \dfrac{1}{\hbar^2}\dfrac{\partial^2 E}{\partial k_i \partial k_j}$, where E and k denote the energy and momentum respectively, and $\hbar$ is the reducd Plank constant. Thus, the flatter (dispersive) the band near the band edges is, the heavier (lighter) the effective mass is. For an traditional semiconductor, such as GaAs and CdTe, the VBM is mostly contributed by anions' p-orbitals, and the CBM is mostly contributed by cations' s- and anions' s-orbitals. Because the high-energy-level s-orbitals are more delocalized than the low-energy-level p-orbitals, the highest valence band is flatter than that of the lowest conduction band, and the effective mass of electron is thus much lighter than that of hole, resulting in a higher electron mobility than that of hole. In comparison, because of the existence of lone-pair Pb s-electrons, the energy dispersion relationship in perovskite is dramatically different. The band edge structure of $MAPbI_3$ is inverted compared to the conventional semiconductors, whose CBM is derived from Pb p-orbitals, and the VBM is a mixture of Pb s- and I p-orbitals. As a result, the Pb^{2+} p-orbital has a much higher energy level than the anion p-orbitals, and the lower conduction band of $MAPbI_3$ is more dispersive. On the other hand, the upper valence band of $MAPbI_3$ is also dispersive due to the strong s-p coupling around the VBM. Based on the density function theory (DFT) calculations, it is found that the effective mass of the hole of $MAPbI_3$ is comparable to that of its electron. The balanced effective mass in $MAPbI_3$ could lead to the ambipolar conductivity [14], which is favored by the p-i-n junction solar cells. Besides that, the quasiparticle self-consistent including the spin-orbital coupling calculations predict a lower effective hole mass, which means that the many-body effect may play a role for small carrier effective mass [15].

One of the superior properties of perovskite absorbers is their high visible light absorption, which makes the photoactive layer as thin as 500 nm [16]. In comparison, the thicknesses of the absorbing layers in the first and second-generation solar cells are 300 μm and 2 μm, respectively. Therefore, the cost of the perovskite solar cell could be much lower than those of conventional Si and GaAs solar cells. The first principle calculation has been employed to elucidate the strong optical absorption of perovskite, and the mechanisms of optical transition of Si, GaAs, and perovskite are depicted schematically in Figure 13.3. The optical absorption of semiconductors is fundamentally determined by two factors, where one is the transition matrix elements between the valence band (VB) states and conduction band (CB) states, and

(a) 1st Generation   (b) 2nd Generation   (c) Perovskite halide

Si p,s Ga s + As s Pb p

E_g^i E_g^d E_g^d

Si p As p Pb s + I p

weak   **moderate**   **strong**

Indirect   direct   direct

p ->p transition   As p-> Ga s transition   I p -> Pb p transition

        As p-> As s transition   Pb s -> Pb p transition

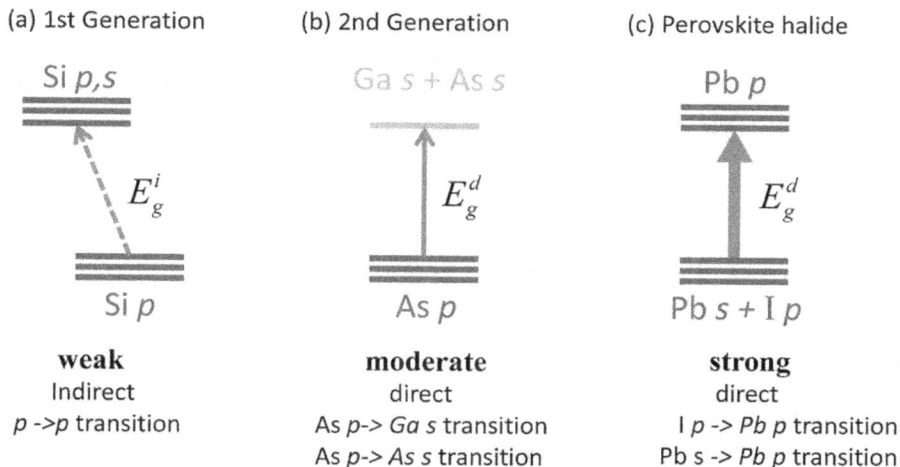

FIGURE 13.3 Schematic of the optical absorption of (a) the first-generation silicon solar cells, (b) the second-generation GaAs solar cells, and (c) the perovskite solar cells.

Source: Reprinted with permission from W.-J. Yin et. al., Journal of Materials Chemistry A 3, 8926–8942 (2015). Copyright 2015 Royal Society of Chemistry.

the other is the joint density of states (JDOS). The former accounts for the probability of each VB to CB transition, while the latter corresponds to the total number of possible photoelectric transitions [17].

For the optical absorption of Si, the transition is from the Si p-orbital to a mixture of its p- and s-orbitals. Because of the indirect bandgap feature, light absorption in Si is much weaker than that of the direct bandgap semiconductors, such as GaAs and perovskite. In addition, although both GaAs and perovskite are direct bandgap semiconductors, their lower parts of the CB are different. For GaAs, the lower part of its CB is derived from the dispersive s-band (Figure 13.3(b)), while it is mainly composed of degenerate Pb p-bands for the perovskite (Figure 13.3(c)), which makes the JDOS much higher and the photoelectric transition in perovskite is stronger. Besides, the upper part of the VB of perovskite is composed of Pb p- and s-orbitals, and the band edge optical transition thus becomes an intra-atomic transition with a high transition probability comparable to that of GaAs. Therefore, the perovskite shows a strong optical transition in the visible range, making it promising for high-efficiency cells.

The conductivity of perovskite could be either p-type or n-type depending on the material grow process, which indicates that MAPbI$_3$ is bipolarly dopable. Therefore, electrons and holes could be separated through an intrinsic p-i-n junction, making the hole transportation material (HTM) free solar cell possible. Note that several reports have shown that carbon materials, such as carbon black, graphene, and carbon nanotubes, can be used in the HTM free architecture [18].

Solution process growth inevitably introduces a lot of point defects in perovskite, such as vacancies, interstitial, and substitutions, however, first principle calculations show that most of the point defects are shallow defects [19]. Compare to other deep level defects, the shallow defects levels are close to band edges, which can donate or accept electrons to bulk bands and are effective for absorbers. It is believed that the character of the shallow defect plays an important role in the defect tolerances of perovskite solar cells.

The perovskite layer used in solar cells is polycrystalline and disordered. Particularly in the mesoporous TiO_2 based architecture, most of the perovskite (around 70%) is confined in the pores and is highly disordered, with tremendous grain boundaries (GBs) in the solution-processed layers. However, the electron beam induced current (EBIC) experiments on as-grown $MAPbI_3$ show that there is no difference between the GB and bulk, indicating negligible charge separation and recombination occurs at grain boundaries [20]. It is still under debate whether the benign grain boundary properties are intrinsic, or arised from other effects such as unintentional defect segregations.

First principal calculations have provided insights into the origin of the benign grain boundary properties. A model system based on I-I wrong bonds in the $CsPbI_3$ grain boundary shows that the antibonding $pp\sigma*$ state is not deep within the bandgap, but below the maximum of the valence band. On one hand, the strong coupling between Pb $6s$- and I $5p$-orbitals significantly increase the VBM level; on the other hand, the large I-I bond length decreases the anion-anion interaction, leading to a shallow defect energy level in the perovskite layer.

Organic cation in hybrid perovskite is noncentrosymmetric, whose dipole moment exhibits spontaneous electric polarization. It is then proposed that ferroelectric domains could exist in bulk perovskite. The alignment of the dipole of the organic cation and the intrinsic lattice distortion should account for the crystal centrosymmetry breaking. Berry phase-based calculation shows that the perovskite has comparable bulk polarization (around 38 $\mu C/cm^2$) to that of conventional ferroelectric oxide perovskite, such as $KNbO_3$ (30 $\mu C/cm^2$). It has been proved that not only the open-circuit voltage (V_{oc}) of the ferroelectric perovskite oxides-based solar cell could be larger than the fundamental bandgap, but also the carrier lifetime and charge separation are improved because of the internal electric field. Therefore, Frost *et al.* proposed that a "ferroelectric highways" model might exist, which could enhance the transportation of electrons and holes [21].

13.3   POLARONS IN HYBRID PEROVSKITES AND THEIR IMPACT ON CARRIER DYNAMICS

Solution-processed lead-halide hybrid perovskite, such as $MAPbI_3$, exhibits relatively low mobilities (1–10^2 cm^2 V^{-1} s^{-1}) but rather long diffusion lengths ($\approx$ μm) and lifetimes ($\approx$ μs) [22], which means a rather low electron-hole recombination rate (10^{-10} cm^3 s^{-1}) should be expected in polycrystalline perovskite layers (*i.e.*, even comparable to that of the pure single crystal inorganic semiconductors). The underlying mechanism of such low charge carrier recombination rate is still under debate, where several models, such as Rashba effect [23], ferroelectricity [24], photons recycling [25], and polarons [6] have been proposed to understand the carrier dynamics in hybrid perovskite. Recently, a common growing consensus is that the lattice interactions lead to the formation of polarons and the resultant carrier transport can be explained by polaron dynamics. In short, the formation of large-size polarons in perovskites efficiently screens the carrier–carrier and carrier–defect scattering, resulting in an ultra-low recombination rate as well as a long lifetime.

DFT calculation has been employed to understand the formation of polarons and their impacts on carrier dynamics. Recently, a tight-binding model has been developed to gain insights into the effects of A-site molecular rotation and inorganic sublattice vibration. As shown in Figure 13.4(a), a large-sized polaron, around 50 Å, formed in a MAPbI$_3$ 48 × 48 × 48 supercell, which spans around 8 primitive cells. Calculation results show that the A-site plays an important role in the formation energy of polarons. The rapid disordered rotation of A-site molecular induces a

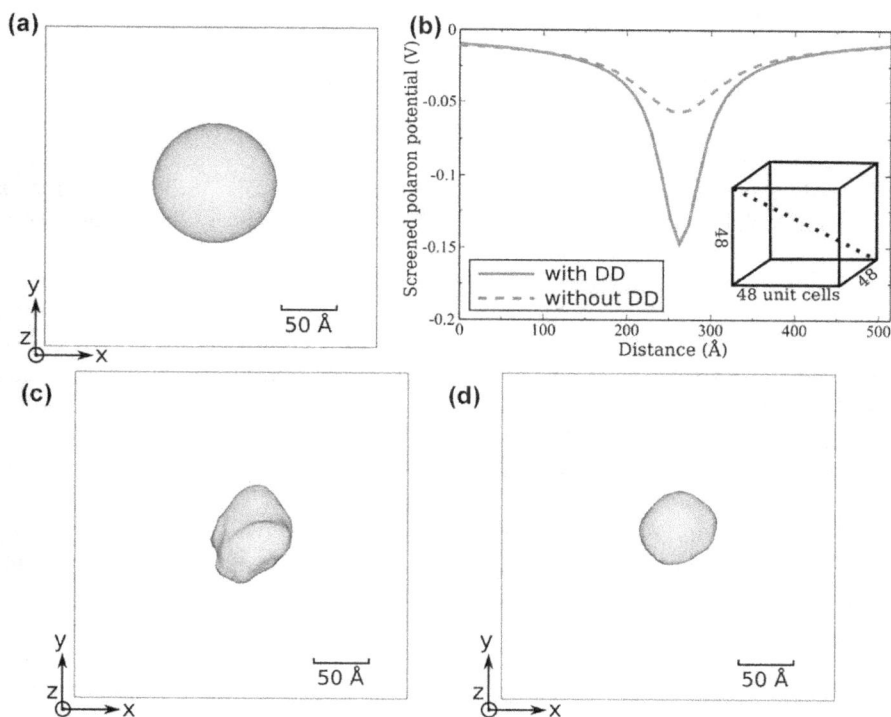

FIGURE 13.4   Large size polaron formed in MAPbI$_3$. (a) The CBM state with large polaron effect, all the atomic are not shown for clearness. (b) The screened polaron polarization potential along the diagonal line of the cubic supercell (the diagonal line is shown in the inset diagram) for the case with and without dynamic disorder (DD) effect. (c) The localized CBM state induced by DD without a large polaron effect. (d) The further localized CBM state under DD and polaron effect.

Source: Reprinted with permission from F. Zheng, et. al., Energy & Environmental Science 12, 1219–1230 (2019). Copyright 2019 Royal Society of Chemistry.

random potential on an inorganic sublattice such as PbI_3^{-1}, leading to the so-called dynamic disorder effects, which further promote wavefunction localized in real space [26]. The calculated formation energy is around -12 meV without the dynamic disorder, compared to -55 meV by including dynamic disorder (Figure 13.4(b)). Localized conduction band minimum could display the size of polaron. The resultant volume of charge density (> 85%) is larger without the polaron effect (Figure 13.4(c)) whereas it becomes smaller under a significant polaron effect and dynamic disorder (Figure 13.4(d)), which means that the formation of polaron could further localize the electrons and holes. Molecular dynamics simulation shows that the inorganic sublattice has two competing effects: the positive side is that the vibration of the sublattice provides an additional driving force to boost the carrier mobility; and the negative side is that large polaron polarization localizes the electrons, reducing its mobility. Eventually, the overall effect is to slow down the electron mobility by roughly a factor of two [6].

13.4 HIGH-EFFICIENCY PEROVSKITE SOLAR CELLS

In 2009, Miyasaka *et al.* reported the first application of hybrid perovskites as the light sensitizers in the liquid electrolyte dye-sensitized mesoscopic TiO_2 solar cells with only a power conversion efficiency (PCE) of 3.8% [27]. In 2012, solid-state perovskite solar cells (PSCs) with PCE ~ 10% are achieved by Snaith and Nam-Gyu Park groups [28,29]. The huge performance improvement is obtained by replacing the liquid electrolyte with a solid hole-conducting material: spiro-OMeTAD (2,2',7,7'-Tetrakis[N, N-di(4-methoxyphenyl)amino]-9,9'-spirobifluorene. After that, the PCE of PSCs is boosted from 9.7% with pure $MAPbI_3$ to a certified record of 25.7% by using $FAPbI_3$ based perovksite in the year 2021, where FA cation stands for CH_5I_2 [30].

The typical structure of PSC is shown in Figure 13.5, where the sensitized mesoscopic type is shown in Figures 13.5(a) and 13.5(c), and the planar thin-film type are shown in Figures 13.5(b) and 13.5(d). The main difference is whether the mesoscopic TiO_2 is used or not. For the sensitized mesoscopic PSCs, the photo-generated electrons are first transported to the TiO_2 layer, and then collected at the electrode and extracted into the circuit. For the planar thin-film PSCs, the perovskite itself acts as a highly efficient, ambipolar charge-conductor. Generally, the electron transportation layer (ETL) is deposited onto an ITO film before the perovskite growth, then the hole transportation layer (HTL) is spin-coated and the metal electrode finally evaporated. This procedure leads to the so-called normal-type, otherwise it is called inverted-type if the ETL and HTL are exchanged.

13.4.1 CRYSTAL STRUCTURE CONSIDERATION

Hybrid perovskite, such as $MAPbI_3$, could change its phase from cubic (α) to tetragonal (β), and then to orthorhombic (γ) by rotating and tilting PbI_6 sub-octahedra as the temperature decreases [31]. These transitions should be considered when designing solar cells because they may change the optoelectronic properties. For example, black-phase α-$FAPbI_3$ is suitable for solar cell but not the yellow δ-$FAPbI_3$, where the

Sensitized perovskite solar cell

(a)

cathode

p-type contact

mp-TiO$_2$ + perovskite

n-type contact

anode

Thin-film perovskite solar cell

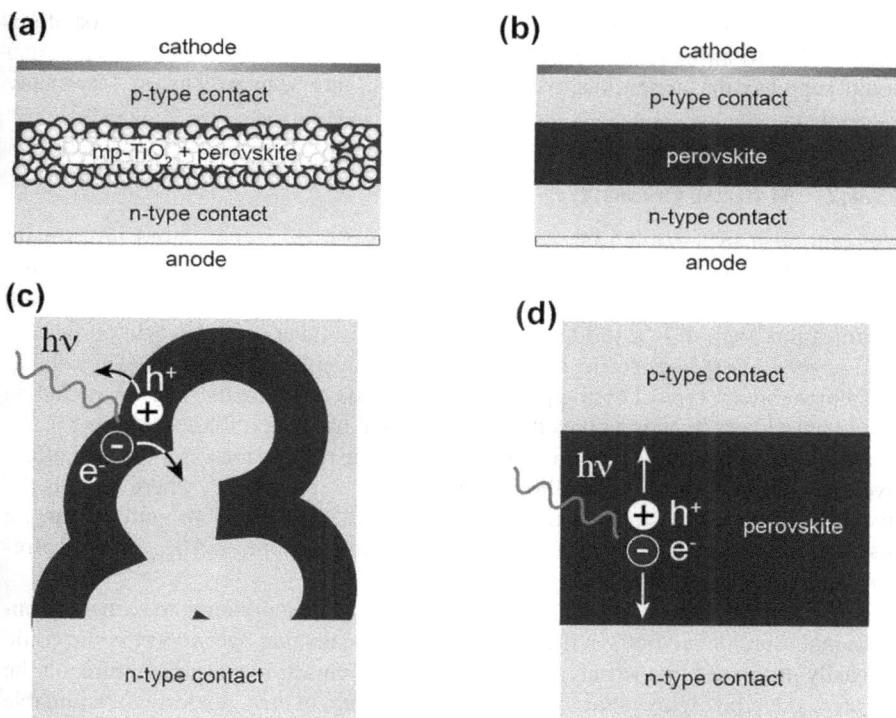

(b)

cathode

p-type contact

perovskite

n-type contact

anode

(c)

hv

h$^+$

$\oplus$

$\ominus$ e$^-$

n-type contact

(d)

p-type contact

hv

$\oplus$ h$^+$

$\ominus$ e$^-$ perovskite

n-type contact

FIGURE 13.5 Typical perovskite solar cell structures. The sensitized perovskite solar cell (a) and the charge separation at the mesoscopic surface (c). The thin-film perovskite solar cell (b) and the charge separation within the perovskite layer (d).

Source: Adopted from https://en.wikipedia.org/wiki/Perovskite_solar_cell, No permission is required.

α to δ transition could easily occur under ambient conditions. For the cubic phase, the noncentrosymmetric organic cation, MA$^+$, must be randomly oriented to satisfy the O_h symmetry, therefore, the cubic phase is stable at room temperature. However, the MA$^+$ becomes more ordering as the temperature decreases, and the PbI$_6$ sub-octahedra rotates to accommodate the ordering of MA$^+$, leading to the tetragonal phase. Further cooling down the MA$^+$ to a completely ordering state, octahedra must tilt out of the plane, resulting in an orthorhombic phase. Besides the temperature, high pressure could also trigger the phase transition, but in general it is not considered in solar cell applications.

Most of the point defects in hybrid perovskite exhibit shallow acceptor/donor features [32], however, the GBs formed during the solution process may create trap states that could decrease the electron diffusion length [33], which is evidenced that the carriers in a single crystal have a longer diffusion length. To gain more insights about the role of the GBs, theoretical calculation and experimental measurements have been employed to study the carriers across the GBs in the time/spatial domain, however, they don't reach consistent and conclusive results. Although many investigations have demonstrated that the GBs could enhance the charge dissociation, thus leading to an increased photocurrent; however, the passivation of GBs, which could suppress the nonradioactive recombination, has been widely used by many researchers.

13.4.2 MATERIAL CHEMISTRY

As mentioned in Section 13.1, the tolerance factor is commonly used to estimate the perovskite structure formability. However, it is a challenge to direct apply the tolerance factor criteria to hybrid perovskite, because the conventional concept assumes that the ion is a solid sphere, but the organic cations in hybrid perovskite are not spherically symmetric, and they form hydrogen bonds with the BX_6 octahedra. Furthermore, typical hybrid perovskite materials based on the large-sized iodide would have lower bond ionicities than conventional oxide perovskites, making it difficult to estimate the effective ionic radii of the constitutive ions. A modified effective ionic radius is proposed and formulated as $r_{Aeff} = r_{mass} + r_{ion}$, where r_{mass} is the distance between the center of mass of the molecule and the atom with the largest distance to the center of mass, excluding the hydrogen atom; and r_{ion} is the corresponding ionic radius of this atom.

The chemical bond of hybrid perovskite is found to be more ionic than that of conventional semiconductors. On the positive side, it means that hybrid perovskite could be easily processed in a variety of polar aprotic solvents at room temperature; on the negative side, the highly polar ionic nature of bonding, in turn, rendering it volnuable to moisture, oxygen, and polar solvents, resulting in a relatively low environmental stability [34]. Nevertheless, the organic A-site cation doesn't contribute directly to the electronic band edge that significantly affects the optoelectronic properties of the perovskite materials, thus providing an opportunity to fine-tune the crystal structure of the materials by using different A-site molecular.

A site(cation)-modified hybrid perovskites. $MAPbI_3$ with a bandgap of 1.57 eV is the first light absorber for photovoltaic device with an MA^+ ionic radius of 217 pm and a tolerance factor of 0.91. It crystallized as a tetragonal perovskite at room temperature and undergo a phase transition to a cubic structure at elevated temperatures ($54 - 57$ °C). At room temperature, dipolar MA^+ cations reorient randomly with a residence time of $\sim$ 14 ps, which stabilized the energetic charge carriers by forming large polarons, leading to long carrier lifetimes [35]. However, the dynamic disorder of the A cation highlights the relatively weak bond between the A cation and PbX_6 octahedra. Consequently, the degradation of $MAPbI_3$ under illumination or at elevated temperatures is typically triggered by the volatilization of the MA cation.

The high PCE of $MAPbI_3$ PSC is limited by its relatively large bandgap and low carrier mobility. After a few years study, $FAPbI_3$ with a lower bandgap of 1.48 eV is proposed as an alternative to overcome the drawbacks of the $MAPbI_3$ perovskite. $FAPbI_3$ crystallizes into a non-perovskite hexagonal yellow phase at room temperature with a bandgap of 2.43 eV [36], and undergoes a structural transition into a pure cubic black phase at about 150 °C, and the cubic phase is maintained even after cooling down to the room temperature. The cubic black phase is found to be metastable at room temperature with a transition activation energy barrier of 0.6 eV, and moisture could significantly accelerate the reverse-phase conversion [37]. Solid-state Nuclear Magnetic Resonance spectroscopy (NMR) measurements show that the reorientation rate of the FA cation is much higher (8.7 ± 0.5 ps) than that of the MA cation (108 ± 18 ps), suggesting a superior charge-carrier stabilization capability. However, the low phase stability of the FAPbI$_3$ perovskite hinders further PCE improvement based on pure FAPbI$_3$ film, with a maximum PCE of ~ 19%.

As a replacement for the volatile A-site organic cations, pure inorganic CsPbI$_3$ with a bandgap of 1.73 eV has been proposed to overcome the stability issue [38]. Cs cation forms a strong chemical bonding with the perovskite lattice, which results in a high thermal stability with the thermal degradation activation energy of 650 ± 90 kJ/mol. However, the small ionic radius of Cs, around 167 pm, leads to a tolerance factor of 0.81, suffering from a low thermodynamic stability at room temperature. A non-perovskite orthorhombic phase with a wide bandgap of 2.82 eV has been found to co-exist in the solution-processed CsPbI$_3$ film. Hydroiodic acid (HI) has been added to improve the phase stability. Besides that, reduced crystal size, such as quantum dot, and I/Br alloy are also helpful in phase stability.

Mixing different A-site cations to tune the tolerance factor has been proved as an effective way to improve the crystallization of perovskite, photo, and thermal stabilities [39,40]. Incorporation of MA (< 50 mol %) into FAPbI$_3$ enables the formation of the cubic perovskite phase at room temperature and significantly enhances the photoluminescence (PL) lifetime of the perovskite film, resulting in an improved photovoltaic performance. However, it is found that the partially replaced FA$_{0.5}$MA$_{0.5}$PbI$_3$ film suffered from much-accelerated degradation compared to that of the pure FAPbI$_3$. This problem has been overcome by the incorporation of relatively small Cs cations. For example, cubic FA$_{0.9}$Cs$_{0.1}$PbI$_3$ phase can be readily formed at low temperature with enhanced stability against moisture and illumination. Inspired by the mixing strategy, a complex system by alloying more kinds of cations, such as Rb$_x$Cs$_y$MA$_z$FA$_{1-x-y-z}$, has been developed to enhance the performances and stabilities of the devices [41]. Furthermore, alloying diverse organic cations like guanidinium (GA) and dimethylammonium (DMA) also exhibited improved stability due to the reduced formation enthalpy (ΔH).

B site(cation)-modified hybrid Perovskites. Most reported high-efficiency solar cells have incorporated Pb-based perovskite materials, which have toxicity issues associated with Pb. Therefore, Sb-based perovskites have been proposed to surpass this problem [42]. Sn has the same valence-shell electron configuration as Pb, however, the ionic radius of Sn^{2+} (69 pm) is much smaller than that of Pb^{2+} (119 pm), which leads to a different band structure and bandgap. For example, the bandgaps of CsSnI$_3$, MASnI$_3$, and FASnI$_3$ were measured to be 1.3 − 1.4 eV. It seems that Sn-based perovskite materials have more favorable bandgaps for single-junction solar cells, however, the reported PCE of Sn-based perovskite (< 12%) is much lower than that of the Pb-based counterpart. The low performance of Sn-based perovskite is attributed to the instability of Sn^{2+} due to its low redox potential, resulting in a high defect density and undesired p-type doping during device fabrication. Partial replacement of Sn^{2+} with Pb^{2+} could effectively suppress the oxidation of Sn^{2+}, but further lower the bandgap down to around 1.25 eV. Recently, a high PCE of 21.1% has been achieved based on a $MA_{0.3}FA_{0.7}Pb_{0.5}Sn_{0.5}I_3$ perovskite by adding metallic Sn into the precursor solution to suppress the formation of Sn^{4+}.

X site (anion)-modified perovskites. Because X-site anion directly contributed to the band edge of the perovskite material, the substitution of X-site anion could significantly alter the electronic properties. The octahedral factor $\mu = r_B / r_X$, where r_B and r_X are the ionic radii of the B-site cation and X-site anion, respectively. Stable octahedral could form when $\mu > 0.41$. With the ionic radius of Pb^{2+} (119 pm), the calculated octahedral factors of $APbCl_3$, $APbBr_3$, and $APbI_3$ are 0.66, 0.61, and 0.54, respectively. As the ionic radius of the halide anion decreases, the bond length decreases accordingly, which leads to an increased wavefunction overlap, therefore an enlarged bandgap is expected [43]. For example, the measured bandgap of $MAPbX_3$ increases from 1.58 eV ($MAPbI_3$) to 2.28 eV ($MAPbBr_3$) to 2.88 eV ($MAPbCl_3$) for various halide anions. In principle, the photovoltaic application requires a lower bandgap, and $MAPbI_3$ seems to be the best choice. However, it has the least structural stability, the partial replacement of I^- with Br^- or Cl^- was, therefore, employed to fine-tune either the optoelectronic property or the chemical stability.

It is found that partially changing the I^- with Br^- could increase the average bond length, which increases the bandgap accordingly. Note that only a small amount of Br can enhance the PCEs owing to the improved fill factor (FF) and V_{oc}. Further increase in Br will decrease the short-circuit current density (J_{sc}) and consequently, the ultimate PCE of the device. A better tolerance factor due to the proper I^-/Br mixing could increase the crystalline of perovskite film, which in turn promotes the charge-carrier transport properties and leads to low series resistance and high FF. Despite the benefits of Br incorporation, the phase homogeneity of the mixed halide perovskite was found to be degraded under light illumination. An alternative way other than I^-/Br mixing is the incorporation of Cl^-, however, measurement shows that only traced amount of Cl residual could be detected after $MAPbI_3$ growth. More interesting thing is that the diffusion length of electrons and holes in Cl-based perovskite is nearly 10 times that of the pure $MAPbI_3$. The roles of the Cl^- ions in iodide-based perovskite films, therefore, have been intensively investigated. It is found that the addition of Cl^- influence the crystallization of the perovskite films. The Cl^- ion in the solution forms intermediate $MAPbI_{3-x}Cl_x$ phases, which are finally converted to $MAPbI_3$ via the volatilization of MACl from the film. The formation of the intermediate phase promotes the growth of large grains with relatively high crystallinity, thereby enhancing the optoelectronic quality of the film.

13.4.3 PROCESS ENGINEERING

The performance of perovskite solar cell is highly dependent on crystallinity and morphology, which are affected heavily by the solution fabrication process. During the solution process, the crystallization of the perovskite film is completed within a few seconds, limiting the effective control over the crystallinity and orientation of the film. In general, the perovskite precursor undergoes a nucleation and growth step to form the crystalline film, which could be effectively modulated by factors such as contact angle, heating, and so on. According to the classical theory, the rate of homogeneous nucleation is described as:

$$\dot{N} = A \cdot \exp\left(-\frac{\Delta G_c}{k_B T}\right)$$

Where A is the preexponential factor, k_B is the Boltzmann constant, and T is the temperature. The crystal-free energy (ΔG_c) can be written as a function of the surface energy γ, molar volume v, and supersaturation of solution S, which yields:

$$\dot{N} = A \cdot \exp\left(-\frac{16\pi\gamma^3 v^2}{3k_B^3 T^3 \left(lnS\right)^2}\right)$$

When a substrate is used as an electron transportation layer, the interface between the substrate and precursor solution could act as nucleation sites, leading to the heterogeneous nucleation, whose critical free energy can be described by introducing a correction term:

$$\Delta G_c^{hetero} = \phi \Delta G_c^{homo}$$

$$\phi = \frac{\left(2+\cos\theta\right)\left(1-\cos\theta\right)^2}{4}$$

where θ is the contact angle of the solution on the substrate. It can be found that the nucleation kinetics can be significantly tuned by the substrate, for example, θ is usually kept at $15° - 20°$ for commonly used SnO_2. The growth stage can be described by a surface reaction controlled or diffusion-controlled process.

One-step process and two-step process. The fabrication of perovskite films can be classified into an one-step or two-step process. For the one-step process, the nucleation and growth stages were completed in a solution that contained all precursor components. While for the two-step process, the perovskite film formed through the step-by-step deposition of each precursor layer, followed by a thermal inter-diffusion process [44].

At an early stage, the conventional one-step process is challenging to achieve high efficiency due to a lack of control of nucleation and growth. For example, for the well-used Dimethylformamide (DMF) solvent system, irregular rods, or needle-like structures with poor coverages on the substrate have been frequently observed [45]. High boiling point and low vapor pressure of DMF severely slow down the nucleation process, leading to a non-uniform grain structure. Solvent engineering, substrate heating, gas blowing, and vacuum treatment have been proposed to quickly remove the solvent during the one-step strategy.

In the solvent engineering process, antisolvent is added to quickly extract solvent from the cast film, which triggers uniform nucleation [46]. Thus far, many antisolvents, including chlorobenzene, toluene, hexane, ethyl acetate, and diethyl ether have been explored. It is found that the antisolvent method is an efficient way to fabricate high-performance hybrid perovskite solar cells, such as $[CsPbI_3]_{0.05}[(FAPbI_3)_{0.85}(MAPbBr_3)_{0.15}]_{0.95}$. Modification of the solvent and the addition of additives have been proved to be an effective way to increase grain size and reduce the defect sites. Compared with the antisolvent method, a simpler way to quickly remove solvent is to apply rapid heating. The heating not only allows for evaporation of the solvent to facilitate nucleation, but also enhances the diffusivity of ions, which yields millimeter-scale large grains. An efficiency of around 20% has been obtained by a hot blade coating process. A gas-blowing or gas-quenching method, generally using inert gas like N_2 or Ar, can also be used to form highly uniform perovskite thin films consisting of densely packed single-crystalline grains. Another effective way to remove the solvent is through vacuum flash-assisted solution processing, which could yield smooth perovskite films of high electronic quality over large areas, the dewetting issue has been effectively suppressed in the vacuum process, leading to the pinhole free homogeneous perovskite films.

The two-step process, also known as the sequential deposition method, involves (i) the deposition of Pb halide or Pb-based adduct onto a substrate, (ii) exposure to a liquid, vapor, or solid of organic salt, and (iii) formation of perovskite film by heat-assisted diffusion reaction. This process allows for better control of the reaction between Pb halide and organic salt, which induces the crystallization of perovskite materials. The detailed deposition process, such as the speed of spin coating, the time socking in the methylammonium iodide (MAI) solution, the substrate heating, could have a strong influence on the grain size and converge of perovskite film that are critical to its performance. Mixed halide materials, such as FA- and MA-based perovskite materials, could also be synthesized using the two-step process, yielding an efficiency as high as 14.9% with superior light-harvesting and charge-collection performance. It is also found that the first intramolecular exchange process, in which the $PbI_2 \cdot DMSO$ film was first deposited, flowed by the deposition of $(CH_5IN_2)FAI$ and post-annealing could enable a high PCE of 20.1%.

Adduct intermediated process. The standard deviation of the average PCE is fairly large, and the intermediate phase during the fabrication process should account for the poor reproducibility of high-efficiency PSCs [47]. Adduct compounds, such as $MAI \cdot PbI_2 \cdot DMSO$ and $MA_2Pb_3I_8 \cdot 2(DMSO)$, have been employed to grow dense uniform perovskite films that is vital for high-efficiency solar cells. The adduct intermediated process yielded a high reproducibility with a small standard deviation.

13.4.4 DEVICE PHYSICS

In terms of the device geometry, PSCs can be classified into mesoscopic type (with the mesoporous nanoparticle layer) and planar type (without the mesoporous nanoparticle layer). Mesoscopic devices utilize high temperature sintered TiO_2 mesoporous layer to extract photo-generated electrons, while planar devices use perovskite itself to dissociate photo carriers, and achieve high efficiency at low temperature that is preferred by flexible solar cells. Besides that, PSCs can also be classified into *p-i-n* or *n-i-p* type based on the direction of charge collection, the device in which electrons are collected at the substrate side is called *n-i-p* type, while the holes are collected at the substrate is named *p-i-n* type.

Electron/hole selective contact. TiO_2 is the first used electron transportation layer as it offers advantages of wide bandgap, appropriate conduction band alignment, and good electron-transport capability, resulting in a PCE > 23%. However, UV-induced photocatalytic activity that causes degradation of device, capacitance issue derived from a high dielectric constant that leads to large hysteresis, and high-temperature sintering show that TiO_2 does present several disadvantages as ETLs. SnO_2 has been proposed as an alternative because of its appropriate band alignment, high electron mobility, wide bandgap, high transmittance, good stability, and easy processing [48]. Furthermore, SnO_2 requires only a much lower temperature less than 200 °C, compared with a high temperature more than 400 °C required for TiO_2, to achieve the same efficiency > 23% without hysteresis. Organic ETLs, such as PCBM, are favorable to flexible PSCs due to their appropriate electrical properties and physical softness. High FF values and negligible current-voltage (I–V) hysteresis have been obtained due to their high electron mobilities and excellent electron-extracting capabilities. However, one drawback of organic fullerene-based ETLs is that they are incompatible with the Ag electrode, and extra interfacial layers are needed to mitigate those constraints.

Organic Spiro-OMeTAD, commonly used in solid-state DSSCs, is the first and most used HTLs in *n-i-p* type devices [49], however, dopants and/or additives such as lithium bis(trifluoromethane) sulfonimide (Li-TFSI) and 4-tert-butylpyridine (tBP) are necessary due to low conductivity ($\sim 10^{-5}$ mS/cm) and low hole mobility ($10^{-5} - 10^{-4}$ cm^2/(V·s)) of pristine Spiro-OMeTAD. Unfortunately, the incorporation of Li-TFSI additives causes serious instability due to their hygroscopic nature. For *p-i-n* devices, PEDOT: PSS has been widely used as HTLs because of its appropriate highest occupied molecular orbital (HOMO) energy level (−5.5 eV), good conductivity, and low-temperature processing, but it also presents highly stability issues due to its hygroscopic and acidic nature. Another alternative is polymeric HTL based on poly [bis (4-phenyl) (2,4,6-trimethylphenyl) amine] (PTAA), which has higher intrinsic hole mobility ($10^{-3} - 10^{-2}$ cm^2/(V·s)) and enhanced stability and could be used in both *n-i-p* or *p-i-n* type cells. High-efficiency cells with poly(3-hexylthiophene) (P3HT) as HTLs without doping have been developed, showing PCEs as high as 22.7% with suppressed hysteresis.

Compared with organic HTLs, the inorganic counterparts have attracted much attention due to their stability. Various inorganic *p*-type semiconducting materials, such as NiO, copper thiocyanate (CuSCN), and copper oxide (CuO_x), have been used as HTLs. The deep valence band, high carrier mobility, good thermal stability, and high transparency make the inorganic HTLs suitable for long stability cells. However, poor physical contact with the perovskite layer has been observed, thus an interface layer like PEDOT: PSS or reduced graphene oxide (rGO) has been adopted to increase the contact.

Interface engineering. A typical PSC consists of four major interfaces, that is electrode/ETL/perovskite/HTL/electrode, whose properties are very important because they directly influence charge extraction/transport/recombination as well as photon transmission. The work function of metal electrode controls the charge transfer at the interface of the charge-transporting layer. For example, low work function (*e.g.*, Al, 4.1 eV) less than the CB (or lowest unoccupied molecular orbital (LUMO) level) of the ETL (below CB of perovskite, −3.9 eV) is required for collection, while a relatively high work function (*e.g.*, Ag, 4.2 eV − 4.7 eV or Au, 5.3 eV) greater than the VB (or HOMO level) of the HTLs (above VB of perovskite, −5.4 eV) is necessary. Surface modification by ultra-thin polymer layer, such as 1 nm–10 nm polyethylenimine ethoxylated (PEIE), can efficiently lower the work function, leading to a significant increase in the FF. A thin layer of bathocuproine (BCP) between PCBM and Ag also shows increased FF for *p-i-n* devices despite the working mechanism has not been fully elucidated.

The interface between perovskite and charge-transporting layers has been mostly investigated. Most of the studies have focused on the oxide-based ETLs/HTLs because the heterogeneous interface generally exhibits poorer physical/electrical contact properties. Implementation of organic layers, such as PEDOT: PSS, triblock fullerene derivative (PCBB-2CN-2C8), graphene quantum dots, or ionic liquid, have been used to enhance electrical contact, leading to increased V_{oc} and FF values along with reduced $I-V$ hysteresis. Atomic doping, such as Y-doped TiO_2, Cl-capped TiO_2, Sb_2S_3 deposited TiO_2, showed an improved J_{sc} and conversion efficiency along with long-term stability.

Defect passivation. Various types of defects including vacancies, interstitials, GBs, interfaces, impurities, and atomic clustering/segregation exist in the solution processed polycrystalline perovskite film, which directly influences the device performance of solar cells, especially the V_{oc} and FF, because the defects generally act as nonradioactive recombination canters. Bulk defects, such as cation/anion vacancies, interstitials, and substituted ions, generally induce shallow energy levels within the bandgap. The addition of potassium ions into $MAPbI_3$ shows reduced $I-V$ hysteresis due to the reduced Frenkel defects. Rb ions are often added to stabilize α-$FAPbI_3$, although the performance of cells was barely improved, the long-term stability has been found greatly enhanced because of the reduced number of vacancies. Furthermore, a divalent cation, methylene diammonium ($^+H_3N-CH_2-NH_3^+$; MDA), has been added to the pure α-FAPbI3 phase to increase thermal stability and improve efficiency. Besides the bulk defects, grain-boundary/surface defects could be efficiently passivated by alkylammonium halides. An excess of MAI has been observed to accumulate between GBs and form highly ionic conducting pathways. Alkylammonium halides with long alkyl chains such as ethylammonium (EA⁺), n-butylammonium (BA⁺), isobutyl ammonium (iso-BA⁺), and guanidinium (GA⁺) have also been used to passivate the interface/surface, most of which improve the device performances and carrier lifetimes by forming two-dimensional (2D) perovskite layers. Recently, phenethylammonium iodide (PEAI), a novel kind of alkylammonium halides bearing benzene moieties, has been shown to exhibit good passivation properties because of the synergetic effect of the π-conjugated structure of the benzene ring. However, its amount is limited due to the insulating properties. Another method to passivate the GBs is cross-linking grains with functional organic molecules, for instance, butylphosphonic acid 4-ammonium chloride (4-ABPACl) form hydrogen bonds (O−H⋯I and N−H⋯I) within the GBs, trimethylolpropane triacrylate (TMTA) greatly improved the long term stability by forming robust continuous polymeric network after thermal annealing, zwitterion is an alternative powerful passivation material as it possesses both positively and negatively charged functional groups.

13.4.5 MORE BEYOND SHOCKLEY-QUEISSER LIMIT

Shockley–Queisser limit sets the maximum efficiency that a single junction solar cell can achieve by assuming that the only loss mechanism is the radiative recombination. Tandem solar cells, hot carrier devices, photo recycling, and concentrator solar cells have been proposed to overcome that limit. For the tandem solar cells, such as a double-junction solar cell, a top cell with a higher bandgap is fabricated onto a bottom cell with a lower bandgap, where high energy photons are absorbed by the top cell and low energy photons are absorbed by the bottom ones. The thermalization loss of the photons' excess energies thus could be minimized, and the highest efficiency of perovskite/silicon tandem solar cells has reached 29.1% recently. Bandgap engineering and performance accurate measurement are the two most important ways to further improve the efficiency of tandem perovskite solar cell.

In addition to the tandem strategy, an alternative way to minimize energy loss from the thermalization of photoexcited charge carriers is to make use of hot carriers, whose excess energies are extracted to energy-selective contacts (ESCs) with narrow DOS before relaxation. It is widely accepted that the halide perovskite has a relatively longer carrier cooling lifetime (>100 ps) than that of the traditional semiconductors (typically < 1 ps), albeit the exact mechanism is still under debate. Several different mechanisms such as the hot-phonon bottleneck, acoustical-optical phonon up-conversion, auger heating, and large polarons have been suggested but they don't reach a conclusive consensus. Although the perovskite materials are promising in hot carrier solar cells, it is rather challenging to develop ESC materials with suitable energy states. Therefore, intensive research efforts are still required to demonstrate efficient hot carrier PSCs.

In principle, the coexist of long charge carrier diffusion paths and the high density of crystalline disorders indicates the possible existence of an active photon recycling process, which can prolong the charge carrier lifetimes beyond their diffusion lengths where energy transport occurs via multiple cycles of charge recombination and reabsorption, and eventually, an increased V_{oc} is thus expected. However, contradictory results have still been reported and further studies may be required to confirm the photon recycling effect.

Lastly, concentrated solar cell strategy is also utilized because the fabrication cost of the active material is greater than that of a concentrator mirror. The attempt to implement hybrid perovskite solar cells under concentrated illumination, *e.g.*, a planar PSC based on $FA_{0.83}Cs_{0.17}Pb_{2.7}Br_{0.3}$, exhibited an increased efficiency from 21.1% under 1 sun to 22.9% under 31 suns. Further increasing the illumination to 128 suns deteriorates the FF thus suppressing the performance improvement.

13.5 HYBRID PEROVSKITE LIGHT EMITTING DIODE

From the fundamental perspective, perovskite-based light-emitting diodes (PeLEDs) and solar cells are the two mirror sides of the optoelectronic process, and they commonly share a similar device structure, that is, anode/HTL/perovskite/ETL/cathode,

(a)

(b)

Perovskite film quality
- Film morphology
- Trap states
- Thermal instability
- Ambient instability
- Ion migration

Surface quality
- Film morphology
- Surface defects

Charge transport layers
- Imbalanced charge injection
- Ambient instability

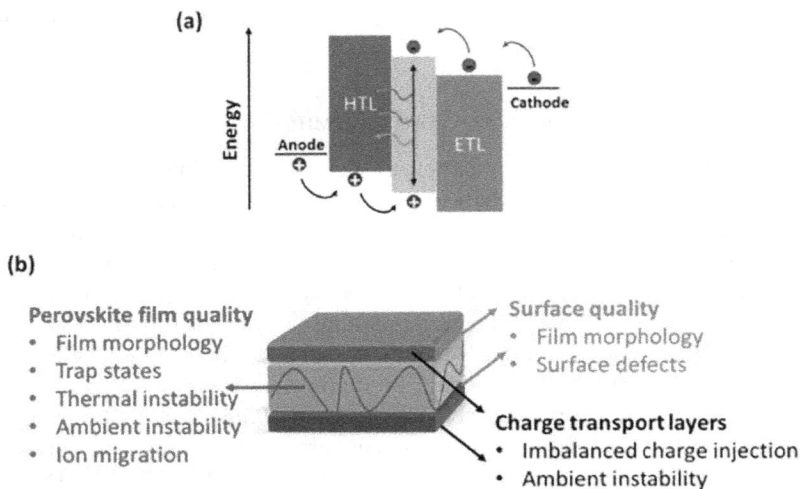

FIGURE 13.6 Perovskite light-emitting diodes. (a) The schematic illustrates a PeLED structure and work mechanism. (b) The main factors that affect the performance, including performance film, surface, and charge transport layers.

Source: Reprinted with permission from Y. Zou, et. al., Materials Today Nano 5, 100028 (2019). Copyright 2019 Elsevier.

as shown in Figure 13.6. The difference between them is that the electron and hole recombined in perovskite for the diode, generating light emission, whereas the electron and hole generated in perovskite for the solar cell, generating electricity instead. Compared with solar cells which are mainly based on 3D perovskites, the low-dimensional counterparts, including 0D perovskite nanocrystals and 2D Ruddlesden-Popper (RP) perovskites, exhibit much better performance in the applications of PeLEDs. The peak external quantum efficiency (EQE) of red, green, and blue PeLEDs have reached 21.3%, 20.31% and 12.3%, respectively, which is compared to that of the quantum dot light emitting diode (QLEDs) [50].

13.5.1 LUMINESCENCE MECHANISM AND UNDERLYING PHOTOPHYSICS

The charge carrier recombination dynamics of 3D perovskites can be quantitatively described by the rate equation $\frac{dn}{dt} = G + k_1 n + k_2 n^2 + k_3 n^3$, where n is the carrier density, G is the generation rate, and k_1, k_2, and k_3 are the coefficients of the first-order monomolecular recombination, second-order bimolecular recombination, and third-order Auger recombination respectively. Monomolecular recombination is the direct consequence of the charge trapping in defect states and dominates over the low carrier density range $< 10^{15}$ cm^{-3} with a typical k_1 of $\sim 10^7$ s^{-1}. Radiative bimolecular recombination generally occurs at an intermediate density regium of 10^{15}–10^{17} cm^{-3} with k_2 rates of $\sim 10^{-10}$ cm^3·s^{-1}, which is usually attributed to the non-Langevin direct band-to-band recombination in the continuum states of the conduction and valence bands. Finally, Auger recombination which takes place at a very high carrier density $> 10^{17}$ cm^{-3} with typical k_3 rates of the order of 10^{-28} cm$^6 \cdot$s^{-1}. The photoluminescence quantum yield (PLQY) is defined as the ratio of the radiative part to total recombination rates, and it is formulated as $\eta_{radiative} = \dfrac{nk_2}{k_1 + nk_2 + n^2 k_3}$, where $\eta_{radiative}$ increases as moving from monomolecular to bimolecular regimes and decreases when Auger recombination takes place.

There are two approaches to simultaneously improve the PLQY and suppress the PL broadening. The first approach is the quantum spatial confinement through structure dimension reduction, which enables the luminescence emission from both excitons and free carriers. Quasi-2D, 1D, and 0D systems have been investigated for light emission, and highly efficient exciton self-trapping has shown below-gap broadband luminescence. The second method is the geometrical reduction of the emitter to obtain tuneable, color-pure emission. Structures with precise tuneable diameters, such as nanocrystals, quantum dots, nanowire arrays, have been fabricated and offer excellent luminous efficiencies.

For light emission application, external quantum efficiency (EQE) is the key parameter and expressed as EQE $= f_{balance} \times f_{e-h} \times \eta_{radiative} \times \eta_{out}$, where $f_{balance}$ stands for the probability of charge injection balance, f_{e-h} is the probability of exciton formation per injected carrier, $\eta_{radiative}$ donates PLQY of emitter material and η_{out} is the light-out coupling efficiency. Extensive efforts have been devoted to improving the PLQY of perovskite film, such as morphology control and composition engineering, near-unit internal PLQY in perovskites has been achieved. Further enhancing the EQE requires the increase of $f_{balance}$ and f_{e-h}, which is in general, based on the optimization of electron or hole injection layers. The internal quantum efficiency (IQE), defined as the product of $f_{balance}$, f_{e-h} and $\eta_{radiative}$, almost reaches the limit, and the light out-coupling management emerges as the effective way to further improve the EQE. Appropriate material engineering techniques can be implemented to boost η_{out} by manipulating the transient dipole moments (TDM) of the materials. A new figure of merit, the alignment constant (ζ), defined by considering the positions and orientations of the emissive TDMs as well as their refractive indices, has been proposed and formulated as $\zeta = \dfrac{sin^2 \varphi_{TDM}}{n_{EML}^4 - sin^2 \varphi_{TDM} \left(n_{EML}^4 - 1 \right)}$, where n_{EML} is the refractive index of the emitting perovskite layer and φ_{TDM} is the angle of the emissive TDM to the film surface. ζ thus has a value between 0 and 1, corresponding to completely horizontal and vertical TDMs respectively. Optimization of these parameters can further improve η_{out} and device performance.

13.5.2 Light Management

By taking photon recycling (PR) into account and assume EQE has a value close to unity, thus we have $EQE_{PR,max} = LEE_0 / \left(LEE_0 + A_{para} \right)$, where LEE_0 is the direct light outcoupling and A_{para} is the parasitic absorption loss. Light can be trapped in the active layer because of the waveguide effect induced by the large refractive index (RI) contrast between the perovskite and the surrounding. It is proposed that the optimal device structure for high η_{out} should, in general, employ the following architecture: transparent electrode/high RI transport layer/perovskite/low RI transport layer/reflective electrode. In addition, optical nanostructure has been demonstrated to further improve the η_{out}. Simulation results show that the patterned LEDs can achieve η_{out} values that are 1.75 times larger than their planar counterparts. Apart from this, resonant nanostructures, such as Au nanospheres and Ag nanorods, have been incorporated into the emitter layer, transport layer, or emitter–transport interface to maximize efficiency.

External out coupler is another way to extract the light in perovskite optoelectronic devices. For example, moth-eye nanostructures and half-ball lens can greatly increase the EQEs and current efficiencies. Besides that, device stacks fitted with 1D nanostructures, such as metallic nanogrids, angle resolving microstructures, and distributed Bragg reflectors, can serve as polarizers, directional emitters, and narrow emissions, respectively.

13.6   STABLE ISSUE OF HYBRID PEROVSKITE

Moisture, thermal and photo instability of perovskite has become the most critical issue limiting the commercialization of PSC devices. Besides that, the degradation of HTLs and ETLs could be detrimental to the overall device performance. In view of this, encapsulation, interfacial modification, perovskite composition engineering, inorganic HTLs, and novel device configurations have been developed to improve long-term stability. However, the stability is still far away from the practical requirement, that is, less than a 10% drop in solar cells performance and a warranty of 20–25 years [34].

The main factors that induce the degradation of perovskite are summarized in Figure 13.7. Water/moisture is considered to be one of the most dominant factors that degrade and destabilize PSCs. The hybrid perovskite like $MAPbI_3$ is highly polarity, which could decompose into PbI_2, CH_3NH_2I, and HI through the water-assisted chemical processes. As a result, the color of $MAPbI_3$ becomes yellow (an indication of PbI_2 formation) and loses power conversion capability rapidly. It is found that the water molecule act as a Lewis base and accept one H^+ from the ammonium, leading to its degradation via intermediate steps. Intermediates like HI and solid PbI_2 are soluble in water, together with the evaporation of organic CH_3NH_2, which, in turn, facilitates the decomposition of perovskite. $MAPbI_3$ shows greater moisture sensitivity than those of $MAPbBr_3$ and $MAPbCl_3$. Mixed I/Br with the composition of $MAPbI_{3-x}Br_x$ exhibits improved stability due to its smaller atom size and more negative charges.

Heating the hybrid perovskite induces phase transformation from low temperature distorted orthorhombic state to the intermediate temperature tetragonal phase, and then, the high temperature cubic phase. Further increase the temperature will lead to decomposition of perovskite into CH_3NH_2 and HI. The possible temperature-induced decomposition path can be written as $CH_3HN_3PbI_3 \rightarrow PbI_2 + CH_3HN_2 \uparrow + HI \uparrow$. It

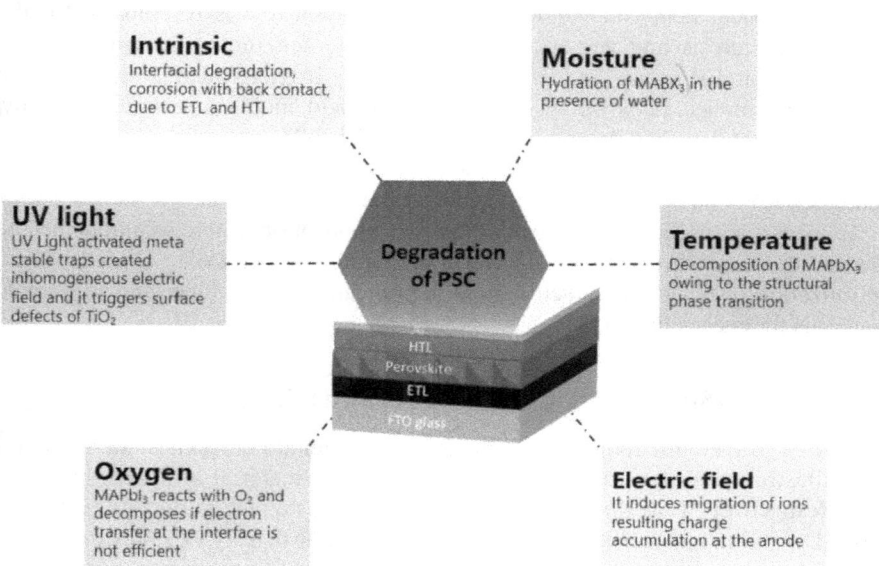

FIGURE 13.7 Summary of perovskite solar cells' degradation by various factors.

Source: Reprinted with permission from Z. H. Bakr et. al., Nano Energy 34, 271–305 (2017). Copyright 2017 Elsevier.

is found that replacing the organic MA with the inorganic Cs could greatly improve the thermal stability, with a decomposition temperature up to 180 °C.

Upon ultraviolet exposure, holes will be generated in compact-TiO_2 (c-TiO_2) or mesoporous-TiO_2 (mp-TiO_2), which reacts with the oxygen adsorbed at oxygen surface vacancies, thereby acting as deep traps and leading to recombination. It shows that the decomposition of $MAPbI_3$ undergoes the following reaction steps:

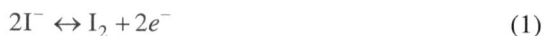

$$2I^- \leftrightarrow I_2 + 2e^- \tag{1}$$

at TiO_2 and $MAPbI_3$ interface, and

$$3CH_3NH_3^+ \leftrightarrow 3CH_3NH_3 \uparrow + 3H^+ \tag{2}$$

$$I^- + I_2 + 3H^+ + 2e^- \leftrightarrow 3HI \uparrow \tag{3}$$

In reaction (1), TiO_2 first becomes excited and takes the electron from the halide ion (I^-), which leads to the decomposition of perovskite, results in the by-product I_2. As the perovskite layer becomes degraded, the equilibrium moves to the right side in reaction (2). Finally, the electron returns from TiO_2 to perovskite and reacts with I^- and H^+ to release HI. Replace TiO_2 with SnO_2 shows better stability since SnO_2 is less active.

The electrical field could induce the ion migration that is responsible for the observed hysteresis and non-steady current output. Ion migration is possible to change the band structure and thus alter the energy barrier which could affect the device performance. In the absence of an electrical field, mobile ions move randomly, and on the contrary, they will accumulate at the HTL/ETL interface when an electrical field is established during the operation of device, which is opposite to the photo-generated field and decreases the efficiency of the device.

To meet the stability requirement of industry application, future research must be focused upon improvements on the intrinsic stability of perovskite absorber layer, optimization of device architecture, and identification and engineering of durable materials for encapsulation.

13.7 DIMENSION REDUCTION IN HYBRID PEROVSKITE

Compared to the bulk counterparts, the low dimensional perovskites are schematically illustrated in Figures 13.8(a) and 13.8(b) for the material level and structural level, respectively. Low dimensional structures, such as quantum dots (0D), nanowires (1D), and nanoplatelets (2D), have shown peculiar density of states due to the

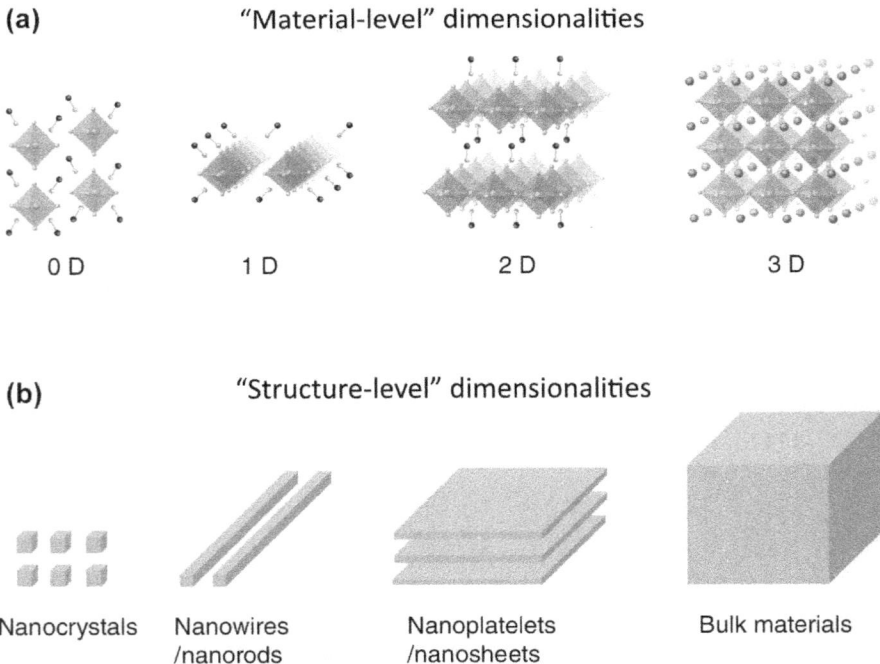

(a)   "Material-level" dimensionalities

0 D   1 D   2 D   3 D

(b)   "Structure-level" dimensionalities

Nanocrystals   Nanowires   Nanoplatelets   Bulk materials
                /nanorods   /nanosheets

FIGURE 13.8   Low-dimension perovskite at the atomic level (a) and the structure level (b).

Source: Adopted from P. Zhu et al., InfoMat 2, 341–378 (2020). No permission is required. Copyright 2020 the Authors.

quantum size effects [51]. In addition, the bandgap of low dimensional perovskite is larger than that of its bulk counterpart, and generally has a higher luminescence. Furthermore, low dimensional perovskite materials are synthesized in single-crystalline states, which exhibits enhanced photo-generated carrier transport properties and stabilities because of reduced defects and grain boundaries. Owing to the anisotropic geometry and large surface-to-volume ratio, the charge transport in low dimensional perovskite could be readily tuned by surface states and environmental factors.

Quasi-2D halide perovskites, with the chemical formula $R(NH_3)_2A_{n-1}B_nX_{3n+1}$, have been synthesized by solution-phase growth and chemical vapor deposition. Meanwhile, the 1D nanowire perovskite can be obtained by solution-phase growth through the self-template-directing and surfactant-directed mechanism. Besides that, vapor-growth like chemical vapor deposition (CVD) has been employed to grow ultra-long perovskite nanowire with controlled orientation. For the 0D nanocrystals, ligand-assisted precipitation, hot injection growth, and template growth are commonly adopted synthesis methods.

Low dimensional perovskite structures have found applications in the fields of solar cells, light-emitting diodes, photodetectors, image sensors, lasers, and memory devices. However, there still exist several disadvantages compared to their bulk counterparts. For example, for the solar cells based on 2D nanosheets, although they have better stability, the efficiencies are much lower. For LEDs, the EQEs of the devices fabricated from 2D or 0D halide perovskites are still much lower than the bulk structures. In the future, many efforts are needed to design novel device structures to fully exploit the unique features of lower-dimensional hybrid perovskite.

13.8 INDUSTRY APPLICATION AND FUTURE OF HYBRID PEROVSKITE

A few companies have started the industry commercialization of perovskite solar cells or the tandem solar cells containing hybrid perovskites. They include Oxford Photovoltaics in UK; CubicPV in US; EneCoat in Japan; Frontier Energy Solution in Korea; Greatcell Energy and Greatcell Solar in Australia; Microquanta Semiconductor in China; P3C in India and Saule Technologies in Poland. Figure 13.9 shows that a large-scale perovskite solar cell with efficiency > 19% can be readily manufactured.

Three key factors that determine the industry application of perovskite solar cells are cost, efficiency, and lifetime (Figure 13.10(a)). Currently, the photovoltaic (PV) market is dominated by silicon based solar cell with a typic PCE of 18% – 20%, whereas the highest reported PCE of hybrid perovskite solar cell has already reached 25.7% in 2021 with small acitve area. However, the reported device area is much smaller than that of the silicon cell. More detailed comparisons between perovskite and silicon PV technologies are shown in Figure 13.10(b). Besides that, the efficiency of the PV modules of silicon (> 10000 cm^2), CIGS (> 10000 cm^2) and CdTe (> 10000 cm^2) are 24.4%, 18.6%, and 19.0%, respectively. In comparison, the largest perovskite module is around 1000 cm^2 with an efficiency of 17.9%, exhibiting the largest cell-to-module efficiency gap which must be solved before practical

FIGURE 13.9   Industrial fabricated large sized perovskite solar cell.

Source: Adopted from https://www.businesswire.com/news/home/20200209005049/zh-CN/, No permission is required.

application. The Levelized cost of energy (LCOE) of perovskite has been calculated with lifetime. Figure 13.10(c) shows that the current perovskite PV has a relative high LCOE around 15 US cents kW^{-1} h^{-1} and a lifetime of about 5 years, which still lags far behind than the competing Si, CIGS and CdTe technologies. Nevertheless, there is little doubt that the LCOE of perovskite PV would become lower if the device lifetime could be extended to more than 25 years. An overall PV market prediction shows that in the year 2030, the silicon PV technology will decrease to 44.8% sharply, and the emergent PV technology such as perovskite maybe increase to more than 40% dramatically, as shown in Figure 13.10(d).

The real outdoor application of perovskite solar cells is seriously limited by their poor device stabilities. The primary factors that affect the device stability could be classified into extrinsic or intrinsic factors. Advanced encapsulation technology can be used to prevent extrinsic factors like moisture and oxygen in ambient air. On the other hand, the intrinsic stability factors, such as (1) the ion dissociation and migration, (2) metal–perovskite reactions, and (3) residual strain, could be solved by devices engineering and novel material synthesis. Several prosperous future research opportunities can be envisioned, such as (1) perovskite material composition engineering, (2) robust transparent conductive oxides development, (3) perovskite module development, and (4) development of a "solar energy harvesting and storage" system.

With the rapid progress made in the academic and industry fields of hybrid perovskite solar cells, we are optimistic that the stability and large-scale issues could

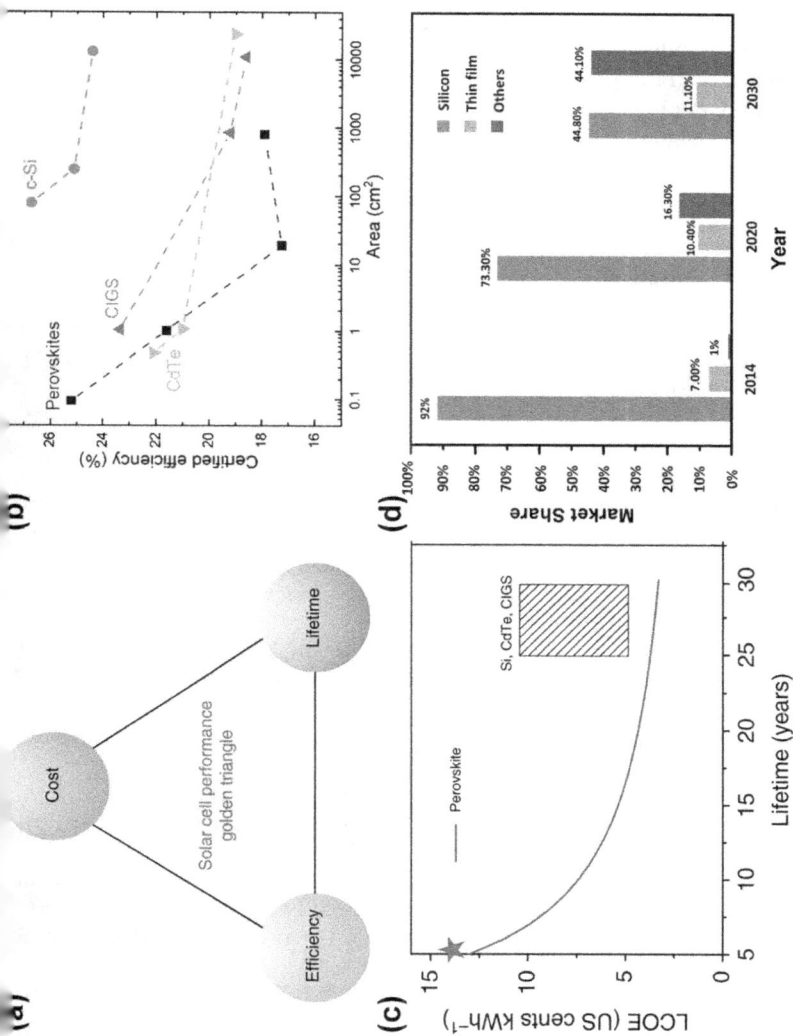

FIGURE 13.10 Industry application of perovskite solar cells. (a) Golden triangle of solar cell performance including cost, efficiency, and lifetime. (b) The detailed comparison between perovskite and silicon technologies. (c) LCOE of perovskite and commercial Si, CdTe and CIGS PV technologies as a function of lifetime, the red star shows a current estimation of perovskite solar cell. (d) Global PV market estimated before the year 2030.

Source: Reprinted with permission from Y. Cheng, et. al., Energy & Environmental Science 14, 3233–3255 (2021). Copyright 2021 Royal Society of Chemistry.

be overcome in the next few years, and the outdoor applications of large-scale PSCs, perovskite solar modules and perovskite-based tandem module will initiate soon. The wide implementation of hybrid perovskite based integrated system and PV panels would accelerate the world's transition from a fossil fuel economy to a renewable green energy economy.

Compared with the fast industrialization of hybrid perovskite based solar cells, the practical application of hybrid perovskite in LEDs is still in its infancy. The main issue is the device stability, which only works less than 50 hours under operation, far behind the commercial inorganic quantum dot based LEDs or the organic LEDs (up to a million hours). For hybrid perovskite LEDs, the high electrical field under operation severely induces ion migration and accelerates the structure degradation, charge accumulation and electrode corrosion, much research effects are required to solve the stability challenge before the industrial applications.

ACKNOWLEDGEMENTS

This work is supported by the Fundamental Research Funds for the Central Universities (2021CDJQY-022) and the Natural Science Foundation of Chongqing (No. cstc2021jcyj-msxmX0270)

BIBLIOGRAPHY

[1] Goldschmidt V. Crystal structure and chemical correlation. *Ber Dtsch Chem Ges* 1927, **60**: 1263–1296.
[2] Li C, Lu X, Ding W, Feng L, Gao Y, Guo Z. Formability of ABX3 (X= F, Cl, Br, I) Halide Perovskites. *Acta Crystallographica Section B: Structural Science* 2008, **64**(6): 702–707.
[3] Yin W-J, Shi T, Yan Y. Unique properties of halide perovskites as possible origins of the superior solar cell performance. *Advanced Materials* 2014, **26**(27): 4653–4658.
[4] Bercegol A, Ramos FJ, Rebai A, Guillemot T, Ory D, Rousset J, Lombez L. Slow diffusion and long lifetime in metal halide perovskites for photovoltaics. *The Journal of Physical Chemistry C* 2018, **122**(43): 24570–24577.
[5] D'Innocenzo V, Grancini G, Alcocer MJP, Kandada ARS, Stranks SD, Lee MM, Lanzani G, Snaith HJ, Petrozza A. Excitons versus free charges in organo-lead trihalide perovskites. *Nature Communications* 2014, **5**(1): 3586.
[6] Zheng F, Wang L-W. Large polaron formation and its effect on electron transport in hybrid perovskites. *Energy & Environmental Science* 2019, **12**(4): 1219–1230.
[7] Meggiolaro D, Ambrosio F, Mosconi E, Mahata A, De Angelis F. Polarons in metal halide perovskites. *Advanced Energy Materials* 2020, **10**(13): 1902748.
[8] Ghosh D, Welch E, Neukirch AJ, Zakhidov A, Tretiak S. Polarons in halide perovskites: A perspective. *The Journal of Physical Chemistry Letters* 2020, **11**(9): 3271–3286.
[9] Miyata K, Meggiolaro D, Trinh MT, Joshi PP, Mosconi E, Jones SC, De Angelis F, Zhu X-Y. Large polarons in lead halide perovskites. *Science Advances* 2017, **3**(8): e1701217.
[10] Walsh A, Payne DJ, Egdell RG, Watson GW. Stereochemistry of post-transition metal oxides: revision of the classical lone pair model. *Chemical Society Reviews* 2011, **40**(9): 4455–4463.
[11] Walsh A, Watson GW. The origin of the stereochemically active Pb(II) lone pair: DFT calculations on PbO and PbS. *Journal of Solid State Chemistry* 2005, **178**(5): 1422–1428.

[12] Wei S-H, Zunger A. Electronic and structural anomalies in lead chalcogenides. *Physical Review B* 1997, **55**(20): 13605–13610.

[13] Xue J, Wang R, Chen X, Yao C, Yang Y. Reconfiguring the band-edge states of photovoltaic perovskites by conjugated organic cations. *Science* 2021, **371**(6529): 636–640.

[14] Giorgi G, Fujisawa J-I, Segawa H, Yamashita K. Small photocarrier effective masses featuring ambipolar transport in methylammonium lead iodide perovskite: a density functional analysis. *The Journal of Physical Chemistry Letters* 2013, **4**(24): 4213–4216.

[15] Brivio F, Butler KT, Walsh A, van Schilfgaarde M. Relativistic quasiparticle self-consistent electronic structure of hybrid halide perovskite photovoltaic absorbers. *Physical Review B* 2014, **89**(15): 155204.

[16] Zhang Y, Seo S, Lim SY, Kim Y, Kim S-G, Lee D-K, Lee S-H, Shin H, Cheong H, Park N-G. Achieving reproducible and high-efficiency (> 21%) perovskite solar cells with a presynthesized FAPbI3 powder. *ACS Energy Letters* 2019, **5**(2): 360–366.

[17] Yin W-J, Yang J-H, Kang J, Yan Y, Wei S-H. Halide perovskite materials for solar cells: a theoretical review. *Journal of Materials Chemistry A* 2015, **3**(17): 8926–8942.

[18] Chen H, Yang S. Carbon-based perovskite solar cells without hole transport materials: the front runner to the market? *Advanced Materials* 2017, **29**(24): 1603994.

[19] Yin W-J, Shi T, Yan Y. Unusual defect physics in CH3NH3PbI3 perovskite solar cell absorber. *Applied Physics Letters* 2014, **104**(6): 063903.

[20] Edri E, Kirmayer S, Henning A, Mukhopadhyay S, Gartsman K, Rosenwaks Y, Hodes G, Cahen D. Why lead methylammonium tri-iodide perovskite-based solar cells require a mesoporous electron transporting scaffold (but not necessarily a hole conductor). *Nano Letters* 2014, **14**(2): 1000–1004.

[21] Frost JM, Butler KT, Brivio F, Hendon CH, van Schilfgaarde M, Walsh A. Atomistic origins of high-performance in hybrid halide perovskite solar cells. *Nano Letters* 2014, **14**(5): 2584–2590.

[22] Buizza LRV, Crothers TW, Wang Z, Patel JB, Milot RL, Snaith HJ, Johnston MB, Herz LM. Charge-carrier dynamics, mobilities, and diffusion lengths of 2D–3D hybrid butylammonium–cesium–formamidinium lead halide perovskites. *Advanced Functional Materials* 2019, **29**(35): 1902656.

[23] Niesner D, Wilhelm M, Levchuk I, Osvet A, Shrestha S, Batentschuk M, Brabec C, Fauster T. Giant rashba splitting in ${\mathrm{CH}}_{3}{\mathrm{NH}}_{3}{\mathrm{PbBr}}_{3}$ Organic-Inorganic Perovskite. *Physical Review Letters* 2016, **117**(12): 126401.

[24] Fan Z, Xiao J, Sun K, Chen L, Hu Y, Ouyang J, Ong KP, Zeng K, Wang J. Ferroelectricity of CH3NH3PbI3 perovskite. *The Journal of Physical Chemistry Letters* 2015, **6**(7): 1155–1161.

[25] Pazos-Outón LM, Szumilo M, Lamboll R, Richter JM, Crespo-Quesada M, Abdi-Jalebi M, Beeson HJ, Vrućinić M, Alsari M, Snaith HJ, Ehrler B, Friend RH, Deschler F. Photon recycling in lead iodide perovskite solar cells. *Science* 2016, **351**(6280): 1430–1433.

[26] Kang J, Wang L-W. Dynamic disorder and potential fluctuation in two-dimensional perovskite. *The Journal of Physical Chemistry Letters* 2017, **8**(16): 3875–3880.

[27] Kojima A, Teshima K, Shirai Y, Miyasaka T. Organometal halide perovskites as visible-light sensitizers for photovoltaic cells. *Journal of the American Chemical Society* 2009, **131**(17): 6050–6051.

[28] Kim H-S, Lee C-R, Im J-H, Lee K-B, Moehl T, Marchioro A, Moon S-J, Humphry-Baker R, Yum J-H, Moser JE. Lead iodide perovskite sensitized all-solid-state submicron thin film mesoscopic solar cell with efficiency exceeding 9%. *Scientific Reports* 2012, **2**(1): 1–7.

[29] Lee MM, Teuscher J, Miyasaka T, Murakami TN, Snaith HJ. Efficient hybrid solar cells based on meso-superstructured organometal halide perovskites. *Science* 2012, **338**(6107): 643–647.

[30] Wu T, Qin Z, Wang Y, Wu Y, Chen W, Zhang S, Cai M, Dai S, Zhang J, Liu J. The main progress of perovskite solar cells in 2020–2021. *Nano-Micro Letters* 2021, **13**(1): 1–18.

[31] Egger DA, Kronik L. Role of dispersive interactions in determining structural properties of organic–inorganic halide perovskites: insights from first-principles calculations. *The Journal of Physical Chemistry Letters* 2014, **5**(15): 2728–2733.

[32] Motti SG, Meggiolaro D, Martani S, Sorrentino R, Barker AJ, De Angelis F, Petrozza A. Defect activity in lead halide perovskites. *Advanced Materials* 2019, **31**(47): 1901183.

[33] Sherkar TS, Momblona C, Gil-Escrig L, Ávila J, Sessolo M, Bolink HJ, Koster LJA. Recombination in perovskite solar cells: significance of grain boundaries, interface traps, and defect ions. *ACS Energy Letters* 2017, **2**(5): 1214–1222.

[34] Wali Q, Iftikhar FJ, Khan ME, Ullah A, Iqbal Y, Jose R. Advances in stability of perovskite solar cells. *Organic Electronics* 2020, **78**: 105590.

[35] Zhu H, Miyata K, Fu Y, Wang J, Joshi PP, Niesner D, Williams KW, Jin S, Zhu X-Y. Screening in crystalline liquids protects energetic carriers in hybrid perovskites. *Science* 2016, **353**(6306): 1409–1413.

[36] Ma F, Li J, Li W, Lin N, Wang L, Qiao J. Stable α/δ phase junction of formamidinium lead iodide perovskites for enhanced near-infrared emission. *Chemical Science* 2017, **8**(1): 800–805.

[37] Lee JW, Kim DH, Kim HS, Seo SW, Cho SM, Park NG. Formamidinium and cesium hybridization for photo-and moisture-stable perovskite solar cell. *Advanced Energy Materials* 2015, **5**(20): 1501310.

[38] Wang K, Jin Z, Liang L, Bian H, Bai D, Wang H, Zhang J, Wang Q, Liu S. All-inorganic cesium lead iodide perovskite solar cells with stabilized efficiency beyond 15%. *Nature Communications* 2018, **9**(1): 1–8.

[39] Pellet N, Gao P, Gregori G, Yang TY, Nazeeruddin MK, Maier J, Grätzel M. Mixed-organic-cation Perovskite photovoltaics for enhanced solar-light harvesting. *Angewandte Chemie* 2014, **126**(12): 3215–3221.

[40] Jeon NJ, Noh JH, Yang WS, Kim YC, Ryu S, Seo J, Seok SI. Compositional engineering of perovskite materials for high-performance solar cells. *Nature* 2015, **517**(7535): 476–480.

[41] Saliba M, Matsui T, Domanski K, Seo J-Y, Ummadisingu A, Zakeeruddin SM, Correa-Baena J-P, Tress WR, Abate A, Hagfeldt A. Incorporation of rubidium cations into perovskite solar cells improves photovoltaic performance. *Science* 2016, **354**(6309): 206–209.

[42] Stoumpos CC, Malliakas CD, Kanatzidis MG. Semiconducting tin and lead iodide perovskites with organic cations: phase transitions, high mobilities, and near-infrared photoluminescent properties. *Inorganic Chemistry* 2013, **52**(15): 9019–9038.

[43] Mosconi E, Amat A, Nazeeruddin MK, Grätzel M, De Angelis F. First-principles modeling of mixed halide organometal perovskites for photovoltaic applications. *The Journal of Physical Chemistry C* 2013, **117**(27): 13902–13913.

[44] Im J-H, Kim H-S, Park N-G. Morphology-photovoltaic property correlation in perovskite solar cells: One-step versus two-step deposition of CH3NH3PbI3. *APL Materials* 2014, **2**(8): 081510.

[45] Eperon GE, Burlakov VM, Docampo P, Goriely A, Snaith HJ. Morphological control for high performance, solution-processed planar heterojunction perovskite solar cells. *Advanced Functional Materials* 2014, **24**(1): 151–157.

[46] Xiao M, Huang F, Huang W, Dkhissi Y, Zhu Y, Etheridge J, Gray-Weale A, Bach U, Cheng YB, Spiccia L. A fast deposition-crystallization procedure for highly efficient lead iodide perovskite thin-film solar cells. *Angewandte Chemie International Edition* 2014, **53**(37): 9898–9903.

[47] Jeon NJ, Noh JH, Kim YC, Yang WS, Ryu S, Seok SI. Solvent engineering for high-performance inorganic–organic hybrid perovskite solar cells. *Nature Materials* 2014, **13**(9): 897–903.

[48] Jiang Q, Zhang X, You J. SnO2: a wonderful electron transport layer for perovskite solar cells. *Small* 2018, **14**(31): 1801154.

[49] Bach U, Lupo D, Comte P, Moser J-E, Weissörtel F, Salbeck J, Spreitzer H, Grätzel M. Solid-state dye-sensitized mesoporous TiO2 solar cells with high photon-to-electron conversion efficiencies. *Nature* 1998, **395**(6702): 583–585.

[50] Kar S, Jamaludin NF, Yantara N, Mhaisalkar SG, Leong WL. Recent advancements and perspectives on light management and high performance in perovskite light-emitting diodes. *Nanophotonics* 2020, **10**(8): 2103–2143.

[51] Hong K, Le QV, Kim SY, Jang HW. Low-dimensional halide perovskites: review and issues. *Journal of Materials Chemistry C* 2018, **6**(9): 2189–2209.

14 Near-Infrared Organic Materials for Biological Applications

Qi-Wei Zhang and Yang Tian

CONTENTS

14.1 Introduction...393
14.2 Main Types of NIR Organic Materials 394
    14.2.1 NIR Materials Based on Small Molecules.....................................394
    14.2.2 NIR Materials Based on Supramolecular Architectures.................397
    14.2.3 NIR Materials Based on Organic Polymers...................................403
    14.2.4 NIR Materials Based on Covalent Organic Frameworks................405
14.3 Bioapplications of NIR Organic Materials 407
    14.3.1 Biological Analytes Sensing 408
    14.3.2 Bioimaging in Vivo...409
    14.3.3 NIR-Based Photodynamic Therapy 411
    14.3.4 NIR-Based Photothermal Therapy..413
    14.3.5 NIR-Light-Triggered Drug Release..414
14.4 Summary and Perspective..416
Acknowledgements..417
Bibliography 418

14.1 INTRODUCTION

Infrared (IR) light refers to the electromagnetic radiation that lies between the upper edge of visible light and the beginning of microwave, which could be roughly divided into three segments, namely, near-infrared (NIR), mid-infrared (MIR) and far-infrared (FIR). More details on IR frequency band are introduced in Chapter 1. In this chapter, we will focus on the NIR region, especially for that ranging from 700 nm to 1700 nm, due to its accessibility by organic molecules, and suitability for biological applications. NIR materials are defined as matters interacting with NIR light, including absorption, sensitization, emission, and reflection under photo, chemical or electrical stimulations. Research on NIR materials is of great importance in the fields of both fundamental science and applied technology, and its contents mainly cover the development of novel molecules and materials, the mechanism of structure-property relationships, and the exploration of new application scenarios. Generally, NIR materials can be divided into three categories: (i) Inorganic materials, including metal nanoparticles,

DOI: 10.1201/9781003202608-14

metal/rare-earth oxide/oxysulfide nanomaterials, 2D materials (like carbon or black phosphorus); (ii) Organic materials, such as π-conjugated small molecules, organic semiconductor polymers, supramolecular systems, and covalent organic frameworks (COFs); (iii) Organic-inorganic hybrid materials, for example, metal-organic complexes or frameworks. In the past few decades, due to the development of fundamental theories and advanced techniques, more and more NIR materials have emerged, which are used not only in traditional industrial fields such as energy and communication, but also in high-tech fields, especially in biological systems, including sensing, imaging and photo-theranostics, due to their low autofluorescence interferences and deep penetration in biological tissues. In this chapter, we will focus on different kinds of NIR organic materials and their biological applications.

14.2 MAIN TYPES OF NIR ORGANIC MATERIALS

In principle, the absorption and emission wavelengths of organic materials depend on the energy gap between the highest occupied molecular orbital (HOMO) and the lowest unoccupied molecular orbital (LUMO). Therefore, strategies that can reduce the HOMO-LUMO energy gaps (HLEG) of molecules to lower energies (< 1.77 eV) can be used to construct NIR organic materials. For example, one can rationally design and modulate the push-pull electron effect of the molecular structure, the conjugation degree of π-systems, the molecular aggregation or assembly state, and so forth. Through these approaches, a series of NIR organic materials have been developed in terms of small molecules, supramolecules, polymers, and COFs.

14.2.1 NIR MATERIALS BASED ON SMALL MOLECULES

Organic chromophores that absorb NIR light are among the most-known NIR materials, which are widely used in heat-absorbs, solar cells, optical disks, photosensitizers, etc. These NIR small molecules usually have delocalized π-conjugated structures, and a straightforward strategy to decrease their HLEGs is to extend the conjugation by coupling extra alkenes/alkynes or phenyl groups. As shown in Figure 14.1,

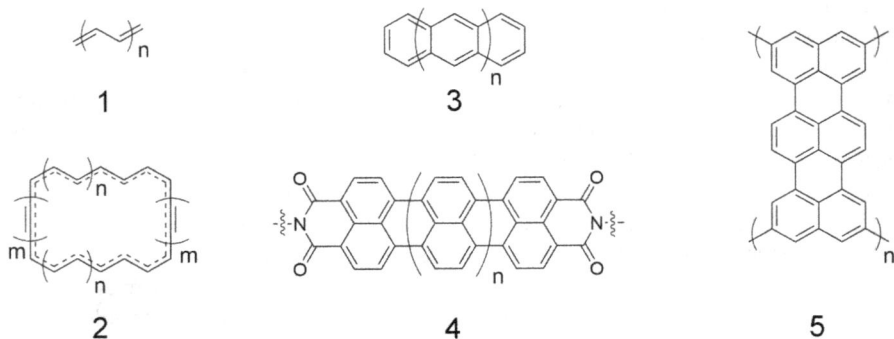

FIGURE 14.1   Representative chemical structures of the π-conjugated compounds.

the most representative π-conjugated compounds are *trans*-polyacetylenes (1), whose absorption wavelengths are positively correlated to the repeated methine units within a certain range, after which the effective conjugated length may hardly increase, thus the absorption wavelength will reach a plateau value around 600 nm. Cyclic polyene-yne derivatives (2) are capable to extend their absorption wavelengths into NIR region (ca. 800 nm) if their conformations are rigid and effectively fixed. Another kind of π-conjugated compounds are acenes (3) that derived by the annulation of benzene, and their HLEGs are reduced by the extension of aromatic rings. However, a plateau maximum absorption wavelength of about 700 nm will be reached, due to the decreased stability at longer conjugation. Therefore, although it is theoretically feasible to obtain NIR absorption by extending the π-conjugations of these compounds, it is practically difficult to realize NIR absorption, especially for wavelength over 800 nm. On the other hand, rylene (4) and its imide (5) derivatives are a series of readily available aryl systems, and their absorption wavelength maxima could be shifted to over 1000 nm by extending the naphthalene units along the long or short axis, as well as peripheral modifications.

Another common and effective strategy to reduce the molecular HLEG is to introduce strong electron donor (D) and electron acceptor (A) pairs into the conjugated structure, forming an electron push-pull effect. There are two possible reasons account for the lowered HLEGs in D-A type molecules. The first one is the existence of two or more resonant structures (D-A ↔ D$^{+=}$A$^-$) in these chromophores. According to the Peierls theorem [1], in a quasi-one-dimensional conjugated system, the HLEG is dependent on the bond length alterations. Because the resonance structures can reduce the bond length alternations between the single and double bonds within the D-A backbones, thus it can effectively reduce the energy gap. The second reason is that the newly formed hybrid molecular orbital by conjugating the D and A groups has a higher HOMO energy level than that of D, and a lower LUMO energy level than that of A, resulting in a relatively small HLEG and red shifted absorption wavelength (Figure 14.2). Moreover, for these D–A type chromophores, the band gap can be finely tuned by modifying the electron-donating/-withdrawing abilities of the D/A units, or modulate their intra-/inter-molecular electronic interactions. For example, polymethines, as a kind of special polyene derivatives with two terminal carbon atoms replaced by D and A groups respectively, usually have longer absorption and emission wavelengths than the corresponding polyenes. As shown in Figure 14.3(a), a typical class of synthetic polymethines is the cyanine dyes, which possess unique charged chromophores and an odd number of methines. Both the terminal D/A groups and the π-conjugated linker between them (numbers of methines) modulate the photophysical properties of the dye [2]. Another kind of NIR dyes with resonance stabilized zwitterionic structures are squaraines. As schemed in Figure 14.3(b), a typical squaraine molecule usually consists of an electron-deficient four-membered ring core as an electron acceptor, and two electron-rich moieties on both sides as electron donors, forming a donor-acceptor-donor (D-A-D) structure. The intramolecular charge transfer character as well as the extended π-conjugation skeleton, endow them with high molar extinction coefficients in the region from visible to NIR light [3].

Owing to the great development on photonics and optronics, high energy lasers thrive in key sectors such as industry and medicine, which further facilitates the advancement

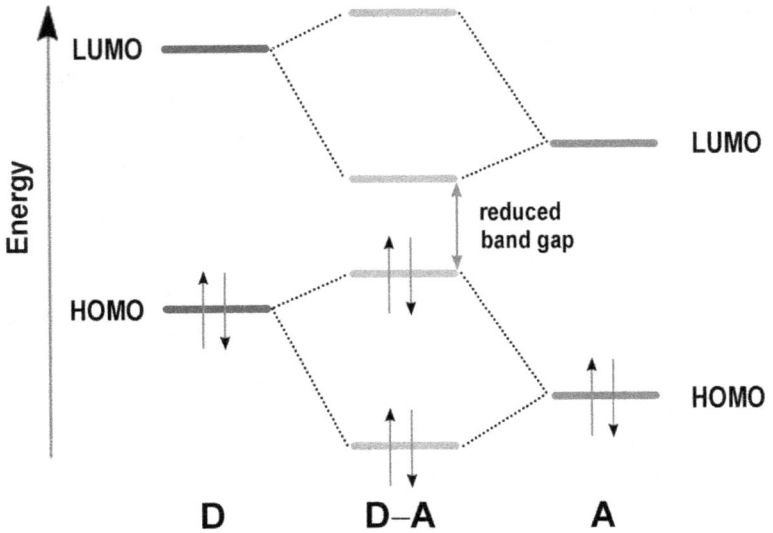

FIGURE 14.2 Schematic illustration of the hybridization of the energy levels of an electron donor (D) and an electron acceptor (A), leading to a D-A type conjugate with a narrowed HOMO-LUMO energy gap.

a) Cyanines

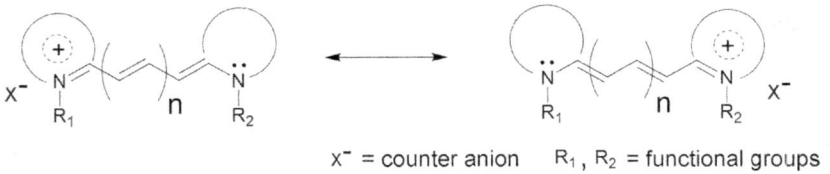

x^- = counter anion R_1, R_2 = functional groups

b) Squaraines

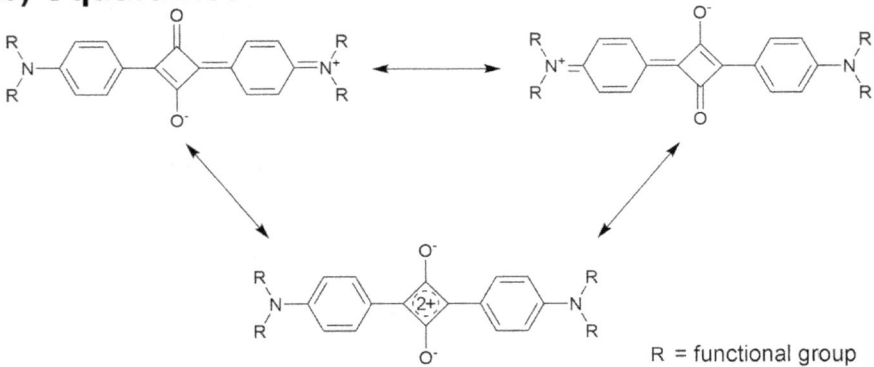

R = functional group

FIGURE 14.3 Possible resonance structures for cyanines (a) and squaraines (b), respectively

of two-photon or even multi-photon technology, because such nonlinear optical phenomenon is only observed under irradiation with high intensity light. Specifically, two-photon or multi-photon absorption refers to the nonlinear optical process that matters simultaneously absorb two or more photons of the same or different frequencies, therefore, this may lead to an absorption wavelength much longer than that of the normal one-photon absorption of a chromophore [4,5]. For example, supposing the absorption wavelength of a dye in the case of one-photon absorption is 405 nm, then in the case of two-photon absorption, the absorption wavelength can reach around 810 nm. Therefore, designing small organic molecules with two-photon absorption features is a very effective approach to construct NIR molecules. For the illustration of the detailed mechanisms and molecular design strategies, readers may refer to the reported reviews [6,7].

14.2.2 NIR MATERIALS BASED ON SUPRAMOLECULAR ARCHITECTURES

In addition to the two strategies described above to achieve NIR organic compounds by covalent synthesis and fabrication, there is also a supramolecular approach for tuning the HLEGs of organic chromophores. Supramolecular chemistry refers to the chemistry beyond the molecules, which employs noncovalent interactions and recognitions to construct well-defined assemblies and nanostructures [8–15]. Compared with traditional synthetic strategies, supramolecular methods have unique advantages in two primary aspects. The first one is the elimination of tedious organic synthesis by using of intermolecular self-assembly approach. The second superiority is the tunability, reversibility and adaptation, which arise from the dynamic and noncovalent nature of the intermolecular interactions. In this section, several typical supramolecular systems, including supramolecular charge transfer complex, aggregates, and organic radicals, will be introduced and discussed for their abilities to lower HLEGs and gain NIR activities.

The organic electron donor and acceptor chromophores can not only be conjugated through covalent bonds, but also interact with each other through non-covalent stacking and form charge-transfer (CT) complexes. This kind of noncovalent-linked supramolecular self-assembly based intermolecular CT process is also called through-space ICT. In such CT complexes, electronic transitions can occur between the HOMO of the D component and the LUMO of the A component, resulting in a significant reduction in HLEG, and a longer wavelength absorption band can be observed. However, most intermolecular D-A stacking is weak, dynamic, and unstable, so the character of CT absorption or emission is faint. To enhance these CT interactions and further decrease the HLEG, a straightforward strategy is to design energy-level matched D-type chromophores with stronger electron-donating abilities and A-type chromophores with stronger electron-withdrawing abilities, but it often means complicated syntheses. Alternatively, another convenient, effective, and more "supramolecular" approach is to confine the donor and acceptor in a nano-sized molecular cavity through the introduction of macrocycles, thereby optimizing the D-A distance and stacking arrangement, by which enhance the association constant. Meanwhile, as the third component, the macrocycle can also generate multiple interactions and binding sites, thereby effectively stabilize the structures of the supramolecular D-A complexes [16–18]. Cucurbit[n]uril (CB[n]) is a kind of widely used macrocyclic molecules, which consist of several (n = 5, 6, 7, 8, . . .) glycoluril units bound together by methylene bridges (Figure 14.4) [19,20]. Considering the relatively larger rigid internal hydrophobic cavity, cucurbit[8]uril (CB[8]) is popularly selected as a host molecule to simultaneously accommodate two aromatic chromophores, for example, an electron donor and an electron acceptor, thus forming a supramolecular ternary complex (Figure 14.4). In such systems, within the cavity of CB[8], the D-A distance is shortened and association is stabilized, resulting in an enhanced and red-shifted CT absorption and emission (in visible or NIR region) as compared with those without CB[8] [21]. For instance, the formation of dyad complexation in water between a phenothiazine compound (9) as a donor and a methyl viologen (6) as an acceptor inside the cavity of CB[8], produces a board of NIR absorption band from 500 nm to 1100 nm, with a maximum at 756 nm, which can be attributed to the intermolecular CT absorption facilitated by the formation of a CB[8]-based 1:1:1 ternary complex [22].

Stacked aggregates in solution or solid states are ubiquitous for π-conjugated planar organic molecules, which could be roughly classified into H-aggregates and J-aggregates based on the spectral shifts. Specifically, the former induces the main absorption peak moving toward a shorter wavelength (blue-shift), whereas the latter shifts to a longer wavelength (red-shift), compared to the corresponding monomers (Figure 14.5(a)). These photophysical transitions could be predicted and rationalized

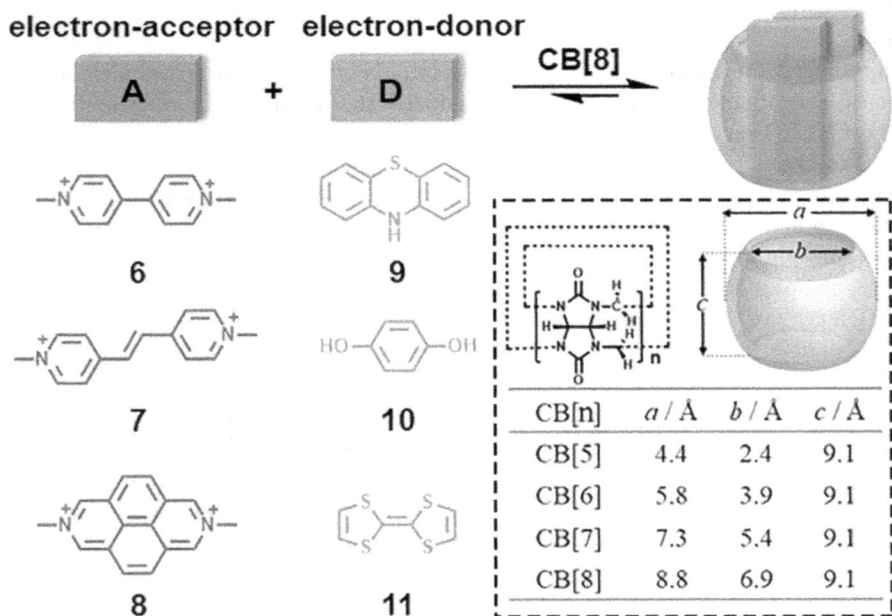

CB[n]	a / Å	b / Å	c / Å
CB[5]	4.4	2.4	9.1
CB[6]	5.8	3.9	9.1
CB[7]	7.3	5.4	9.1
CB[8]	8.8	6.9	9.1

FIGURE 14.4 Structures of the cucurbit[n]urils and some typical electron donor and acceptor pairs that can form ternary supramolecular complexes with cucurbit[8]uril. The counteranions were omitted for clarity.

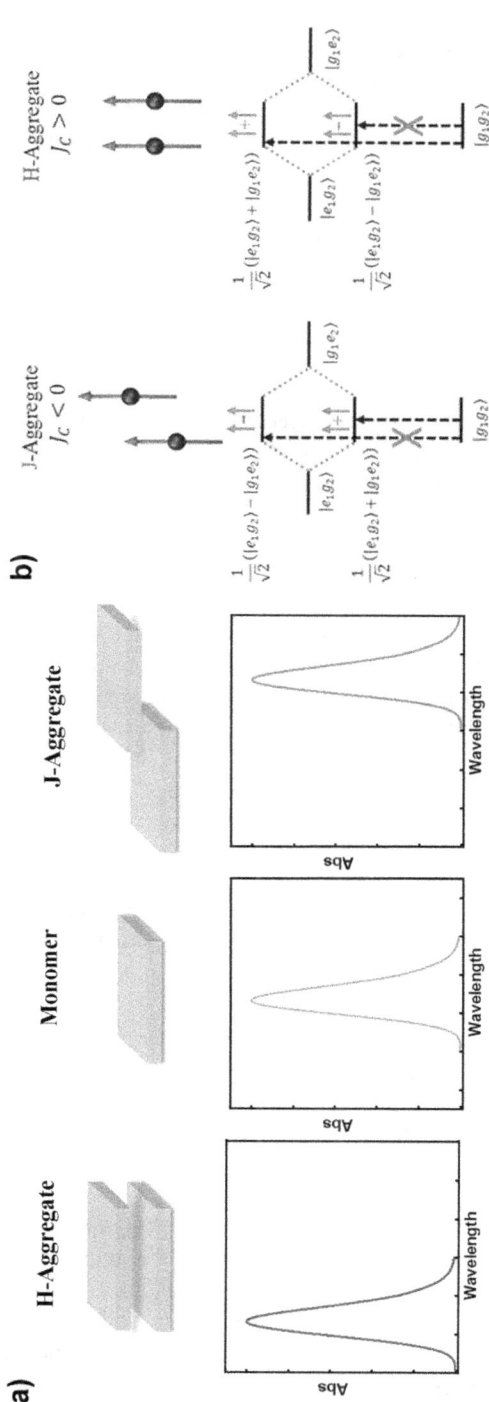

FIGURE 14.5 (a) Schematic illustrations of the H-aggregate, monomer and J-aggregate, and the corresponding absorption spectra shifts. (b) Energy level and possible transition diagrams for the J-aggregate and H-aggregate dimers.

Source: Reprinted with permission from Hestand, N.J., et al. *Chem. Rev.* 118, 7069–7163, 2018. Copyright 2018, American Chemical Society.

by Kasha's theory [23]. As illustrated in Figure 14.5(b), in a simplified two-molecule aggregated model, two split delocalized excited states are formed due to the Coulomb coupling (J_C), which contain an in-phase and an out-of-phase linear combinations of two locally excited states, with the former shifted by J_C and the latter shifted by $-J_C$. Transitions only occur between the ground state and the states with in-phase transition dipole moments (the bottom and up bands of the split excited states in J-aggregate and H-aggregate, respectively.) [24]. As a result, the J-aggregation provides a good approach to red-shift the absorption wavelength of organic dyes from visible light region at their monomer state to NIR region at J-aggregated states. For normal organic chromophores, decrease the energy gap between the excited and ground states will increase the internal conversion due to the increased overlap between the lowest vibrational level of the excited state and the high vibrational level of the ground state, which is known as the energy gap law. However, J-aggregates may provide an approach to circumvent this law. Because the transition dipole moments of chromophores in J-aggregate are coupled together and delocalized over the aggregate by Coulomb interactions, resulting in a lower reorganization energy as well as a suppressed vibrational relaxation rate. According to this mechanism, J-aggregation may provide a physical compensation within the framework of the energy gap law, and thus often displays strong emission, making it a promising strategy for constructing NIR emitting materials.

There are several approaches for structural modifications of organic dyes to realize the formation of largely slipped head-to-tail or shifted plate arrangements (J-aggregates), while avoid the face-to-face stacking (H-aggregates). The first way is to employ noncovalent interactions at specific positions, such as hydrogen bonding, to direct the head-to-tail stacking. Another way is to introduce steric hindrances or electrostatic repulsive groups to restrain the H-aggregation. For example, Börjesson and co-workers developed a quaterrylene-based superradiant J-aggregate that exhibits strong NIR absorption and emission [25]. As shown in Figure 14.6(a), the bay-functionalization of perylene with two C_6-chains generates a twisted plane, thus breaking the symmetry and suppressing the face-to-face π-π stacking. Distinct absorption spectra were observed when the dihexylquaterrylene dye was dissolved in toluene and $C_2H_2Cl_4$, which were assigned to the monomer and J-aggregate states, respectively (Figure 14.6(b)). Due to the strong coupling between dye molecules within the aggregated state, the absorption spectrum transformed from a typical rylene vibronic pattern to a narrower and more intense peak with its maximum significantly red-shifted (~ 180 nm) from visible to NIR regime. On the other hand, owing to the delocalization of the transition dipole moments in the J-aggregate, the reorganization energy of the system is reduced, leading to a slower rate of internal conversion and higher fluorescence quantum yield.

Since the first report on the triphenylmethyl radical [26], the discipline of organic radical chemistry has advanced tremendously. Such compounds always have narrow HLEGs, thus are promising potential NIR materials. However, most of the organic radicals are transient radical anions or cations generated by reduction or oxidation of π-conjugated organic compounds, so they are generally unstable and highly reactive. There are two strategies to stabilize these organic radicals, namely the covalent approach and noncovalent (supramolecular) approach. The former adopts the

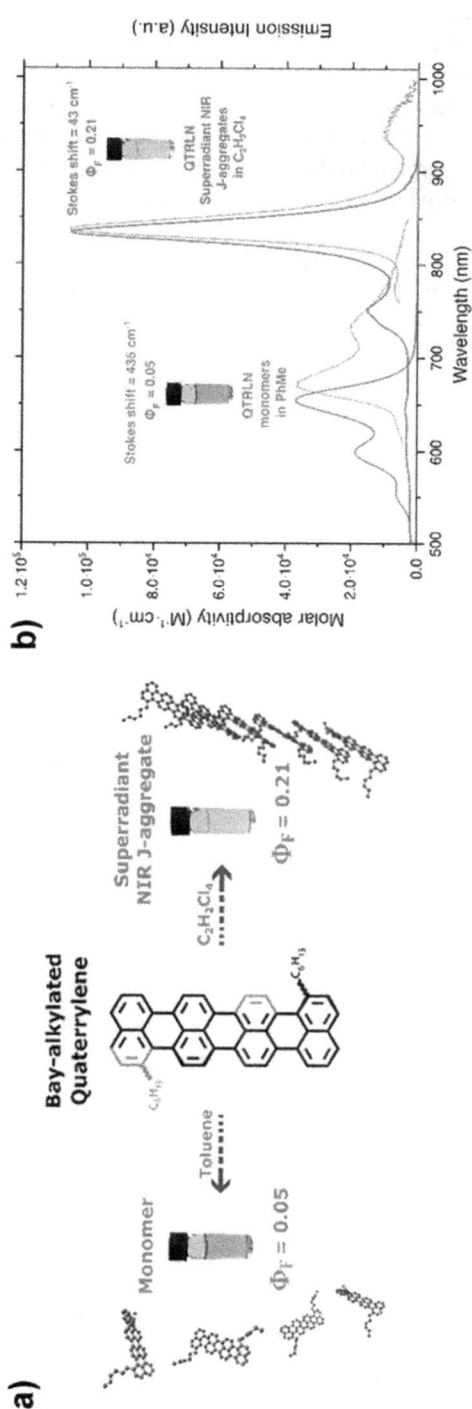

FIGURE 14.6 (a) Illustration of the dihexylquaterrylene in different aggregation states. (b) Absorption and emission spectra of dihexylquaterrylene recorded in toluene (blue) and $C_2H_2Cl_4$.

Source: Reprinted with permission from Cravcenco, A., et al. *J. Am. Chem. Soc.* 143, 19232–19239, 2021. Copyright 2021, American Chemical Society.

modification of parent molecules by substituents with strong electron-withdrawing and electron-donating abilities to stabilize the radical anions and cations, respectively [27,28]. The latter approach utilizes macrocycles, cages or nano frameworks to encapsulate organic radicals into their cavities, thus physically shield them from interactions with the surrounding environment or reagents, which has been proved quite effective in stabilizing organic radicals [29,30]. On the other hand, the special spin-spin affinity between organic radicals in turn provides a new kind of driving force for the supramolecular self-assembly. Based on these synergistic advantages, supramolecular radical chemistry has become a research hotspot, and a series of supramolecular radical materials with distinct self-assembled architectures have been reported [31,32]. For example, as shown in Figure 14.7(a), Wang and co-workers developed a dumbbell-like viologen derivative (12), which could be transformed to the corresponding [2]rotaxane (R12) by slipping the macrocycle CB[7] [33]. Through a series of measurements including ^{1}H NMR, ESR and absorption, they confirmed that the rotaxane R12 exhibited good reversibility during chemical or electrochemical

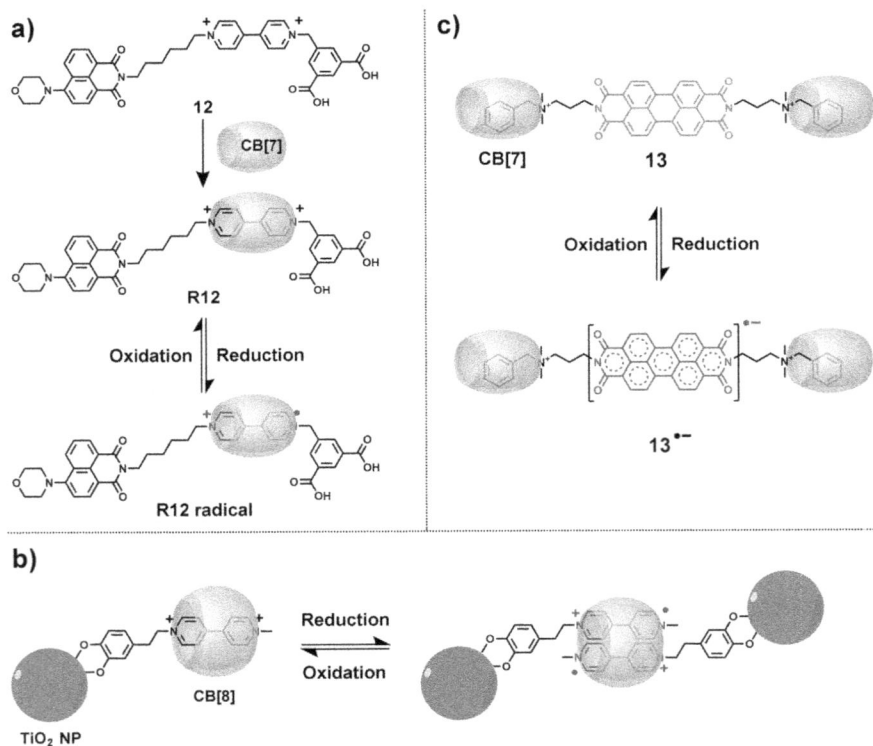

FIGURE 14.7 (a) The formation of a [2]rotaxane (**R12**) by slipping of CB[7] onto a dumbbell molecule **12**, and the redox processes. (b) The redox-indued self-assembly/disassembly of viologen-functionalized TiO$_2$ nanoparticles with the aid of the macrocycle CB[8]. (c) CB[7] stabilized perylene diimide anion radical. The counteranions were omitted for clarity.

reduction/oxidation cycles, and the cationic radical form of the supermolecule could keep stable, while compound 12 would decompose during reduction due to the instability of the radical. These results clearly confirmed that the macrocycle CB[7] could stabilize the organic radical by host-guest encapsulation. Another example of using radical-cation interaction to direct the intermolecular and inter-nanoparticle self-assembly was reported by Qu, Tian, et al. [34]. As illustrated in Figure 14.7(b), the authors shown that the self-assembly/disassembly between methyl viologen-functionalized TiO_2 nanoparticles (NP) can be regulated by chemical or photochemical redox reactions with the aid of the macrocycle CB[8]. Besides, reversible self-assembly/disassembly switches are also accompanied by the photocatalytic activity transitions of the TiO_2 NPs. Xu, Zhang, et al, reported a supramolecular complex (**13**) containing a perylene diimide chromophore and two CB[7] macrocycles, which could be selectively reduced to the corresponding radical anion from (**13$^{•-}$**) by E. coli (Figure 14.7(c)) [35]. In this supramolecular system, the bulky CB[7] macrocycles at the ends of the perylene diimide derivative can effectively diminish the face-to-face π-π stacking of the chromophore, thus restrain the quenching of the radical anion. As a result, stable radicals with three absorption peaks in the NIR region (735 nm, 782 nm and 820 nm, respectively) could be obtained, which showed good bacterial inhibition due to the photothermal effect of 13$^{•-}$ under an NIR laser (808 nm) irradiation.

These examples have demonstrated the convenience and effectiveness of supramolecular strategies in constructing and stabilizing NIR organic materials by lowering the HLEGs of organic chromophores through the formation of supramolecular ICT complexes, J-aggregates, and radicals. The reversible and dynamic nature of noncovalent interactions may further bring new advanced NIR materials with smart stimuli-responsive functions.

14.2.3   NIR MATERIALS BASED ON ORGANIC POLYMERS

Organic polymers are drawing much attention in fabrication of NIR materials due to their tunable absorbance and emission wavelength, multi-site modifiability, high stability, and good biocompatibility/biodegradability. Basically, NIR organic polymers may refer to polymers that are inherently NIR-active, or aggregated materials in which the NIR-inert polymers act as shells for NIR molecules. In this section, we restrict our attention to the former, and will briefly discuss the classes of conjugated semiconducting polymers, especially the D–A-type polymers, coordination polymers, and NIR-dye-branched polymers, etc.

Semiconducting polymers, such as polyaniline, polypyrrole, polydopamine, etc., are common conjugated organic polymers with NIR absorption (Figure 14.8(a)) [36,37]. Polyaniline is kind of good photothermal agent, in which the imide groups can react with acids and transform to the corresponding salts, which results in a red-shift of the absorption to NIR region. Polypyrrole is obtained from the pyrrole monomer by oxidative polymerization, which shows strong absorption in the NIR region and good photostability. As a result, it is widely used in bioelectric and biomedical fields. Polydopamine is an important component of melanin that is widely present in the human body, and it can be degraded naturally, so it has excellent biocompatibility. These conjugated polymers can be further converted to nanoparticles by

a)

Polyaniline   polypyrrole   polydopamine

b)   **D-A type polymer**   **D-π-A type polymer**

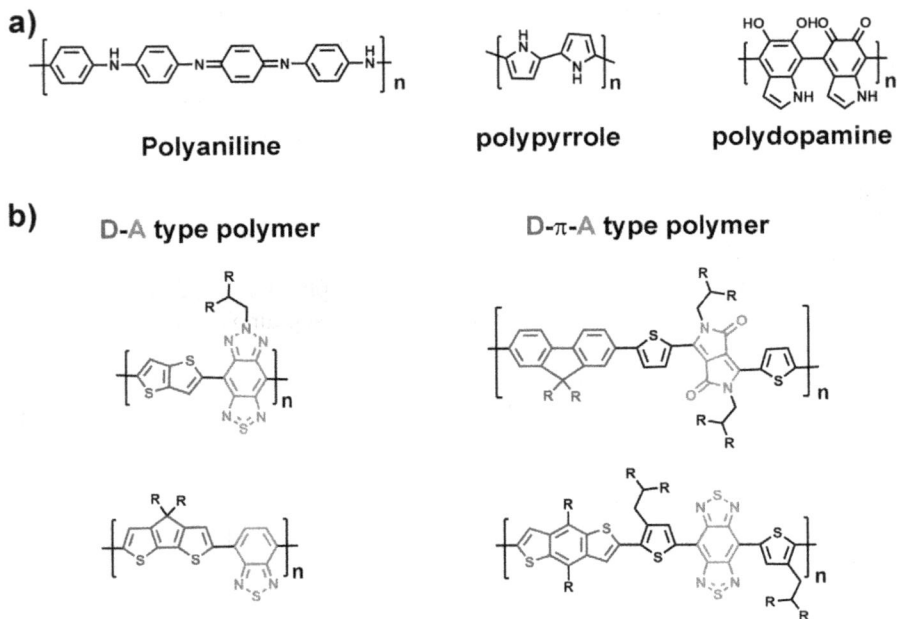

FIGURE 14.8 (a) Chemical structures of polyaniline, polypyrrole, and polydopamine. (b) Structures of some representative examples of D-A type polymer and D-π-A type polymer.

nanoprecipitation in the organic/aqueous solvents system, driven by the collapse of hydrophobic polymer chains when transformed from organic solvent to water. These nanoparticles usually exhibit better photostability and photothermal property, therefore have become a new class of nanomaterials with long absorption wavelengths, high absorption coefficients and tunable dimensions. What's more, a series of anticancer drugs can be loaded inside these nanoparticles due to the hydrophobic effect and π-π stacking interaction. On top of that, their surfaces can be further coated with silica or other materials to achieve extended functions, endowing them with good application prospects in biological fluorescence/photoacoustic imaging, photothermal therapy (PTT), and stimuli-responsive drug delivery.

The previous sections have mentioned that the introduction of electron donor and acceptor groups into small organic molecules can effectively enhance the ICT processes and lower the HLEGs, and this strategy can be further extended to the construction of conjugated D-A type polymers. In these polymers, the electron donor, electron acceptor and π-spacer all play important roles, and they jointly determine the photophysical properties of the conjugated polymers. For the electron donors, introduction of electron-rich subunits (like fused arenes) or atoms (such as those with lone-pair electrons) can increase the electron-donating ability. While for the electron acceptors, fusing heavy atoms (like selenium, tellurium, etc.) can enhance the electron-withdrawing ability and lead to red-shifted absorption owing to the decreased ionization potential and aromaticity, as well as the increased bond length in the acceptor, described as the "atomistic band gap engineering" by Seferos et al. [38]. Besides, extending the conjugation and planarity of the molecule, as well as reducing the steric hindrance of adjacent groups can also lead to longer absorption/emission wavelength. Figure14.8(b) presented some typical D-A type or D-π-A type conjugated polymers, which have been used as NIR fluorescent probes, photothermal agents, etc. [39–41]. For the syntheses of these polymers, typical synthetic procedures such as the Stille polycondensation and the Suzuki coupling reactions are widely used for the polymers with thiophene and phenyl repeat units, respectively. However, the development of new synthesis methods with more variability, mild reaction condition and broad adaptability remains an urgent matter.

Another kind of polymeric NIR material is coordination polymer, which is usually formed by crosslinking between metal ions and some natural polyphenol molecules through multivalent coordination bonds [42]. Tannic acid and gallic acid are two kinds of typical natural phenols that can be extracted from plants. A series of metal ions including ferric, ruthenium, and vanadium ions, can quickly be chelated by the digallolyl groups of tannic acid to generate polymeric networks, which absorb NIR light [43]. What's more, these metal-tannic acid coordination polymers can adhere to the surfaces of templates, such as nan-vesicles, and maintain their photophysical properties. For example, the ferric-tannic acid network on a polymeric nano-vesicle template (PNV@Fe-TA) possesses photoacoustic, magnetic and NIR fluorescence multimodal imaging features, as well as photothermal conversion ability. Therefore, this kind of coordination polymer is a promising NIR material with inherent good biocompatibility, and suitable for imaging-guided disease therapies. In addition to these approaches, another direct way to obtain NIR polymers is to append existing NIR active small molecular dyes as side chains to the main chain polymer backbone. This strategy is a combination of small molecule dye chemistry and polymer chemistry. The skeleton of the polymer will affect the chemical environment of the branched dyes to some extent, however, the intrinsic NIR photophysics are mainly derived from the branched small molecular dyes, which have been discussed in the previous sections.

14.2.4   NIR MATERIALS BASED ON COVALENT ORGANIC FRAMEWORKS

The development from small molecules to polymers has not only brought new architectures and functions, but also greatly improved the processability and practicality of organic materials. However, amorphous polymers or networks still have limitations in structural control and characterization, which may, to some extent, restrict their advanced applications. In this context, Côté, Yaghi and co-workers pioneered a class of porous crystalline materials, named covalent organic frameworks (COFs) [44]. COFs can be constructed by linking small molecular building blocks through dynamic covalent bonds, such as B-O-B, B-O-C, B-O-Si, C=N, N=N, etc., and predictable control of the crystalline structures (3D frameworks) can be possibly achieved by the deft selection of building blocks and appropriate reaction conditions. Since the first discovery, this kind of conspicuous functional organic materials have grown tremendously, and have been used in various fields, including sensing, adsorption, separation, catalysis, optoelectronics, biomedicine, etc., due to their desirable features like crystallinity, structural modularity, porosity, stability, versatility, and biocompatibility [45,46].

As a structural extension of small molecules, strategies that are suitable for constructing NIR small molecules can also be applied to COFs materials. For example, Xie and co-workers developed a series of electron donor and acceptor (D-A) involved COFs in mild conditions [47]. These synthetic D-A COFs exhibit good colloidal stabilities and possess uniform spherical nano morphologies with their sizes flexibly adjustable by changing the amount of catalyst, which further influence the absorption spectra. Furthermore, the absorption band can be broadened and red-shifted to the second NIR bio-window, by changing the electron-donating ability of the donor units in COF (Figure 14.9(a)). This strategy can be applied to other aromatic constituents and linkage modes, and build more advanced NIR D-A COF materials. On the other hand, planar NIR dyes can be modified with reactive groups and used as building blocks or appendants for COFs. Porphyrin (Por), phthalocyanine (Pc) and boron-dipyrrin (BODIPY) derivatives are planar π-delocalized macrocycles that may absorb or emit visible-to-NIR light, which make them fascinating building blocks for constructing NIR COFs (Figure 14.9(b)) [48–51]. For example, Pang, et al., developed a nanoscale porphyrin-conjugated COF (Por COF) though a solution-based aging method at room temperature (Figure 14.9(b)) [49]. The UV-VIS-NIR spectrum of this Por COF-dispersed aqueous solution exhibits broad absorption peaks

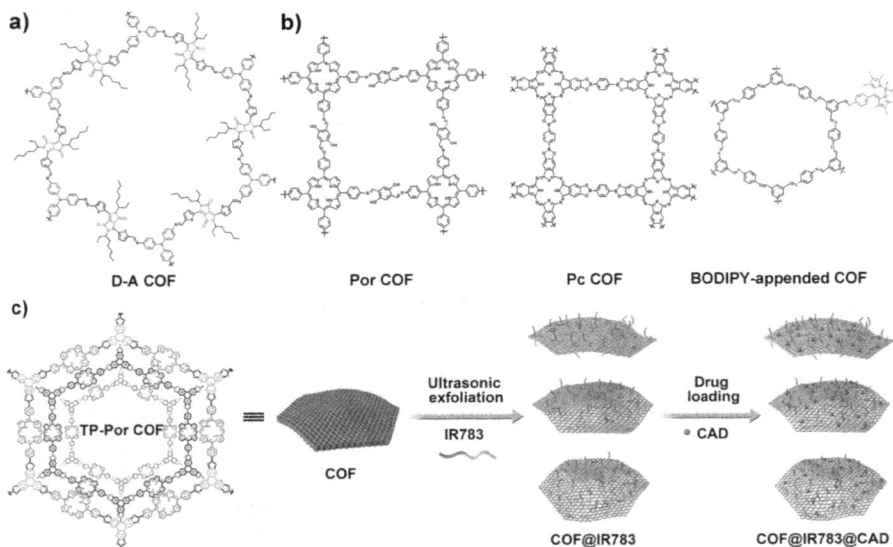

FIGURE 14.9 Structures of some representative COFs: (a) D–A type COF, (b) porphyrin-based COF (Por COF), phthalocyanine-based COF (Pc COF), and BODIPY-appended COF. (c) Schematic illustration of the fabrication process for the nanocomposite COF@IR783@CAD.

Source: Reprinted with permission from Wang, K., et al. *ACS Appl. Mater. Interfaces* 11, 39503–39512, 2019. Copyright 2019, American Chemical Society.

from 400 nm to over 800 nm. What's more, these COF-based nanoparticles possess high photothermal conversion and singlet oxygen generation efficiencies when irradiated by 808 nm and 650 nm lasers, respectively. In addition to the direct use as NIR materials, COFs can also be used as carriers for adsorbing NIR small molecules due to their porous properties. As shown in Figure 14.9(c), Tian and co-workers proposed a porphyrin-based COF that can adsorb an NIR cyanine dye (IR783) and form a nanocomposite (COF@IR783) [52]. This nanocomposite exhibits better dispersibility and stability in water than the COF itself, and the nanosized morphology as well as the negatively charged nature also endow it with improved blood circulation, enhanced permeability, and retention (EPR) effect, and PTT performance. What's more, this dye-adsorbed COF material can further act as a carrier for an antitumor prodrug (cis-aconityl-doxorubicin, CAD) and form a multifunctional antitumor agent (COF@IR783@CAD).

14.3   BIOAPPLICATIONS OF NIR ORGANIC MATERIALS

NIR materials refer to matters that interact with NIR light, and the interactions may include absorption, sensitization, and emission, which can be employed for photoconversion, photoreaction and optical imaging, respectively. NIR light generally penetrates deeper than UV light ($\lambda < 400$ nm) and visible light ($\lambda \approx 400$–700 nm) in biological tissues, due to their weaker absorptions and scatterings by bio-matters and water [53]. In biology, the NIR bands with deeper penetration depths of biological tissues are often called as "biological windows", which mainly include three regions: the first biological window (NIR-I, 700 nm [sometimes may start from 650 nm]–900 nm), the second biological window (NIR-II, 1000 –1350 nm), and the third biological window (NIR-III, 1550 –1870 nm) (Figure 14.10) [53,54]. On the other

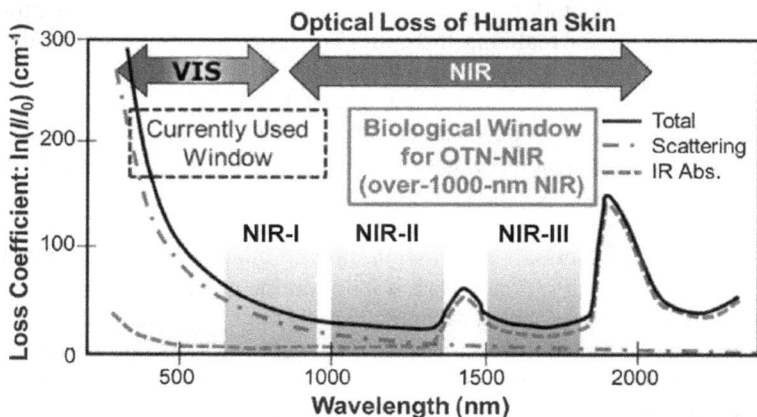

FIGURE 14.10   Optical loss of human skin and the NIR-I, NIR-II and NIR-III biological windows.

Source: Reprinted with permission from Hemmer, E., et al. *Nanoscale* 5, 11339–11361, 2013. Copyright 2013, Royal Society of Chemistry.

hand, biological matters have low autofluorescence in NIR region, which may provide higher signal-to-background ratio (SBR) for optical sensing. Therefore, a series of NIR organic materials that are active at these biological windows have been developed and their biological applications are explored. In this section, the discussions will be focused on applications including biological analytes sensing, bioimaging, photodynamic therapy, photothermal therapy, and photo-triggered drug release.

14.3.1 Biological Analytes Sensing

Optical sensing technology is a remarkable visualization modality for investigations of biological analytes and processes, due to its tremendous potential for obtaining qualitative and quantitative biological information with high sensitivity and temporal-spatial resolution in vivo, without invasion and avoid complicated operations. Specifically, organic chromophore-based fluorescent probes can be rationally designed to meet high selectivity and reversibility. A representative fluorescent probe consists of a recognition site (receptor), a fluorescent signal readout site (indicator), and a linkage between these two blocks (spacer). When the analyte is recognized by the receptor, the interactions between them can lead to a variation in fluorescence of the indicator transferred through the spacer. The output signal could be fluorescence wavelength, intensity, polarity, and lifetime [55,56], while the changes could be spectra shifts, quenching (turn-off), activation (turn-on), ratiometric, etc. [57–62]. The response mechanisms may include photo-induced electron transfer (PeT), intramolecular charge transfer (ICT), Förster resonance energy transfer (FRET), excited state intramolecular proton transfer (ESIPT), aggregation-induced emission (AIE) and so forth [63–70]. As mentioned above, NIR light has unique advantages in biological systems, therefore, NIR fluorescent dyes have become promising candidates for indicators in fluorescent probes. These NIR fluorescent probes possess deep tissue penetration, low photodamage, low autofluorescence background, etc., thus providing a vital choice for realizing the optical imaging-based biomedical diagnosis and surgery.

To date, a series of NIR fluorescent probes based on quantum dots, carbon nanomaterials, rare-earth doped nanoparticles, etc. have been studied extensively. However, NIR organic fluorescent probes are more fascinating because of their defined molecular structures, easy derivatization, high fluorescence quantum yield, and most importantly, low bio-toxicities. As a representative, cyanine dyes are quite attractive as NIR absorption and emission units in biological probes (labeling and sensing), in which the most famous one is indocyanine green (ICG), which has been approved as a fluorescent contrast for human use by the United States Food and Drug Administration (FDA) and European medicines agency (EMA) [71]. So far, a variety of biological analytes and parameters can be monitored by NIR fluorescent probes, which include metal ions, reactive oxygen species (ROS), reactive nitrogen species (RNS) reactive sulfur species (RSS, like biothiols and H_2S), pH, gas, enzymes, amino acids, temperature, viscosity, etc., providing molecular tools for deciphering biological activities or pathogenesis in vitro or in vivo [72–74]. For example, Zhang and co-workers synthesized a series of pentamethine cyanine NIR fluorophores with significant anti-quenching feature, as well as superior photostability [75]. These dyes absorb and emit light at wavelengths beyond 1000 nm, whose intensities are responsive to pH variations, thus affording noninvasive ratiometric NIR-II fluorescence sensing of pH which is suitable for sensing pH fluctuations in deep tissues in vivo. Another kind of widely used NIR fluorophores for molecular probes are BODIPY derivatives. Their photophysical properties can be dramatically changed by different substitutions on their conjugated core, therefore, BODIPY-based probes can sensitively respond to external reagents like H_2S, GSH, etc., in fluorescence "off-on" or ratiometric modes, which can be potentially used to monitor the redox homeostasis of biological systems [76].

14.3.2 BIOIMAGING IN VIVO

Compared with traditional clinical imaging methods, such as computed tomography (CT), positron emission tomography (PET), magnetic resonance imaging (MRI), as well as thermal and ultrasound imaging, fluorescence biological imaging technology is a more remarkable visualization modality due to its instantaneity, versatility, accuracy, and safety. With the development of a series of advanced NIR fluorescent dyes, especially those targetable and activatable NIR fluorophores, considerable progress has been made in fluorescence bioimaging technology. Because normal downconversion NIR-I fluorophores may require visible light for excitation, so the penetration of the incident light is limited. Whereas, for longer wavelength light (like NIR-II or NIR-III), the excitation wavelengths could fall in the NIR region, ensuring better fluorescence imaging depths, and thus has better application prospects.

To date, various NIR fluorophores including some commercially available dyes such as ICG, IR-26, IR-783, IR-806, IR-1061, etc., have been successfully used for bioimaging, which may include precise organ/tissue imaging, tumor imaging, dynamic metabolism imaging, etc. For instance, Smith group synthesized a tetralactam macrocycle appended with six copies of a bone-targeting ligand [77]. By taking advantage of the host-guest self-assembly strategy, a linear fluorescent PEGylated squaraine dye could thread into two macrocycles and form a pseudorotaxane architecture. This pre-assembly strategy not only endowed the squaraine dye with better photostability, red-shifted absorbance and emission wavelength, targeting ability, but also provided targeting ability. Imaging results showed that there was no fluorescence attenuation at the bone sites after 24 hours in a living mouse, indicating the high targeting ability and chemical stability of this supramolecular NIR probe (Figure 14.11(a)). Tang and co-workers developed s series of semiconducting polymer nanoparticles (SPNs) that integrated planar π-conjugated groups and twisted units into one polymer system, thus achieved strong absorption coefficient and high aggregated-state fluorescence quantum yield [78]. Through this rational molecular design strategy, much brighter NIR-II (1300–1400 nm) emission was obtained for blood vessel imaging in vivo (Figure 14.11(b)). Another important issue in medicine is the imaging of tumor tissue, which demands fluorescent probes with the abilities to selectively label cancerous cells. A series of characteristic and abnormal metabolisms have been observed in tumor tissues, such as hypoxia, low pH, high energy-demand, high vasculature, over expression of biological receptors, and enhanced mitochondria activity, etc. [79]. Therefore, NIR fluorescent molecules that

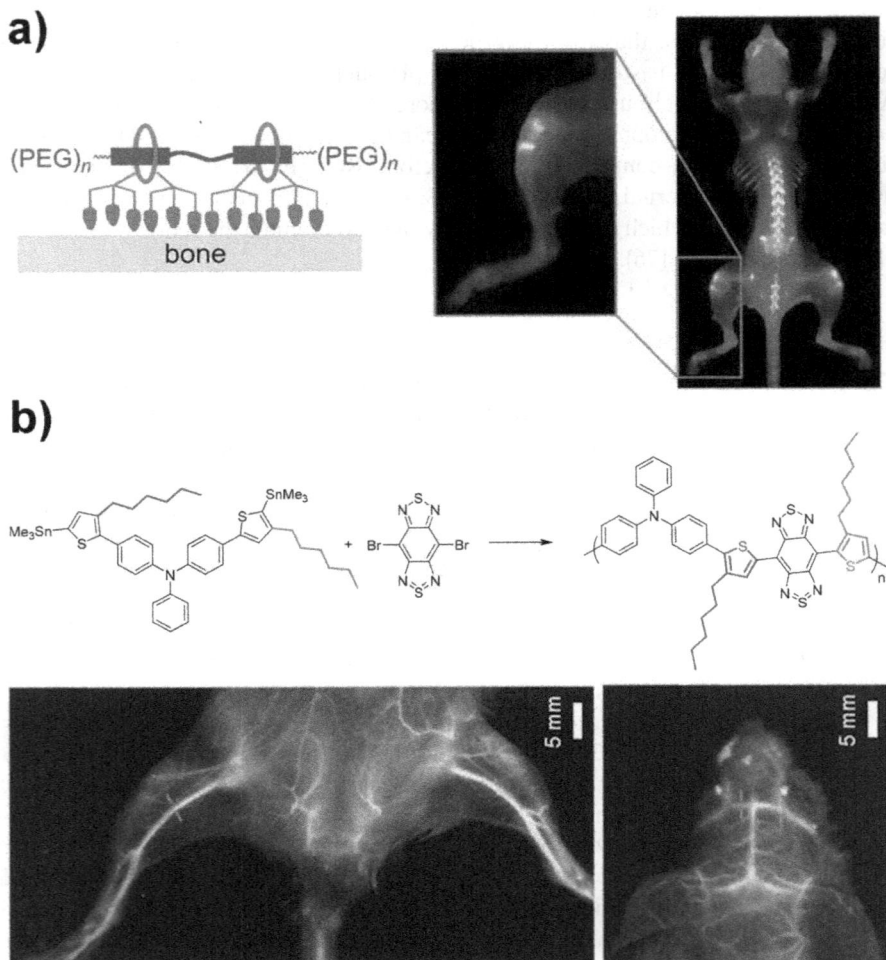

FIGURE 14.11 (a) Schematic illustration of the structure of the pre-assembled [3]pseudoro-taxane targeted on the surface of bone with appended targeting ligands [left], and the representative fluorescence imaging for bone of a skinless mouse intravenously injected with the supramolecular dye [right].

Source: Reprinted with permission from Peck, E., et al. *Bioconjugate Chem.* 27, 1400–1410, 2016. Copyright 2016, American Chemical Society.

(b) Synthetic route to the conjugated polymer [top], and NIR-II fluorescent imaging (under a 1319 nm LP filter) of blood vessels in the cerebral cortex and hindlimb of a mouse preinjected with the polymer dye.

Source: Reprinted with permission from Liu, S., et al. *J. Am. Chem. Soc.* 142, 15146–15156, 2020. Copyright 2020, American Chemical Society.

can specifically target or monitor these factors can be used for tumor delineation and imaging-guided surgery. Considering the better practical clinical applications, wash-free NIR fluorescence imaging is attracting more and more attention, because it can greatly simplify the preprocessing, shorten the detection time, as well as minimize the potential background interference. From this perspective, NIR fluorophores with AIE characters (AIEgens) should be promising candidates for this wash-free fluorescence imaging technology, because they are non-emissive in dilute solution, but emit intensely in the targeted and aggregated sites [80].

There is another biological imaging modality based on NIR materials has received increasing attention, that is, photoacoustic imaging (PAI), which combines the advantages of optical and ultrasonic imaging, and possesses high spatial resolution and contrast. For example, compared to the traditional ultrasonic imaging modality, PAI can distinguish between targets by using different molecular markers or contrasts with distinct absorption wavelengths; compared to CT, there is no radioactive injury or cumulative effects for PAI; compared to MRI, it can be used for real-time and dynamic imaging; compared to fluorescence imaging, ultrasonic signal obtained in PAI has less tissue scattering and deeper tissue penetration (up to 5 cm and 6 cm in clinical applications) [81]. In PAI, the ultrasound signal is generated by the photoacoustic (PA) effect. Specifically, when the pulsed laser irradiates the contrast agents, it can absorb the photo energy, leading to transient thermal elastic expansion, and further forms an ultrasonic wave (photoacoustic signal), which can be detected and imaged by external ultrasound transducer. Therefore, the photoacoustic contrast agents play crucial role in PAI, which are typically NIR materials, including metal nanoparticles, 2D carbon materials, organic dyes, etc. Compared with inorganic materials, organic materials have fine and predictable structure and absorption spectra, lower toxicity, rapid metabolism, thus are arousing increasing research interest. There are several principles for designing organic NIR PA agents: (i) have fast non-radiative decays; (ii) exhibit high excitation coefficients in the NIR region; (iii) possess various molecular motor units in structure; (iv) have low emission quantum yield [82]. Since these organic molecules usually exhibit photothermal conversion properties, they are often used simultaneously in photothermal therapy. Examples can be found in Section 14.3.4 that follows.

14.3.3 NIR-Based Photodynamic Therapy

Cancer has become one of the major diseases threatening human health and is a leading cause of death, which accounts for around 10 million deaths in 2020 (nearly one sixth of all deaths) [83]. The traditional cancer treatments include surgery, chemotherapy, radiotherapy, etc. However, surgical treatment has great trauma and high recurrence rate, and chemotherapy suffers from non-specificity and high toxicity to normal tissues, while radiotherapy may have cumulative radiation injury. All these facts have brought serious and nonnegligible side effects to patients. In recent years, phototherapy, such as photodynamic therapy (PDT), is gaining more and more attention due to the high accuracy and minimal invasiveness during treatment. PDT is evolved from early discoveries about the phototoxicity of some organic dyes, and is defined as an oxygen-involved photoreaction of a photosensitizer (PS) under light irradiation a specific wavelength, generating cytotoxic ROS (such as singlet oxygen, $^1O_2$) and oxidizing crucial cellular proteins, eventually resulting in tumor ablation [84,85]. There are three indispensable elements, namely PS, oxygen and light, and the absence of any one will not produce toxicity to the biological systems. This brings great temporal and spatial precision to the treatment, which can be controlled by light. Figure 14.12 illustrates the photoreaction processes and mechanism of PDT. Firstly, the PS molecule absorbs light to transition from the ground state to the singlet excited state ($^1PS^*$), which then undergoes intersystem crossing (ISC) to form a triplet excited state ($^3PS^*$). This relatively long-lived excited state can react with biomolecules to form reactive free radical anions or cations, which further react with oxygen to generate superoxide anion radicals ($O_2^{\bullet-}$), hydroxyl radicals ($\bullet OH$), hydrogen peroxides (H_2O_2), and so forth (Type-I). Alternatively, $^3PS^*$ can directly transfer the energy to oxygen molecules and generate $^1O_2$ (Type-II). In most cases, these two reaction pathways coexist during PDT, and the generated ROS may cause direct damage to tumor cells, or destroy the blood vessels to block oxygen and nutrients supply to tumor. Besides, activation of antitumor specific immune system by PDT may also aid the ablation of tumor [86].

PSs play a central role in PDT, and their properties fundamentally affect the ROS generation. NIR-absorbing organic materials are promising candidates for PSs, due to their deeper tissue penetration and lower light-associated damage. Besides, the localization, concentration and metabolism of the PSs together influence the therapeutic effect. Therefore, great efforts have been devoted for the development of NIR-active PSs to enhance the clinical effectiveness and practicability of PDT. For example, porphyrins and phthalocyanines are traditional PSs, whose absorption wavelengths

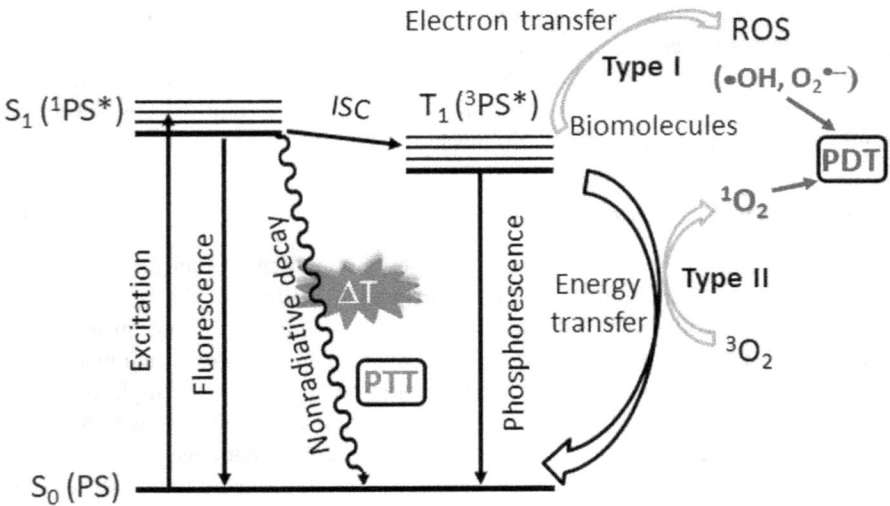

FIGURE 14.12 Schematic illustration of the photophysical and photochemical mechanisms for PDT and PTT.

can be shifted to NIR by using modification strategies mentioned in the previous sections. Longer triplet excited state lifetime and higher $^1O_2$ generation quantum yield can be achieved by the complexation with some metal ions. Besides, some metal, like Mn^{2+} and Gd^{3+}, chelated porphyrins or phthalocyanines can also be used as magnetic resonance imaging (MRI) contrast agents [87]. However, the problems of poor solubility and strong aggregation, as well as slow clearance in vivo, have to be overcome in the future. Cyanines are another kind of promising dyes not only be used in fluorescence imaging but also in PDT as PSs. The strategy of heavy-atom effect can be utilized to cyanines to increase their rates of ISC thus enhancing the generation of $^1O_2$ [88]. Although these significant developments of PSs have led to great progress in PDT, the application of PDT is still limited due to some intrinsic drawbacks. One of them is that, PDT is dependent on the presence of oxygen and it will be gradually depleted as the irradiation progresses, which further aggravate the hypoxia of tumors and deteriorated the treatment effect of PDT. Therefore, the combination of oxygen-carrier molecules with NIR PSs, covalently or noncovalently, would be an effective strategy to improve the comprehensive therapeutic effect of PDT [89].

14.3.4 NIR-Based Photothermal Therapy

In addition to the PDT, there is another attractive phototherapy for cancer treatment, named photothermal therapy (PTT). PTT can kill the cancer cells by using the photothermal effect of photothermal agents (PTAs) that convert the absorbed light energy into heat and rapidly raise the local temperature of tumor tissues, inducing the apoptosis of cancer cells [90,91]. Compared with other therapies, PTT possesses a series of advantages, such as light controlled precision treatment, minimized damage to the surrounding healthy tissues, convenient operation, low toxicity, and avoiding drug resistance [92]. Different from the PDT, PTT does not rely on oxygen during treatment, therefore it could be used for hypoxic tumors. There are two indispensable elements in PTT, namely, PTAs and light. Due to the crucial role of PTAs, considerable effort has been devoted to the design and construction of advanced PTAs. Generally, ideal PTAs should absorb strong NIR light in biological windows, selectively accumulate in tumor tissues, and have high photothermal conversion efficiency (PCE) [93]. To date, most of the PTAs are based on inorganic materials, such as noble metal nanomaterials, metal sulfide oxide materials, carbon nanomaterials, and Mxenes [94]. Although these inorganic materials possess favorable absorbance features, high PCEs, and good photostability, organic materials may win out in terms of better biocompatibility, biodegradability, and easier functionality.

As shown in Figure 14.12, a ground state organic molecule absorbs a photon and transfers to an excited state, which rapidly undergoes internal conversion to the lowest vibrational level of the first excited singlet state (S_1). This state can decay back to the ground state radiatively (emitting fluorescence), or non-radiatively (generating heat). Otherwise, the S_1 state is also possible to undergo ISC to a triplet state (typically T_1), which could be employed for PDT (see Section 14.3.3), or decay back to the ground state radiatively (emitting phosphorescence), or non-radiatively (generating heat). From this molecular photoelectronic mechanism, it can be seen that photons absorbed by molecules can generate heat through non-radiative deactivation processes, including intramolecular rotations and vibrations, which can be used to increase the temperature of the system and be used in PTT. In recent years, NIR-active organic PTAs have received increasing attention, which function by absorbing photons from NIR light.

From the perspective of energy conservation, in order to make NIR dyes have better photothermal conversion performance, it is necessary to prohibit the fluorescence, phosphorescence and ROS generation, while enhance the non-radiative transitions as much as possible. To date, a series of traditional NIR organic dyes, such as cyanines, porphyrins, phthalocyanines, diketopyrrolopyrroles, etc., have been reported to have PTT effect by some rational structural modifications [94–96]. A straightforward method to convert an emissive molecule to a non-emissive PTA, is to introducing flexible, rotatable and vibratile groups to the conjugated structure to directly boost the intramolecular motion. Besides, other strategies, such as employing photophysical quenching processes like PeT, aggregation, nanoengineering and supramolecular-engineering may also inhibit the radiative pathway [97,98]. For example, Liu and co-workers synthesized a borondifluoride bridged azafulvene molecule (BAF4) with a D-A-D structure and active intramolecular motions. This molecule exhibits strong NIR-II absorption but negligible fluorescence emission [99]. By assembling with Pluronic F-127 matrix, the BAF4 nanoparticles (BAF4 NPs) were formed, showing a very high PSE value of 80% by irradiation with a 1064 nm laser (0.75 W cm^{-2}), which was much higher than most of the reported inorganic or organic PTAs, providing excellent PTT effect for tumor ablation. What's more, the BAF4 NPs also showed PA signal under 910 nm pulsed laser irradiation, thus could be used as a PA agent to guide the PTT in vivo, further enhancing the accuracy of treatment (Figure 14.13). As a result, PAI-guided PTT or NIR fluorescence imaging-guided PTT (by regulating the balance of an NIR dye between its radiative and non-radiative decays), may hold great potential in imaging-guided phototherapy for tumors.

14.3.5 NIR-Light-Triggered Drug Release

Precision medicine not only asks for matching the right drug with the specific patient, but also delivering the right dosage to the right site at the right time. Based on this idea, a new concept of on-demand drug delivery (or smart drug-delivery) system has emerged, which is able to explicitly control over where, when, and how much dosage by using internal or external stimuli. To date, a series of stimuli have been used individually or synergistically, such as temperature, pH, redox, biomarkers, electric/magnetic field, ultrasonic, light, and so forth [100]. Among these possible stimuli, the light-input, especially that in the NIR region, is the most attractive modality, due to its cleanness, non-invasiveness, remote control, high spatial and temporal resolution [101]. For example, the drugs can be automatically released from in vivo medical devices simply by shining a light on the skin. In principle, the light-triggered drug release systems consist of a photo-sensitive motif, and a drug (or pre-drug) motif. The former undergoes photophysical or photochemical processes after absorbing light energy, including photothermal conversion, photocleavage reaction, photogeneration of free radicals or ROS, etc.

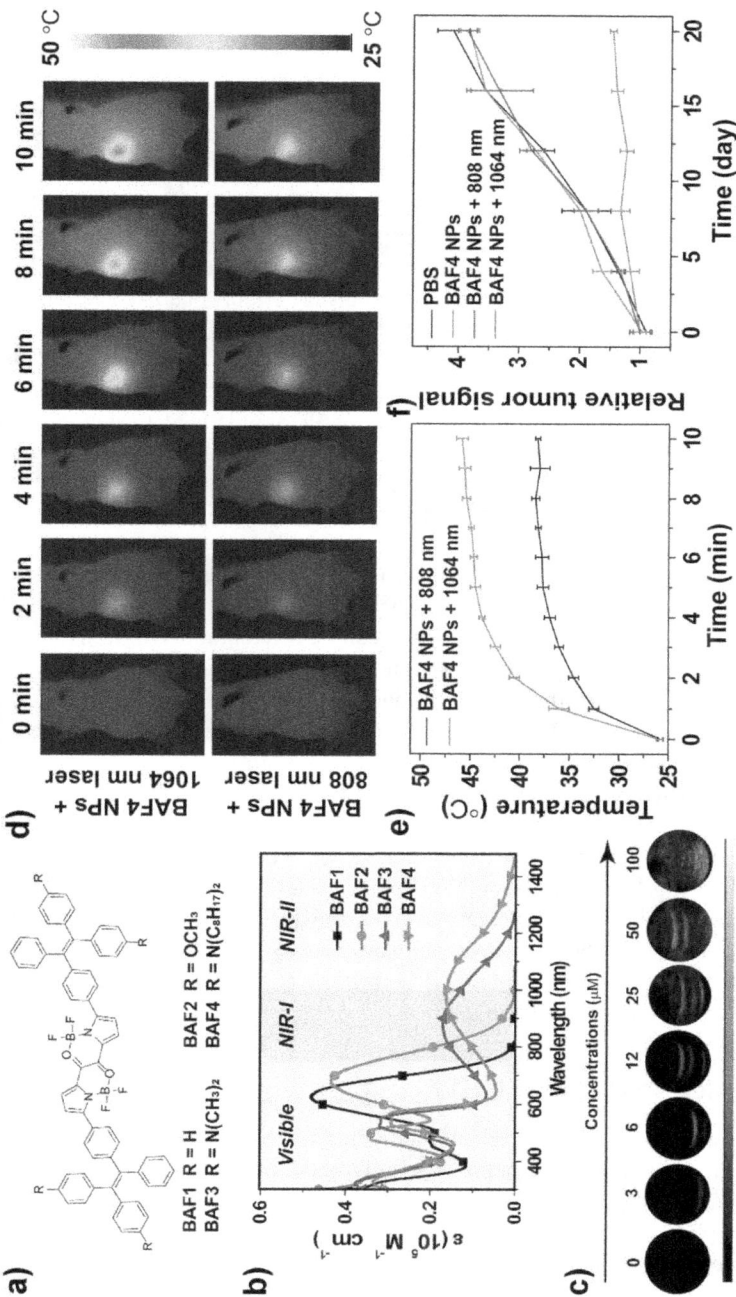

FIGURE 14.13 (a) The chemical structures and (b) absorption spectra in dichloromethane of compounds BAF1–4. (c) In vitro PA images of aqueous dispersions of BAF4 NPs at different concentrations under 910 nm laser irradiation. (d) Infrared thermal imaging of orthotopic liver tumor mice treated with BAF4 NPs under 808 or 1064 nm laser (0.75 W cm⁻²) irradiation for different time. (e) The corresponding temperatures of tumor-site as a function of irradiation time. (f) Relative orthotopic liver tumor luminescence levels of the mice at different time under different treatments.

Source: Reprinted with permission from Jiang, Z., et al. *Angew. Chem. Int. Ed.* 60, 22376–22384, 2021. Copyright 2021, Wiley-VCH GmbH.

which induce the decomposition of nanocarriers or cleavage of the chemical linkage between the photosensitizer and drug, thereby releasing the drug at the desired site in a controlled manner [102].

The most widely used photophysical process for drug release is the photothermal mode. As noted above, PTAs can rapidly increase the temperature under NIR light irradiation, which further breaks the structure of drug-encapsulated nanoassemblies, leading to the release of therapeutic agents. For example, organic NIR PTAs combined with a thermal-sensitive phospholipid with an appropriate phase transition temperature, forming nanoparticles in aqueous medium, offers a suitable platform for drug encapsulation, and the photothermal induced gel-lipid phase transition can effectively release the encapsulated small molecular drugs [103]. Another quite delicate strategy to employ photothermal effect for drug release is the NIR photothermal-induced blasting, reported by Jia, Liu, and co-workers [104]. They constructed a nanodrug by coating the doxorubicin (an anti-cancer drug) and NH_4HCO_3 dual loaded nanoparticle with a polydopamine (PDA) film. When irradiated by NIR light, the photothermal effect of PDA will lead to decomposition of NH_4HCO_3 which generates carbon dioxide (CO_2) and ammonia (NH_3) gases, thus resulting in a "bomb-like" release of doxorubicin in situ. On the other hand, the photochemically triggered drug delivery method facilitates the release of encapsulated or linked therapeutic agents through cleavage of covalent bonds between the carrier and drug. There may be two kinds of approaches for this photochemically triggered release method. The first and direct one is to introduce photo-cuttable covalent bonds as linkers, which can be broken under light irradiation, such as the o-nitrobenzyl systems and some easter bonds. However, due to the high energy required to break the covalent bonds, in most cases, UV irradiation is needed, which greatly limits their bio-applications in situ. Recently, some new NIR-light cleavable systems have been developed, demonstrating promising potential for further photo-triggered drug release [105,106]. The second and indirect approach is based on the photosensitization-induced oxidation, whose mechanism is the same as PDT. In this approach ROS, such as $^1O_2$, $\bullet OH$, $O_2^{\bullet-}$, H_2O_2, etc., are firstly generated during NIR irradiation, which can oxidate some kinds of molecular components like lipids and cholesterols, resulting in structural disruption of the nanocarrier and release the encapsulated drug. Due to the same photoreaction mechanism with PDT, NIR-light triggered drug release is often used in the chemo and photodynamic combination therapy.

14.4 SUMMARY AND PERSPECTIVE

Organic molecules have become excellent candidates for NIR materials, and their structure-property relationships are being gradually understood and well employed. What's more, the convergence of organic chemistry, supramolecular chemistry, polymer science, phonology, analytical chemistry, and nanotechnology, has greatly facilitated the progress of NIR organic materials with various advanced and tunable properties. However, despite the current success, there is still a long way to go in terms of new molecular designs and practical applications in industry, defense, energy, space, and biomedicine, etc.

Generally, there are two directions for future research on NIR organic materials, namely, new structures, and better properties. First, for the structure, currently used NIR molecules mostly rely on a limited number of classical primitives, including cyanines, porphyrins, phthalocyanines, perylenes, BODIPYs, etc. Therefore, designing more novel molecular structures to fundamentally enrich the sources of NIR materials, and developing new rules or theories to guide molecular design, will be a challenging but valuable task. Second, for biological applications, the biosecurity is the first and foremost issue for the organic NIR materials. As a result, the biocompatibility, metabolic capacity, and long-term biodegradability for new NIR materials should be taken into account and inspected carefully. In addition, as an application of biosensing and imaging in vivo, NIR-II light is superior to NIR-I in terms of deeper tissue penetration depth and lower scattering. Therefore, developing NIR-II fluorophores, especially those activatable NIR-II fluorescent probes, are particularly desirable, which may provide powerful tools for monitoring and imaging analytes of interest in vivo with high SBR. Another direction for development of new organic NIR imaging materials is the construction of NIR phosphorescent materials, which can further increase the SBR for bioimaging, owing to their large Stokes shifts, and especially long emission lifetime (typically in the time domain of microseconds to seconds), which can minimize or even eliminate the short-lived autofluorescence background interference through a time-gated imaging technology. For phototherapy applications, NIR materials with integrated functions and therapeutic effects (or called all-in-one materials) will be the future direction. For example, combining of PDT and PTT, chemotherapy, and PDT/PTT, in a single NIR material, can not only achieve a complementary synergistic effect in tumor ablation, but also effectively prevent the generation of drug resistance. Moreover, NIR materials with imaging-guided phototherapy, and signal-feedback phototherapy, may contribute to the accurate theranostics for a series of diseases. To sum up, although some challenges still need to be addressed, NIR organic materials have a promising future for widespread use in smart and precision medicine.

It is also worthy to mention that there are enormous opportunities by combining the NIR organic materials with cutting-edge nanotechnologies, triggering a whole range of paradigm-shift technologies and applications. For example, the NIR organic fluorescent probes could be bonded to nanophotonic structures to engineer their emission intensities and directivities (see Chapter 6), which have been proposed for various emergent applications such as single molecule sensing, point-of-care system (POC), lab-on-chip, early diagnosis, and so forth. Researchers are advised to pay special attention to these multi-disciplinary directions, and explore possibility for revolutionary applications.

ACKNOWLEDGEMENTS

This work is supported by the Shanghai Pujiang Program (19PJ1402800), the National Natural Science Foundation of China (NSFC, 21904040, 21635003, 21827814, 21811540027), and the Fundamental Research Funds for the Central Universities.

BIBLIOGRAPHY

[1] Peierls, R. E. 1955. *Quantum Theory of Solids*. Oxford University Press, Oxford.

[2] Bilici, K.; Cetin, S.; Aydındogan, E.; Yagci Acar, H.; Kolemen, S. 2021. Recent Advances in Cyanine-Based Phototherapy Agents. *Front. Chem.* 9: 707876.

[3] Ilina, K.; MacCuaig, W. M.; Laramie, M.; Jeouty, J. N.; McNally, L. R.; Henary, M. 2019. Squaraine Dyes: Molecular Design for Different Applications and Remaining Challenges. *Bioconjugate Chem.* 31: 194–213.

[4] Wu, W.; Zheng, T.; Tian, Y. 2020. An Enzyme-Free Amplification Strategy Based on Two-Photon Fluorescent Carbon Dots for Monitoring miR-9 in Live Neurons and Brain Tissues of Alzheimer's Disease Mice. *Chem. Commun.* 56: 8083–8086.

[5] Li, W.; Liu, Z.; Fang, B.; Jin, M.; Tian, Y. 2020. Two-Photon Fluorescent Zn^{2+} Probe for Ratiometric Imaging and Biosensing of Zn^{2+} in Living Cells and Larval Zebrafish. *Biosens. Bioelectron.* 148: 111666.

[6] Zhang, Q.; Tian, X.; Zhou, H.; Wu, J.; Tian, Y. 2017. Lighting the Way to See Inside Two-Photon Absorption Materials: Structure–Property Relationship and Biological Imaging. *Molecules.* 10: 223.

[7] Pascal, S.; David, S.; Andraud, C.; Maury, O. 2021. Near-Infrared Dyes for Two-Photon Absorption in the Short-Wavelength Infrared: Strategies Towards Optical Power Limiting. *Chem. Soc. Rev.* 50: 6613–6658.

[8] Lehn, J. M. 1995. *Supramolecular Chemistry: Concepts and Perspectives*. John Wiley & Sons, Weinheim.

[9] Qu, D.-H.; Wang, Q.-C.; Zhang, Q.-W.; Ma, X.; Tian, H. 2015. Photoresponsive Host-Guest Functional Systems. *Chem. Rev.* 115: 7543–7588.

[10] Zhang, Q.; Qu, D.-H.; Wu, J.; Ma, X.; Wang, Q.; Tian, H. 2013. A Dual-Modality Photoswitchable Supramolecular Polymer. *Langmuir.* 29: 5345–5350.

[11] Zhang, Q.; Qu, D.-H.; Ma, X.; Tian, H. 2013. Sol-Gel Conversion Based on Photoswitching between Noncovalently and Covalently Linked Netlike Supramolecular Polymers. *Chem. Commun.* 49: 9800–9802.

[12] Zhang, Q.; Yao, X.; Qu, D.-H.; Ma, X. 2014. Multistate Self-Assembled Micro-Morphology Transitions Controlled by Host-Guest Interactions. *Chem. Commun.* 50: 1567–1569.

[13] Li, H.; Zhang, H.; Zhang, Q.; Zhang, Q.; Qu, D.-H. 2012. A Switchable Ferrocene-Based [1]Rotaxane with an Electrochemical Signal Output. *Org. Lett.* 14: 5900–5903.

[14] Stoffel, S.; Zhang, Q.-W.; Li, D.-H.; Smith, B. D.; Peng, J. W. 2020. NMR Relaxation Dispersion Reveals Macrocycle Breathing Dynamics in a Cyclodextrin-Based Rotaxane. *J. Am. Chem. Soc.* 142: 7413–7424.

[15] Wang, Q.; Zhang, Q.; Zhang, Q.-W.; Li, X.; Zhao, C.-X.; Xu, T.-Y.; Qu, D.-H.; Tian, H. 2020. Color-Tunable Single-Fluorophore Supramolecular System with Assembly-Encoded Emission. *Nat. Commun.* 11: 158.

[16] Song, Q.; Jiao, Y.; Wang, Z.; Zhang, X. 2016. Tuning the Energy Gap by Supramolecular Approaches: Towards Near-Infrared Organic Assemblies and Materials. *Small.* 12: 24–31.

[17] Li, D.; Feng, Z.; Han, Y.; Chen, C.; Zhang, Q.-W.; Tian, Y. 2022. Time-Resolved Encryption via a Kinetics-Tunable Supramolecular Photochromic System. *Adv. Sci.* 9: 2104790.

[18] Sun, R.; Zhang, Q.; Wang, Q.; Ma, X. 2013. Novel Supramolecular CT Polymer Employing Disparate Pseudorotaxanes as Relevant Monomers. *Polymer.* 54: 2506–2510.

[19] Barrow, S. J.; Kasera, S.; Rowland, M. J.; del Barrio, J.; Scherman, O. A. 2015. Cucurbituril-Based Molecular Recognition. *Chem. Rev.* 115: 12320–12406.

[20] Zhang, Q.; Tian, H. 2014. Effective Integrative Supramolecular Polymerization. *Angew. Chem. Int. Ed.* 53: 10582–10584.

[21] Ko, Y. H.; Kim, E.; Hwang, I.; Kim, K. 2007. Supramolecular Assemblies Built with Host-Stabilized Charge-Transfer Interactions. *Chem. Commun.*: 1305–1315.

[22] Sun, S.; Gao, W.; Liu, F.; Fan, J.; Peng, X. 2010. Study of an Unusual Charge-Transfer Inclusion Complex with NIR Absorption, and Its Application for DNA Photocleavage. *J. Mater. Chem.* 20: 5888–5892.

[23] Kasha, M.; Rawls, H. R.; El-Bayoumi, M. A. 1965. The Exciton Model in Molecular Spectroscopy. *Pure Appl. Chem.* 11: 371–392.

[24] Hestand, N. J.; Spano, F. C. 2018. Expanded Theory of H- and J-Molecular Aggregates: The Effects of Vibronic Coupling and Intermolecular Charge Transfer. *Chem. Rev.* 118: 7069–7163.

[25] Cravcenco, A.; Yu, Y.; Edhborg, F.; Goebel, J. F.; Takacs, Z.; Yang, Y.; Albinsson, B.; Börjesson, K. 2021. Exciton Delocalization Counteracts the Energy Gap: A New Pathway toward NIR-Emissive Dyes. *J. Am. Chem. Soc.* 143: 19232–19239.

[26] Gomberg, M. 1900. An Instance of Trivalent Carbon: Triphenylmethyl. *J. Am. Chem. Soc.* 22: 757–771.

[27] Chen, Y.; Li, J.; Zhao, Y.; Zhang, L.; Tan, G.; Zhu, H.; Roesky, H. W. 2021. Stable Radical Cation and Dication of a 1,4-Disilabenzene. *J. Am. Chem. Soc.* 143: 2212–2216.

[28] Pushkarevsky, N. A.; Chulanova, E. A.; Shundrin, L. A.; Smolentsev, A. I.; Salnikov, G. E.; Pritchina, E. A.; Genaev, A. M.; Irtegova, I. G.; Bagryanskaya, I. Y.; Konchenko, S. N.; Gritsan, N. P.; Beckmann, J.; Zibarev, A. V. 2019. Radical Anions, Radical-Anion Salts, and Anionic Complexes of 2,1,3-Benzochalcogenadiazoles. *Chem.-Eur. J.* 25: 806–816.

[29] Huang, B.; Mao, L.; Shi, X.; Yang, H.-B. 2021. Recent Advances and Perspectives on Supramolecular Radical Cages. *Chem. Sci.* 12: 13648–13663.

[30] Ouari, O.; Bardelang, D. 2018. Nitroxide Radicals with Cucurbit[n]urils and Other Cavitands. *Isr. J. Chem.* 58: 343–356.

[31] Barnes, J. C.; Fahrenbach, A. C.; Cao, D.; Dyar, S. M.; Frasconi, M.; Giesener, M. A.; Benítez, D.; Tkatchouk, E.; Chernyashevskyy, O.; Shin, W. H.; Li, H.; Sampath, S.; Stern, C. L.; Sarjeant, A. A.; Hartlieb, K. J.; Liu, Z.; Carmieli, R.; Botros, Y. Y.; Choi, J. W.; Slawin, A. M. Z.; Ketterson, J. B.; Wasielewski, M. R.; Goddard, W. A.; Stoddart, J. F. 2013. A Radically Configurable Six-State Compound. *Science.* 339: 429–433.

[32] Zheng, X.; Zhang, Y.; Cao, N.; Li, X.; Zhang, S.; Du, R.; Wang, H.; Ye, Z.; Wang, Y.; Cao, F.; Li, H.; Hong, X.; Sue, A. C. H.; Yang, C.; Liu, W.-G.; Li, H. 2018. Coulombic-Enhanced Hetero Radical Pairing Interactions. *Nat. Commun.* 9: 1961.

[33] Qian, Z.; Huang, X.; Wang, Q. 2017. Stabilizing Benzyl Viologen Radical Cation by Cucurbit[7]uril Rotaxanation. *Dyes Pigments.* 145: 365–370.

[34] Zhang, Q.; Qu, D.-H.; Wang, Q.-C.; Tian, H. 2015. Dual-Mode Controlled Self-Assembly of TiO2 Nanoparticles Through a Cucurbit[8]uril-Enhanced Radical Cation Dimerization Interaction. *Angew. Chem. Int. Ed.* 54: 15789–15793.

[35] Yang, Y.; He, P.; Wang, Y.; Bai, H.; Wang, S.; Xu, J.-F.; Zhang, X. 2017. Supramolecular Radical Anions Triggered by Bacteria in Situ for Selective Photothermal Therapy. *Angew. Chem. Int. Ed.* 56: 16239–16242.

[36] Zhang, Y.; Wang, Y.; Yang, X.; Yang, Q.; Li, J.; Tan, W. 2020. Polyaniline Nanovesicles for Photoacoustic Imaging-Guided Photothermal-Chemo Synergistic Therapy in the Second Near-Infrared Window. *Small.* 16: 2001177.

[37] Theune, L. E.; Buchmann, J.; Wedepohl, S.; Molina, M.; Laufer, J.; Calderón, M. 2019. NIR- and Thermo-Responsive Semi-Interpenetrated Polypyrrole Nanogels for Imaging Guided Combinational Photothermal and Chemotherapy. *J. Control. Release*. 311–312: 147–161.

[38] Gibson, G. L.; McCormick, T. M.; Seferos, D. S. 2012. Atomistic Band Gap Engineering in Donor–Acceptor Polymers. *J. Am. Chem. Soc.* 134: 539–547.

[39] Pu, K.; Mei, J.; Jokerst, J. V.; Hong, G.; Antaris, A. L.; Chattopadhyay, N.; Shuhendler, A. J.; Kurosawa, T.; Zhou, Y.; Gambhir, S. S.; Bao, Z.; Rao, J. 2015. Diketopyrrolopyrrole-Based Semiconducting Polymer Nanoparticles for In Vivo Photoacoustic Imaging. *Adv. Mater.* 27: 5184–5190.

[40] Zhang, C.; Zeng, Z.; Cui, D.; He, S.; Jiang, Y.; Li, J.; Huang, J.; Pu, K. 2021. Semiconducting Polymer Nano-PROTACs for Activatable Photo-Immunometabolic Cancer Therapy. *Nat. Commun.* 12: 2934.

[41] Dai, Y.; Du, W.; Gao, D.; Zhu, H.; Zhang, F.; Chen, K.; Ni, H.; Li, M.; Fan, Q.; Shen, Q. 2022. Near-Infrared-II Light Excitation Thermosensitive Liposomes for Photoacoustic Imaging-Guided Enhanced Photothermal-Chemo Synergistic Tumor Therapy. *Biomater. Sci.* 10: 435–443.

[42] Du, C.; Wu, X.; He, M.; Zhang, Y.; Zhang, R.; Dong, C.-M. 2021. Polymeric Photothermal Agents for Cancer Therapy: Recent Progress and Clinical Potential. *J. Mater. Chem. B.* 9: 1478–1490.

[43] Liu, T.; Zhang, M.; Liu, W.; Zeng, X.; Song, X.; Yang, X.; Zhang, X.; Feng, J. 2018. Metal Ion/Tannic Acid Assembly as a Versatile Photothermal Platform in Engineering Multimodal Nanotheranostics for Advanced Applications. *ACS Nano.* 12: 3917–3927.

[44] Côté, A. P.; Benin, A. I.; Ockwig, N. W.; O'Keeffe, M.; Matzger, A. J.; Yaghi, O. M. 2005. Porous, Crystalline, Covalent Organic Frameworks. *Science.* 310: 1166–1170.

[45] Ding, S.-Y.; Wang, W. 2013. Covalent Organic Frameworks (COFs): From Design to Applications. *Chem. Soc. Rev.* 42: 548–568.

[46] Dong, Y.-B.; Guan, Q.; Zhou, L.-L.; Li, W.-Y.; Li, Y.-A. 2019. Covalent Organic Frameworks (COFs) for Cancer Therapeutics. *Chem.-Eur. J.* 26: 5583–5591.

[47] Xia, R.; Zheng, X.; Li, C.; Yuan, X.; Wang, J.; Xie, Z.; Jing, X. 2021. Nanoscale Covalent Organic Frameworks with Donor: Acceptor Structure for Enhanced Photothermal Ablation of Tumors. *ACS Nano.* 15: 7638–7648.

[48] Jiang, D. 2020. Covalent Organic Frameworks: An Amazing Chemistry Platform for Designing Polymers. *Chem.* 6: 2461–2483.

[49] Shi, Y.; Liu, S.; Liu, Y.; Sun, C.; Chang, M.; Zhao, X.; Hu, C.; Pang, M. 2019. Facile Fabrication of Nanoscale Porphyrinic Covalent Organic Polymers for Combined Photodynamic and Photothermal Cancer Therapy. *ACS Appl. Mater. Interfaces.* 11: 12321–12326.

[50] Guan, Q.; Fu, D.-D.; Li, Y.-A.; Kong, X.-M.; Wei, Z.-Y.; Li, W.-Y.; Zhang, S.-J.; Dong, Y.-B. 2019. BODIPY-Decorated Nanoscale Covalent Organic Frameworks for Photodynamic Therapy. *iScience.* 14: 180–198.

[51] Zhang, Q.; Elemans, H. J. A. A. W.; White, P.; Nolte, R. J. M. 2018. A Manganese Porphyrin-α-Cyclodextrin Conjugate as an Artificial Enzyme for the Catalytic Epoxidation of Polybutadiene. *Chem. Commun.* 54: 5586–5589.

[52] Wang, K.; Zhang, Z.; Lin, L.; Hao, K.; Chen, J.; Tian, H.; Chen, X. 2019. Cyanines-Assisted Exfoliation of Covalent Organic Frameworks into Nanocomposites for Highly Efficient Chemo-Photothermal Tumor Therapy. *ACS Appl. Mater. Interfaces.* 11: 39503–39512.

[53] Smith, A. M.; Mancini, M. C.; Nie, S. 2009. Second Window for in Vivo Imaging. *Nat. Nanotechnol.* 4: 710–711.

[54] Hemmer, E.; Venkatachalam, N.; Hyodo, H.; Hattori, A.; Ebina, Y.; Kishimoto, H.; Soga, K. 2013. Upconverting and NIR Emitting Rare Earth Based Nanostructures for NIR-Bioimaging. *Nanoscale*. 5: 11339–11361.

[55] Wu, Z.; Liu, M.; Liu, Z.; Tian, Y. 2020. Real-Time Imaging and Simultaneous Quantification of Mitochondrial H2O2 and ATP in Neurons with a Single Two-Photon Fluorescence Lifetime-Based Probe. *J. Am. Chem. Soc.* 142: 7532–7541.

[56] Ge, L.; Tian, Y. 2019. Fluorescence Lifetime Imaging of p-tau Protein in Single Neuron with a Highly Selective Fluorescent Probe. *Anal. Chem.* 91: 3294–3301.

[57] Li, W.; Fang, B.; Jin, M.; Tian, Y. 2017. Two-Photon Ratiometric Fluorescence Probe with Enhanced Absorption Cross Section for Imaging and Biosensing of Zinc Ions in Hippocampal Tissue and Zebrafish. *Anal. Chem.* 89: 2553–2560.

[58] Liu, Z.; Wu, P.; Yin, Y.; Tian, Y. 2019. A Ratiometric Fluorescent DNA Nanoprobe for Cerebral Adenosine Triphosphate Assay. *Chem. Commun.* 55: 9955–9958.

[59] Liu, Z.; Wang, S.; Li, W.; Tian, Y. 2018. Bioimaging and Biosensing of Ferrous Ion in Neurons and HepG2 Cells upon Oxidative Stress. *Anal. Chem.* 90: 2816–2825.

[60] Liu, Z.; Tian, Y. 2021. Recent Advances in Development of Devices and Probes for Sensing and Imaging in the Brain. *Sci. China Chem.* 64: 915–931.

[61] Liu, Z.; Jing, X.; Zhang, S.; Tian, Y. 2019. A Copper Nanocluster-Based Fluorescent Probe for Real-Time Imaging and Ratiometric Biosensing of Calcium Ions in Neurons. *Anal. Chem.* 91: 2488–2497.

[62] Huang, H.; Tian, Y. 2018. A Ratiometric Fluorescent Probe for Bioimaging and Biosensing of HBrO in Mitochondria upon Oxidative Stress. *Chem. Commun.* 54: 12198–12201.

[63] Hang, Y.; Boryczka, J.; Wu, N. 2022. Visible-Light and Near-Infrared Fluorescence and Surface-Enhanced Raman Scattering Point-of-Care Sensing and Bio-Imaging: A Review. *Chem. Soc. Rev.* 51: 329–375.

[64] Zhang, Q.-W.; Li, D.; Li, X.; White, P. B.; Mecinović, J.; Ma, X.; Ågren, H.; Nolte, R. J. M.; Tian, H. 2016. Multicolor Photoluminescence Including White-Light Emission by a Single Host–Guest Complex. *J. Am. Chem. Soc.* 138: 13541–13550.

[65] Li, D.; Hu, W.; Wang, J.; Zhang, Q.; Cao, X.; Ma, X.; Tian, H. 2018. White-Light Emission from a Single Organic Compound with Unique Self-Folded Conformation and Multistimuli Responsiveness. *Chem. Sci.* 9: 5709–5715.

[66] Li, D.; Han, Y.; Jiang, Y.; Jiang, G.; Sun, H.; Sun, Z.; Zhang, Q.-W.; Tian, Y. 2022. Achieving Adjustable Multifunction Based on Host–Guest Interaction-Manipulated Reversible Molecular Conformational Switching. *ACS Appl. Mater. Interfaces.* 14: 1807–1816.

[67] Sedgwick, A. C.; Wu, L.; Han, H.-H.; Bull, S. D.; He, X.-P.; James, T. D.; Sessler, J. L.; Tang, B. Z.; Tian, H.; Yoon, J. 2018. Excited-state intramolecular proton-transfer (ESIPT) based fluorescence sensors and imaging agents. *Chem. Soc. Rev.* 47: 8842–8880.

[68] Guo, Z.; Park, S.; Yoon, J.; Shin, I. 2014. Recent Progress in the Development of Near-Infrared Fluorescent Probes for Bioimaging Applications. *Chem. Soc. Rev.* 43: 16–29.

[69] Ge, L.; Liu, Z.; Tian, Y. 2020. A Novel Two-Photon Ratiometric Fluorescent Probe for Imaging and Sensing of BACE1 in Different Regions of AD Mouse Brain. *Chem. Sci.* 11: 2215–2224.

[70] Tian, X.; Murfin, L. C.; Wu, L.; Lewis, S. E.; James, T. D. 2021. Fluorescent Small Organic Probes for Biosensing. *Chem. Sci.* 12: 3406–3426.

[71] Du, Y.; Liu, X.; Zhu, S. 2021. Near-Infrared-II Cyanine/Polymethine Dyes, Current State and Perspective. *Front. Chem.* 9: 718709.

[72] Li, C.; Chen, G.; Zhang, Y.; Wu, F.; Wang, Q. 2020. Advanced Fluorescence Imaging Technology in the Near-Infrared-II Window for Biomedical Applications. *J. Am. Chem. Soc.* 142: 14789–14804.

[73] Li, D.; Pan, J.; Xu, S.; Fu, S.; Chu, C.; Liu, G. 2021. Activatable Second Near-Infrared Fluorescent Probes: A New Accurate Diagnosis Strategy for Diseases. *Biosensors.* 11: 436.

[74] Mei, Y.; Zhang, Q.-W.; Gu, Q.; Liu, Z.; He, X.; Tian, Y. 2022. Pillar[5]arene-Based Fluorescent Sensor Array for Biosensing of Intracellular Multi-Neurotransmitters through Host–Guest Recognitions. *J. Am. Chem. Soc.* 144: 2351–2359.

[75] Wang, S.; Fan, Y.; Li, D.; Sun, C.; Lei, Z.; Lu, L.; Wang, T.; Zhang, F. 2019. Anti-Quenching NIR-II Molecular Fluorophores for in Vivo High-Contrast Imaging and pH Sensing. *Nat. Commun.* 10: 1058.

[76] Xu, G.; Yan, Q.; Lv, X.; Zhu, Y.; Xin, K.; Shi, B.; Wang, R.; Chen, J.; Gao, W.; Shi, P.; Fan, C.; Zhao, C.; Tian, H. 2018. Imaging of Colorectal Cancers Using Activatable Nanoprobes with Second Near-Infrared Window Emission. *Angew. Chem. Int. Ed.* 57: 3626–3630.

[77] Peck, E. M.; Battles, P. M.; Rice, D. R.; Roland, F. M.; Norquest, K. A.; Smith, B. D. 2016. Pre-Assembly of Near-Infrared Fluorescent Multivalent Molecular Probes for Biological Imaging. *Bioconjugate Chem.* 27: 1400–1410.

[78] Liu, S.; Ou, H.; Li, Y.; Zhang, H.; Liu, J.; Lu, X.; Kwok, R. T. K.; Lam, J. W. Y.; Ding, D.; Tang, B. Z. 2020. Planar and Twisted Molecular Structure Leads to the High Brightness of Semiconducting Polymer Nanoparticles for NIR-IIa Fluorescence Imaging. *J. Am. Chem. Soc.* 142: 15146–15156.

[79] Haque, A.; Faizi, M. S. H.; Rather, J. A.; Khan, M. S. 2017. Next Generation NIR Fluorophores for Tumor Imaging and Fluorescence-Guided Surgery: A Review. *Biorg. Med. Chem.* 25: 2017–2034.

[80] Huang, X.; Zhang, R.; Chen, C.; Kwok, R. T. K.; Tang, B. Z. 2021. Wash-Free Detection and Bioimaging by AIEgens. *Mater. Chem. Front.* 5: 723–743.

[81] Du, J.; Yang, S.; Qiao, Y.; Lu, H.; Dong, H. 2021. Recent Progress in Near-Infrared Photoacoustic Imaging. *Biosens. Bioelectron.* 191: 113478.

[82] Li, C.; Liu, C.; Fan, Y.; Ma, X.; Zhan, Y.; Lu, X.; Sun, Y. 2021. Recent Development of Near-Infrared Photoacoustic Probes Based on Small-Molecule Organic Dye. *RSC Chem. Bio.* 2: 743–758.

[83] Ferlay, J.; Colombet, M.; Soerjomataram, I.; Parkin, D. M.; Piñeros, M.; Znaor, A.; Bray, F. 2021. Cancer Statistics for the Year 2020: An Overview. *International Journal of Cancer.* 149: 778–789.

[84] Lucky, S. S.; Soo, K. C.; Zhang, Y. 2015. Nanoparticles in Photodynamic Therapy. *Chem. Rev.* 115: 1990–2042.

[85] Celli, J. P.; Spring, B. Q.; Rizvi, I.; Evans, C. L.; Samkoe, K. S.; Verma, S.; Pogue, B. W.; Hasan, T. 2010. Imaging and Photodynamic Therapy: Mechanisms, Monitoring, and Optimization. *Chem. Rev.* 110: 2795–2838.

[86] Liu, Y.; Meng, X.; Bu, W. 2019. Upconversion-based Photodynamic Cancer Therapy. *Coord. Chem. Rev.* 379: 82–98.

[87] Xue, X.; Lindstrom, A.; Li, Y. 2019. Porphyrin-Based Nanomedicines for Cancer Treatment. *Bioconjugate Chem.* 30: 1585–1603.

[88] Atchison, J.; Kamila, S.; Nesbitt, H.; Logan, K. A.; Nicholas, D. M.; Fowley, C.; Davis, J.; Callan, B.; McHale, A. P.; Callan, J. F. 2017. Iodinated Cyanine Dyes: A New Class of Sensitisers for Use in NIR Activated Photodynamic Therapy (PDT). *Chem. Commun.* 53: 2009–2012.

[89] Shen, Z.; Ma, Q.; Zhou, X.; Zhang, G.; Hao, G.; Sun, Y.; Cao, J. 2021. Strategies to Improve Photodynamic Therapy Efficacy by Relieving the Tumor Hypoxia Environment. *NPG Asia Mater.* 13: 39.

[90] Lv, S.; Miao, Y.; Liu, D.; Song, F. 2020. Recent Development of Photothermal Agents (PTAs) Based on Small Organic Molecular Dyes. *ChemBioChem.* 21: 2098–2110.

[91] Liu, J.; Qu, Y.; Zheng, T.; Tian, Y. 2019. A Dual-Mode Nanoprobe for Evaluation of the Autophagy Level Affected by Photothermal Therapy. *Chem. Commun.* 55: 9673–9676.

[92] Yan, C.; Zhang, Y.; Guo, Z. 2021. Recent Progress on Molecularly Near-Infrared Fluorescent Probes for Chemotherapy and Phototherapy. *Coord. Chem. Rev.* 427: 213556.

[93] Jung, H. S.; Verwilst, P.; Sharma, A.; Shin, J.; Sessler, J. L.; Kim, J. S. 2018. Organic Molecule-Based Photothermal Agents: An Expanding Photothermal Therapy Universe. *Chem. Soc. Rev.* 47: 2280–2297.

[94] Liu, Y.; Bhattarai, P.; Dai, Z.; Chen, X. 2019. Photothermal Therapy and Photoacoustic Imaging via Nanotheranostics in Fighting Cancer. *Chem. Soc. Rev.* 48: 2053–2108.

[95] Zhu, H.; Cheng, P.; Chen, P.; Pu, K. 2018. Recent Progress in the Development of Near-Infrared Organic Photothermal and Photodynamic Nanotherapeutics. *Biomater. Sci.* 6: 746–765.

[96] Zhang, Y.; Zhang, S.; Zhang, Z.; Ji, L.; Zhang, J.; Wang, Q.; Guo, T.; Ni, S.; Cai, R.; Mu, X.; Long, W.; Wang, H. 2021. Recent Progress on NIR-II Photothermal Therapy. *Front. Chem.* 9: 728066.

[97] Liu, S.; Li, Y.; Kwok, R. T. K.; Lam, J. W. Y.; Tang, B. Z. 2021. Structural and Process Controls of AIEgens for NIR-II Theranostics. *Chem. Sci.* 12: 3427–3436.

[98] Kwon, N.; Kim, H.; Li, X.; Yoon, J. 2021. Supramolecular Agents for Combination of Photodynamic Therapy and Other Treatments. *Chem. Sci.* 12: 7248–7268.

[99] Jiang, Z.; Zhang, C.; Wang, X.; Yan, M.; Ling, Z.; Chen, Y.; Liu, Z. 2021. A Borondifluoride-Complex-Based Photothermal Agent with an 80% Photothermal Conversion Efficiency for Photothermal Therapy in the NIR-II Window. *Angew. Chem. Int. Ed.* 60: 22376–22384.

[100] Linsley, C. S.; Wu, B. M. 2017. Recent Advances in Light-Responsive on-Demand Drug-Delivery Systems. *Ther. Deliv.* 8: 89–107.

[101] Pan, P.; Svirskis, D.; Rees, S. W. P.; Barker, D.; Waterhouse, G. I. N.; Wu, Z. 2021. Photosensitive Drug Delivery Systems for Cancer Therapy: Mechanisms and Applications. *J. Control. Release.* 338: 446–461.

[102] Tao, Y.; Chan, H. F.; Shi, B.; Li, M.; Leong, K. W. 2020. Light: A Magical Tool for Controlled Drug Delivery. *Adv. Funct. Mater.* 30: 2005029.

[103] Lai, Y.; Zhu, Y.; Xu, Z.; Hu, X.; Saeed, M.; Yu, H.; Chen, X.; Liu, J.; Zhang, W. 2020. Engineering Versatile Nanoparticles for Near-Infrared Light-Tunable Drug Release and Photothermal Degradation of Amyloid β. *Adv. Funct. Mater.* 30: 1908473.

[104] Li, M.; Sun, X.; Zhang, N.; Wang, W.; Yang, Y.; Jia, H.; Liu, W. 2018. NIR-Activated Polydopamine-Coated Carrier-Free "Nanobomb" for In Situ On-Demand Drug Release. *Adv. Sci.* 5: 1800155.

[105] Weinstain, R.; Slanina, T.; Kand, D.; Klán, P. 2020. Visible-to-NIR-Light Activated Release: From Small Molecules to Nanomaterials. *Chem. Rev.* 120: 13135–13272.

[106] Shrestha, P.; Dissanayake, K. C.; Gehrmann, E. J.; Wijesooriya, C. S.; Mukhopadhyay, A.; Smith, E. A.; Winter, A. H. 2020. Efficient Far-Red/Near-IR Absorbing BODIPY Photocages by Blocking Unproductive Conical Intersections. *J. Am. Chem. Soc.* 142: 15505–15512.

Index

A

absorber, 66, 108, 194, 366
absorption, 12, 62, 87, 129, 154, 187, 209, 242, 365
absorption surface, 59
absorption volume, 59
active chiral metamaterials, 194
active materials, 139, 240, 292
actuator, 329
AlN, 329
amino acids, 334
anapole, 157
angular momentum, 187
anisotropic magnetoresistance (AMR), 266
anisotropy, 208
anormalous Hall effect (AHE), 264
antiferromagnet (AFM), 262
antiferromagnetic resonance, 262
antiferromagnetic spin pumping, 273
anti-solvent, 375
asymmetric emission, 198
asymmetric excitation, 198
asymmetric transmission effect, 183
authentication, 317

B

bio-chemical, 7
biological imaging, 409
biological sensing, 408
biology, 7, 407
biomolecules, 334
black-body radiation, 4
black phosphorus, 216–218
bottom-up, 213
boundary condition, 23, 56
bounded-state-in-continuums (BIC), 163
broadband wave plate, 193
bulk inversion asymmetry, 226

C

carbon materials, 297, 298
carbon nanotubes, 298
Casimir force, 166
cell traction force, 340
charge current, 269
charge density, 24
charge density wave, 76
charge transport, 378

chemical vapor deposition (CVD), 64, 158, 211, 297, 385
chemical vapor transport, 211
chiral beam splitter, 193
chiral biosensor, 197
chirality, 179, 180
chirality transfer, 184
chiral metamaterials, 180
chiral mirror, 194
chiral motors, 195
chiral Purcell factor, 199
chiral thermal effect, 200
chiroptical effect, 180–182
chloride, 378
circular conversion dichroism, 183
circular dichroism, 180
circular photogalvanic effect (CPGE), 224, 272
circular polarization detection, 224–227
circular polarized light, 183
colossal magnetoresistance (CMR), 266
communication, 12, 15, 19
compressive strain, 337
conduction band offset, 345
conductive polymer, 299–302
conductive polymeric composite, 299
conservative field, 25
continuity equation, 24
converse piezoelectric effect, 329
copper nanowires, 293
core-shell nanowires, 341
coupling effects in piezoelectric material, 335–337
covalent organic frameworks, 405
crystal structure, 68, 216
current density, 24

D

data storage, 12
degrees of freedom, 208
dewetting, 159
dielectric material, 154–158
dielectric permittivity, 43
diffraction, 168
dipole, 155, 163, 167
direct piezoelectric effect, 329, 330
directionality, 167
dispersion relation, 34, 48
dispersion spatial, 45
dispersion temporal, 45
DNA origami, 190

donor–acceptor, 395
drug release, 414

E

eigenmode, 23, 34
eigenmode resonance, 34
eigenmode theory, 34
electric dipole emitter, 163
electric dipole mode, 155
electric dipole radiation, 263
electric displacement field, 40, 332
electric field, 24
electric susceptibility, 42
electroluminescence, 337
electromagnetic wave, 1, 84, 121, 164, 181
electromagnetic wave manipulation, 123–128
electromechanical, 329, 330, 332
electronic skin, 290
electron-phonon interaction, 74
encryption, 317
enantioselective synthesis, 185
energy density, 27
evanescent wave, 186
exciton, 48, 64, 210
extinction, 155
extrinsic chirality, 180

F

Fabry-Pérot (F-P), 348
faraday effect, 261, 265
far field, 90, 138, 167, 185
femtosecond (fs) laser, 267
ferrimagnet (FiM), 262
ferrimagnetic resonance, 263
ferromagnet (FM), 262
ferromagnetic resonance, 263
flexible devices, 299, 340, 376
flexible dye-sensitized solar cells, 306
flexible electrochromic display, 312
flexible electronics, 291
flexible inorganic solar cells, 306
flexible LED, 312
flexible optical security label, 316
flexible organic solar cells, 306
flexible perovskite solar cells, 306, 370–379
flexible photochromic display, 312
flexible photodetector, 309–311
flexible photovoltaic cell, 306–308
flexible substrate, 287–291
flexible transparent electrode, 302–305
fluorescence, 90, 163, 190
flux growth, 211
force density, 30
formamidinium, 389

G

GaN, 329
gas blow, 375
Ge-Sb-Te (GST), 240
ghost imaging, 271
giant magnetoresistance (GMR), 266
glancing angle deposition, 190
glycine, 334
gold nanowires, 296
graphene, 62–63, 209, 297
GST composition, 240
GST crystallography, 245, 252

H

Helmholtz decomposition, 25, 54
heterointerface, 341
heterostructure, 213
hexagonal, 66, 219, 332
hole transport, 367
hotspot, 159
hydrogels, 289

I

induced chiroptical effect, 184, 197
information security, 316
infrared (IR), 13–15, 274
infrared detectors, 309
InGaN, 337
in-plane, 216
interference, 157
intrinsic chirality, 180
inverse Hall effect effect (ISHE), 261
inverse Rashba-Edelstein effect (IREE), 272
iodide, 372
ionic liquid, 378

K

Kerker condition, 157
Kramers-Kronig relations, 43, 46

L

laser, 4, 170, 348–352
ligands, 385
light-emitting diode (LED), 4, 170, 337–340
lightening, 12
light-matter interactions, 107, 123, 209, 353
linear dichroism, 217
linear polarization detection, 215
local response approximation, 46
localized surface plasmon resonance, 87
longitudinal field, 26
loss, 37, 87, 122, 154, 188, 243

M

macroscopic Maxwell's equations, 41
magnetic, 262–265
magnetic dipole emitter, 163
magnetic dipole mode, 155
magnetic dipole radiation, 263
magnetic field, 40
magnetic heterostructures, 273, 276, 277
magnetic induction field, 24
magnetic permeability, 43
magnetic susceptibility, 42
magnetic tunnel junction (MTJ), 261
magnetization field, 39
magneto-optic effect, 265
magneto-optic Kerr effect (MOKE), 265
material relation, 41
Maxwell stress tensor, 31
mechanical deformation, 344
mechanoluminescence (ML), 356
medium homogeneous, 38
medium inhomogeneous, 52
medium piecewise homogeneous, 59
meta-device, 122, 129–142
metal nanomesh, 303
metal nanowire, 292–296
metamaterial, 88, 172, 123–125
metasurface, 88, 172, 126–128, 355
methylammonium lead trihalide, 388
microcavity, 348
micro-nano devices, 358
micro-nano structures, 155, 333
microscopic Maxwell's equations, 24
micro-sphere lithography, 190
microstructure, 294
Mie resonance, 155
moiré patterns, 210
molecular beam epitaxy
    (MBE), 211
momentum density, 31
momentum transfer equation, 31
multi-junction, 366, 379
multiple quantum wells, 337
multipole, 155

N

nanofabrication, 95–98
nanowire network, 293, 303
near-band edge (NBE)
    emission, 351
near field, 83, 130, 160, 185
near-infrared (NIR), 13, 394
night vision, 15
nitrogen-vacancy center (NV), 163
non-centrosymmetric crystal, 334
nonlinear chiroptical effect, 182

nonlinearity, 168
non-radiative, 164
nucleation and growth, 245

O

OLED, 314
optical activity, 180
optical communication, 12, 215
optical force, 164–166
optical frequency, 2–12
optical physical unclonable function, 321
optoelectronic, 10, 62, 90, 170, 209, 287,
    330, 365
optomechanical resonator, 353
opto-spintronics, 261
orbital angular momentum, 208
organic-inorganic halide perovskite,
    344, 389
oscillator, 17

P

paper substrate, 291
particle accelerator, 17
passivation, 372
PEDOT, PSS, 299
permittivity, 43, 154
perovskite, 223, 332, 363
phase change material, 139, 239
phonon, 48
photoconductive effect, 214
photocurrent, 208
photodetector, 63, 90, 214, 309–311, 344, 385
photodynamic therapy, 411
photogating effect, 215
photography, 7
photoinduced inverse spin Hall effect, 278
photolithography, 7
photoluminescence (PL), 337
photon, 35
photonic-crystal spintronic THz
    emitter, 270
photonic devices, 142, 170, 193, 354
photonic Green's function, 74–75
photoresponsivity, 214
photothermal therapy, 413
photothermoelectric effect, 215
photovoltaic effect, 214
physical vapor deposition, 211
piezoelectric charge coefficient, 332
piezoelectric effect, 330–332
piezoelectricity, 330
piezoelectric materials, 332–334
piezoelectric strain coefficient, 332
piezoluminescence, 336

piezophotonics, 330, 335
piezophototronics, 330, 335
plasmon, 48, 83, 122, 153
plasmonic, 180, 209, 240, 344
plasmonic application, 99–109
plasmonic material, 88–94
p-n junction, 221, 336
polariton, 48
polarization, 214
polarization field, 39
polarized light, 208
polaron, 372
polyaniline (PANi), 300, 301
polydimethylsiloxane (PDMS), 287
polyethylene naphthalate (PEN), 291
polyethylene terephthalate (PET), 291
polyimide (PI), 288
polypyrrole (PPy), 300
polyurethane (PU), 391
polyvinylidene difluoride (PVDF), 334
power density, 27
poynting theorem, 28
poynting vector, 28
pressure sensor, 296
purcell effect, 164, 199
PZT, 332

Q

quadrupole, 155
quantum cascade laser (QCL), 14
quantum dot (QD), 163, 314, 378
quantum emitter, 163
quantum optics, 12

R

racemic chiral structure, 197
Raman imaging, 319
Raman scattering, 161
Rashba-Edelstein effect (REE), 265
recycling, 368
refractive index, 239, 241
remote sensing, 12
resonant bonding, 242
rhenium disulfide, 219

S

scattering, 161, 167
Schottky barrier height, 336
security, 18, 171, 316
self-assembled microsphere
      monolayer, 190
semiconductor nanowires, 296

sensor, 16, 63, 89, 165, 197, 271, 290,
      332, 385
shell, 180
silver nanowires, 292
soft materials, 286, 377
solar blind, 7
solar cells, 336
solar irradiance, 4
solenoidal field, 25
spectroscopy, 11, 16, 19
spin current, 264
spin Hall effect (SHE), 261, 264
spin light emitting diodes
      (Spin-LED), 266
spin-orbit coupling, 264
spin-orbit interaction, 200, 215
spin-orbit torque (SOT), 261
spin photodiode, 276
spin pumping, 265
spin Seebeck effect, 265
spin-to-charge conversion, 272
spin-transfer torque (STT), 261
spintronic, 262, 263
spintronic THz emitter, 163, 269
spin valve, 262
spontaneous emission, 163, 198
spoof surface plasmon polaritons, 88
stability, 372
sterilization and disinfection, 6
stimulated Raman scattering (SRS), 163
stokes parameter, 195
strain, 289, 293, 331
strain-induced piezopotential, 336
stress, 332
stretchable electronics, 291
structural colors, 317
structural inversion asymmetry, 225
super-chiral field, 200
supramolecules, 394
surface charge, 57
surface current, 57
surface-enhanced Raman spectroscopy
      (SERS), 162, 317
surface plasmon polariton, 84
surface plasmon resonance, 84, 87
synthesis, 210

T

tactile sensor, 293
tandem, 379
tensile strain, 341
terahertz (THz), 16–19, 74, 276, 345
textile, 290
theranostics, 394

thermography, 15
third-generation semiconductor, 344
THz detector, 277, 312
THz nano-oscillator, 273
THz pulse emission, 267
THz spintronics, 262
THz time-domain spectroscopy (THz-TDS), 279
time reversal symmetry, 226
time-temperature transformation
    diagram, 246
top-down, 211
topological insulator, 76, 226, 262
topological phase transitions, 76
toroidal mode, 157
toxic, 373
tracking, 15
transition metal dichalcogenides, 209
transverse field, 26
trivalent lanthanide rare-earth ion, 163
trochoidal dichroism, 186
trochoidal spin, 186
tunnel magnetoresistance (TMR), 266
twisted multilayer graphene, 210
two-dimensional material, 209, 378

U

ultrafast demagnetization, 262, 267
ultraviolet (UV), 2–7
uniformity, 375
UV photodetector, 309

V

vacuum, 375
van der Waals, 208
virtual photon, 35
visible, 8–12
volatile phase change material, 255
volume current, 59

W

water-solubility, 382
wavefront shaping, 130–133
weyl semimetal, 227
whispering gallery mode (WGM), 163, 351
wireless communication, 19
wurtzite, 332

X

X-ray, 1

Y

yield, 381
Young's modulus, 290

Z

ZnO, 292, 329
ZnO nanowires, 292

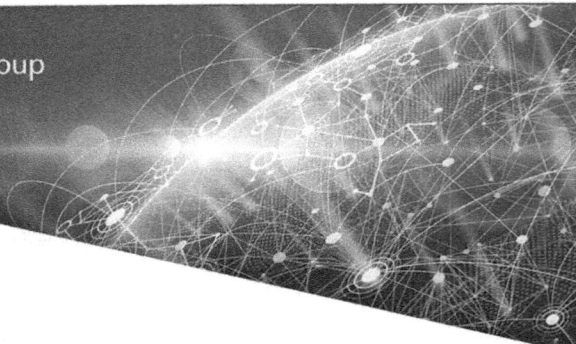

For Product Safety Concerns and Information please contact our EU
representative GPSR@taylorandfrancis.com
Taylor & Francis Verlag GmbH, Kaufingerstraße 24, 80331 München, Germany

www.ingramcontent.com/pod-product-compliance
Lightning Source LLC
Chambersburg PA
CBHW060745220326
41598CB00022B/2331

*9 7 8 1 0 3 2 0 6 5 0 7 6 *